Automotive Steering, Suspension, and Wheel Alignment

First Edition

By Chek-Chart Publications, a Division of H.M. Gousha

Catherine M. Godwin, *Editor*
Janette E. Kok, *Managing Editor*

HarperCollins*CollegePublishers*

Acknowledgments

In producing this series of textbooks for automobile technicians, Chek-Chart has drawn extensively on the technical and editorial knowledge of the nation's carmakers and suppliers. Automotive design is a technical, fast-changing field, and we gratefully acknowledge the help of the following companies and organizations who provided us with help and information, so that we could create the most up-to-date text and illustrations possible. These companies and organizations are not responsible for changes in procedures or specifications made by the carmakers or suppliers, or for any errors or omissions in the instructions or illustrations contained in this book or in any other Chek-Chart product:

American Isuzu Motors, Inc.
American Racing Equipment
Balco, Snap-On Tools Corporation
Central County Occupational Center Program (CCOC/P), San Jose
Chrysler Corporation
CR Service, SKF Manufacturing Company
Evergreen College, San Jose
Ferrari North America
Ford Motor Company
General Motors Corporation
  Buick Motor Division
  Cadillac Motor Car Division
  Chevrolet Motor Division
  Oldsmobile Division
Honda Motor Company, Ltd.
Hunter Engineering Company
Hyundai Motor Company
Jaguar Cars, Ltd.
Laguna Seca Raceway
Mazda Motor Corporation
Mitsubishi Motors Corporation
Nissan Motor Company, Ltd.
OTC, a Division of SP^x Corporation
Porsche Cars North America, Inc.
Saab-Scania AB
Suzuki Motor Corporation
Tires Unlimited, San Jose
Toyota Motor Corporation
Volkswagen America, Inc.
Yokohama Tire Corporation
Yugo America, Inc.

The comments, suggestions, and assistance of the following reviewers were invaluable:

Tom McCormack, Evergreen College, San Jose CA
Dwight Tracy, Santa Rosa Junior College, Santa Rosa CA
Lawrence Meicher, Madison Area Technical College, Madison WI
Clem Childress, Fischer Technical Institute, Virginia Beach VA
Royce Foster, Indiana Vocational Technical College, Lafayette IN

At Chek-Chart, Daniel L. Doornbos managed the production of this book. Original art and photographs were produced by John Badenhop, Dave Douglass, Gerald A. McEwan, F. J. Zienty, Catherine M. Godwin, and Janette E. Kok. The project is under the direction of Roger L. Fennema.

Library of Congress Cataloging-in-Publication Data

Automotive steering, suspension, and wheel alignment / by Chek-Chart Publications, a division of H.M. Gousha; Janette E. Kok, editor. — 1st ed.
p. cm.
Includes index.
Contents: v. 1. Classroom manual — v. 2. Shop manual.
ISBN 0-06-500167-2 (set)
1. Automobiles—Steering-gear—Juvenile literature.
2. Automobiles—Springs and suspension—Juvenile literature.
3. Automobiles—Wheels—Alignment—Juvenile literature. I. Kok, Janette E. II. Chek-Chart Publications (Firm)
TL259.A96 1993
629.24—dc20 92-30385
CIP
95 9 8 7 6 5 4 3 AC

Contents

On the Cover:
Front — The Hunter D111 digital wheel aligner, courtesy of Hunter Engineering Company.
Rear — The Toyota Camry front suspension, courtesy of Toyota Motor Corporation; and the Dodge Stealth 2WD rear suspension, courtesy of Chrysler Corporation.

Introduction to Automotive Steering, Suspension, and Wheel Alignment

Automotive Steering, Suspension, and Wheel Alignment is part of the HarperCollins/Chek-Chart Automotive Series. The package for each course has two volumes, a *Classroom Manual* and a *Shop Manual*. Other titles in this series include:

- Automatic Transmissions and Transaxles
- Automotive Brake Systems
- Automotive Heating, Ventilation, and Air Conditioning
- Automotive Electrical and Electronic Systems
- Automotive Engine Repair and Rebuilding
- Engine Performance, Diagnosis, and Tune-Up
- Fuel Systems and Emission Controls.

Each book is written to help the instructor teach students to become competent and knowledgeable professional automotive technicians. The two-manual texts are the core of a complete learning system that leads a student from basic theories to actual hands-on experience.

The entire series is job-oriented, especially designed for students who intend to work in the car service profession. A student will be able to use the knowledge gained from these books and from the instructor to get and keep a job in automotive repair or maintenance. Learning the material and techniques in these volumes is a giant leap toward a satisfying, rewarding career.

The books are divided into *Classroom Manuals* and *Shop Manuals* for an improved presentation of the descriptive information and study lessons, along with representative testing, repair, and overhaul procedures. The manuals are to be used together: the descriptive material in the *Classroom Manual* corresponds to the application material in the *Shop Manual*.

Each book is divided into several parts. Instructors will find the chapters to be complete, readable, and well thought-out. Students will benefit from the many learning aids included, as well as from the thoroughness of the presentation.

The series was researched and written by the editorial staff of Chek-Chart and was produced by HarperCollins*Publishers*. For over 60 years, Chek-Chart has provided car and equipment manufacturers' service specifications to the automotive service field. Chek-Chart's complete, up-to-date automotive data bank was used extensively to prepare this textbook series.

Because of the comprehensive material, the hundreds of high-quality illustrations, and the inclusion of the latest automotive technology, instructors and students alike will find that these books will keep their value over the years. In fact, they will form the core of the master technician's professional library.

How To Use This Book

Why Are There Two Manuals?

Unless you are familiar with the other books in this series, *Automotive Steering, Suspension, and Wheel Alignment* will not be like any other textbook you've used before. It is actually two books, the *Classroom Manual* and the *Shop Manual*. They have different purposes and should be used together.

The *Classroom Manual* teaches you what you need to know about automotive steering and suspension systems, wheels and tires, and wheel alignment theory. The *Classroom Manual* will be valuable in class and at home, for study and for reference. You can use the text and illustrations for years to refresh your memory — not only about the basics of automotive steering and suspension, wheels and tires, and wheel alignment but also about related topics in automotive history, physics, and technology.

In the *Shop Manual*, you learn test procedures, troubleshooting, and how to repair the systems and parts you read about in the *Classroom Manual*. The *Shop Manual* provides the practical, hands-on information you need to work on automotive chassis systems. Use the two manuals together to understand fully how the systems work and how to fix them when they don't work.

What's in These Manuals?

These key features of the *Classroom Manual* make it easier for you to learn and to remember what you learn:

- Each chapter is divided into self-contained sections for easier understanding and review. This organization shows you clearly which parts make up which systems, and how various parts or systems that perform the same task differ or are the same.
- Most parts and processes are fully illustrated with drawings or photographs. Important topics appear in several different ways, to make sure you can see other aspects of them.
- Important words in the *Classroom Manual* text are printed in **boldface type** and are defined on the same page and in a glossary at the end of the manual. Use these words to build the vocabulary you need to understand the text.
- Review questions are included for each chapter. Use them to test your knowledge.
- Every chapter has a brief summary at the end to help you review for exams.
- Every few pages you will find sidebars — short blocks of ''nice to know'' information — in addition to the main text.

The *Shop Manual* has detailed instructions on test, service, and overhaul procedures for modern components and current chassis systems. These are easy to understand and usually have step-by-step explanations to guide you through the procedures. The *Shop Manual* contains:

- Helpful information that tells you how to use and maintain shop tools and test equipment
- Safety precautions
- Clear illustrations and diagrams to help you locate trouble spots while you learn to read service literature
- Test procedures and troubleshooting hints that will help you work better and faster
- Tips the professionals use that are presented clearly and accurately
- A sample test at the back of the manual that is similar to those given for Automotive Service Excellence (ASE) certification. Use this test to help study and prepare yourself when you are ready to be certified as a steering and suspension expert.

Where Should I Begin?

If you already know something about automotive chassis systems and know how to repair them, you will find that this book is a helpful review. If you are just starting in car repair, then the book will give you a solid foundation on which to develop professional-level skills.

Your instructor will design a course to take advantage of what you already know and what facilities and equipment are available to work with. You may be asked to read certain chapters of these manuals out of order. That's fine. The important thing is to really understand each subject before you move on to the next.

Study the vocabulary words in boldface type. Use the review questions to help you understand the material. When you read the *Classroom Manual*, be sure to refer to your *Shop Manual* to relate the descriptive text to the service procedures. And when you are working on actual car systems and components, look back to the *Classroom Manual* to keep the basic information fresh in your mind. Working on such a complicated piece of equipment as a modern car isn't always easy. Use the information in the *Classroom Manual*, the procedures in the *Shop Manual*, and the knowledge of your instructor to help you.

The *Shop Manual* is a good book for work, not just a good workbook. Keep it on hand while you're working on equipment. It folds flat on the workbench and under the car, and can stand quite a bit of rough handling.

When you perform test procedures and overhaul equipment, you will need a complete and accurate source of manufacturer's specifications, and the techniques for pulling computer codes. Most auto shops have either the carmaker's annual shop service manuals, which lists these specifications, or an independent guide, such as the Chek-Chart *Car Care Guide*. This unique book, with ten-year coverage, is updated each year to give you service instructions, capacities, and troubleshooting tips that you need to work on specific cars.

PART ONE

General Service Operations

Chapter

1

Safety Precautions, Shop Practices, and Special Tools

Automotive chassis systems are easy to service and overhaul, once you understand how they are built and how they work. If you have a good grasp of the fundamentals presented in the *Classroom Manual*, you should not have difficulties learning chassis system service.

This chapter introduces a number of safety precautions for chassis work. Among the most important of these are the procedures used to lift cars and properly support them before you begin work. After the safety section, some shop practice tips are provided to help you work more efficiently. Finally, this chapter also covers the unique tools used by the chassis technician. Like any specialized area of automotive repair, chassis work requires special tools in addition to the usual assortment of basic hand tools.

SAFETY PRECAUTIONS

Safety is important, both to you and to the people who work around you. No repair job is worth losing your life, or suffering a permanent, crippling injury, or even loss of time and income resulting from a preventable accident. Even so, technicians who follow poor work practices take these risks regularly, even when the correct procedures require little or no extra time or effort. By following a few common-sense safety practices, you can minimize your chances of being involved in an accident or causing accidental injury to others.

Some basic safety precautions for working on any automotive system include the following:

- Tie back long hair, tuck in loose clothing, and button long sleeves or wear a short-sleeved shirt, especially if you are going to work near moving parts
- Remove rings, watches, or jewelry, especially if you are going to work near moving parts or electrical components
- Wear safety glasses or goggles when there is a risk of foreign objects or liquids getting into your eyes; this can happen whenever you work under the car, or when you use almost any type of power tool
- If you are not sure how to use any tool, machine, or test equipment, ask your instructor or supervisor to show you how to use it safely
- Keep floors, aisles, and your work area clear of tools, parts, and materials
- Do not carry sharp tools, such as screwdrivers, in your pockets; carry them in your hand with the sharp or cutting edge downward.

Emergencies

In case of emergencies in the shop, know the location of all of the following:

- First aid supplies, the telephone, and emergency numbers
- Fire extinguishers and fire alarm switches
- Emergency equipment such as eyewashes and showers
- Emergency exits.

Fire

Most automotive shops contain a number of flammable substances, and the space for storage and work may be limited. To avoid starting fires:

- Do not bring any source of spark or flame near a vehicle's fuel tank, fuel lines, or fuel filler neck
- Do not smoke in the shop or while you are working
- Do not use gasoline as a cleaning solvent
- Keep spray cans and containers of flammable materials away from heat sources.

Solvents and Chemicals

The solvents, cleaning solutions, lubricants, and other chemicals found in automotive shops require safe handling. Besides the danger of fire, already mentioned, these substances usually are poisonous. The fumes alone of many chemicals used for automotive service can cause sickness. Avoid endangering your health by following these precautions:

- Make sure your shop is well-ventilated, regardless of the weather
- Always read the instructions and warnings accompanying a new product before using it; follow all the precautions the manufacturer recommends
- Use rubber gloves and safety glasses when working with chemicals
- Use plastic or rubber aprons to protect your skin and clothing, and change out of any chemical-soaked clothing as quickly as possible
- Do not splash cleaning solvent as you put parts into a parts washer
- Clean up spills immediately
- Wash your hands with soap and water or with hand cleaner after handling chemicals, especially before eating, drinking, smoking, or using the toilet
- Do not use gasoline, solvent, or thinner to clean your hands
- Dispose of all used chemicals properly — never pour them down sink drains, toilets, or waste water drains.

Operating Vehicles in the Shop

When working on a car, you must drive it into and out of the shop and onto and off lifts. To do chassis work, you reach parts from under the car, through the passenger compartment or trunk, or through the engine compartment. Often you must start the engine briefly while the car is on a lift, for example, to activate the power steering. Observe the following safety rules:

- Drive no faster than five miles per hour (eight kilometers per hour) into and out of the shop area
- Ensure adequate ventilation or connect the exhaust pipe to shop exhaust ducts before running the engine; engine exhaust contains carbon monoxide, which can poison you gradually within hours
- Before starting the engine, set the parking brake, chock the drive wheels, and put the transmission or transaxle into neutral, for a manual, or park, for an automatic
- Turn off the engine before working under the hood
- If you must run the engine while working under the hood, use extreme caution and make sure to keep yourself and your tools out of the way of moving parts
- Keep in mind that electric radiator fans can start at any time if the engine has been running recently, even with the ignition off; disconnect the fan power lead before working near it
- In the engine compartment and under the car, keep away from exhaust components; catalytic converters in particular stay hot long after the engine is stopped.

Electrical Safety

Some power steering systems and suspension systems use electronic controls and electrical components. To work safely with electrical parts:

- Make sure your hands, the floor, and your entire work space are dry before touching any electrical switches or plugs or using any electrical equipment
- Remove the battery ground cable when working on or near electrical components
- Keep flames and sparks away from the battery at all times; the battery gives off

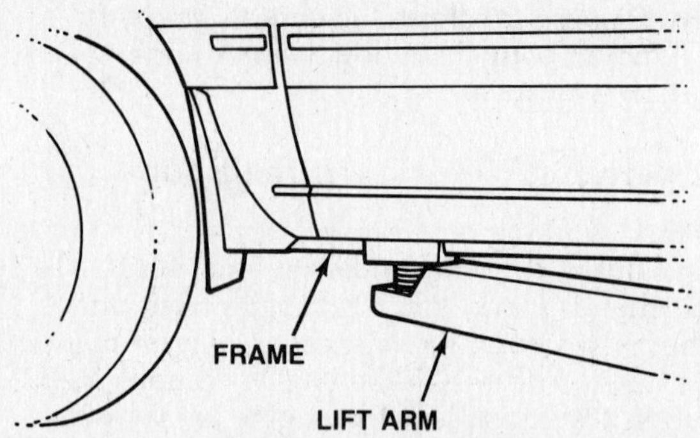

Figure 1-1. A frame-contact lift raises the frame of the car and lets the suspension, wheels, and tires hang freely.

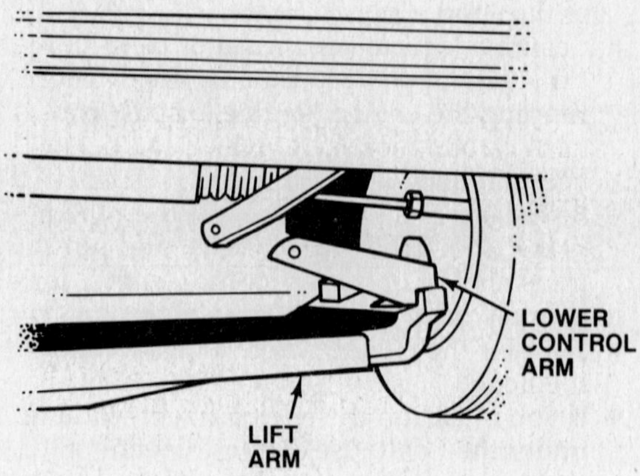

Figure 1-2. The adapters of a suspension-contact lift fit under the suspension lower control arms, so the suspension stays compressed when the car is lifted.

Figure 1-3. The technician drives a vehicle onto a runner lift, and the lift raises it from under the wheels.

Figure 1-4. With an alignment lift, the technician raises the vehicle on runners. Pneumatic jacks in the lift allow lifting one or both ends off the runners.

explosive hydrogen gas, and if ignited can explode and spew sulfuric acid throughout the shop

- Never use the battery top as a tool tray; a metal tool touching both terminals will melt and can cause an explosion.

Chapter Three of this *Shop Manual* provides more information about how to safely and effectively service electrical devices and electronic systems.

Using Lifts

Lifts raise a vehicle to a convenient working height. There are three basic types of lifts:

- Frame-contact
- Suspension-contact
- Runner.

A frame-contact hoist lifts the vehicle by the frame, figure 1-1, letting the wheel and tire assemblies hang to the lowest point of suspension travel. A suspension-contact lift has claws that fit onto the suspension lower control arms, figure 1-2, so the wheels and tires do not hang free. A runner lift raises the car by the drive-on runners under the tires, with the weight on the wheel and tire assemblies, figure 1-3. Alignment racks generally are equipped both as runner and frame-contact lifts. The technician uses the runner lift to raise the car and measure the alignment angles, then the built-in jacks to lift one or both ends of the vehicle, figure 1-4, as needed, to work on making the adjustments.

A sense of caution is important when lifting vehicles. Even small, ''light-weight'' cars often weigh over 2,000 pounds (900 kg), and you should be mindful of the danger every time you place yourself underneath a ton of steel

RWD TRUCK

SUSPENSION-CONTACT POINTS

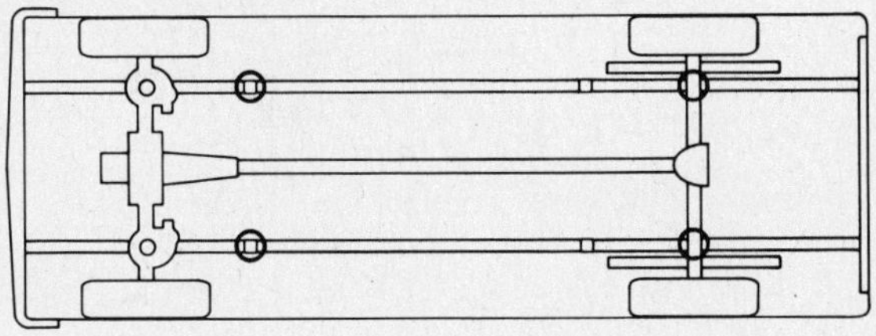

FRAME-CONTACT POINTS

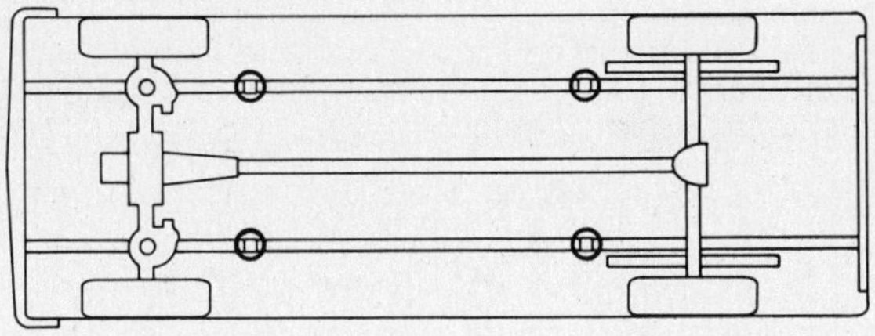

JACK POINTS

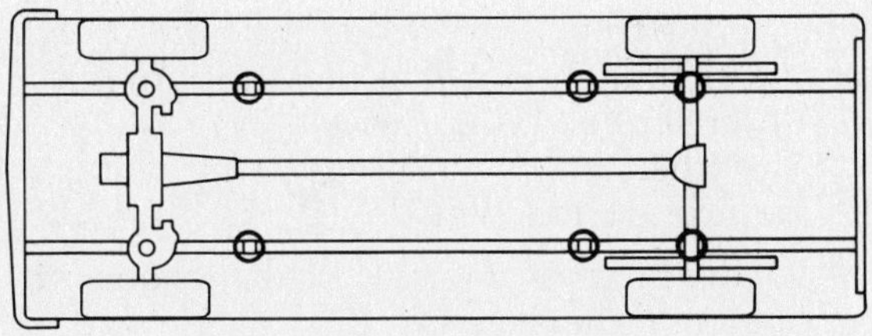

Figure 1-5. Use the manufacturer's recommended lift points to avoid damaging the chassis.

balanced on a lift. If a vehicle is lifted improperly or if removing components changes the weight distribution, a car or truck can fall off a lift and injure or kill people.

When using a frame-contact or suspension-contact lift, use the lift points designated by the manufacturer, figure 1-5. You can find these in the manufacturer's shop manual and in some aftermarket publications. If you lift at any other point, you are likely to damage the frame or platform, which could make the vehicle less crashworthy. For some cars, only frame-contact points are specified, figure 1-6. Put lift adapters only at those points. Raise the lift until the lift pads or claws are just under the car, position them at the lift points, and raise the lift until the pads or claws contact the lift points. Check for secure contact before lifting the vehicle the rest of the way.

FWD CAR

FRAME-CONTACT POINTS

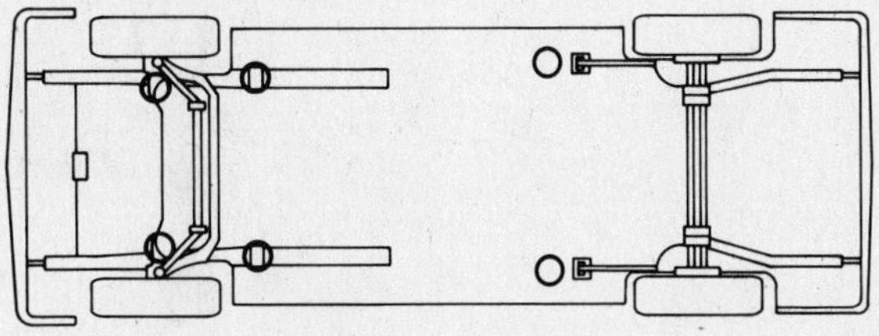

JACK POINTS

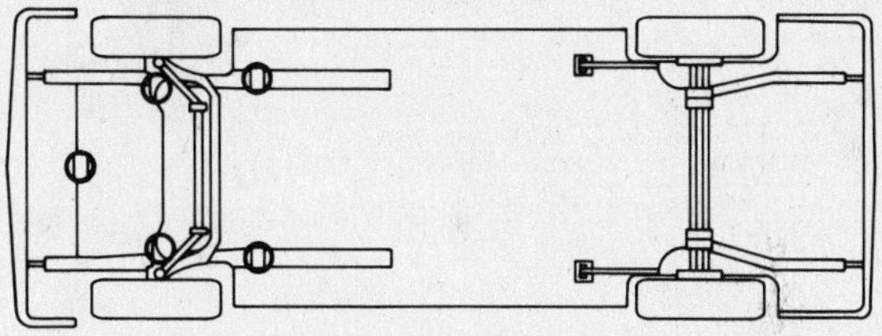

Figure 1-6. On some cars, there are no suspension-contact lift points. Lift at the frame-contact points or under the wheels.

When you use a runner lift, make sure the vehicle cannot roll. Set the parking brake, put the transmission in neutral for a manual, or park for an automatic, and block the wheels, figure 1-7.

Once a vehicle is on a lift, think through the procedures you want to perform in terms of their effect on the distribution of weight. This is particularly important with frame-contact and suspension-contact lifts. If you are planning to remove a component from the front or rear axle, use a jack to support the opposite end of the car. Place the jack under a sturdy undercar component located on the bumper side of the wheel. In other words, if you are going to remove a rear-axle component, place a jack under the front end of the car, between the front wheels and the bumper. That way, the car or truck cannot start to tilt toward the opposite end when weight is removed from the end at which you are working.

Some manufacturers also specify chaining the vehicle frame to the lift at the same end from which you are going to remove a component. That is, chain the rear of the car to the lift if you plan to remove a rear component, and chain the front to the lift if you are going to remove a part from the front end. That way, if the jack under the opposite end gets jolted or knocked over, the car still will not fall off the lift.

Figure 1-7. Make sure that a car on a runner lift cannot roll forward or back. Set the parking brake and block the wheels.

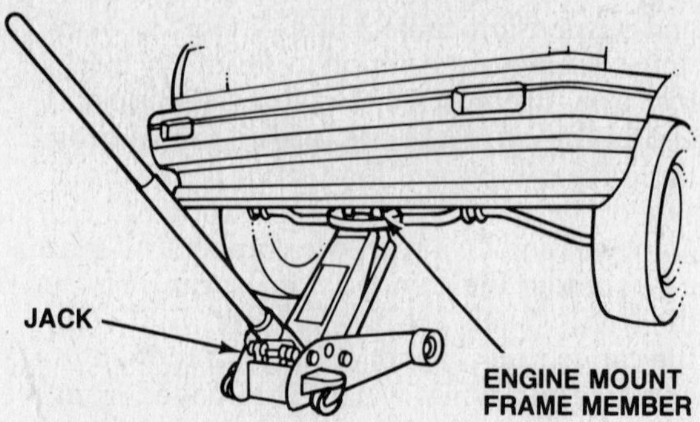

Figure 1-8. For some jobs, you can lift one end of the car with a jack and secure it with safety stands. Never work underneath a car supported only by a jack.

For example, if you are going to drop a solid rear axle out of a rear-wheel-drive (RWD) car, the car will become considerably more front-heavy. A jack under the front of the car will

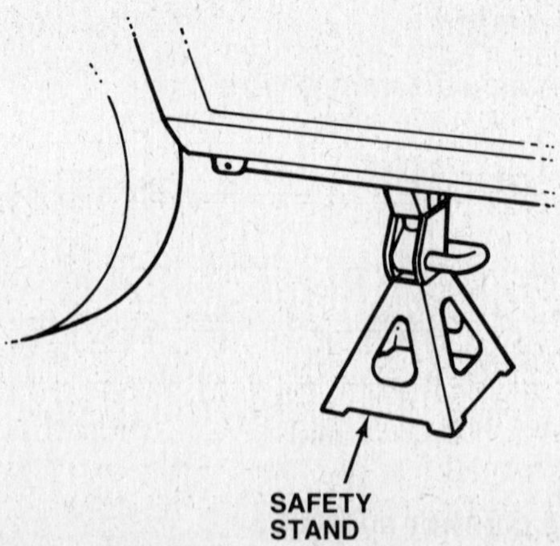

Figure 1-9. After lifting a car with a jack, secure it with safety stands. Make sure the safety stands are rated to hold the vehicle's weight.

support the weight so the car does not tilt forward. Chaining the rear of the car to the lift provides further security. For front-wheel-drive (FWD) vehicles, keep in mind that probably 60 percent of the weight is already at the front, and that removing rear-axle components may make the rear practically weightless compared to the front. Supporting the front of the car and chaining the rear to the lift is an easy way to provide for your own safety.

Some other common-sense rules for using lifts include the following:

- Follow the manufacturer's instructions for operating the lift
- Inspect the lift daily, and do not use it if any part of it needs repair
- Make sure no extensions or other lift components will obstruct the vehicle as you drive it on and off the lift
- Do not lift a vehicle that weighs more than the lift's capacity
- Never lift a vehicle with a person inside
- Do not try to override any of a lift's safety devices.

Jacks and Safety Stands

Although most shops have lifts, many simple chassis jobs — particularly changing wheel and tire assemblies — can be done by lifting one end of a car and putting it on safety stands. In this case, you can raise the car with a hydraulic jack, figure 1-8, and place safety stands under the frame, figure 1-9, or under strong suspension or drivetrain components. Make sure the jack and safety stands are in good condition and that they are warranted to support as much weight as the vehicle will place

Figure 1-10. Before detaching the leaf spring in this car, the technician supported the rear axle to keep it from falling.

on them. Never work on a vehicle supported only by a jack, and never use a jack to move one end of a car.

Check the stability of the vehicle's position on the safety stands before you go underneath it. Be aware that safety stands are not as safe nor as stable as a lift. It is possible to knock a vehicle off stands by bumping into it with enough force, and this would crush the person underneath, causing severe injury or death.

Bumper jacks were once popular, but most manufacturers do not allow using them with late-model cars. Never use the jack that came with the car as a support when you perform under-car work. Those jacks are suitable for changing a wheel and tire assembly in a roadside emergency situation, but they are not strong. You would be foolish to trust your life to one by crawling under a vehicle supported by one.

Supporting Undercar Components

A suspension system uses a number of links that together support the car body and locate the wheels. Because the links act as a system, removing or disconnecting one suspension part can cause another part to drop out of place. For example, sometimes a shock absorber or leaf spring limits the downward travel of a suspension or axle. If you remove the shock or spring, the axle and wheels will drop, causing damage to the vehicle, particularly the brake hydraulic lines, and possible injury to the person under the car.

Supporting the axle or suspension components with safety stands before removing the parts that hold them in place prevents this accident, figure 1-10. An alternative is to raise the car only enough to lift the weight off the wheels, while allowing the tires to still touch the ground. In this case, the ground supports the wheels and axle or suspension, and limits the downward travel. The disadvantage of this method is the cramped space in which you must work.

Compressing Springs

Suspension springs are constantly under pressure. When the weight of the frame and body rest on the spring, they are carrying their "static load", as explained in Chapter Eight of the *Classroom Manual*. The tension in a loaded suspension spring provides a lot of potential force. With torsion bars and longitudinally mounted leaf springs, spring tension releases gradually as you release the spring from its mount or as you lower the component on which it rests, but other springs, especially coil springs, fly out of place as soon as their upper and lower seats are separated enough, releasing many pounds of force against whatever they encounter. If this happens, it causes considerable damage and may injure or kill people in the shop.

A spring compressor, figure 1-11, keeps a coil spring from releasing its tension when it is freed from its seat. With a spring compressor, the technician removes the spring under pressure and slowly releases the tension after spring removal. Spring compressors also are necessary for installing springs.

Some transverse leaf springs require a special compressor for removal from the rear axle. Since transverse leaf springs are relatively uncommon, these are usually special manufacturer's tools.

Compressor plates

Compressor plates use two plates, a threaded rod, and a nut to compress a coil spring. The two circular plates push the coils together. The threaded rod fits through the lower plate and threads into the upper plate. The nut threads onto the rod below the lower plate to secure that end to the lower plate. The technician turns the rod to thread it through the upper plate, which pulls the upper plate toward the lower plate, compressing the coils between the two plates.

Spring clips

Spring clips are useful for keeping a coil spring compressed while you remove it, but they cannot do the compressing. To use them, jack up the component on which the lower end of the spring seats. This compresses the spring. Put the clips over the coils. Long clips can usually secure five coils, and short ones can secure four. Adjust the jack height, as needed, to

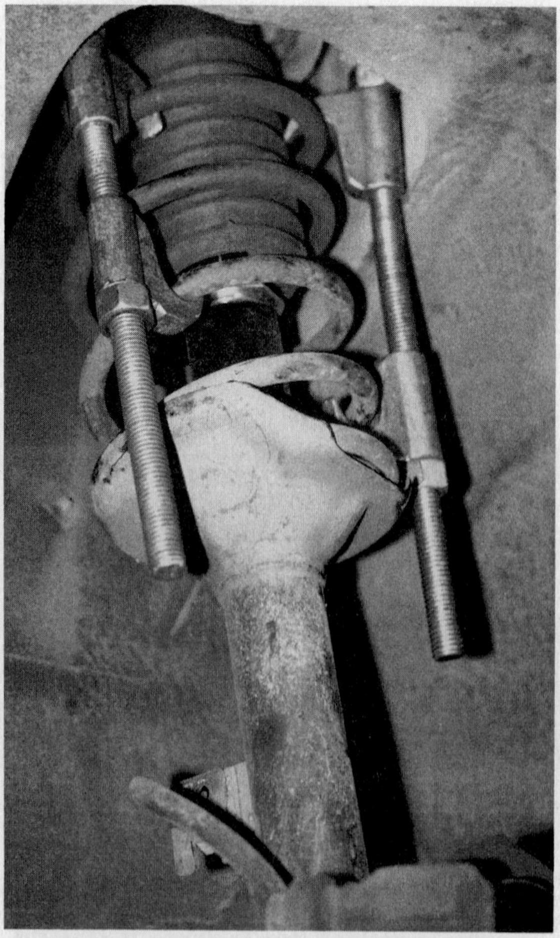

Figure 1-11. A spring compressor keeps a coil spring compressed so it will not injure anyone when removed. An unsecured spring releases hundreds of pounds of force.

position the clips. Slowly lower the jack to release pressure from the spring. The clips should fit snugly on the coils, keeping the spring compressed. As you lower the jack, check the clips to be sure they are secure.

The clips keep the spring compressed after you remove it from the vehicle. If you need to remove the clips while the spring is off the car, use compressor plates.

Integral-spring strut compressors

Most struts have a coil spring as part of their construction. Many chassis shops have pneumatic compressors for integral-spring struts, figure 1-12. With these, follow the manufacturer's instructions for mounting the strut in the press and compressing the spring. If a strut compressor is not available, you can use a coil spring compressor to compress the spring in a strut, but it will have to have forks that can fit around the shock cartridge instead of plates.

Figure 1-12. Some struts contain coil springs, and you must use special equipment to compress and remove the spring from the strut.

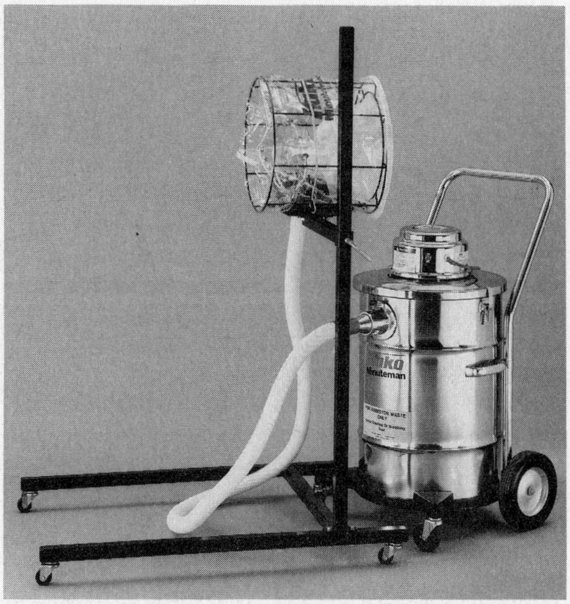

Figure 1-13. Special brake vacuums allow technicians to clean brake friction assemblies without the danger of inhaling asbestos fibers.

Mounting and Dismounting Tires

Mounting tires onto wheels and dismounting them are everyday procedures for the chassis technician. After doing this job hundreds of times, it is easy to become careless, but there are certain precautions the technician should remember each time he or she mounts a tire:

- Do not stand directly over a tire while you are inflating it to seat the beads; if by any chance it exploded, it would injure you — stand as far off as possible
- Do not have your hand near the wheel flange while seating the beads
- Never exceed the maximum tire pressure stated on the sidewall when you are seating the beads; overinflating may forcefully rupture the tire beads or the wheel rim, and you could be injured
- Do not drop or bounce an inflated tire; you may lose control of it and cause damage.

There is also one important caution for dismounting a tire. Sometimes car owners repair a flat tire that has had a puncture by filling it with a sealant/inflator out of a pressurized container. Some of these products contain explosive chemicals, and if the technician strikes a spark while scraping the bead-breaking tool around the wheel rim, the tire can explode. Technicians have been severely injured by this happening. If you are going to dismount a tire that contains sealant, bleed the air out, inflate it, and bleed it out again. Do this a number of times to purge the chemicals out of the tire before you dismount it.

Brake Service Safety

Chassis technicians often perform minor brake system service, such as topping off the brake fluid or inspecting and changing brake pads or shoes. While doing these tasks, keep safety in mind.

The main danger of doing a lot of brake work is that many brake pads and shoes contain asbestos, which if inhaled causes lung disease. If pads or shoes do contain asbestos, tiny particles of it will be present in any dust that comes from the brake friction assembly at each wheel, so it is important not to scatter this dust into the air. In particular, never use an air hose to blow brake dust off the wheel or brake assembly. Doing so fills the air with asbestos particles, and you will not be able to avoid inhaling them.

Some shops have special brake vacuums, figure 1-13, that have a High-Efficiency Particulate Air (HEPA) filter that traps asbestos fibers. Technicians can use this type of vacuum to clean the brake friction assembly before working on it. Another method of cleaning the friction assembly is to use brake cleaning solvent in a brake washer or from a spray can. The solvent keeps asbestos fibers out of the air. If you ever use a brush to clean a part of a brake friction assembly, make sure the part is wetted with solvent, so that the brush will not scatter asbestos fibers into the air. Never scrape or brush a dry friction assembly. If brake dust gets on the floor of your shop, wet sweep it or clean it with a vacuum that has a HEPA filter. Do not dry sweep it, as this too will release asbestos fibers into the air.

Brake solvent, like other solvents discussed earlier, also requires special handling. In particular, avoid inhaling the fumes and never smoke near it. Wash your hands of brake cleaning solvent and of brake dust before eating, drinking, smoking, or using the toilet.

Finally, polyglycol brake fluid is also a powerful solvent. Do not let it contact your skin, and always wear safety glasses while you work with it. Be careful not to spill any brake fluid on the car's finish, as it will remove the paint in seconds. If you do spill it, do not wipe it across the paint with a cloth, as this will increase the damage. Instead, rinse it away with lots of clean water.

Figure 1-14. Use a steering wheel puller to remove the steering wheel from the steering shaft. Never pound on the steering wheel, or you may collapse the shaft.

SHOP PRACTICES

Good shop practices are any habits you develop that help you do your work more safely and efficiently. Organizing your work area and keeping it clean may seem like a bother, but in the long run it will save you trouble by making your work easier. The practices recommended in the following paragraphs are simple common sense. It is not a complete list — when you have read it, see if you can think of other shop practices that will help you do a first-class job.

Organize your work area. This keeps unneeded movements to a minimum, which increases your efficiency and makes repair procedures less tiring. Keep your work bench clean and free of unnecessary tools, old parts, and other clutter. This provides more useful space to work in and reduces the possibility of losing small parts.

Keep your tools clean and dry. A clean tool is less likely to slip off the part on which you are working or out of your hand. Put your tools back in their place after each use so you will always know where they are when you need them. Keep the shop floor and your work area clean of spilled fluids, lubricants, or cleaners. Have clean shop towels on hand to wipe up spills.

Provide yourself with enough light to do your work correctly and without eyestrain. If you use an extension light, make sure it has a safety shield so you do not burn yourself on the bulb.

Have plastic bags or other suitable containers available for saving old parts to be returned to the customer. Never dispose of old parts until the repairs are finished and the customer has been given the option of taking the parts home.

Keep up-to-date shop manuals, technical service bulletins, and other specification sources available for use. Never guess — refer to the proper service literature whenever you are uncertain about a specification or procedure.

SPECIAL TOOLS

Many chassis repair jobs can be done with the usual assortment of hand tools found in any professional garage. The chassis technician also uses a number of special tools and equipment. Some of these special tools are also used in other types of automotive service and repair, while others are specific to chassis work. The following paragraphs discuss both types of special tools.

Steering Wheel Pullers

A steering wheel puller is used to remove the steering wheel from the steering shaft, figure 1-14. This tool consists of a slotted plate with a center hole and bolt, and an assortment of anchor bolts that fit through the slots into holes in the steering wheel. A similar tool, which can also be used as a steering wheel puller in some cases, is a harmonic balancer puller.

To use a steering wheel puller, place the shaft of the center bolt on top of the steering shaft (having already removed the steering wheel trim and center bolt), and choose two anchor bolts that fit into the holes in the steering wheel. These bolts go through the slots in the plate and secure the plate to the steering wheel. When you use a wrench to thread the center bolt through the plate, the plate will move away from the steering shaft and the anchor bolts will pull the steering wheel along with them. Eventually, the tool will pull the steering wheel off the shaft.

Tie Rod Adjuster

Varying the length of the tie rods in the steering linkage varies the toe-in measurement. Many technicians use a vise grip to make this adjustment, but a special tool for it, figure 1-15, helps turn the threaded rod or sleeve to

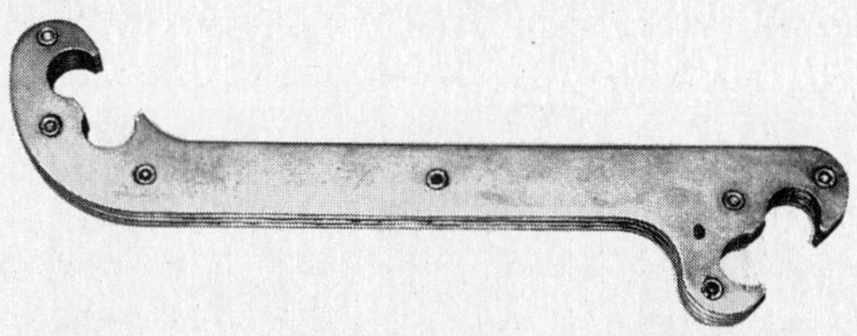

Figure 1-15. A special tie rod adjuster allows the technician to turn the adjusting sleeve or rod without damaging the tie rod.

adjust tie rod length without damaging the tie rod. The end of the adjuster loops over the tie rod and, as the technician pulls down on the tool, it grips the rod or sleeve and turns it.

Snapring Pliers

Snaprings are flat, circular pieces of spring steel used to hold seals in place, locate bearings, and retain mounting plates. There are two types of snaprings: internal and external. Each requires a specialized type of pliers, figure 1-16, for removal although some combination tools can handle both.

External snaprings require pliers that pull the two eyes apart, releasing the snapring from its groove. Internal snaprings require the two eyes to be squeezed together, releasing the ring from its groove.

Because the eyes of different snaprings vary in size, some snapring pliers come with removable, interchangeable ends. They can be used on a variety of snaprings. Alternatively, some technicians prefer to purchase a complete set of pliers with different size ends.

Torque Wrenches

In chassis repair, it is often essential that a nut or bolt be tightened a specific amount. Manufacturers typically make torque recommendations for the through-bolts of suspension bushings — which must always be tightened with the vehicle on the ground and the weight on the suspension — and for lug nuts. On any chassis part, tightening beyond the specifications, "for good measure", can stretch the fastener beyond its elastic limit, strip the threads, or distort the parts being fastened together. A common mistake of many chassis technicians is to tighten wheel lug nuts with an impact wrench, going far beyond specifications and warping the brake rotor or drum as a result.

Using a torque wrench ensures that you will tighten fasteners properly. Torque wrenches are also used to measure steering gear tolerances and to make wheel bearing adjustments.

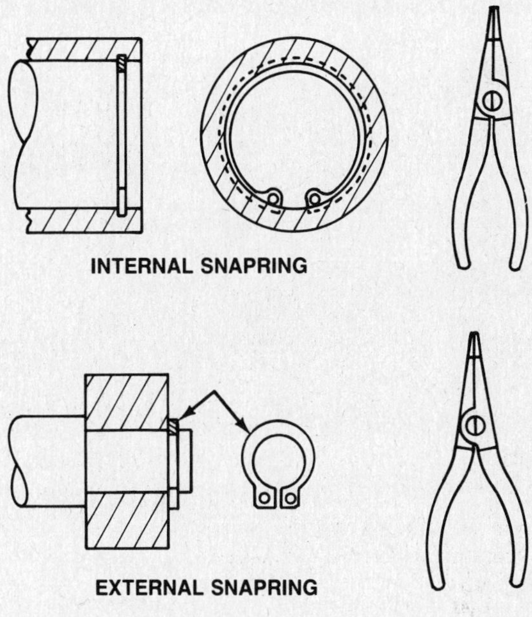

Figure 1-16. Use snapring pliers to remove or install internal or external snaprings.

A torque wrench has a scale on it that indicates the amount of tightening force being applied to a nut or bolt. The scale may be graduated in inch-pounds (in-lbs), foot-pounds (ft-lbs), Newton-meters (Nm), kilogram-meters (kg-m), or a combination of these values.

There are three basic kinds of torque wrenches, figure 1-17:

- Beam
- Dial
- Micro-adjusting.

With a beam-type torque wrench, a pointer indicates the torque on a scale near the handle. A needle and scale built into the wrench body indicates the torque on a dial-type wrench. With the micro-adjusting wrench, the technician presets the torque by rotating the wrench handle until it aligns with the desired value on the body of the wrench. When the preset level of tightness is reached, the mechanism inside the wrench overcenters and "clicks". The technician can feel the mechanism's release through the handle of the wrench, as well as hearing it click. The micro-adjusting wrench is sometimes called a "click-type" torque wrench.

Parts Washers

Chassis shops use parts washers to clean disassembled components such as steering gears that are being rebuilt or wheel bearings that need to be cleaned before packing with grease.

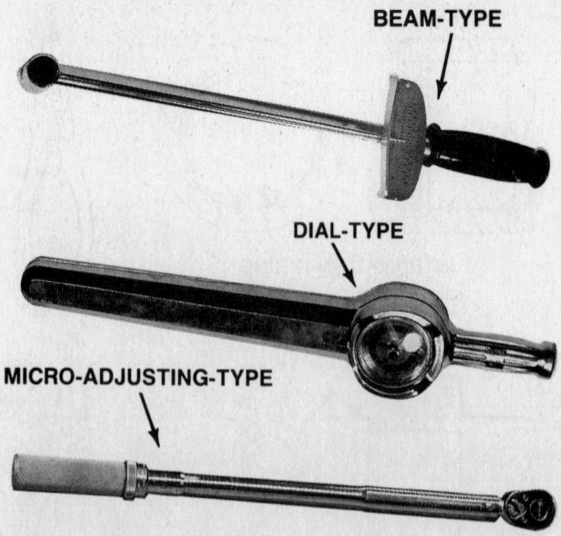

Figure 1-17. Each type of torque wrench — beam, dial, and micro-adjusting — measures twisting force so the technician will not overtighten or undertighten fasteners.

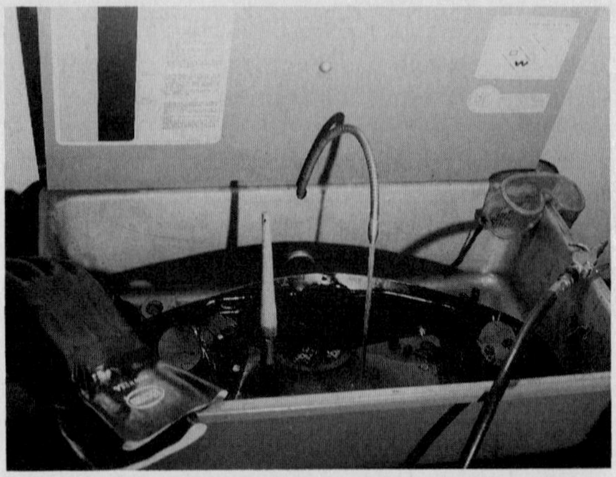

Figure 1-18. A parts washer lets the technician direct solvent over disassembled components to clean them.

A typical parts washer has a steel tank to hold the parts and a hose to direct a stream of solvent over the parts, figure 1-18. A small electric pump circulates the solvent, and a filter removes large contaminants.

Air Compressors

An air compressor is found in almost every automotive repair shop. Compressed air directed through a nozzle is a common method of drying parts after cleaning them in a parts washer. Most air coming from a compressor contains traces of water and oil, so do not use a compressor to dry brake parts. Petroleum oil

Figure 1-19. A grease gun and cartridges are basic lubrication equipment for a chassis technician.

on brake components damages the brake system seals. If you use compressed air to dry bearings you have cleaned, be careful not to spin the balls or rollers. Doing so will damage the bearing because there is no lubricant in the assembly.

The most common use a chassis technician has for pressurized air is to inflate tires. Air hoses for tires have pressure indicators to let you know how much pressure is in the tire, but always double-check with your own tire pressure gauge. The gauges on shop air hoses are used so much that they are not always correct.

Grease Guns

A grease gun, figure 1-19, is among a chassis technician's most frequently used tools. Loaded with a grease cartridge, it is used to dispense grease through a fitting into steering and suspension ball joints and driveline U-joints. Sometimes the joint has a grease plug, and you must remove the plug and install a fitting. To use a grease gun, place the nozzle on the grease fitting and squeeze the handle gently, figure 1-20. Some shops use bulk grease dispensers, with the gun attached by a tube to a large grease container. Do not apply the grease with too much pressure, or you may damage the seal on the part being lubricated.

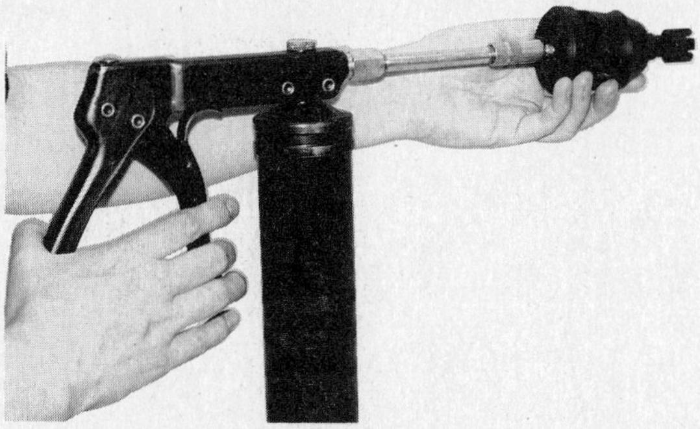

Figure 1-20. To grease undercar components through a grease fitting, put the grease gun nozzle onto the fitting and squeeze the gun applicator.

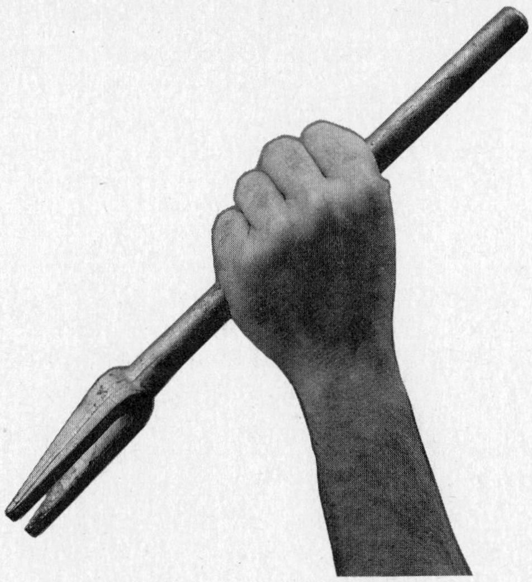

Figure 1-21. A pickle fork forces the top and bottom of the ball joint housing apart.

Lubricating the chassis grease points is a frequent, routine service. Since the late 1980s, however, there has been an industry-wide trend to use permanently sealed chassis joints, which have no fittings and cannot be lubricated — never try to grease a permanently sealed component. In spite of the trend, there are still many chassis lube points: on older cars, on some new cars, and on replacement parts.

Pickle Forks

A pickle fork, figure 1-21, is a long-handled, pronged tool used to break apart steering linkage and suspension ball joints. The prongs are wedged so the thin edge can fit between the two pieces of a ball joint housing. As you

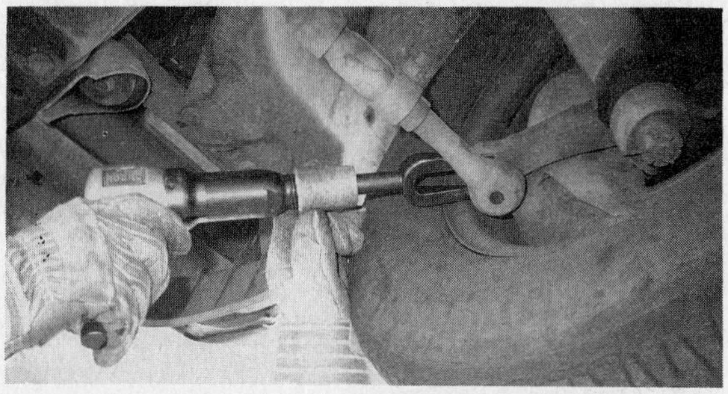

Figure 1-22. A pickle fork extension on a pneumatic impact hammer allows quick disassembly of ball joints.

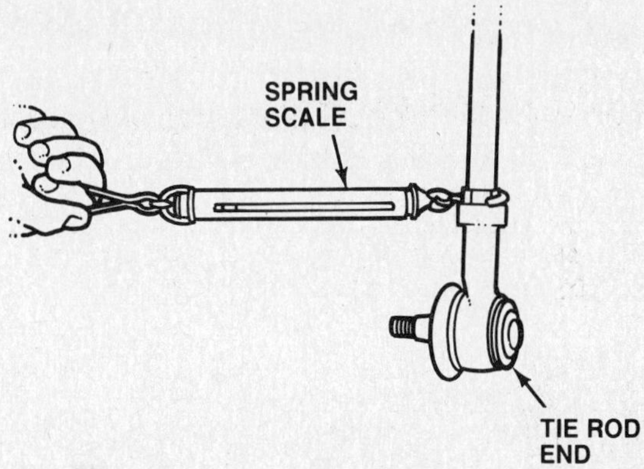

Figure 1-23. Technicians use a spring scale to check the swing resistance of a tie rod in a rack and pinion steering system.

hammer the handle, the wedge pries the housing apart. Some pickle forks are extensions that can be put on pneumatic hammers, figure 1-22, so air pressure can do the work. Be aware that pickle forks ruin the joint dust cover. They are best used for removing joints you intend to discard.

Spring Scale

A spring scale measures force. Its primary use in a chassis shop is for checking tie rod swing resistance in a rack and pinion steering gear. The check is performed with the steering gear and tie rods out of the car. Hook one end of the scale over the tie rod and pull the other end and measure how much force it takes to move the tie rod, figure 1-23. The scale is in Newtons (N) or pounds (lbs) of force.

Power Steering Service Tools

A power steering system requires special tools to service the hydraulic components and the parts that run them. Flare nut wrenches are used on the hydraulic fittings, pressure gauges

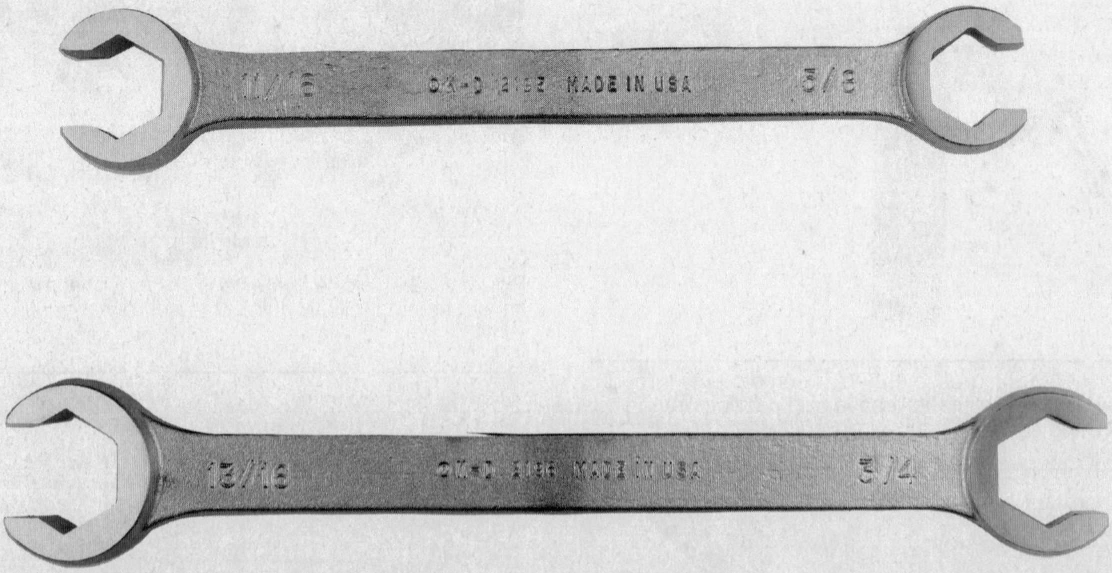

Figure 1-24. Flare nut wrenches are open at one side to fit over hydraulic lines, and they are designed to distribute torque onto each side of the wrench to avoid distorting hydraulic fittings.

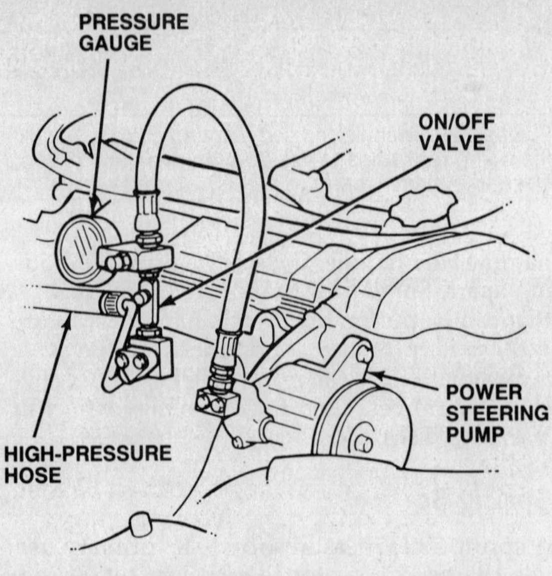

Figure 1-25. A power steering pressure gauge installs between the pump and the high-pressure hose to measure pump output.

measure the hydraulic pressure in the system, special pullers allow removal of the power steering pump pulley, and drive belt tension gauges measure the tension of the belt that drives the pump pulley.

Flare nut wrenches

Flare nut wrenches are used on hydraulic connections and fittings. These tools are essentially box-end wrenches with one side of the box opening cut away so the wrench can be passed over hydraulic lines, figure 1-24.

Hydraulic fittings have relatively thin walls and are easily damaged if turned with an open-end wrench, because open-end wrenches place all the torque on just two flats of a six-sided fitting. A fitting that is distorted even a small amount will leak. Flare nut wrenches distribute the force more evenly, to prevent distortion.

Pressure gauges

A pressure gauge for a power steering system has high-pressure fittings that allow it to be installed between the pump and the high-pressure hose, figure 1-25. A valve on the gauge opens and closes the gauge to system pressure. The gauge dial indicates the fluid pressure from the pump in pounds per square inch (psi) or kilopascals (kPa).

Pulley pullers

A special puller enables the technician to remove the pulley from the power steering pump, figure 1-26. A threaded rod fits into the center hole of the pulley, butting up against the pulley shaft. A pair of clamps is attached to a nut on the shaft, and they fit over the flange around the pulley center hole. The technician secures the clamps to the pulley then threads the nut up the rod, and the clamps pull the pulley up with the nut.

Drive belt tension gauges

Rubber-and-fabric V-belts must be tensioned correctly for efficient output. To resist slipping,

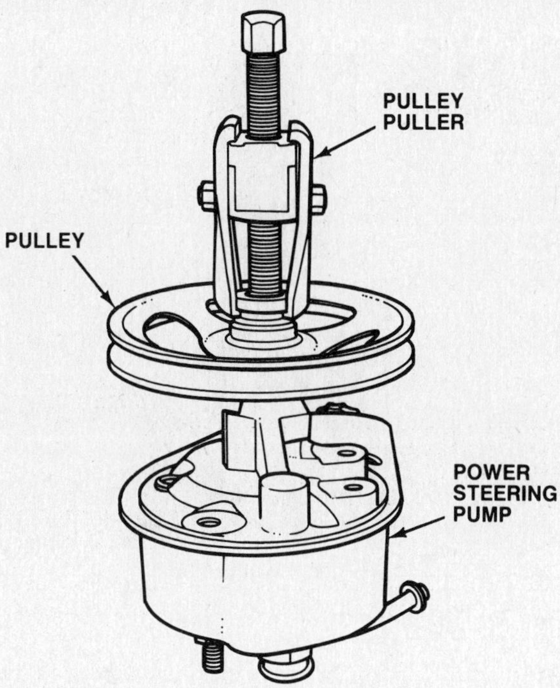

Figure 1-26. As the puller nut threads up the rod, its attachments pull the pump pulley free of the pump.

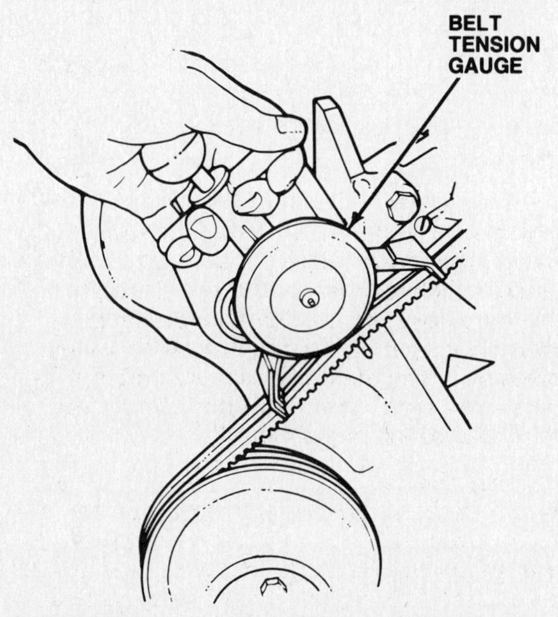

Figure 1-27. A belt tensioning gauge tells the technician whether the drive belt is too tight or loose, which would interfere with the proper functioning of the power steering pump.

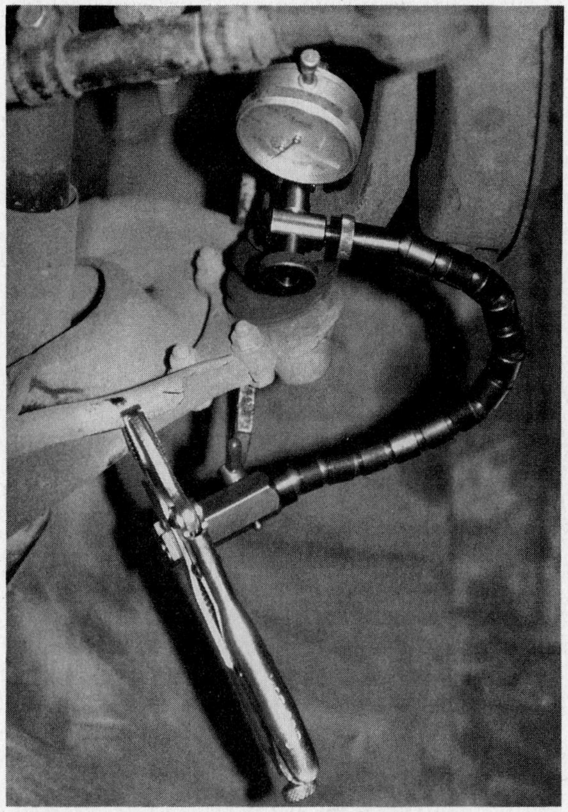

Figure 1-28. This dial indicator is positioned to measure ball joint play. Dial indicators measure movement in small increments that the technician cannot usually detect without the tool.

there must be enough friction between the sides of the belt and the V-shaped groove in the pulley. If the tension is too low, the belt will slip and the driven pulley will not turn at the required speed. In belt-driven steering pumps, this reduced efficiency can interfere with the operation of the power steering system.

The best method for gauging correct belt tension is to use a drive belt tensioning gauge, figure 1-27. This gauge measures the force necessary to deflect a given length of the belt a small distance and will indicate whether the tension is correct. Follow the directions that accompany the tool to use it correctly.

Serpentine belts used on newer cars typically have built-in tensioning devices and require only periodic inspection for wear and correct operation.

Dial Indicators

Chassis technicians use dial indicators to check ball joint play, wheel bearing play, and wheel and tire runout. A dial indicator, figure 1-28,

Figure 1-29. A holddown cone secures the wheel and tire assembly in the tire changing machine.

Figure 1-30. A bead-breaking attachment separates the tire beads from the wheel rim.

registers movement in increments of thousands of an inch or of a millimeter. The dial usually has a stand for placing it near the component whose movement will be measured, and an extension from the ball joint touches the face of the component. The extension may

Figure 1-31. A lever attachment pulls the beads over the wheel rim to remove the tire from the wheel.

have a little wheel at the end if it is for checking wheel and tire runout, to roll along the surface of the wheel or tire as it spins. When used to check the wheel bearings or ball joints, the dial indicator registers how much these parts move in and out. When measuring wheel or tire runout, it measures irregularities in the rotation of the part.

Tire Changing Machines

Special machines are required to mount tires to wheels and to dismount them. Tire changers have a center shaft that fits through the center hole of the wheel. Various sizes of holddown cones thread onto this shaft to secure the wheel and tire assembly into the tire changer, figure 1-29. Wheels on some late-model cars do not have a center hole, and for these you must use a rim clamp. Deflate the tire before removing it from the wheel. Use the bead breaking attachment of the tire changer to separate the beads of the deflated tire from the wheel rim, figure 1-30. Another attachment, a lever that fits over the center shaft, pulls the beads over the rim, figure 1-31. Place the lever over the shaft and fit the end under the top tire bead. Operate the tire changer according to the manufacturer's instructions to spin the wheel and tire assembly. As it spins, the lever lifts the bead out of the wheel rim. Repeat for the lower bead to completely separate the tire from the wheel.

The same lever works to put the beads into the wheel rim when mounting a tire. Once the beads are inside the wheel drop center, loosen

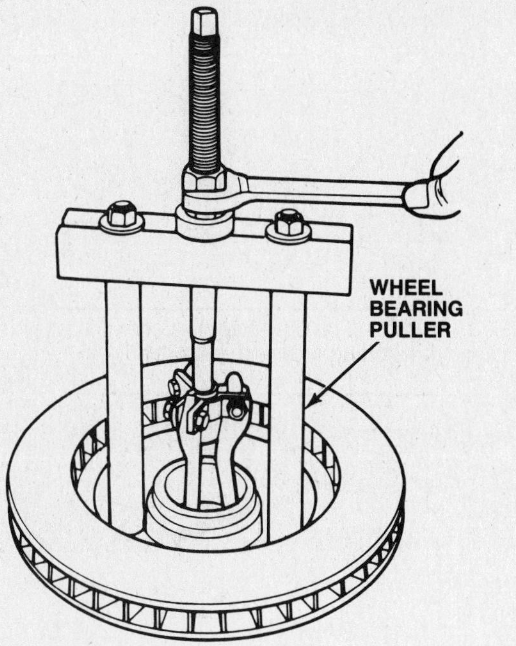

Figure 1-32. As the rod threads up out of the puller, its attachments pull the wheel bearing out of the hub.

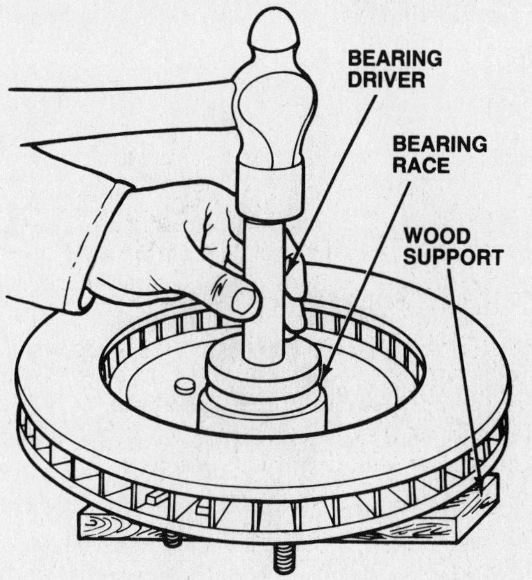

Figure 1-33. When the technician strikes the driver with a hammer, the force is evenly applied to the wheel bearing to drive it straight into the hub.

the holddown cone, then slowly inflate the tire while it is still secured in the tire changer until the beads seat in the rim. They make a loud noise when they seat. Never overinflate a tire to seat the beads. If the beads do not seat by about 30 psi (210 kPa), deflate the tire, lubricate the beads, and try again. If you overinflate, the tire may explode and cause severe injury or death.

Wheel Bearing Tools

Wheel bearing service is part of a chassis technician's jobs. Some wheel bearings are nonserviceable, meaning they cannot be disassembled for lubrication or inspection. If these wheel bearings fail, the technician simply replaces them, often as part of an entire hub assembly. However, some wheel bearings, particularly front wheel bearings in RWD vehicles, are serviceable. Technicians can remove them from the wheel hub, clean them, inspect them, pack them with grease, and reinstall them into the hub. A few special tools help in this task, wheel bearing pullers for removing the bearing, drivers for putting it back into the hub, and bearing packers for greasing it quickly.

Wheel bearing pullers and drivers

Wheel bearing pullers and drivers are removal and installation tools. A wheel bearing puller works like the other pullers we have looked at in this chapter. Two studs hold a plate away from the wheel hub while a central threaded rod attaches to the bearing and pulls it out of the hub as the technician threads the rod out of the puller, figure 1-32.

A bearing driver applies equal pressure to the entire bearing as the technician drives it into the hub with a hammer, figure 1-33. With a driver, the technician can drive the bearing straight into the hub and is less likely to install it at an angle.

Wheel bearing packers

A wheel bearing packer, figure 1-34, uses pressure to quickly force grease into a wheel bearing. Using a bearing packer is a much quicker and cleaner way to lubricate wheel bearings than packing the grease in by hand. Make sure all traces of old grease are cleaned out of the bearing before adding new grease.

Seal Pullers and Drivers

A seal puller, figure 1-35, extends into hard-to-reach places and hooks onto seals so a technician can pull the seals out. Seal drivers are special tools that push an entire seal into place, figure 1-36, applying even force to the whole seal to avoid stretching, twisting, or distorting the seal. Chassis technicians may use seal pullers and drivers in a number of situations, most commonly to remove and install wheel bearing or driveline seals.

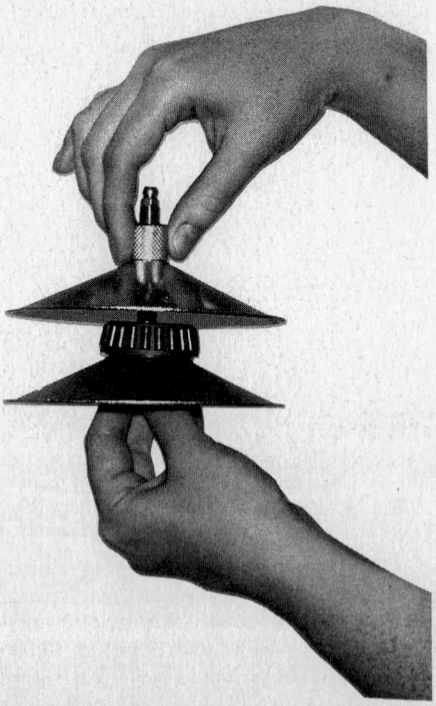

Figure 1-34. A wheel bearing packer is a neat, quick way to fill a wheel bearing with grease. Apply grease through the fitting on the packer.

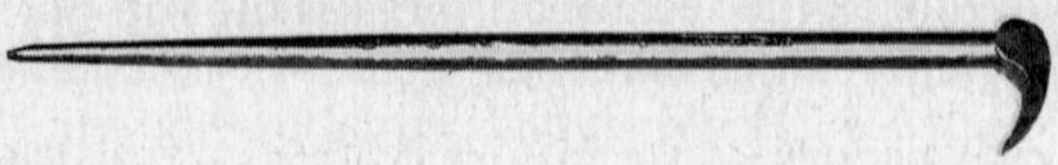

Figure 1-35. A seal puller, or "lady finger", extends into hard to reach places to remove seals.

Wheel Balancers

Wheel and tire assemblies that are balanced have their weight evenly distributed around the hub — static balance — and side-to-side — dynamic balance. Imbalanced wheels cause steering and handling problems and increase tire wear. Technicians can use one of a number of machines to balance wheels and tires. The machines indicate where more weight is needed to balance the assembly, and the technician applies wheel weights, hammer-on clip-type or adhesive-backed, to the wheel where needed.

The three most common types of wheel balancers are:

- Bubble balancer
- Spin balancer
- On-the-car balancer.

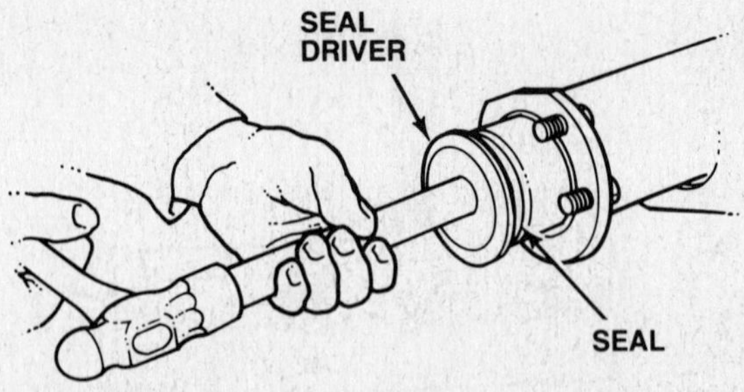

Figure 1-36. A seal driver applies even force to a seal to avoid distorting it during installation.

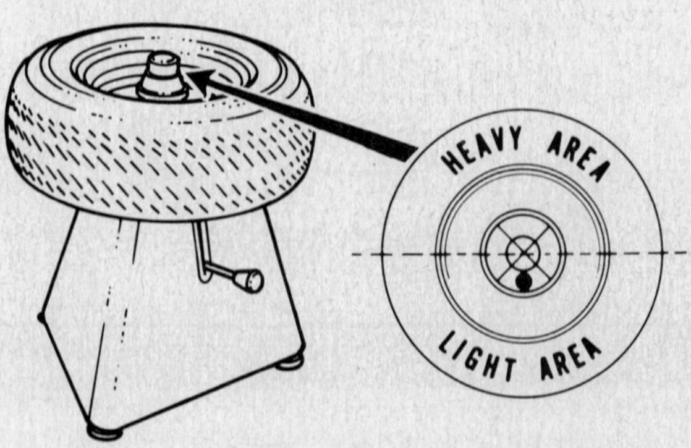

Figure 1-37. A bubble balancer can measure only static wheel balance. It indicates whether the wheel weight is evenly distributed around the hub.

Of these three, the most widely used is the spin balancer, which can check both static and dynamic balance of the wheel and tire assembly.

Bubble balancers

Bubble balancers are an old-fashioned means of correcting static balance. A glass capsule on the center shaft has fluid in it with a bubble at the top. When the bubble is in the center of the capsule, the weight around the shaft is equally distributed. When a technician puts a wheel and tire assembly on the shaft, figure 1-37, the bubble indicates whether the weight is evenly distributed around the hub — static balance. If the tire is not balanced, the technician lays weights on the wheel until it balances, then marks the positions of the weights. After taking the wheel and tire assembly off the shaft, the technician hammers or sticks the weights to the wheel, splitting the weight evenly between the front and back of the wheel to avoid changing dynamic balance. The drawback of a bubble balancer is that it cannot measure dynamic balance, so the technician cannot correct it.

Figure 1-38. A spin balancer can check both static and dynamic wheel balance. A computer display indicates where to place wheel weights.

Figure 1-39. Some spin balancers have the driveshaft on the side.

Figure 1-40. Apply a weight where the balancer display indicates. Clip-type wheel weights that hammer onto the rim are used for steel wheels.

Spin balancers

Spin balancers are the most common wheel balancers because they are the simplest to use and correct both static and dynamic imbalance. The wheel and tire assembly mounts onto a drive shaft on top or on the side of the balancer, figures 1-38 and 1-39. Some top-mount spin balancers double as tire changers because a wheel and tire assembly mounts into it in the same way, and it can spin the assembly for tire removal or mounting. Computer controls enable it to operate as a balancer.

The technician secures the wheel and tire assembly in the balancer and keys in information. Most balancers need the following information:

- Wheel diameter
- Wheel width
- Distance from wheel to balancer cover.

The distance from the wheel to the balancer cover approximates wheel offset. Next, the technician closes the cover and turns on the balancer. The machine spins the wheel and tire assembly and senses its balance. When it stops spinning the assembly, its screen displays a diagram of the wheel and indicates where to place wheel weights and how heavy the weights should be. The technician adds the weights, figure 1-40, then checks the balance again to see if it is correct.

On-the-car balancers

An on-the-car wheel balancer checks balance as the wheel and tire assembly spins while still installed on the car. This type of balancer allows compensation for imbalance in the wheel hub or brake rotor as well as in the wheel and

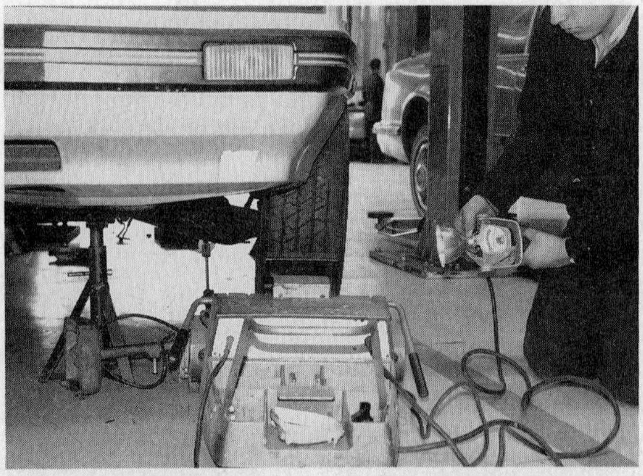

Figure 1-41. An on-the-car balancer allows compensation for hub and brake rotor imbalance as well as wheel and tire imbalance.

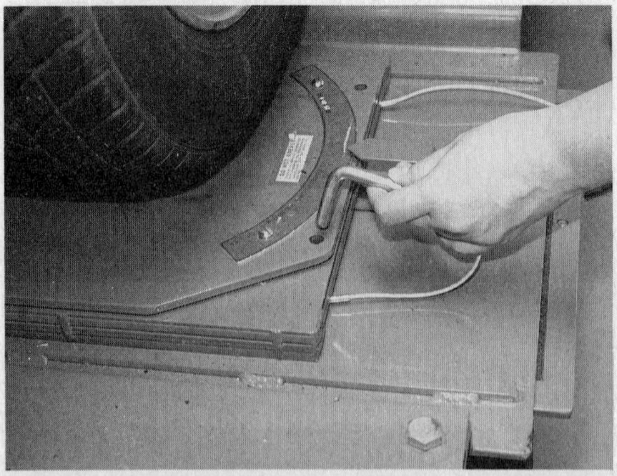

Figure 1-42. Lock the turntables before driving a car on or off an alignment rack.

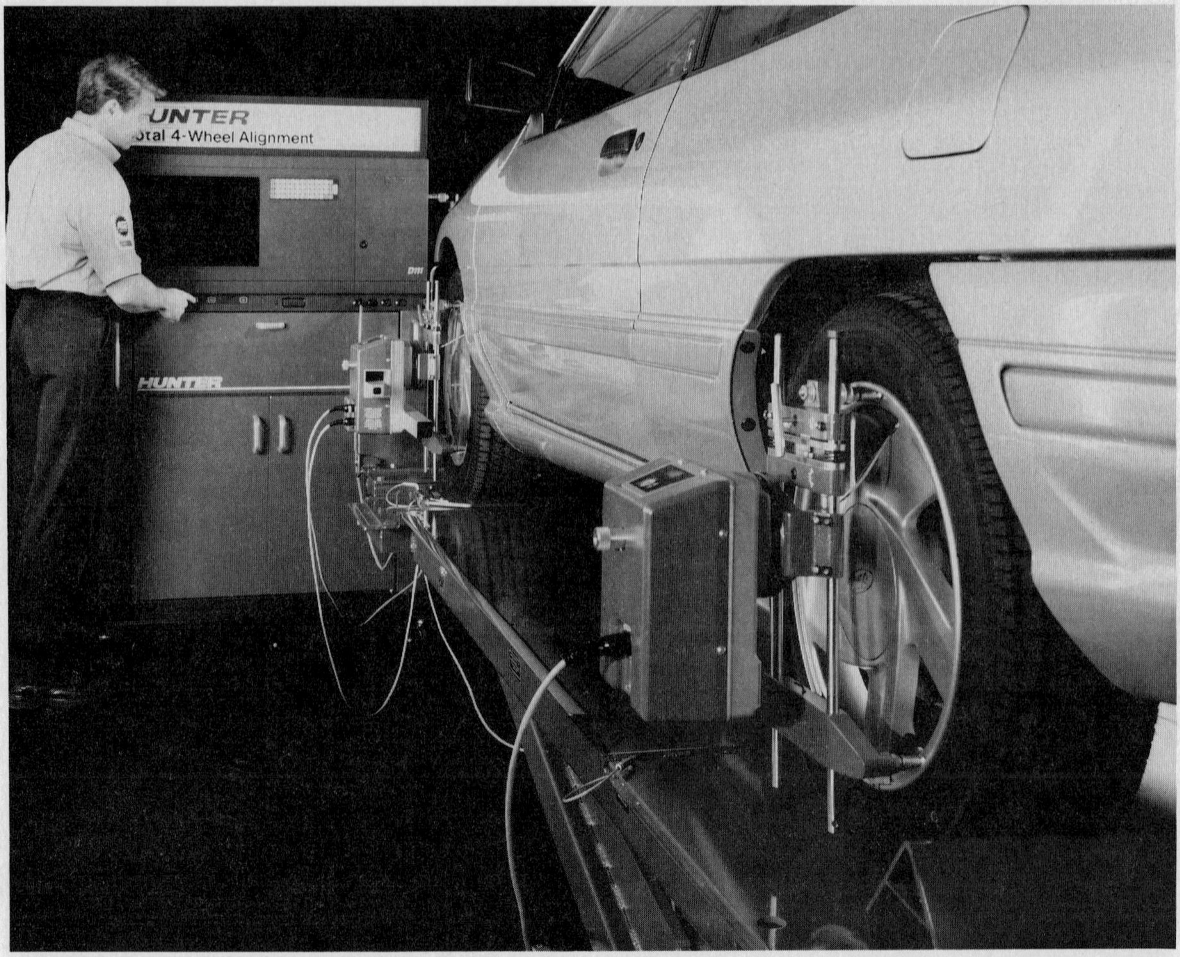

Figure 1-43. The complete four-wheel alignment is the best method of ensuring good handling and maximum tire life.

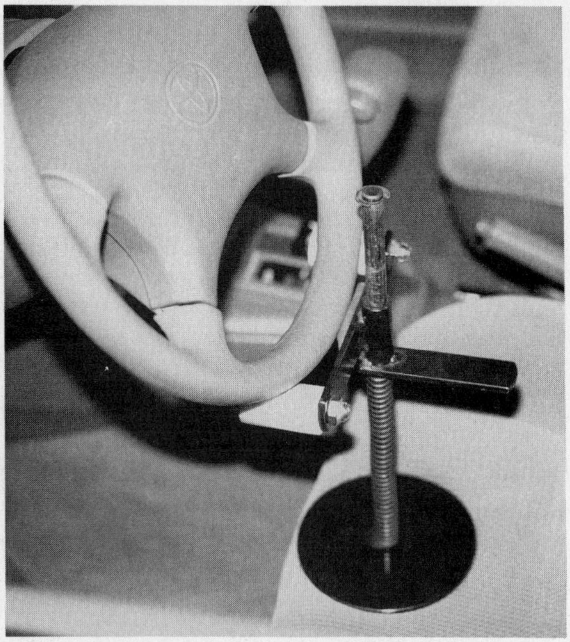

Figure 1-44. A steering wheel lock keeps the steering wheel straight during the toe adjustment. This prevents come-backs when the customer notices a cocked steering wheel.

tire assembly itself. However, whenever you rotate the wheels, you must re-balance them if they have been balanced on the car.

On-the-car balancing uses a motorized drum on which the tire rests and that spins the wheel and tire assembly, an electronic vibration sensor that mounts on the wheel, and a meter that points at the sensor and flashes a strobe when it receives a signal from the sensor indicating a lack of balance, figure 1-41. The technician either makes a white chalk mark on the tire or uses the tire valve as a reference. When the strobe flashes, the technician takes note of the position of the reference mark. When the wheel and tire assembly stops spinning, he or she turns the assembly until the reference mark is in the same position as when the strobe flashed, at which point the heavy part of the assembly is at the bottom and the light part is at the top. The technician adds weights at the top of the wheel, dividing the weight evenly between the front and back of the wheel.

Wheel Alignment Equipment

Wheel alignment positions the wheel and tire assemblies to promote good handling and minimize tire wear. An alignment rack has turntables that the vehicle's front wheels rest on, so the technician can turn the wheels side-to-side to check toe out on turns and caster. Pins that fit through the turntables, figure 1-42, lock them in place while the vehicle is driven on and off the rack. Most modern four-wheel alignment equipment uses four wheel-mounted sensors to gather information and computers to interpret the data, figure 1-43. The alignment measurements, the specifications for the vehicle, and the methods of adjusting alignment are displayed on a screen. Older equipment often is capable only of checking front-wheel alignment and uses a combination of bubble indicators, measuring rods, and other manual methods of measurement. Specifications and adjustment methods are found in service literature, either the manufacturer's shop manual or an aftermarket publication.

Steering wheel lock

A steering wheel lock keeps the steering wheel centered during toe adjustments. It braces the steering wheel against the driver's seat, figure 1-44, so that the feedback through the steering system will not move the steering wheel when the technician is adjusting the tie rods.

Brake lock

A brake lock works similarly to a steering wheel lock. It braces the brake pedal against the steering wheel or driver's seat to keep the brakes applied during a procedure.

Chapter

2 Chassis Problem Diagnosis

The first step in any repair procedure is to determine what is causing the problem. To remove and replace parts, hoping each time that it will solve the problem, without being sure of the source of the problem, is the mark of an unskilled technician. On the other hand, if you, as a knowledgeable technician, gather all the information you can obtain about a vehicle and then think logically about how the systems work, you can narrow down the possibilities and make a reasonable decision about what type of repair is needed.

The benefits of good diagnosis and the drawbacks of poor procedures are many. The most important benefit is providing good service for customers. If you make a correct diagnosis and repair, your customer pays for only the materials and services necessary to make the car work well. When a technician diagnoses incorrectly and makes numerous repairs, the customer pays for goods and services he or she did not need, and this is not fair business. Moreover, performing unnecessary repairs is no more profitable for your shop than doing the job right the first time, because you receive the same pay for your work and materials whether the work is necessary or not.

In the long run, unnecessary repairs are bad for business, for two reasons. One, the customer who pays for them is unlikely to return and very likely to tell his or her friends not to come to you for repairs, and, two, the time you spend on repeated attempts to fix one customer's car could be better spent performing correct, efficient repairs on several customers' vehicles. The more customers you attract and keep, the more secure your position is.

Now that we have seen *why* correct diagnosis is good, let us look at *how* to diagnose problems in the chassis.

GATHERING INFORMATION

To correctly diagnose problems, an automotive technician must possess two things:

- Knowledge
- Understanding.

The technician must *know* exactly what the problem is and *understand* the workings of the various chassis systems well enough to be able to figure out a likely cause for the problem. The sources of knowledge, in this case, are the:

- Customer
- Visual inspection
- Road test.

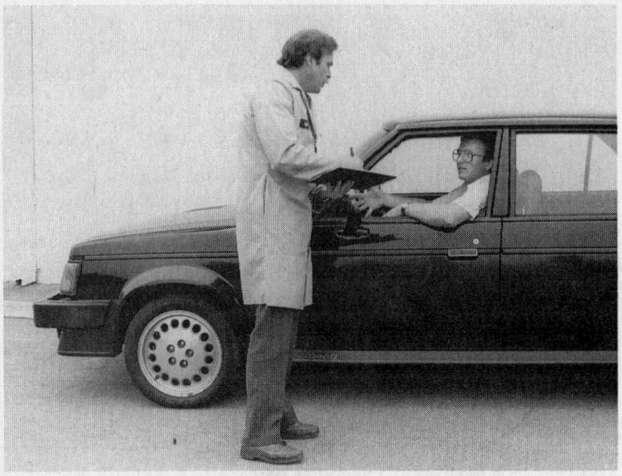

Figure 2-1. Talk to the customer, if at all possible, to get a clear picture of what the problem is.

The understanding comes into play when you analyze the information you gather. A technician's understanding of how a car operates comes from study, training, and experience. It is this understanding that makes the difference between an automotive technician and any other person.

While you are gathering information from the customer, the inspection, and the road test, there are two questions you should keep in mind:

1. What is the symptom?
2. What is the car doing when the symptom occurs?

When you analyze the information to make a diagnosis, there are two more questions to think about:

3. What specific components are affected by what the car is doing at that time?
4. Which of these components is most likely to cause this symptom?

When you have answered all four of these questions, you should have a pretty good idea of what part is causing the problem symptom. If possible, verify your diagnosis by inspecting or testing the part before beginning your repair.

The Vehicle Owner

Start with talking to the customer, figure 2-1. This person has the most experience with the vehicle, and he or she has paid at least enough attention to the problem to seek a repair. In addition to providing you with information, time spent listening attentively to the customer establishes a good relationship with him or her.

Use your conversation to find out what the symptom is and when it happens. Maybe the customer notices at a particular corner on the way to work — is this the only left turn on that drive? Or does the problem occur during braking? Maybe the problem occurs once the car is on the freeway and up to cruising speeds. Most customers do not know a lot of technical jargon, but they are knowledgeable about their own driving routine and habits. Use your own technical knowledge to relate what the customer tells you to what is happening in the car under those circumstances.

After speaking with the customer, turn to the inspection and road test.

Visual Inspection

It is better to inspect the car before driving it, just in case some part is so severely damaged that driving the car would be dangerous or would cause further damage. Your inspection may also turn up problems that are just beginning or of which the customer is unaware. A complete inspection includes checking the following:

- Tire wear patterns
- Tire pressure
- Wheel condition
- Ride height
- Springs or torsion bars
- Struts or shocks
- Bump stops
- CV joint boots
- Rack and pinion steering system boots
- Steering linkage
- Suspension links and bushings
- Ball joints.

Tire wear patterns can indicate alignment problems, incorrect tire pressures, or improper driving habits. Figure 2-2 shows what problems the wear patterns may indicate. Bring tire pressure up to the specified pressure if it is low, but remember that driving on the tires increases the pressure. If the car has been sitting in the shop for a few hours, the pressure readings will be correct, but if the customer just drove in from a place more than a mile away, the pressure reading will be higher because the tires will be hot. Look at the wheels and see if the rims are bent or damaged. If the car is riding low or sags at one corner, the springs are worn out. If the vehicle has adjustable torsion bars, you may be able to compensate for the wear. Test the shocks or struts by bouncing each corner of the car vigorously and then letting go. If the car body bounces more than twice after you release it, the shock or strut at that wheel is worn out.

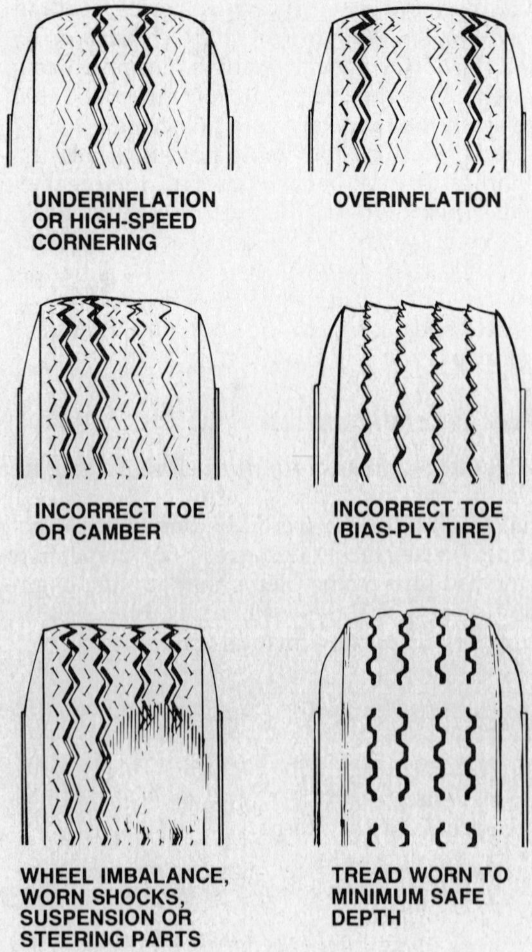

Figure 2-2. Check the tires to see if you can pinpoint a problem by identifying the wear pattern.

Next look at the undercar components, figure 2-3. A slight oil film on the shock cartridge is normal, but look for heavier fluid leaks. Look at the bump stops to see if damage there shows that the suspension has been bottoming out. Examine the constant velocity (CV) joint boots on front-wheel-drive (FWD) and independent-rear-suspension, rear-wheel-drive (RWD) cars, looking for tears, cracks, or lubricant leaks. Check the boots on a rack and pinion steering gear for the same problems.

Grip one piece of steering linkage at a time and try to move it. Sloppy movement indicates worn linkage ball joints or tie rod ends. Look at the suspension links — including control arms, strut rods, and antiroll bars — and see if any are noticeably bent or damaged. Inspect the bushings for wear or damage. Push and pull on the links to see if bushings are worn down.

Figure 2-3. Put the vehicle on a lift and inspect the undercar components for obvious damage or wear.

To check a suspension ball joint for excessive clearance, grasp the arm connected to it and try to move it side-to-side or up-and-down. Unload a load-carrying ball joint to check it, as described in Chapter Nine of this *Shop Manual*.

Road Test

Unless the visual inspection makes it obvious what the problem is, take the car out for a road test. Even if you have a good idea of the problem, a road test can still be helpful, since it can either confirm your guess or show you something about how the car works that might change your mind. During the road test, make sure you drive the car under a variety of conditions. If you test drive for chassis problems frequently, it is a good idea to chart out a course that enables you to:

- Accelerate and decelerate
- Cruise straight ahead at various speeds
- Corner both to the left and right
- Drive on both smooth and bumpy surfaces.

If possible, have the car owner in the car with you during the test drive, so he or she can point out the symptom causing the complaint. If you notice and correct a different problem from the one the car owner wants repaired, you still will not have a satisfied customer.

Notice what kind of a problem it is. Is it a noise or vibration? Where does it seem to come from — the front or rear of the car, or from one side or another? Is it a handling problem, such as wander, pull, or lead? Is it a ride problem? Different components cause different types of problems when they fail or wear down, so get a clear idea of the nature of the problem.

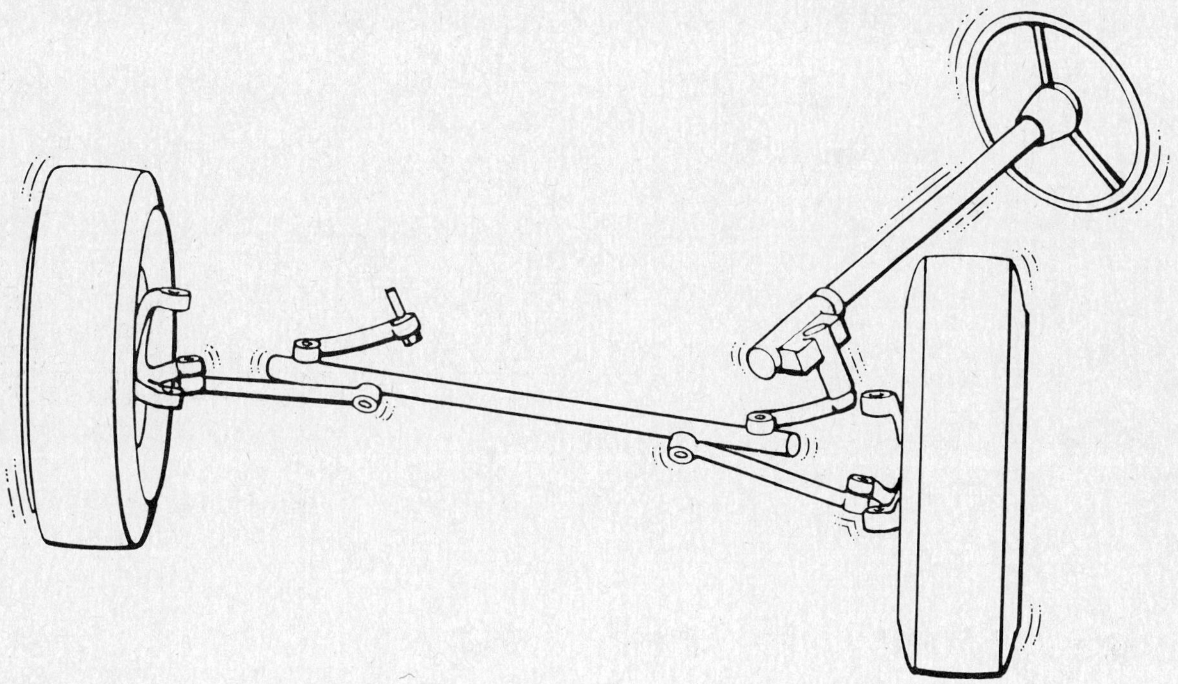

Figure 2-4. Vibrations felt through the steering wheel may come from the wheel and tire assemblies through the steering linkage.

It is crucial to pay attention not only to the symptom but also under what conditions it occurs. Notice whether its occurrence is related to:

- Acceleration or deceleration
- Engine speed
- Vehicle speed
- Straight-ahead driving
- Cornering
- Road surface conditions.

Different systems — steering, suspension, drivetrain — are affected by different conditions, so noting the conditions helps you decide which system is causing the problem. Having gathered all this information, you need to analyze it to determine the cause of the symptom.

ANALYZING THE INFORMATION

In analyzing the problem, first consider the type of symptom with which you are dealing. Certain chassis parts make typical noises or vibrate in typical ways when they break or wear out. Vibrations that come through the steering wheel often are transmitted from the wheels and tires through the steering system, figure 2-4, and may indicate runout due to damaged components or to a lack of balance. Vibrations coming through the floor are more likely to come from the drivetrain. General front-end vibrations could come either from the engine or from a FWD powertrain.

Worn drivetrain parts often clunk when the driver accelerates or decelerates, as the gears or shafts suddenly move after excessive clearance creates a lag time. Worn CV joints, figure 2-5, may knock continuously during straight-ahead driving, and outer CV joints click repeatedly during turns when they are worn, as the balls knock against the cage. A rumble can be either the noise of damaged wheel bearings or CV joints, or the sound of overinflated tires. Squeals usually come from rubber components such as drive belts or tires. Loose suspension or steering links, or exhaust parts, may rattle or knock rapidly as the car drives over a bumpy surface. Simple lack of lubrication may make ball joints and bushings squeak during suspension travel, figure 2-6, but if the lubricant has been gone for any significant length of time, the part is probably damaged.

Problems that occur only during acceleration and deceleration usually originate in the drivetrain, figure 2-7. Likely problems are failed or worn gears, bushings, or bearings in the transmission or in the differential or final drive. Another possibility is worn-out universal joints

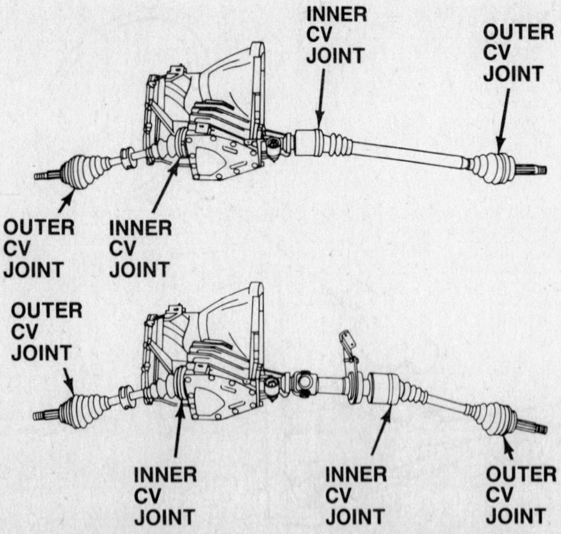

Figure 2-5. Constant velocity joints make noise when they are worn. They may knock or vibrate according to road speed, or the outer CV joints may click during turns.

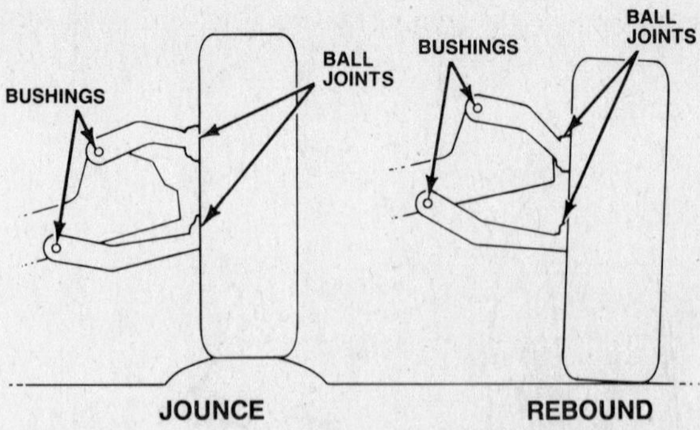

Figure 2-6. Suspension pivots may squeak during wheel travel if they are not lubricated.

(U-joints), figure 2-8, or CV joints. If a FWD car pulls to one side during acceleration and to the opposite side during deceleration, figure 2-9, it has torque steer. To correct torque steer, check for:

- Unequal ride height side-to-side
- Cocked engine, transaxle, or engine cradle
- Heavy load in the rear, pulling the front of the car up
- Collision damage to drivetrain components.

Look at the car to see if one side is riding lower than the other. If it is low on one side, it probably needs new springs or shocks. Check the manufacturer's specifications for the location of the engine, transaxle, and engine subframe mounting hardware, and see if any of these are incorrectly installed. Heavy loads in the rear of the vehicle may alter the axle shaft angles by pulling the front end of the car up. If the owner usually keeps heavy objects in the rear, you may want to recommend heavy-duty rear springs or shocks. First, check if these are legal for the vehicle. Finally, ask the customer if the car has been in a collision, or check for signs of collision damage to the suspension, frame, and sheet metal. If none of these problems are present, but the car still has torque steer, it may simply be a characteristic of the design, to which the owner will have to become accustomed.

Problems related to engine speed most likely occur in the engine compartment and may involve loose or damaged engine mounts, loose or broken accessory pulleys, or worn or improperly tensioned drive belts. Engine vibration is aggravated by damaged engine mounts. Replacing the old mounting hardware with new parts helps isolate the engine from the body and reduces the amount of vibration transmitted to the passenger compartment. Examine the drive belt pulleys for damage and replace them as necessary. Also, check belt tension, as described in Chapter Six of this *Shop Manual*, and correct it if possible. Inspect the drive belts for wear, glazing, or deterioration.

If the symptom occurs during straight-ahead driving or is related to vehicle speed, the problem could be in the drivetrain or wheels, or in the suspension or steering system. Some possibilities are:

- Imbalanced driveshaft or axle shafts
- Worn or failed U-joints or CV joints
- Wheel or tire runout or imbalance
- Dragging brake pads or shoes
- Worn or failed wheel bearings
- Improperly inflated tires
- Tires that interfere with the underbody or suspension
- Mis-aligned wheels
- Worn control arm bushings
- Worn or damaged steering system parts.

As mentioned earlier, each type of problem creates a slightly different symptom, with the driver feeling drivetrain problems through the floor, and problems with the steering system or wheels through the steering wheel. If the driveshaft or an axle shaft is bent or damaged, its rotation will be irregular and cause problems. Replace the damaged parts. Sometimes

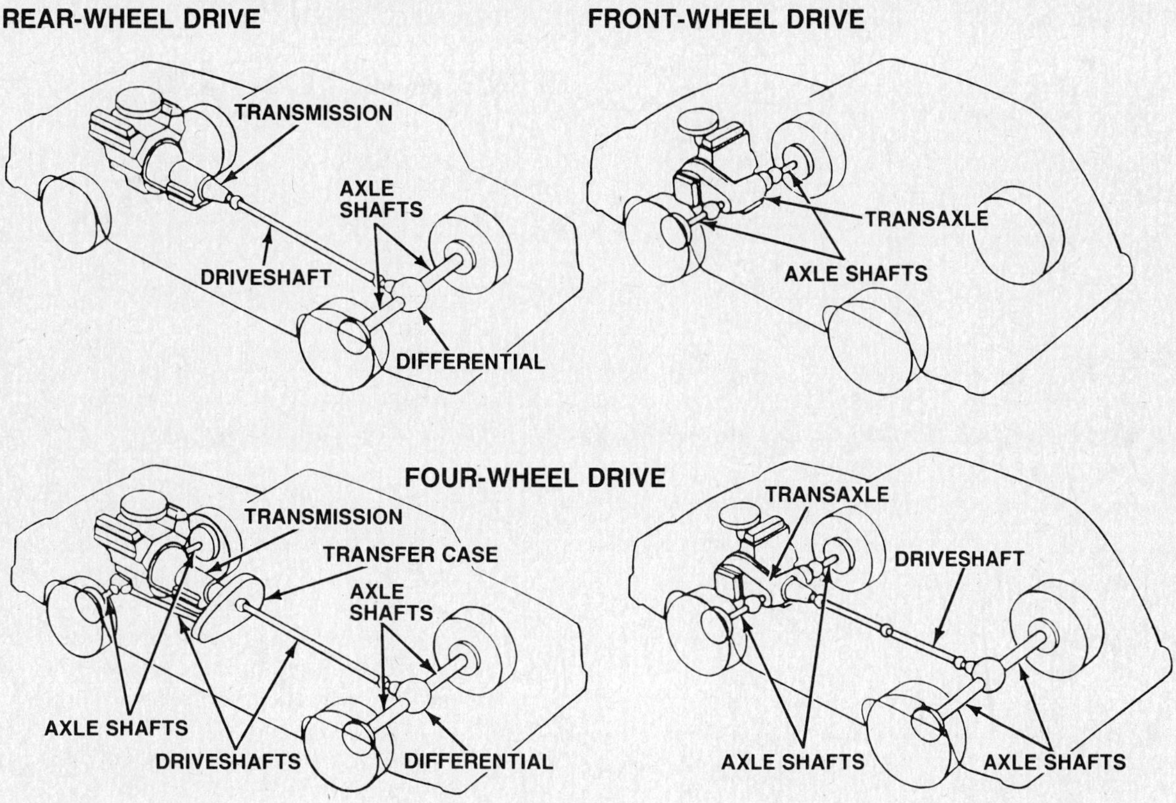

Figure 2-7. Drivetrain problems usually show up during acceleration or deceleration, when the rate of drivetrain operation changes.

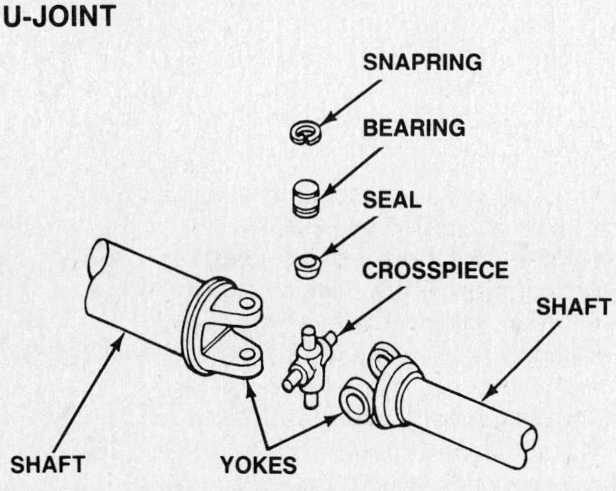

Figure 2-8. The U-joints in a drivetrain can cause vibrations when they wear out.

drivetrain shafts are thrown out of balance by large amounts of dirt clinging to them and changing the weight distribution. In this case, cleaning up the shafts can correct the problem. Inspect the U-joints and CV joints as described

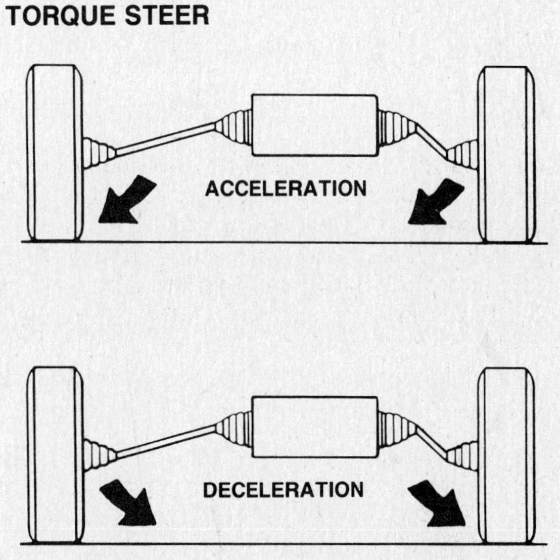

Figure 2-9. Torque steer can be a problem in some FWD vehicles. See if there is some reason, such as a cocked engine sub-frame, that the axle shafts are not at a proper angle.

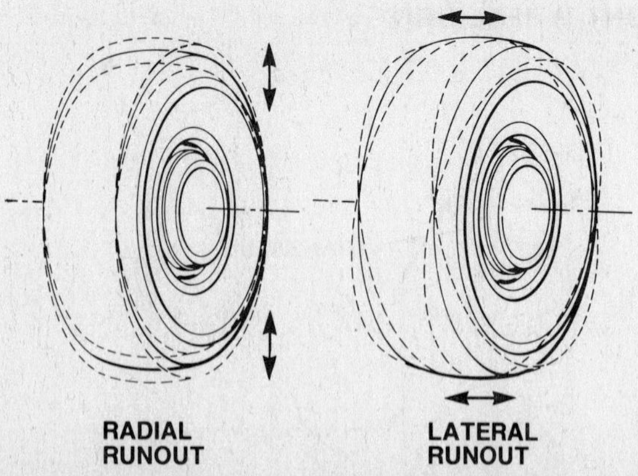

Figure 2-10. Radial runout causes tramp and lateral runout makes the wheels wobble, causing shimmy. Lack of balance is one common cause of runout.

in Chapter Ten of this Shop Manual if the vibration or noise seems to indicate them as the trouble source. Wheel runout or imbalance, figure 2-10, can create a general front-end shimmy or a tramping noise and sensation. If the brake pads or shoes drag on one wheel, it makes the driving feel uneven or unstable. Worn-out wheel bearings allow the wheels to wobble, and they can make a rumbling noise. Improperly inflated tires, especially underinflated tires, make the car feel unstable. Incorrect inflation also wears out tires more quickly. Sometimes car owners try to ''upgrade'' their cars by changing the size of the wheels and tires, but if they choose a size that makes the assembly interfere with the underbody, they create problems. When the wheels are out of alignment, steering often feels unstable. Wear in the suspension or steering system allows sloppy movement of parts that can cause both noise and handling problems.

Sometimes a car constantly leads or pulls to one side. Lead is a less severe condition, in which the car drives toward one side if the driver lets go of the steering wheel. A car pulls if the driver has to actively steer to one side in order to go straight. Causes of lead or pull can be:

- Unequal-size front tires
- Unequally inflated front tires
- Dragging brake pads on one wheel
- Wheels that are out of alignment.

Unequal size or inflation of the front tires, or a grabbing brake pad, makes the tires rotate at slightly different speeds, so that one wheel will tend to be dragged around the other, creating lead or pull. Unequal camber causes the same problem, while unequal caster creates

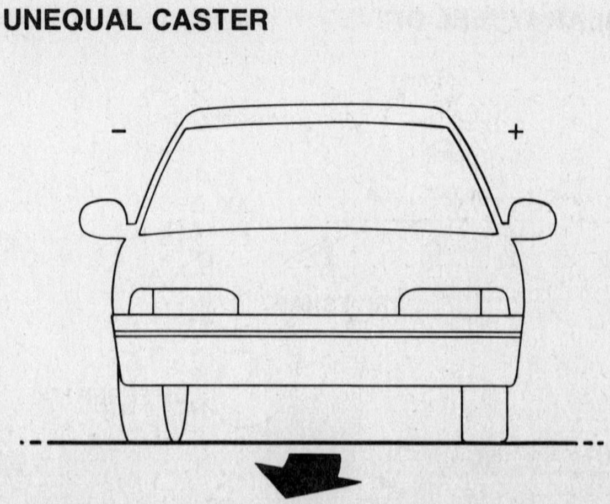

Figure 2-11. Unequal caster is one of the possible causes of pull or lead. The car will tend to drive toward the side with less caster.

pull by allowing one wheel to be deflected more easily, figure 2-11. If none of these are the problem, try switching the front tires side-to-side. If the car pulls or leads in the opposite direction, it may be that the tire has a faulty belt, especially if it is a radial tire. In this case, a new tire or set of tires will eliminate the problem. In a power-assisted steering system, it may be that the steering gear control valve is stuck or tends to push to one side, figure 2-12. This applies more hydraulic pressure to one side than another and tends to turn the car to one side.

Problems related to cornering can be in the steering system, front wheels, or FWD drivetrain and may involve one of the following:

- Loose power steering pump pulley or drive belt that is loose
- Worn or failed steering gear
- Bent or worn tie rods or tie rod ends
- Bent steering arm
- Worn, misadjusted, or failed front wheel bearings
- Worn or improperly inflated front tires, or front tires that interfere with the underbody or suspension
- Worn or failed outer CV joints (FWD).

The driver of a car with a power-assisted steering system, figure 2-13, can suspect problems in the power steering when steering wheel movement becomes stiff and difficult. Check the fluid level and condition first. If the level is low, look for system leaks. Sometimes the pour point depressants in the fluid wear out after some years and steering is stiff in cold weather. In this case, replace the fluid. Other possible causes of stiff steering include low tire

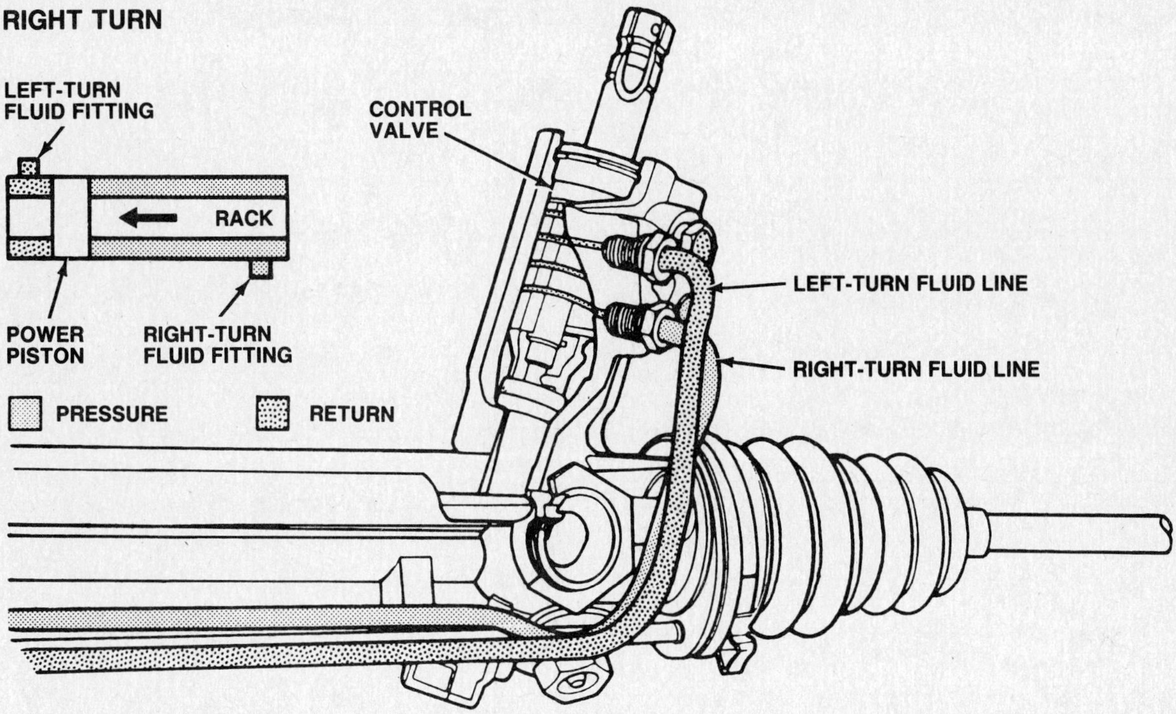

Figure 2-12. If the steering gear control valve sticks in its right- or left-turn position, it will constantly apply hydraulic pressure to one side of the power piston and tend to turn the steering system in that direction.

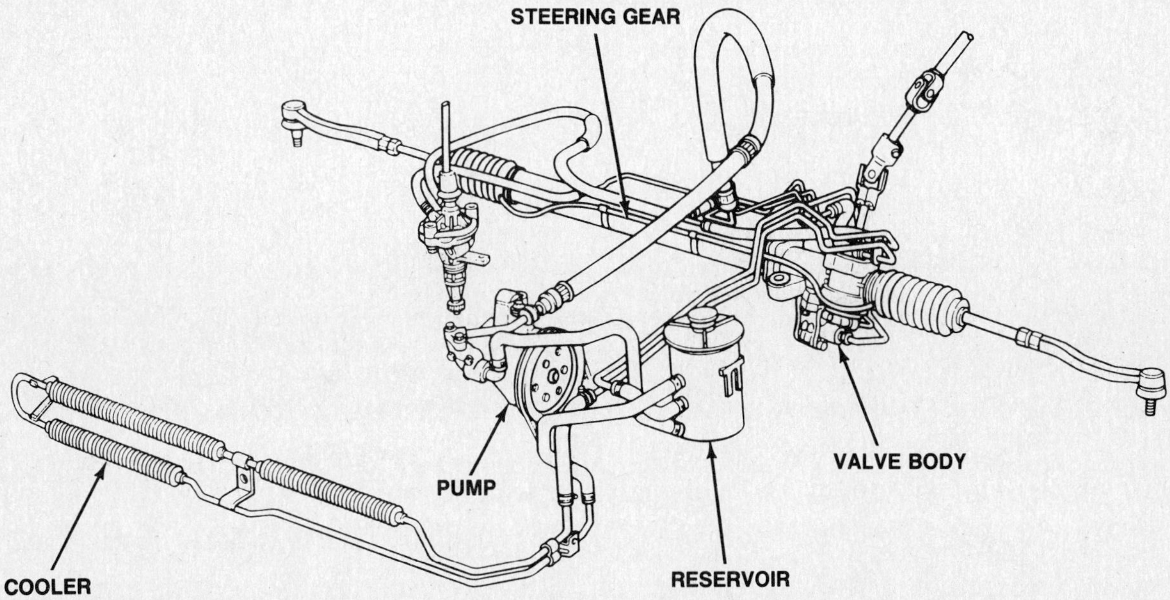

Figure 2-13. Power steering system malfunctions, especially fluid leaks and fluid deterioration, reduce the hydraulic assist and make it more difficult to turn the steering wheel.

pressures, which makes the tires adhere more to the pavement and resist changes in direction, and binding in the steering linkage. Lubricate the linkage if it needs it, but if it has not been lubricated according to the manufacturer's service intervals, the joints may be damaged and need to be replaced.

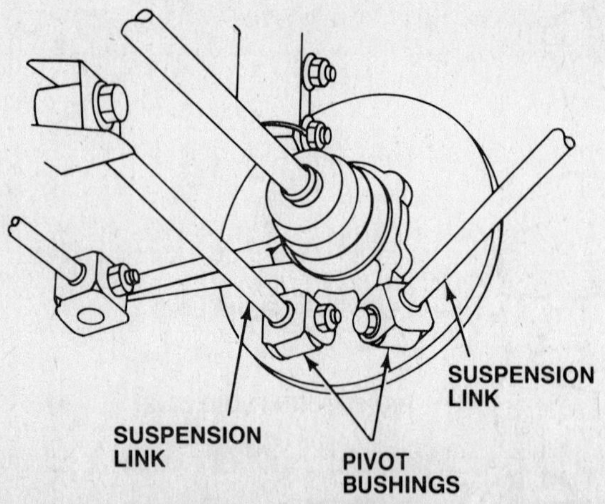

Figure 2-14. If suspension pivot bushings become loose or wear out, they allow the suspension links to vibrate and knock, especially on rough road surfaces.

Problems occurring only on rough road surfaces are most likely in the suspension or steering system, or the exhaust system, and involve one of the following:

- Loose or worn steering or suspension system bushings or bolts
- Suspension ball joints, steering linkage ball joints, or tie rod ends that have too much play
- Worn, damaged, or improperly mounted springs, struts, or shocks
- Exhaust system components that are loose and knock against the underbody or ground.

If suspension pivot bushings, figure 2-14, are loose or damaged, they will allow sloppy movement that can create noise and vibration over rough road surfaces. Tighten loose bolts to manufacturer's specifications with the weight of the vehicle on the wheels. If the bushings are damaged, they must be replaced. Worn ball joints in the suspension or steering linkage also require replacement. Springs, struts, or shocks, figure 2-15, that are worn out allow the car to bounce excessively, making the ride unpleasant. Damaged bump stops indicate that worn springs are allowing the suspension to bottom out. Finally, other undercar components, especially exhaust system parts, are prone to work loose and knock or vibrate when the car is on a bumpy road. Tighten or replace the mounting hardware for these parts.

By the time you are through with your inspection and road test, you will be ready to make an intelligent diagnosis of the car's problem, having narrowed down your choices to the most likely component. For extra help, look at a troubleshooting chart in the manufacturer's service literature. Figure 2-16 shows a general chassis troubleshooting chart. The charts in a manufacturer's shop manual may resemble this one, or they may be flow charts.

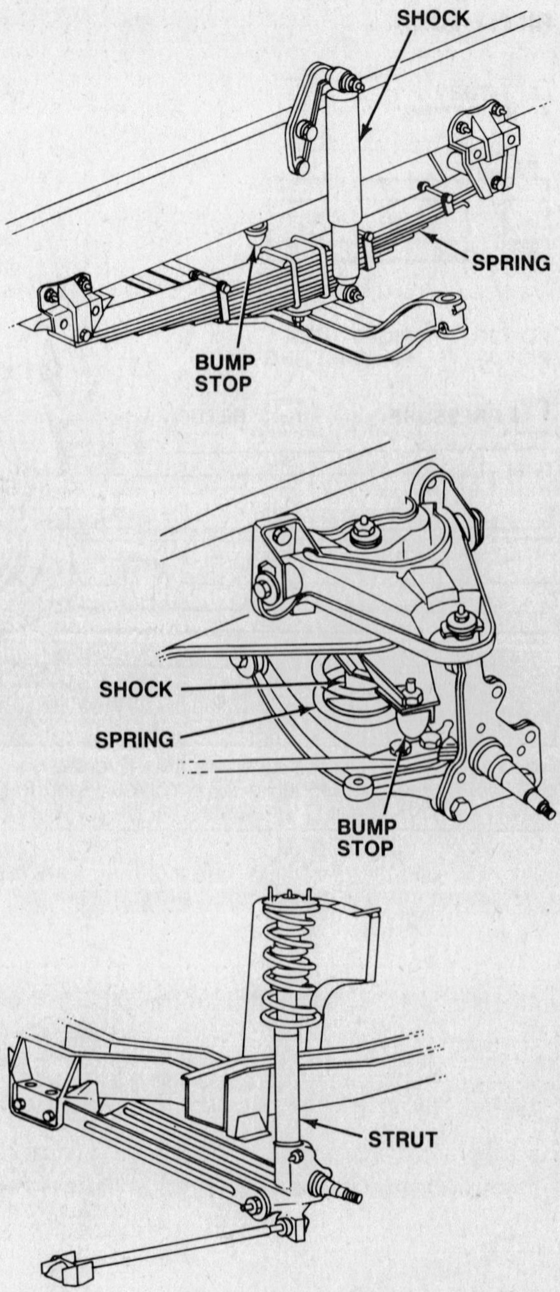

Figure 2-15. Shocks, springs, and struts control bouncing, and ride quality deteriorates when these parts are worn or damaged.

SYMPTOM	POSSIBLE CAUSE	SOLUTION
	Related to Acceleration or Deceleration	
Front end noise	1. Loose or worn front wheel bearings 2. Loose or worn bushings at control arms or shock mounts 3. Worn coil spring isolator(s) 4. Transaxle malfunction, FWD or 4WD	1. Adjust or replace 2. Tighten or replace 3. Replace 4. Service transmission
Clunking noise	1. Worn inner CV joint 2. Faulty transmission gears or bearings 3. Worn or damaged engine or transmission mounts	1. Replace 2. Service transmission 3. Replace
Rumbling noise	1. Worn front wheel bearings 2. Excessive gear clearance in transmission, differential, or transaxle	1. Replace 2. Service transmission, differential, or transaxle
	Related to Engine Speed	
Front end noise	1. Loose or damaged engine accessories or engine mounts 2. Loose or worn drive belt	1. Tighten or replace 2. Adjust or replace
	Related to Vehicle Speed	
Vibrations	1. Unequal tire inflation 2. Unequal tire wear 3. Imbalanced wheels 4. Wheel or tire runout 5. Worn, damaged, or misinstalled driveline joints 6. Imbalanced driveshaft or axle shafts	1. Inflate to correct pressure 2. Rotate or replace 3. Balance wheels 4. Determine cause; correct or replace faulty part(s) 5. Inspect U-joints and/or CV joints; correct problems or replace parts 6. Determine cause; correct or replace faulty part(s)
Rumbling noise	1. Worn or damaged wheel bearings	1. Replace
Thumping or rattling noise	1. Loose or worn control arm bushings, strut upper mount, or antiroll bar bushings and brackets	1. Tighten or replace
	Occurring during Straight-Ahead Driving	
Leading or pulling to one side	1. Unequal tire inflation 2. Incorrect camber or caster 3. Faulty belt in radial tire 4. Faulty power steering gear control valve 5. Dragging brake shoe or pad 6. Broken or weak spring 7. Loose or worn suspension bushings	1. Inflate to correct pressure 2. Perform wheel alignment 3. Switch front tires side-to-side; if direction of pull changes, replace faulty tire 4. Repair or replace steering gear 5. Service brake system 6. Replace 7. Tighten or replace

Figure 2-16. A typical chassis troubleshooting table.

SYMPTOM	POSSIBLE CAUSE	SOLUTION
Occurring during Straight-Ahead Driving		
Wander	1. Incorrect tire inflation 2. Incorrect toe 3. Loose or worn wheel bearings 4. Loose or worn control arm bushings 5. Binding in steering shaft coupling or steering gear, or in front strut upper bearing 6. Steering gear needs adjustment 7. Broken or weak spring 8. Damaged shock or strut	1. Inflate to correct pressure 2. Perform wheel alignment 3. Adjust or replace 4. Tighten or replace 5. Lubricate, repair, or replace 6. Adjust 7. Replace 8. Replace
Front wheel shimmy	1. Imbalanced wheel(s) 2. Incorrect caster 3. Uneven tire wear 4. Loose or worn suspension bushings 5. Worn suspension or steering linkage ball joints	1. Balance wheels 2. Perform wheel alignment 3. Rotate or replace 4. Tighten or replace 5. Replace
Thumping or rattling noise	1. Loose or worn control arm bushings, strut upper mount, or antiroll bar bushings and brackets 2. Loose or damaged steering gear mounting bolts or studs	1. Tighten or replace 2. Tighten or replace
Tire squeal	1. Brake drag 2. Insufficient clearance for tires 3. Wheels severely out of alignment	1. Service brakes 2. Install wheel and tire assemblies that meet manufacturer's specifications 3. Perform wheel alignment
Occurring during Cornering		
Clicking noise	1. Worn or damaged outer CV joint, FWD or 4WD	1. Replace
Increased steering effort	1. Low tire pressure 2. Insufficient lubricant in steering shaft coupling, steering gear, or steering linkage ball joints 3. Damaged steering shaft coupling, steering gear, or steering linkage ball joints 4. Low power steering fluid level 5. Loose power steering pump drive belt 6. Faulty power steering pump 7. Excessive caster	1. Inflate to correct pressure 2. Lubricate 3. Replace 4. Top off; check system for leaks 5. Adjust 6. Confirm by pressure test; replace 7. Perform wheel alignment
Excessive steering wheel freeplay	1. Steering gear needs adjustment 2. Worn or loose steering shaft coupling, steering gear mounting bolts, or steering linkage ball joints 3. Worn steering gear internal components 4. Loose or worn front wheel bearings	1. Adjust 2. Tighten or replace 3. Replace steering gear 4. Adjust or replace

Figure 2-16. A typical chassis troubleshooting table (continued).

SYMPTOM	POSSIBLE CAUSE	SOLUTION
	Occurring during Cornering	
Steering wheel vibrations during low-speed maneuvering	1. Improper tire inflation 2. Air in the power steering system, new or recently repaired power steering system 3. Engine vibration 4. Loose tie rod ends 5. Faulty drive belt tensioner 6. Overcharged air conditioning system	1. Inflate to correct pressure 2. Advise customer to wait and see if condition disappears after several weeks of driving 3. Service engine 4. Replace 5. Replace 6. Check air conditioning pump pressure
Steering wheel does not recenter after turns	1. Improper tire inflation 2. Incorrect wheel alignment 3. Mispositioned or binding steering shaft coupling(s) 4. Binding steering shaft bearings 5. Binding in steering gear 6. Insufficiently lubricated steering linkage ball joints	1. Inflate to correct pressure 2. Perform wheel alignment 3. Reposition or replace 4. Replace 5. Adjust steering gear; repair or replace if needed 6. Lubricate; replace if damaged
Hissing noise during sharp cornering	1. Normal pressure relief valve operation	1. No correction; instruct owner not to keep steering wheel at full lock for more than 5 seconds
Squealing noise under hood	1. Loose drive belt	1. Adjust tension
Whining or growling noise under hood	1. Low power steering fluid level 2. Power steering hoses touching body or frame 3. Faulty power steering pump	1. Top off; check system for leaks 2. Reposition 3. Replace pump, flush power steering fluid, and refill system
Rumbling noise	1. Loose or worn front wheel bearings	1. Adjust or replace
Scrubbing or knocking noise	1. Insufficient clearance for tires 2. Incorrect steering gear installed	1. Install wheel and tire assemblies that meet manufacturer's specifications 2. Install correct steering gear
Tire squeal	1. Low tire pressure 2. Insufficient clearance for tires 3. Worn tie rod ends or bent steering arm 4. Wheels out of alignment	1. Inflate to correct pressure 2. Install wheel and tire assemblies that meet manufacturer's specifications 3. Replace 4. Perform wheel alignment
	Related to Road Surface	
Harsh ride	1. Worn or damaged springs, shocks, or struts 2. Loose or damaged suspension bushings	1. Replace both same-axle springs, shocks or struts 2. Tighten or replace
Rattling noise	1. Loose or worn control arm bushings, shock or strut mounts, or antiroll bar bushings and brackets 2. Loose or damaged steering gear mounting bolts or studs 3. Loose exhaust system components	1. Tighten or replace 2. Tighten or replace 3. Tighten or replace mounting hardware
Squeaks	1. Insufficient lubricant in suspension or steering bushings or ball bearings 2. Damaged shocks or struts	1. Lubricate; replace if damaged 2. Replace

Figure 2-16. A typical chassis troubleshooting table (continued).

TIRES
PASSENGER CAR

Tire sizes and tire pressures represent the car manufacturer's recommendations (for normal speeds).
†Sustained high speeds: **Numeric and Alphanumeric** sizes, add 4 psi (do not exceed 32 psi for load range "B", 36 psi for load range "C", 40 psi for load range "D"); **P-Metric** sizes, add 3 psi (do not exceed 35 psi for Standard Load, 41 psi for Extra Load "XL" tire).
European metric sizes have maximum inflation pressure of either 32 or 36 psi, depending on construction. Refer to maximum inflation limit stamped on tire sidewall.
*For minimum load pressure, refer to tire decal. •Cars equipped with Trailer Tow Option, refer to tire decal.

MAKE YEAR	MODEL	ORIGINAL EQUIPMENT SIZE Std.	Optional	PRES-SURE†• Max. Load* F-R	RIM WIDTH
DODGE Continued					
1988-83	600 ex. convertible	P185/70R14		26-32	5½-JJ
1987-85	Daytona ex. Turbo	P185/70R14	P195/70R14	35-35	5½-JJ
	Daytona Turbo	P205/60HR15		35-35	6-JJ
			P225/50VR15	35-35	6½-JJ
	Shelby Charger	P205/50VR15		35-35	6-JJ
1987-84	Aries	P175/80R13		35-35	5-JB
			P185/70R14	35-35	5½-JJ
1986	Conquest	P215/60R15		28-28	6½-JJ
	W/Intercooler: Front	205/55VR16		28	7-J
	Rear	225/50VR16		28	8-J
1986-85	Omni ex. GLH	P165/80R13	P175/75R13	35-35	5-JB
	Omni GLH	P195/50R15		35-35	6-JJ
	Charger ex. 2.2, Shelby	P165/80R13		35-35	5-JB
			P195/60R14	35-35	5½-JJ
	Charger 2.2	P195/60R14		35-35	5½-JJ
	600 Convertible	P195/60VR15		26-29	6-JJ
1985	Conquest	P195/70R14		28-28	6-J
			P215/60VR15	28-28	6½-JJ
1984	Conquest	P195/70R14		27-27	6-JJ
			P215/60R15	27-27	6½-JJ
	Daytona ex. Turbo	P185/70R14		35-35	5½-JJ
	Daytona Turbo	P195/60R15		35-35	6-JJ
	Omni	P165/80R13		35-35	5-JB
	Charger ex. Shelby	P195/60R14		35-35	5½-JJ
	Shelby Charger	P195/50R15		35-35	6-JJ
	Diplomat	P205/75R15		29-29	7-JJ
1984-83	Colt ex. Vista	155SR13		31-31	4-,4½-J
			P175/70R13	24-24	4½-J
1984-82	400	P185/70R14		29-29	5½-JJ
1983	Aries	P175/75R13	P195/70R13	35-35	5-JB
			P185/70R14	35-35	5½-JJ
	Challenger	195/70SR14		24-24	5½-JJ
	Diplomat	P195/75R15	P205/75R15	29-29	5½-,7-JJ
	Mirada	P195/75R15		29-29	5½-JJ,7-JJ
			P205/75R15	29-29	7-JJ
			P215/75R15	29-29	7-JJ
	Omni ex. Charger 2.2	P175/75R13		35-35	5-JB
			P195/60R14	35-35	5½-JJ
	Charger 2.2	P195/60R14		35-35	5½-JJ
EAGLE					
1992	Summit: 3-dr.	P155/80R13		29-29	4½-JB
	4-dr.	P155/80R13		29-29	5-JB
	Summit ES: 3-dr.	P175/70R13		29-29	4½-JB
	4-dr.	P175/70R13		29-29	5-JB
	Talon: FWD	P205/55R16		29-26	6-JJ
			P205/55VR16	29-26	6-JJ
	Turbo	P205/55VR16		29-26	6-JJ
	4WD	P205/55VR16		32-29	6-JJ
	Summit Wagon FWD: MT	P185/75R14		30-30	5½-JJ
	AT & 4WD	P205/70R14		30-30	5½-JJ
1992-90	Premier ES	P205/60R15		30-30	6-JJ
	Premier LX	P195/70R14		30-30	5½-JJ
			P205/6[illegible]	30-30	6-JJ
1991	Summit	P155/[illegible]		[illegible]-30	5-JB
	Summit ES				5-JB

MAKE YEAR	MODEL	ORIGINAL EQUIPMENT SIZE Std.	Optional	PRES-SURE†• Max. Load* F-R	RIM WIDTH
FORD CROWN VICTORIA, LTD Continued					
1986-85	LTD wagon	P195/75R14	P205/70R14	28-35	5-,5½-JJ
1985	Crown Victoria	P215/75R14		31-35[1]	6½-JJ
			P205/75R15	31-35	6-,6½-JJ
			P215/70R15	31-35	6-,6½-JJ
1984-83	LTD ex. Crown Victoria	P185/75R14	P195/75R14	30-30	5-,5½-JJ
			P205/70R14	35-35	5½-JJ
	Crown Victoria	P215/75R14	P225/75R14	31-35	6½-JJ
			P205/75R15	31-35	6-,6½-JJ
	Wagon ex. Country Squire	P185/75R14	P195/75R14	28-35	5-,5½-JJ
	Country Squire	P215/75R14	P225/75R14	27-35	6½-JJ
			P205/75R15	27-35	6-,6½-JJ
1 Wagon, 27-35 **2** Aluminum wheels, 6½-JJ					
FORD ESCORT, EXP					
1992	LX sedan	P175/65R14		32-32	5-JJ
	LX-E	P185/60R14		32-32	5½-JJ
	GT	P185/60R15		32-32	5½-JJ
	All others	P175/70R13		32-32	5-JJ
1991	GT	P185/60HR15		32-32	6-JJ
1991-89	All ex. GT	P175/70R14		32-32	5-,5½-JJ
1990-89	GT	P195/60R15		30-30	6-JJ
1988	Escort ex. Pony, GT	P165/80R13		30-30	4½-,5-JB
	Pony	P175/80R13		35-35	4½-JJ
	EXP	P185/70R14		30-30	5½-JJ
1988-86	GT	P195/60HR15		30-30	6-JJ
1987	Escort ex. GT	P165/80R13		30-30	4½-,5-JB
			P175/80R13	35-35	5-JB
	EXP ex. GT	P185/70R14		30-30	5-JJ
1986	All ex. GT	P165/80R13		30-30	4½-,5-JB
			P175/80R13	35-35	5-JB
1985-84	All ex. EXP	P165/80R13	P175/80R13	35-35	4½-,5½-JB
			P165/80R13	30-30	5-,5½-JB
			165/70R365‡	30-30	135 mm
			185/65R365‡	28-28	135 mm
	EXP	P165/80R13		30-30	5-,5½-JB
			165/70R365‡	30-30	135 mm
			185/65R365‡	28-28	135 mm
1983	All ex. EXP	P165/80R13		35-35[1]	4½-,5-JB
			165/70R365‡	30-30	135 mm
	EXP	P165/80R13		30-30	5-JB
			P175/80R13	30-30	5½-JB
			165/70R365‡	30-30	135 mm
1 Wagon, 30-30					
FORD FAIRMONT					
1983	Fairmont ex. wagon	P175/75R14	P185/75R14	35-35	5-,5½-JJ
			190/65R390‡	28-28	150 mm
FORD FESTIVA					
1992	GL	165/70SR12		32-32	4½-J
1992-90	L			32-3[illegible]	4-[illegible]
1991-90	L Plus, LX				

Figure 2-17. This specifications chart provides information about the tire sizes and rim widths allowed on each car model, as well as the recommended inflation pressure.

Troubleshooting charts provide helpful pointers, especially for beginners, but they are no substitute for knowledge and understanding of chassis systems. Use them to aid you, not direct you, in your diagnosis. Whenever possible, confirm your diagnosis by closely examining the part before removing it for repair or replacement. The chapters in this *Shop Manual* contain detailed descriptions of how to examine and repair the specific parts of each chassis system.

When you make repairs or adjustments, refer to the manufacturer's specifications for part numbers, torque values, size and capacity, and other information, figures 2-17 and 2-18. Follow these precisely, for two reasons — first, to do the job right and, second, to avoid liability if you make an incorrect repair. If a customer

WHEEL NUT TORQUE

MAKE	YEAR	MODEL	TORQUE Ft/lbs
EAGLE			
	1992-90	Talon	85-100
	1992-89	Summit	65-80
	1991	Premier	90
	1990-89	Premier	63
FORD			
	1992-90	All ex. Escort, Festiva, Probe	85-105
		Escort	65-87
	1992-89	Probe	65-87
	1992-88	Festiva	65-87
	1989-83	All ex. Festiva, Probe	85-105
GEO			
	1992-90	Storm	87
	1992-89	Prizm	76
		Metro	45
	1989	Spectrum	65
		W/aluminum wheels	87
HONDA			
	1991-84	Civic, CRX, Accord, Prelude	80
	1983	Civic	50-65
		Prelude, Accord	72-88
HYUNDAI			
	1992-90	All	65-80
	1989-86	Steel wheels	50-57
		Aluminum wheels	65-72
INFINITI			
	1991-90	Q45	72-87
		M30	76-90
ISUZU			
	1992-90	Stylus	87
	1992-88	Impulse	87
	1989-88	I-Mark: W/steel wheels	65
		W/aluminum wheels	87
	1987-85	I-Mark FWD: W/steel wheels	60
		W/aluminum wheels	87
	1987-84	Impulse	80-94
	1985-83	I-Mark RWD: W/steel wheels	50
		W/aluminum wheels	90
MERCURY			
	1992-91	All ex. Capri, Tracer	85-105
		Capri, Tracer	65-87
	1990	All	85-105
	1989-88	Tracer	65-87
	1989-83	All ex. Tracer	85-105
MERKUR			
	1989-88	Scorpio	52-73
	1989-85	XR4Ti	75-101
MITSUBISHI			
	1992-91	All ex. Eclipse, 3000GT	65-80
		Eclipse, 3000GT	87-101
	1990-87	All	65-80
	1986-84	All w/aluminum wheels	65-80
	1984-83	All w/steel wheels	51-58
	1983	All w/aluminum wheels	58-73
NISSAN			
	1992-90	All	72-87
	1989-86	All ex. 200SX	72-87
		200SX	87-108
	1985-83	All ex. 1986-83 pickup w/steel wheels	58-72
OLDSMOBILE			
	1992-90	All	100
	1989	All ex. Calais	100
		Calais	80
	1988-87	Ciera, Custom Cruiser, Firenza	100
		Calais, Cutlass, Delta 88, Ninety-Eight	80
	1986	Cutlass Supreme	80
		Custom Cruiser: W/1/2″ studs	100
		W/7/16″ studs	80
		Others, FWD	100
	1985	Cutlass Supreme, Delta 88, Calais	80
		Others	100
	1984	Cutlass ex. Ciera	80
	1984-83	Omega, Ciera, Firenza	100
		Custom Cruiser, Ninety-Eight	100
		88 ex. Custom Cruiser	80
		Toronado	100
	1983	Cutlass ex. Ciera	80
		With aluminum wheels	90
PEUGEOT			
	1992-89	405	
PONTIAC			
	1992-90	All ex. LeMans	100
		LeMans	65
	1989	Bonneville, Firebird, Safari	100
		Grand Am	100
		Grand Prix	103
	1989-88	LeMans	65
		Sunbird, 6000	100
	1988	Bonneville, Firebird	80
	1988-87	Firefly	43
		Grand Prix, Firebird	80
	1988-86	Fiero	100
		Sunburst	65
	1987	Bonneville, Grand Am, Safari	100
	1987-83	1000	80
		2000, 6000, Phoenix, Sunbird	100
	1986	Parisienne: W/1/2″ studs	100
		W/7/16″ studs	80
	1986-85	Firefly	45
	1986-84	Bonneville, Grand Prix, Firebird	80
	1985-84	Fiero	80
		Parisienne: W/3/4″ nut	80
		W/13/16″ nut	100
	1983	Bonneville, Grand Prix, Firebird	80
		W/aluminum wheels	100
PORSCHE			
	1990	911 Carrera 4	94
	1990-87	924S	96
	1988-83	944	96
		911, 928	96
	1983	924: Steel wheels	80
		Alloy wheels	96
RENAULT			
	1989-88	Medallion	63
	1987	GTA	65
	1987-85	Alliance, Encore	63
	1986-83	Fuego, 18i	60
	1984-83	Alliance, Encore	55
SAAB			
	1992-90	900, 9000	80-90
	1989-87	900	70-90
	1989-86	9000 w/aluminum wheels	75-90
	1986-83	99, 900	65-80
SATURN			
	1992-91	All	100

Figure 2-18. Torque values are important to avoid damaging parts. Tightening the wheel lug nuts correctly makes the wheel and tire assembly secure without damage to the rotor or hub.

can prove that your work did not meet manufacturer's specifications, you are legally responsible for the resulting equipment failure. Many repair shops maintain a library of manufacturer's service literature, and there are aftermarket publishers (such as Chek-Chart) who gather specifications from all the manufacturers into single publications for the general service station technician's convenience.

When you are done with your repair, road test the car again to make sure you have eliminated the problem. No matter how careful you are, you cannot always make a correct diagnosis the first time. If your repair has not eliminated the trouble symptom, re-evaluate the problem and make a new diagnosis. By being as accurate as possible, both in your diagnosis and your repair work, you will reduce the chances of error and increase your customer's satisfaction.

Chapter

3

Electrical and Electronics Basics

In most cases, repairing a steering and suspension system does not involve work with the vehicle's electrical system. However, some vehicles have electronically controlled suspension and steering systems. These systems use sensors that detect driving conditions, process the information using a computer, and then actuate steering or suspension components with electric, hydraulic, or pneumatic actuators. To repair these systems, you must know how to retrieve trouble codes from the vehicle's suspension computer and perform simple electrical tests.

If an electronically controlled steering or suspension system is not operating properly, an indicator on the dash will illuminate, figure 3-1, informing the driver that the system needs repair. Like an engine computer, the suspension computer will store trouble codes that a trained service technician can retrieve to determine which components or circuits are malfunctioning.

When you service an electronically controlled system, be sure to have the manufacturer's service manual on hand. Although this chapter contains basic information that will help you service electronically controlled steering and suspension systems, it cannot replace the service manual. This chapter includes information on:

- Safety
- Organized troubleshooting
- Using electrical test devices
- Understanding basic terminology and electrical tests.

The last section includes a brief example of how you might troubleshoot a pneumatically actuated, electronically controlled suspension system.

SAFETY

General safety is covered in Chapter One of this *Shop Manual*. When working on a vehicle's electrical system, however, there are additional precautions. Carelessness will result in injury to yourself and others, as well as costly damage to the vehicle's electrical system and electronic components. Observe the following precautions in addition to the general safety guidelines discussed in Chapter One of this *Shop Manual* when testing or servicing a vehicle's electrical system:

- Disconnect the battery negative terminal when working on or near the electrical system, unless a test procedure specifies otherwise. On most batteries, the negative or

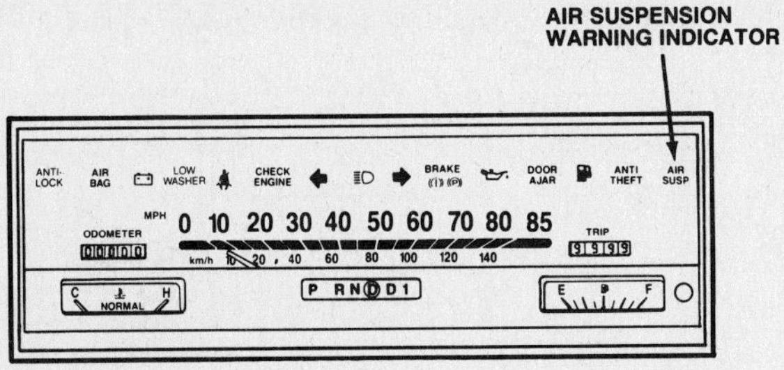

Figure 3-1. If the electronically controlled steering or suspension system malfunctions, an indicator on the dash will illuminate.

ground terminal is marked with a minus (–) sign, and the positive terminal is marked with a plus (+) sign. Sometimes, the battery positive cable and terminal are covered with red insulation.

- Be sure to reconnect battery cables to their proper terminals. Connecting them backward will reverse polarity and damage the alternator and other components that contain diodes.
- When charging a battery, connect the charger to the battery terminals before plugging the charger into the electrical outlet. Also, disconnect the vehicle's battery cables from the battery, and disconnect the cable between the battery and alternator to prevent voltage surge and damage to the electrical system.
- When disconnecting a wire from an electrical component, pull from the connector rather than the wire. Use a small screwdriver with an insulated handle to loosen the connector, if needed.
- Do not ground or apply battery voltage directly to electronic components, unless specified in a test procedure. Applying 12 volts will damage some components.
- When servicing an electronically controlled system, be sure that the ignition is off before disconnecting or connecting the computer wiring harness.
- Do not remove grease from electrical connectors. This special grease deters corrosion.
- Handle the battery carefully. Sulfuric acid in the electrolyte can cause severe skin burns and eye injury. Also, it can damage the vehicle's paint. Be sure to wear safety gloves and glasses when working with a battery.
- Because the battery sometimes generates explosive hydrogen gas, keep sparks, open flames, and smoking materials away from the battery. Charge the battery only in well-ventilated areas.
- Remove any jewelry, such as wristwatches and rings, before servicing a battery or doing any work near electrical systems to prevent arcing and burns.
- Avoid causing sparks while connecting and disconnecting the battery. Gasoline vapors in the engine compartment may ignite if the area is not well ventilated.

ORGANIZED TROUBLESHOOTING

In most cases, an indicator on the dash will illuminate or blink if a vehicle is having problems with an electronically controlled steering or suspension system. From that point, you must determine what is causing the problem. Complex automotive troubleshooting must be approached in a logical, organized manner. To help you develop this habit, use this 10-step checklist when troubleshooting electrical and electronic systems:

1. Ask the Owner or Driver

If possible, talk to the person who was operating the vehicle when the problem first occurred. Many shops have a service writer who talks to customers and prepares work orders for the technician. If you cannot talk to the customer, check the work order for answers to the following questions:

- Were other circuits and accessories being used when the problem occurred?
- Did the customer notice any sparks, odors, noises, or other unusual signs?
- Has the problem occurred before? If so, what repairs were made?

2. Know the System

Review appropriate chapters in the *Classroom* and *Shop Manuals* for information on how suspension systems operate. Use the vehicle manufacturer's service manual to obtain circuit diagrams, service procedures, and specifications. Complete and accurate specifications are the tools that will tell you how the system *should* operate. Without them, your test equipment is useless. The best sources for electrical specifications are the manufacturer's service manuals.

3. Operate the System

Next, see how the system operates. This usually includes a test drive where you have the opportunity to check the vehicle in various operating conditions. For example, if you are checking an electronically controlled suspension system, drive the vehicle on bumpy and

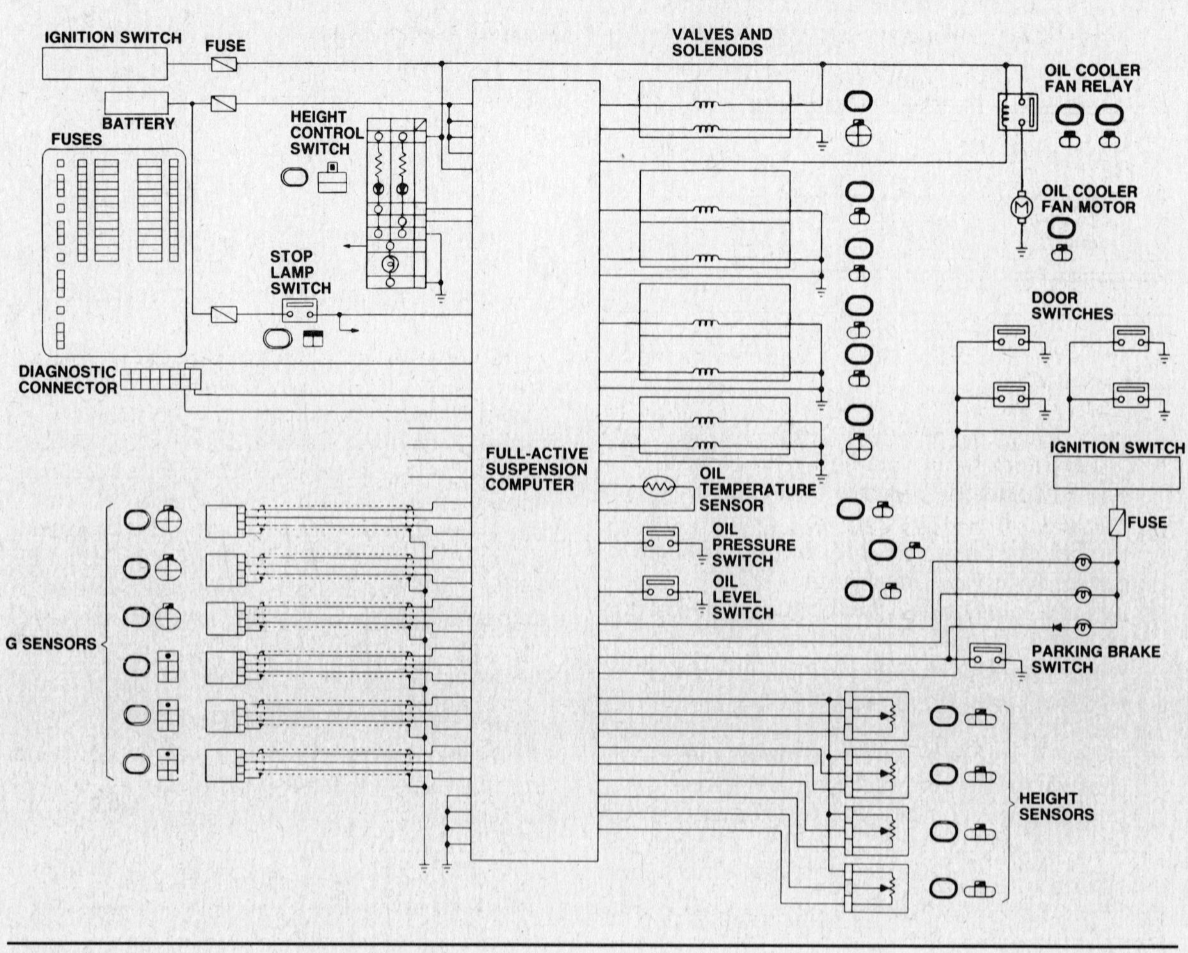

Figure 3-2. This is the electrical circuit diagram for the Infiniti Q45 full-active suspension system. It shows you the power source, switches, fuses, and grounds.

flat roads. Turn a few corners to see how the suspension system responds. Also, check the suspension at various settings. For example, if the suspension system has SOFT, NORMAL, and SPORT suspension system settings, drive the vehicle at all three settings, noting any handling problems.

4. List the Possible Causes

After the initial inspection, list things that you think may be causing the problem. For example, by looking at the electrical system diagram and understanding how the system operates, you may see other possible causes. The list will help you organize your thoughts when you begin testing the system. Most importantly, do not overlook obvious electrical problems. For example, if the vehicle has low-current problems, check the battery and alternator before testing other components for circuit resistance.

5. Isolate the Problem Circuit

You must isolate the circuit that is causing the electrical problem. If your list of possible causes includes several circuits, narrow the possibilities by performing spot checks in each circuit or by using the trouble codes stored in the vehicle's computer. Instructions for retrieving trouble codes are in the manufacturer's service manual. Some examples are given later in this chapter.

6. Know the Problem Circuit

After determining which circuit is causing the problem, study the wiring diagram or schematic for that circuit in detail, figure 3-2. Understanding how the circuit operates is necessary if you are to troubleshoot efficiently. Use a circuit diagram to determine:

- Where the circuit receives battery current
- Which switches and relays control current flow
- What type of circuit protection is used
- Where the circuit is grounded.

7. Test Systematically

Direct current (DC) automotive circuits receive battery current at one end and connect to ground at the other. Because of the way current flows through a DC circuit, the most logical way to test it is to start at one end and work your way to the other. Use these general guidelines if the problem circuit has many components:

- If one electrical part is defective in a circuit with many parts, start the test at the defective part.
- If all of the electrical parts in the circuit are not operating properly, check ground. Then check along the circuit, starting at the point where the circuit gets power.

For example, if the suspension's oil cooler fan motor is not operating properly, figure 3-2, start electrical tests at the motor, checking power and ground. However, if the flow control valves, the fan relay, *and* the oil cooler fan motor are not working, check ground. Then test the circuit, starting at the power source.

In complex systems, such as electronically controlled steering and suspension systems, it is best to test one sub-system at a time. (The oil cooler fan motor and relay may be considered a sub-system.) Doing this narrows down possible causes to a specific sub-system. Then you can perform more detailed tests to isolate the faulty component.

8. Verify Your Findings

After troubleshooting leads you to the defective component, check your results. For example, if you trace the problem to a relay, remove it, and test it separately. If the relay works outside of the circuit, then perhaps the problem was caused by a loose ground or power connection. Checking, cleaning, and tightening connections may keep you from unnecessarily replacing a part that is not defective.

9. Repair

Make repairs only after you have determined the defect. It may be tempting to replace components one at a time, checking to see which replacement part fixes the system. However, this does not always fix the problem. For example, if a short circuit damages an actuating motor, and you replace the motor to fix the problem, the motor will likely be damaged again due to the electrical short. Instead, repair the short circuit, and then replace the motor. If not, the vehicle will likely be back in the shop for repairs in a matter of days.

10. Test Your Repair

This is the final step in any troubleshooting procedure. If you operate the circuit and the problem is gone, then you have succeeded. Never release the vehicle to the customer until after you have checked your work.

BASIC ELECTRICAL TEST DEVICES

There are a number of electrical test devices that you should be familiar with before testing an electronically controlled steering or suspension system. The test equipment described in this section includes:

- A 12-volt test lamp
- A self-powered test lamp
- Jumper wires
- Standard electrical test meters, including the ammeter, voltmeter, ohmmeter, and multimeter
- A volt-amp tester
- Vehicle manufacturers' special testers.

12-Volt Test Lamp

A 12-volt test lamp may simply be a light bulb connected to a pair of wires, or it may be a special tool that has a probe on one end, a wire with an alligator clip on the other end, and a light bulb in the handle, figure 3-3. When you connect one test lead to a good ground and touch the probe or the other test lead to a 12-volt power source, the light bulb comes on. Two common tests are the voltage-seeking and ground-seeking tests.

A voltage-seeking test determines whether voltage is present at a particular point in a circuit. This test can be used to locate an open in a circuit, figure 3-4:

1. Clip one of the test lamp wires to a known chassis ground.
2. Beginning at the point nearest the power source, touch the probe or other test lamp wire to various points along the circuit.
3. If the lamp lights at one test point but not at the next, the circuit is open between those two points.

A ground-seeking test determines whether a conductor, switch, or other electrical component is connected to ground. For example, this test can be used to check the ground switch of an oil pressure warning lamp circuit:

1. Disconnect the oil pressure warning lamp wire from the ground switch.
2. Turn the ignition switch on, but do not start the engine.

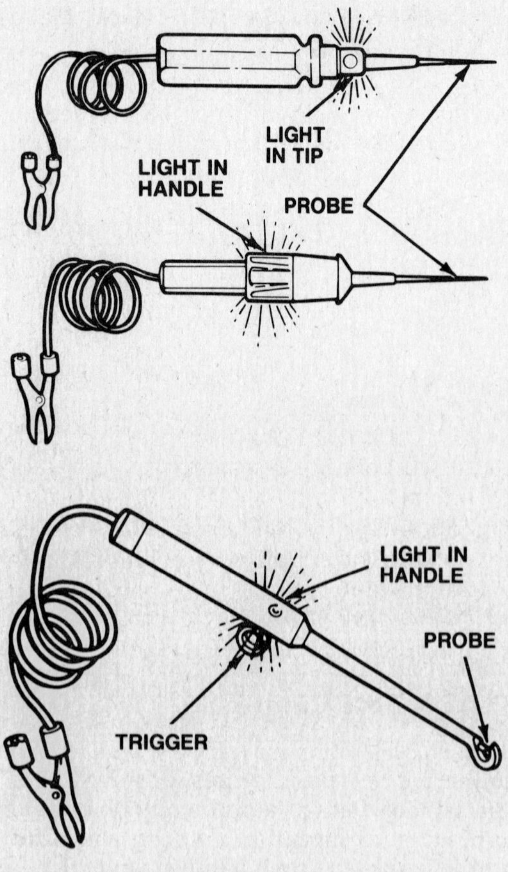

Figure 3-3. To use a 12-volt test lamp, connect the wire to a known ground and touch the test probe at various points along the circuit. The lamp will illuminate if current is flowing in the circuit.

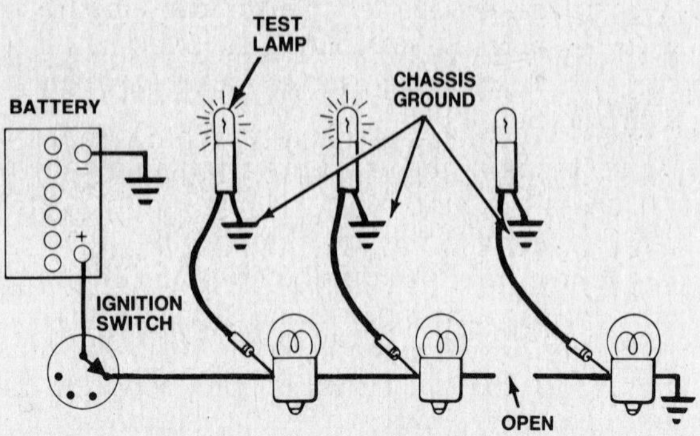

Figure 3-4. You can use a 12-volt test lamp for voltage-seeking tests. In this example, the third test lamp did not light, which indicates a lack of current due to an open in the circuit.

3. Clip one of the test lamp wires to the ground switch wire, and touch the other test lamp wire to the engine block, figure 3-5. If the test lamp and the warning lamp

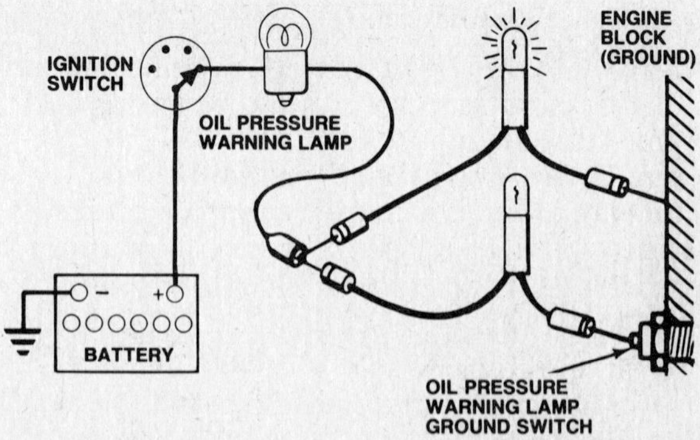

Figure 3-5. This ground-seeking test with a 12-volt test lamp indicates that the oil pressure warning lamp ground switch is defective. When the test lamp is touched to the engine block, which is a good ground, the lamp illuminates.

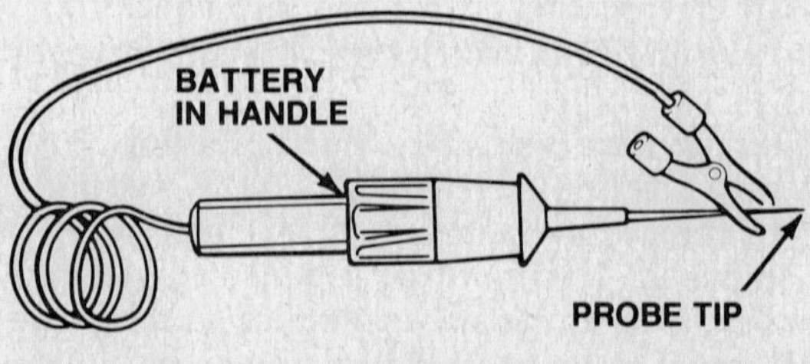

Figure 3-6. This test lamp contains its own power source. Check for continuity in a component by clipping the power source lead to one end of the component and the ground lead to the other end. If the lamp lights, there is a complete circuit, or continuity.

light, then the switch works when properly grounded. This means that the grounding switch may be defective.

4. Keep the test lamp wire clipped to the ground switch wire. Touch the other test lamp lead to the ground switch, figure 3-5. If the test lamp and the warning lamp do not light, the ground switch is defective.

Self-Powered Test Lamp

A self-powered test lamp, sometimes called a continuity tester, is a light bulb and battery that are connected in series with a pair of test wires, figure 3-6. Because this type of test lamp has its own power source, any components that you test with it must be disconnected from any other voltage source. Also, this test lamp cannot be used to check components with high internal resistance, because its power source is too low to complete high resistance circuits. Do not use this for electronic components, such as integrated circuits.

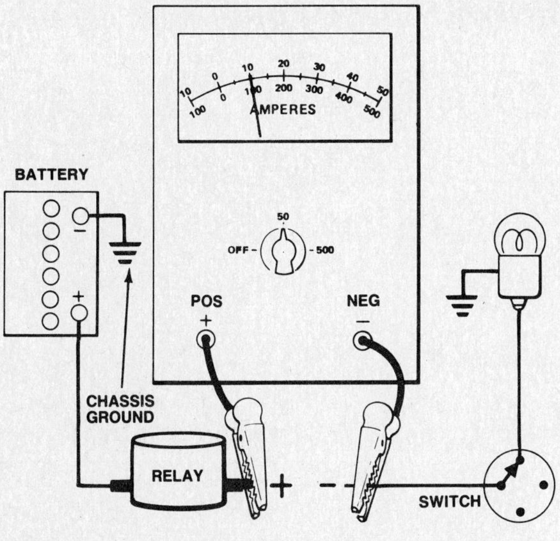

Figure 3-7. The ammeter is connected in series to read the amount of current flowing through a circuit. Connect the ammeter positive lead to the battery side of the circuit, and the negative lead to the ground side.

To use this test lamp when checking for continuity, clip the test lamp leads to both ends of the device being checked. If the lamp lights, the circuit is complete, and there is electrical continuity through the component.

Jumper Wire

A jumper wire is simply a length of flexible wire with an alligator clip at each end. Although it is not a true test instrument, you can use a jumper wire to bypass switches, connectors, and other nonresistive electrical components. If the circuit operates when you bypass the switch or other nonresistive component, then you can deduce that the switch is defective.

Never use a jumper wire to bypass a component with high internal resistance because you may overload the circuit. Also, never use a jumper wire to ground a hot lead. Some tool manufacturers sell a special jumper wire that includes an inline fuse or circuit breaker to protect circuits if you make an improper connection.

Standard Electrical Test Meters

Amperage, voltage, and resistance are three values that you will check when troubleshooting electrical circuits. These values are measured using three basic test meters:

- Ammeters, which measure current
- Voltmeters, which measure voltage
- Ohmmeters, which measure resistance.

At one time, all electrical test meters were analog. Analog meters use a needle moving across a fixed scale to indicate the electrical value being measured. Now digital meters are available that can read amps, volts, and ohms out to three or four digits to the right of the decimal point. This accuracy is needed when checking and adjusting some electronic components that operate at low voltages.

Nearly all digital test meters have high impedance, which is a measurement of electrical resistance in a circuit using alternating current. Many vehicle manufacturers specify that you use a high-impedance (10-megohm or greater) voltmeter when testing computerized electronic systems to avoid damaging the circuitry.

Ammeter

An ammeter measures the amount of current flowing through a circuit in units called amperes, or amps. An ammeter can be used to detect open circuits because an open circuit does not allow current flow.

Always connect an ammeter in series with the circuit being tested because you want all of the current that normally flows through the circuit to flow through the ammeter, figure 3-7. Because ammeters have very little resistance to current flow, they will not affect normal circuit operation. If an ammeter reading is lower than specifications, then there are points of resistance in the circuit.

Voltmeter

Connect a voltmeter in parallel with a circuit when checking voltage. Use a voltmeter to:

- Measure the voltage available at a circuit's power source by connecting the voltmeter's positive lead to the battery positive terminal and the negative lead to the ground terminal.
- Measure the voltage available at any point along a circuit by connecting the positive lead anywhere in the circuit and connecting the negative lead to chassis ground.
- Measure voltage drop by reading the voltage available at two points along a circuit. Then subtract these voltages to calculate voltage drop, figure 3-8.

Voltmeters have very high internal resistance to current flow, so when they are connected in *parallel* with a circuit, only a small amount of current flows through the meter. The rest

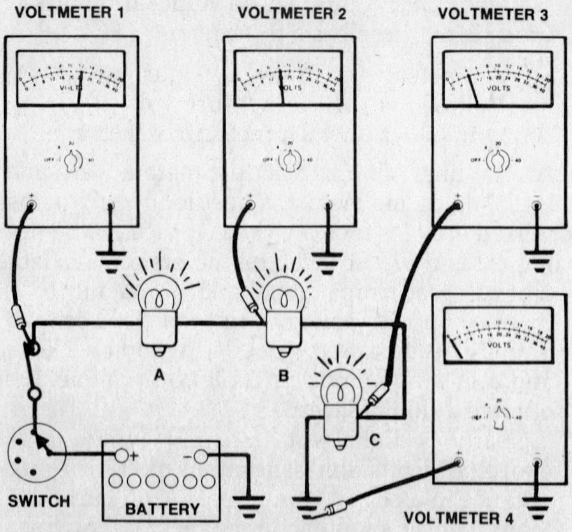

Figure 3-8. After taking voltage readings at various points along a circuit, subtract the values to calculate the voltage drop. For example, the first voltmeter reads 12 volts and the second reads 8 volts, so the voltage drop is 4 volts across the first light bulb.

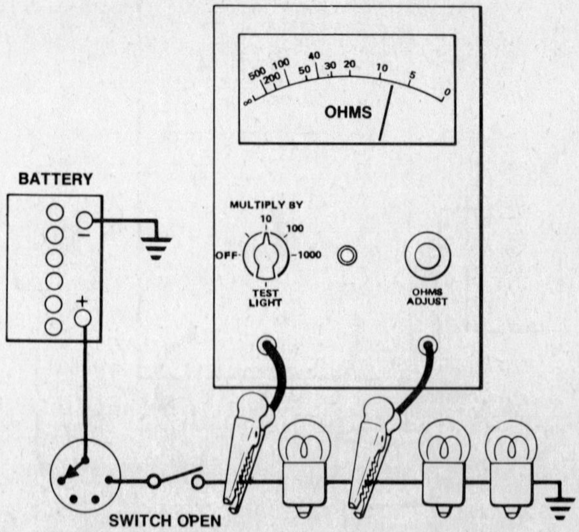

Figure 3-9. An ohmmeter has its own power source. Be sure to disconnect a circuit's power source by disconnecting the battery or opening a switch before checking resistance with an ohmmeter.

flows through the circuit, and the circuit can operate normally. If you connect a voltmeter in *series*, current will flow through the meter, and only a small amount will continue through the rest of the circuit. Often this prevents the circuit from operating normally.

Ohmmeter

An ohmmeter measures the resistance of an electrical component in units called ohms. It has its own power source, usually an internal battery, which makes it different from the ammeter or voltmeter. Use an ohmmeter to check resistance when current *is not* flowing through the circuit. Current flow from an outside source can permanently damage an ohmmeter, especially since it has its own power source. As a precaution, remove the part that you are checking from the circuit or disconnect the power source, such as the battery ground cable, before using an ohmmeter.

Because the ohmmeter's power supply may vary depending on the age of the battery that powers it, you must calibrate the meter before measuring circuit resistance. Set the ohmmeter to the range setting that you will use, and touch the test leads together. If the meter does not read 0, turn the calibration knob until the scale reads 0 ohms. Then the ohmmeter is ready for use.

To measure the resistance of a component, such as a light bulb, open the switch to cut power to the bulb, figure 3-9. Touch the ohmmeter test leads to the wires on either side of the bulb, and read the resistance in ohms.

Multimeters

Multimeters combine the functions of an ammeter, voltmeter, and ohmmeter. A digital volt-ohmmeter (DVOM) is the most common type of multimeter, and most technicians who perform electrical work eventually invest in a good digital multimeter.

To use one of these meters, simply turn the dial to select amps, volts, or ohms, figure 3-10. Then test the circuit using one of the procedures described earlier.

Volt-Amp Tester

Volt-amp testers are used to check automotive charging systems, starting systems, and batteries. These testers are sometimes called CSB (charging-starting-battery) analyzers. A typical volt-amp tester consists of a voltmeter, ammeter, and carbon pile that serves as a variable resistor. You can select the desired charging, starting, and battery tests by setting the knobs and switches on the meter, figure 3-11. On many volt-amp testers, the voltmeter has at least two ranges, one for 12-volt and another for 24-volt systems. The ammeter has more than one range for measuring various levels of battery load, starter current draw, and charging system output. The carbon pile's resistance to current flow is controlled with a knob on

Figure 3-10. A DVOM (digital volt-ohmmeter) is a good all-purpose tool for electrical testing, because it reads amps, volts, and ohms.

the front of the meter. This meter uses heavy-gauge leads to carry large amounts of current.

Vehicle Manufacturers' Special Testers

Special testers simplify tune-up and trouble-shooting procedures on computer-controlled systems by identifying defective circuits and components by running a diagnostic routine. Virtually all late-model vehicles have a diagnostic connector in the engine compartment or under the dash that hooks to a special tester.

Some special testers are sold only to dealership service departments by vehicle manufacturers, and you will not have access to them. However, many testers are manufactured for aftermarket sale to independent repair shops. These special testers will help you retrieve trouble codes and diagnose computer-controlled systems.

On some vehicles, trouble codes can be retrieved without the use of a special tester. Instead, you can use an analog voltmeter to read

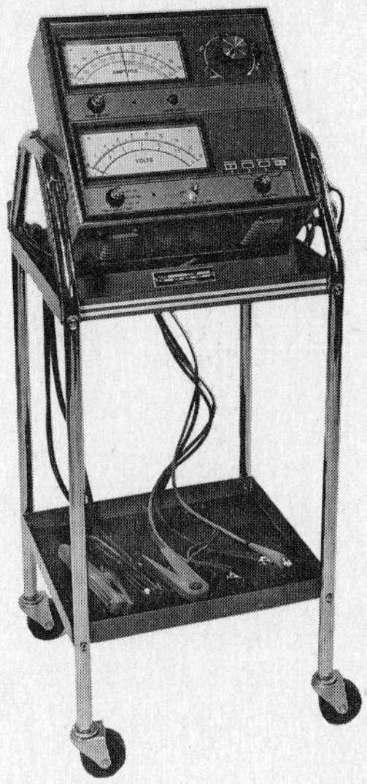

Figure 3-11. The volt-amp tester, also known as the CSB (charging-starting-battery) analyzer, is a common piece of electrical test equipment in an automotive repair shop.

trouble codes after jumping two terminals on the diagnostic connector. Trouble code retrieval for some of the following manufacturers' vehicles are described in this section:

- Chrysler
- Ford
- General Motors
- Nissan/Infiniti
- Mazda
- Honda/Acura
- Mitsubishi
- Toyota/Lexus.

Note: The vehicle's battery must be charged in order to retrieve trouble codes. If you must connect a charger to the battery in order to maintain sufficient current while performing diagnostic tests, be sure to observe correct polarity when hooking up the charger. Internal diodes are installed in many of these circuits, and reversing the battery positive and ground cables will destroy the diodes.

Chrysler

Most 1984 and newer Chrysler vehicles have self-diagnostic capability. A special tester, called DRB or DRB-II, attaches to a diagnostic connector in the engine compartment to retrieve information about the engine and other computer-controlled systems.

However, to read the trouble codes for an electronic suspension system such as the one

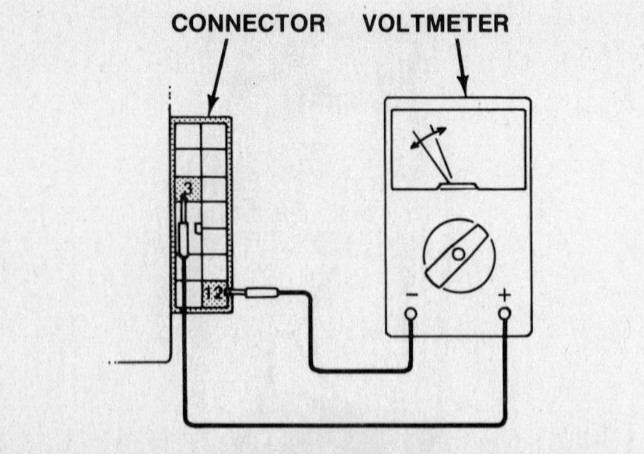

Figure 3-12. To retrieve trouble codes for the Dodge Stealth's electronically controlled suspension, connect a voltmeter to terminals 3 and 12 of the suspension system's diagnostic connector.

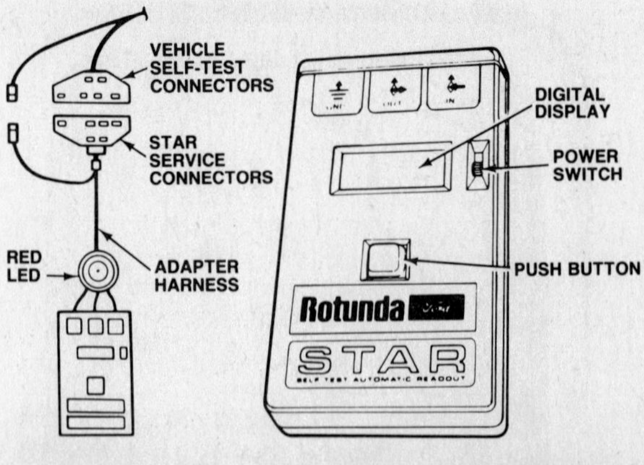

Figure 3-13. This special tester, called STAR (Self-Test and Automatic Readout), is used to retrieve the trouble codes on a number of late-model Ford vehicles.

on the Dodge Stealth, you do not need a special test meter, figure 3-12. Instead, while the ignition switch is off, connect a voltmeter's positive lead to the number 3 terminal and negative lead to the number 12 terminal of the suspension system diagnostic connector. Turn the ignition switch on, and count the needle sweeps on the voltmeter to obtain trouble codes. For example, if a trouble code ''42'' is stored, the needle will sweep across the scale four times, pause, and then sweep across two times.

Record the codes on a sheet of paper, turn the ignition switch off, and disconnect the voltmeter test leads. After making repairs, erase trouble codes by disconnecting the battery cables for more than ten seconds.

Ford

Since 1980, most Ford vehicles with Microprocessor Control Unit (MCU) fuel-metering control or Electronic Engine Control (EEC) systems have had self-diagnostic capabilities. EEC-I, EEC-II, EEC-III, and early MCU systems require dedicated testers unique to each system that plug inline between the electronic control assembly and the vehicle wiring harness. Used in conjunction with a DVOM, these testers monitor all pertinent EEC signals and are designed to reduce the total number of steps required to locate a faulty component. EEC-IV systems and late-model MCU systems have a self-test connector in the engine compartment to which you can connect a special system tester, called the STAR (Self-Test and Automatic Readout) or Super STAR-II tester, figure 3-13.

The STAR and Super STAR testers can be used to retrieve trouble codes for the computer-controlled suspension system used on some Ford models, such as the Thunderbird Super Coupe and Cougar XR-7. With the ignition switch off and the suspension set in the AUTO mode, connect the tester to the ARC (Automatic Ride Control) diagnostic connector, which is in the engine compartment next to the passenger side shock absorber. Turn the tester on, and put the TEST button in the down position. Start the engine, and put the tester in TEST mode, HOLD mode, and back to the TEST mode. This initiates the self-diagnostic test. Trouble codes will blink on the tester if the suspension system is faulty. Use the manufacturers' service manual for a listing of the trouble codes, what they mean, and the recommended repair procedures.

General Motors

General Motors vehicles that use the Computer Controlled Catalytic Converter (C-4) or Computer Command Control (CCC) engine control systems have a test connector under the instrument panel for self-diagnostics. This connector, called either the Assembly Line Communications Link (ALCL) or Assembly Line Diagnostic Link (ALDL), has a test terminal that you must ground to retrieve trouble codes. Trouble codes for engine systems as well as power steering and electronically controlled suspensions, if equipped, are displayed by a flashing CHECK ENGINE or SERVICE ENGINE SOON lamp on the instrument panel. On these diagnostic connectors, test and ground terminal locations vary from year to year, figure 3-14.

One of General Motors' electronically controlled suspension systems, called CCR (Computer Command Ride) 1200, is available as an

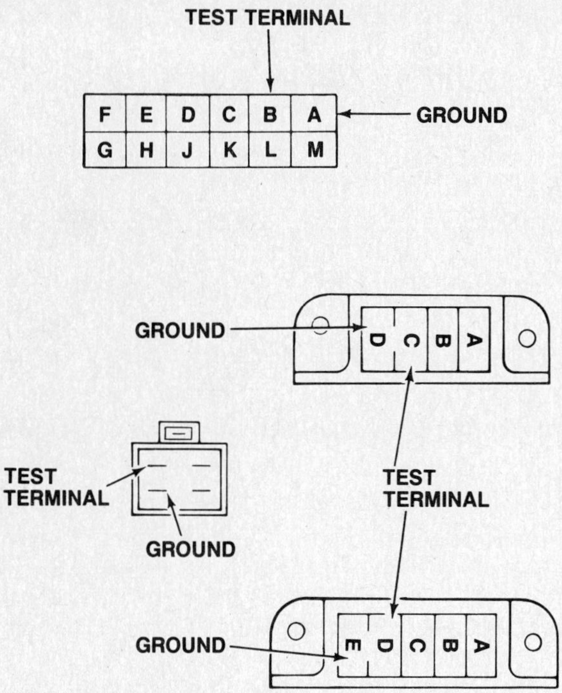

Figure 3-14. General Motors' ALDL and ALCL diagnostic connectors vary from year to year. Follow the vehicle manufacturer's service instructions for the correct terminals to jump when retrieving trouble codes.

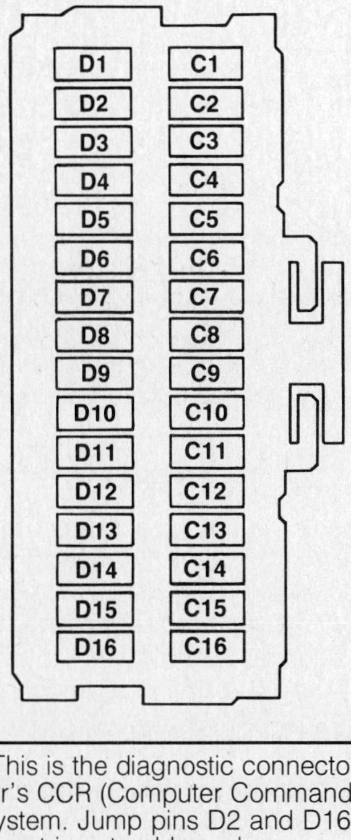

Figure 3-15. This is the diagnostic connector for General Motor's CCR (Computer Command Ride) suspension system. Jump pins D2 and D16 with a jumper wire to retrieve trouble codes.

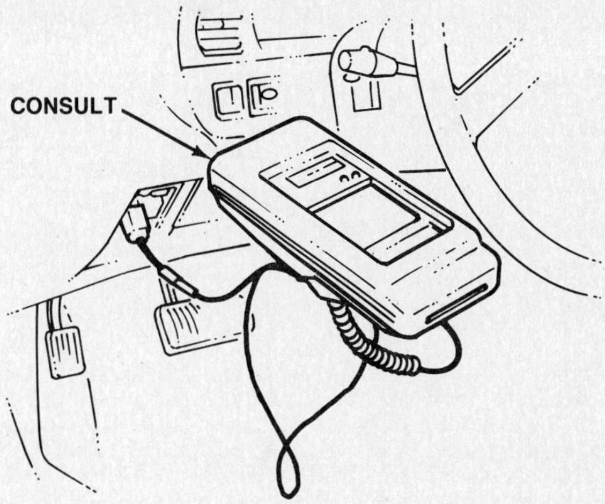

Figure 3-16. The Nissan/Infiniti CONSULT diagnostic computer plugs into a connector in the fuse box on the driver's side of the dash.

option on the Cadillac DeVille. This system uses a separate suspension computer, rather than using the engine computer for ride control data processing. To obtain trouble codes from this system, you can use one of two methods. You can ground pin C of the ALDL connector, or you can jump pins D2 and D16 at the suspension system control module connector, figure 3-15. Then count the number of times the "Service CCR" light flashes on the dash to obtain trouble codes. After retrieving the codes, clear the computer's memory by jumping pins A and C of the ALDL connector. You must plug in and remove the jumper wire at pin C three times to erase codes.

Nissan/Infiniti

Nissan/Infiniti uses a tester called CONSULT to diagnose a number of electronically controlled systems. CONSULT is a small, hand-held computer. It comes with program cards containing test data for the various electronically controlled systems, as well as instruction manuals on how to properly diagnose each system.

CONSULT has the ability to perform an electronically controlled suspension system self-test. Plug CONSULT's connector into the vehicle's diagnostic connector, which is located in the fuse box at the lower left side of the instrument panel, figure 3-16. Start the engine, and press CONSULT's start button. Press the active suspension button and the self-diagnostics button. CONSULT will analyze the circuits in the computer-controlled suspension system and show diagnostic codes

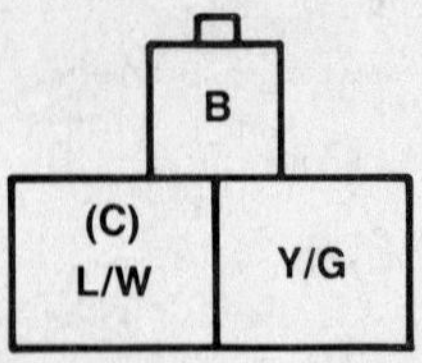

Figure 3-17. You can retrieve trouble codes from Mazda's computer-controlled suspension system, called AAS (Auto Adjusting Suspension), by using a voltmeter at the diagnostic connector.

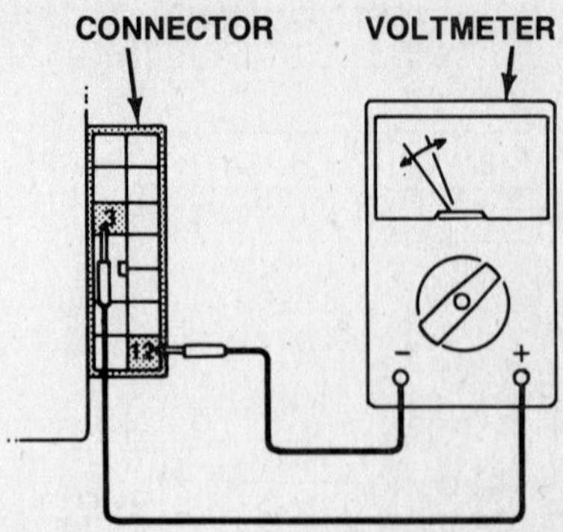

Figure 3-18. Hook up a voltmeter at terminals 3 and 12 of the diagnostic connector to retrieve trouble codes from Mitsubishi's suspension control computer.

on the display panel. From this information you can use a DVOM and the test procedures in the manufacturer's service manual to isolate any electrical problems.

Mazda

The computer-controlled suspension system used on some Mazda vehicles, such as the MX-6 or 626, is called AAS (Auto Adjusting Suspension). To perform a self-diagnostic test, you will need an analog (needle-type) voltmeter. Connect the voltmeter to the C (L/W) terminal of the test connector in the engine compartment and ground, figure 3-17.

To obtain trouble codes, you must simulate a number of driving conditions. After each simulation, you must turn off the ignition to reset the diagnostic tester. For example, put the steering wheel in the straight-ahead position, turn on the ignition, and count the needle sweeps to obtain the trouble codes. Then turn the ignition off to reset the system, and turn it on again. Switch the ride control button in the passenger compartment from normal to sport or from sport to normal. Again, check the voltmeter for trouble code output by counting the needle sweeps. Turn the ignition off, and then on again. With the vehicle on a chassis roller, turn the front wheels at 15 mph (24 km/h), and check for trouble code output. Use the manufacturer's service manual to interpret trouble codes.

Honda/Acura

Until the introduction of the 1992 Prelude, Honda and Acura suspension systems were entirely mechanically operated.

Some of their "high-end" vehicles use variable assist power steering, however. This system uses a computer to change the amount of power steering assist, depending on vehicle speed. Troubleshooting procedures for this system require the manufacturer's service manual and a DVOM, rather than a computerized self-diagnostic test.

Mitsubishi

Some luxury Mitsubishi vehicles, such as the 3000 GT, use computer-controlled suspension and steering systems, which are diagnosed using the MUT (Multi-Use Tester). This tester retrieves trouble codes from the computer and can be used to simulate various driving conditions.

A light on the dash will blink when the computer has trouble codes stored in memory. With the ignition off, hook up the MUT, which plugs into two connectors — one under the dash and the other at the center console. Turn on the ignition and MUT to read trouble codes. See Mitsubishi's service manual for a trouble code table and additional diagnostic tests to help pinpoint the problem.

Rather than using MUT, you can use a voltmeter to retrieve trouble codes, figure 3-18, as on the Dodge Stealth. With the ignition off, connect the voltmeter positive lead to the number 3 terminal of the diagnostic connector and the negative lead to the number 12 terminal. Turn the ignition on, and count the needle sweeps on the voltmeter to read trouble codes. Erase the codes by disconnecting the voltmeter from the diagnostic connector and disconnecting the battery terminals for ten seconds.

Toyota/Lexus

Some Lexus and Toyota vehicles have computer-controlled steering and suspension systems. To retrieve trouble codes from these systems, turn the ignition switch on, and jump terminals Tc and E_1 at the diagnostic connector, figure 3-19. The diagnostic test will begin when you jump the terminals. Count the blinks at the dashboard indicator light to determine the trouble codes. Trouble code charts are in the manufacturer's service manual.

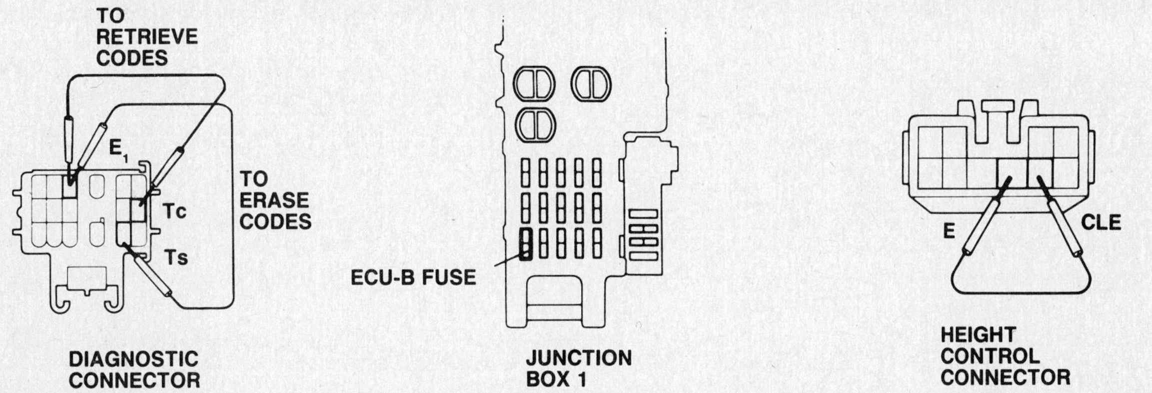

Figure 3-19. After retrieving trouble codes from the suspension computer on a Toyota or Lexus vehicle, be sure to erase them by removing the ECU-B fuse or by jumping terminals at the diagnostic and height control connectors.

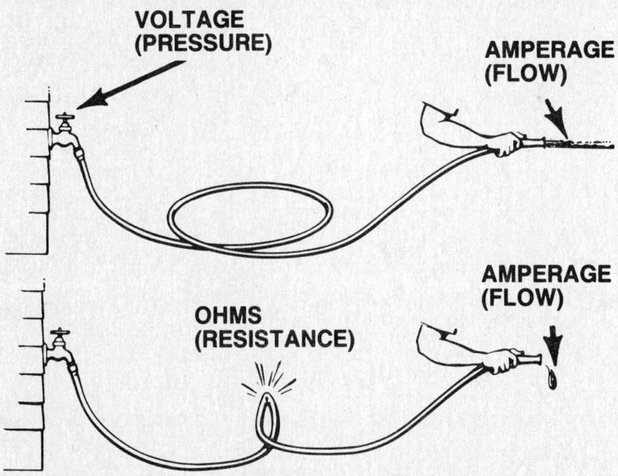

Figure 3-20. The three basic terms used to describe the flow of electricity in a circuit — amps, volts, and ohms — can be compared to the flow of water through a hose.

After retrieving the codes and making repairs, erase trouble codes by removing the ECU-B fuse from junction box 1, or by jumping terminals E_1 and Ts at the diagnostic connector and terminals E and CLE at the height control connector.

BASIC TERMINOLOGY AND ELECTRICAL TESTS

Although computer-controlled systems appear complicated, they are simply groups of electrical circuits. After identifying defective circuits by retrieving trouble codes, perform simple electrical tests for voltage, amperage, and resistance to check components. Understanding electrical terms and being able to perform simple tests will improve your ability to repair complicated computer-controlled systems. In this section we will discuss:

- Basic electrical terminology
- How to check fuses
- How to test and service the battery
- How to inspect and test the charging system.

Because the battery and alternator or generator supply power, they should not be overlooked when troubleshooting a vehicle's electrical system.

Basic Electrical Terminology

In general, three terms are used to describe the flow of electricity in a circuit:

- Amperes, or amps, which are units of current flow
- Volts, which are units of electrical pressure or electromotive force that pushes current through a circuit
- Ohms, which are units of resistance to current flow.

For instructional purposes, the flow of electricity through a circuit can be compared to the flow of water through a hose, figure 3-20.

- The amount of water that flows out of a hose nozzle (measured in liters or gallons) is similar to the amount of current that flows through a circuit (measured in amperes).
- The amount of water that flows through the hose is partly regulated by water pressure (measured in psi). The higher the water pressure, the higher the flow. The amount of current flow through a circuit is similarly regulated by electrical pressure (measured in volts). The higher the voltage, the higher the current flow.

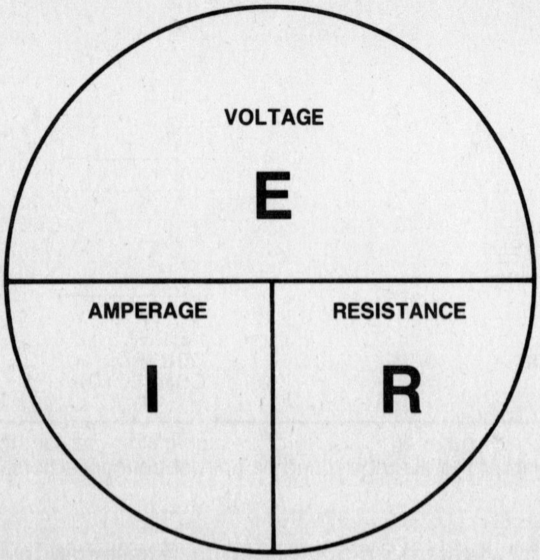

Figure 3-21. Use this diagram to remember Ohm's Law. Cover the unknown value, and then multiply or divide the remaining two.

- The amount of water flowing through a hose is also regulated by resistance, which may be due to the size of the hose or any kinks in it. The higher the resistance, the lower the water flow. The amount of current flowing through a circuit is similarly regulated by electrical resistance (measured in ohms). The higher the resistance, the lower the current flow.

Amps, volts, and ohms have a simple mathematical relationship, called Ohm's Law, which you will use while troubleshooting electrical systems. Ohm's Law states that current flow in a circuit varies directly with voltage and inversely with the resistance of the circuit. In other words, if voltage increases, current flow also increases. However, if resistance increases, current flow decreases.

An easy way to remember Ohm's Law is shown in figure 3-21. If you cover the unknown value, you will reveal the formula used to find it. (In this figure, amperes is abbreviated ''I'' for intensity, voltage as ''E'' for electromotive force, and resistance as ''R''.) If you know a circuit's voltage (E) and resistance (R), cover the unknown value, which is amperage (I). To solve for amperage, divide voltage (E) by resistance (R). Again, if you know two of the three electrical values, you can find the third one by applying Ohm's Law.

The principles of Ohm's Law are used every day in automotive troubleshooting. For example, suppose you want to know if a resistor in a circuit is defective. You could connect an ammeter to the positive and negative leads of the circuit and measure current flow, which, for example, turns out to be 8 amps. Then you would use a voltmeter to measure voltage drop across the resistor, which turns out to be 0.5 volts. With these two values, you can use Ohm's Law to determine the resistor's resistance.

Use the chart in figure 3-21, and cover the R, which is the unknown value. This leaves a formula of E/I, or voltage divided by amperage. So plugging the meter readings into the formula, you get 0.5 divided by 8 = 0.0625 ohms, which is the resistance of the resistor. Compare this value with the manufacturer's specifications to find out if the resistor is good.

A ''leak'' in an electrical circuit is a lot like a leak in a water hose, figure 3-22. If the hose leaks and water escapes, both the quantity and pressure of the water flowing through the hose is reduced. In an electrical circuit, loose connections, corroded terminals, and broken wire insulation will allow ''leaks'', resulting in current and voltage reductions. If you test a circuit and determine that current and voltage are less than specified by the manufacturer, then you should find the ''leak''.

Checking Fuses

It is important that you check fuses before starting an electrical system repair. Fuses, fusible links, and circuit breakers keep excessive current flow from damaging components in a circuit, figure 3-23. In most cases, excessive current flow overheats the wires and other components in a circuit, which can lead to electrical fires. A fuse, fusible link, or circuit breaker prevents this by creating an open in the circuit and stopping current flow. Ratings on a fuse indicate the maximum current that can flow through it, in amps. If a fuse blows, be sure to determine the cause of the excess current. Repair the problem before you replace the fuse, or the fuse will likely blow again.

Replace a blown fuse with a good one that has the correct amperage rating. For example, if an 8-amp fuse blows, replace it with a new 8-amp fuse. If you replace it with a 12-amp fuse, then components in the circuit may be damaged if current exceeds 8 amps. If you replace it with a 6-amp fuse, the fuse may blow during normal circuit operation.

A circuit breaker is similar to a fuse in that it opens the circuit when current exceeds the circuit breaker's maximum amperage setting. However, it differs in that it will allow current

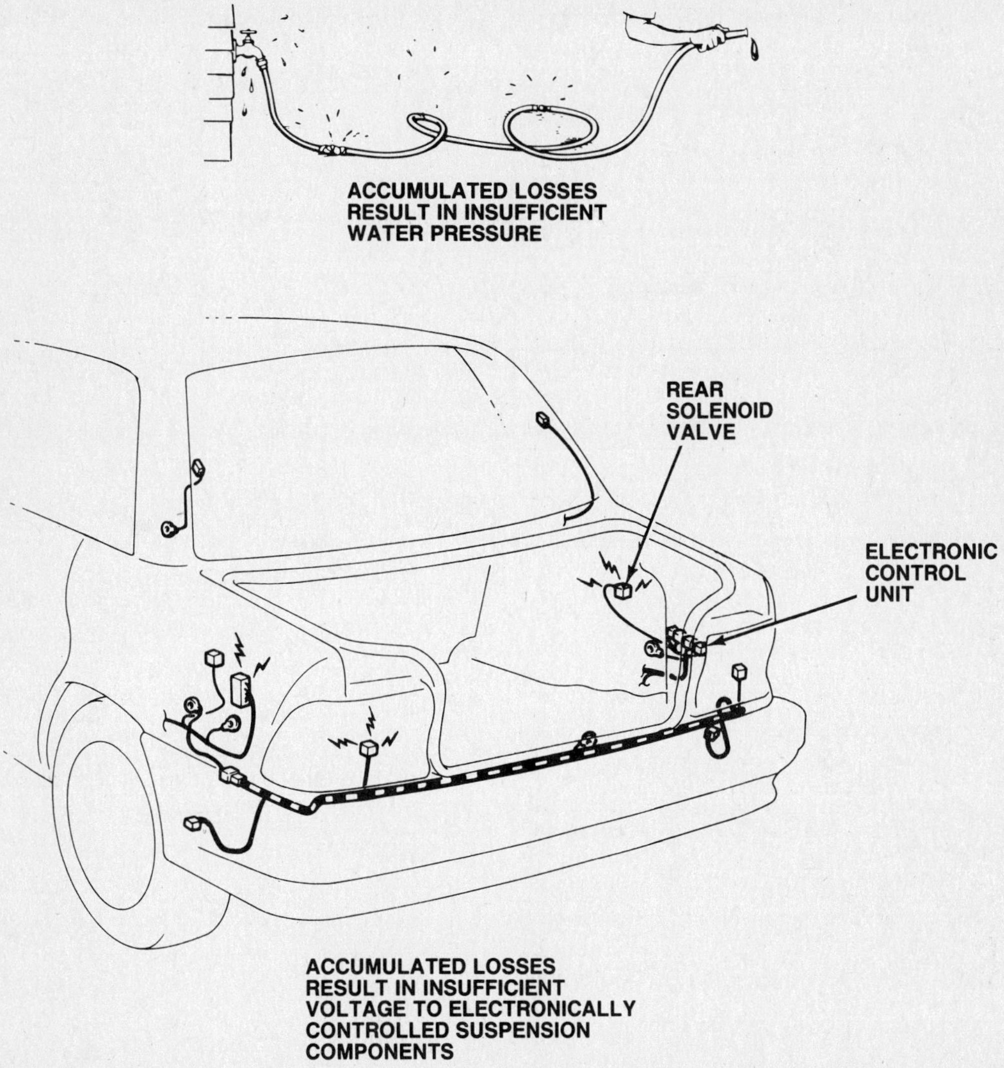

Figure 3-22. Leaks in an electrical circuit are comparable with leaks in a hose. In an electrical circuit, the leaks will result in current and voltage reductions.

to flow in the circuit again once amperage returns to normal. Some circuit breakers have a manual reset button that you must push before the circuit breaker will allow normal operation.

Fuses are located in a fuse box or panel, which is under the dash or in the engine compartment. Some vehicles have separate fuse boxes for various systems. For example, an electronically controlled suspension system may have its own fuse box that is separate from the engine control fuse box. Check the vehicle manufacturer's service manual for fuse box locations, and the types of fuses used in each. Some electrical problems can be repaired by replacing blown fuses or by installing the correct fuse. If you are not sure which fuse to install in a fuse box, check on the top and sides of the box. Often the correct fuse amperage rating is recorded somewhere on the fuse box or panel.

Battery Tests and Service

Check the condition of the battery before starting major electrical repairs. A battery that is producing insufficient current may hamper circuit operation.

Contrary to what many people believe, a car battery does not contain any electricity. Instead, it converts chemical energy to electrical energy. Four types of batteries in use today are:

- Conventional
- Low-maintenance

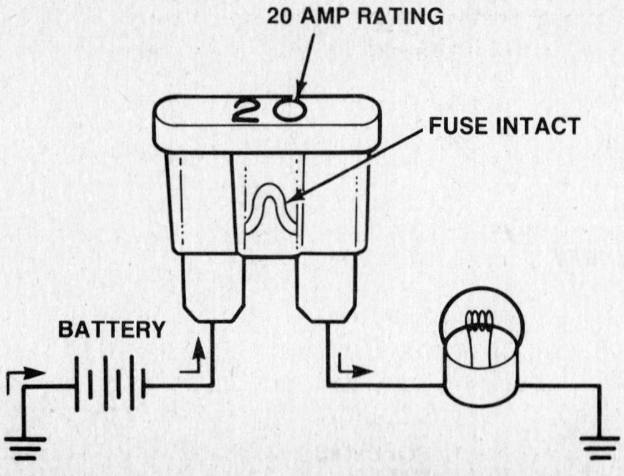

Figure 3-23. This fuse has a 20 amp rating. If current exceeds 20 amps, the fuse will "blow", creating an open in the circuit, which stops current flow. The accessory, a light bulb, is protected from excess current.

- Semi-sealed maintenance-free
- Sealed maintenance-free.

The most commonly used batteries are either low-maintenance or semi-sealed maintenance-free batteries. Low-maintenance batteries have filler caps that allow you to add water to the cells. Semi-sealed maintenance-free batteries usually have small gas vents that prevent pressure buildup in the case. A low-maintenance battery requires that water be added much less often than a conventional battery. A maintenance-free battery will never need additional water during its lifetime.

The liquid inside a battery, electrolyte, is a solution of sulfuric acid and water and has a specific gravity reading of approximately 1.260 when the battery is fully charged. As a battery produces current, it discharges, and the acid in the electrolyte changes to water. This reduces the specific gravity of the electrolyte because there is less of the heavier acid in the mixture. You can measure the specific gravity of battery electrolyte with a hydrometer. If the battery is not fully charged, the specific gravity will be less than 1.260.

Battery testing

To completely and accurately determine the condition of a battery, you must perform two tests: the state-of-charge and capacity (load) tests.

State-of-charge tests

There are a few ways to test a battery's state-of-charge. Always check state-of-charge before performing the capacity test, which is described later in this section.

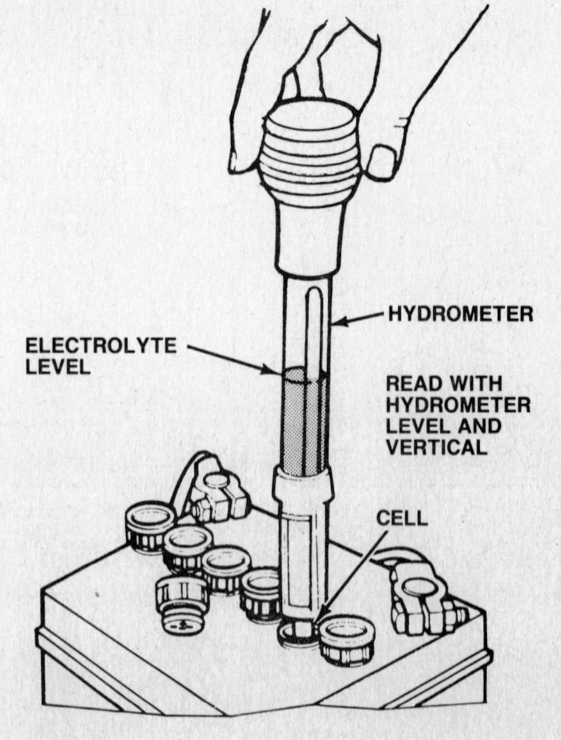

Figure 3-24. Checking the battery electrolyte with a hydrometer is one way to determine state-of-charge. If the specific gravity reading is less than 1.225, charge the battery.

A hydrometer determines the specific gravity of the battery's electrolyte, which is a measure of state-of-charge. Insert the hydrometer into a cell at either end of the battery, figure 3-24. Hold the hydrometer vertically and draw in enough electrolyte to float the indicator. Gently shake or tap the hydrometer to keep the float from sticking to the side of the tube. If the hydrometer contains a thermometer, draw electrolyte several times to stabilize the temperature reading. If the electrolyte level is too low to sample, add water to the cells, and charge the battery at a high rate for *at least* 10 minutes to mix the water and electrolyte before testing. Hold the hydrometer at eye level and read the specific gravity on the indicator. Return the electrolyte to the same cell from which it was taken, and then repeat the test for the other cells. A battery should be recharged if the specific gravity drops below 1.225. A variation of more than 50 points (.050) between any two cells indicates a bad battery that should be replaced.

Maintenance-free batteries used in many late-model cars have a visual state-of-charge indicator built into the top of the case, figure 3-25. A colored plastic ball (usually green, red,

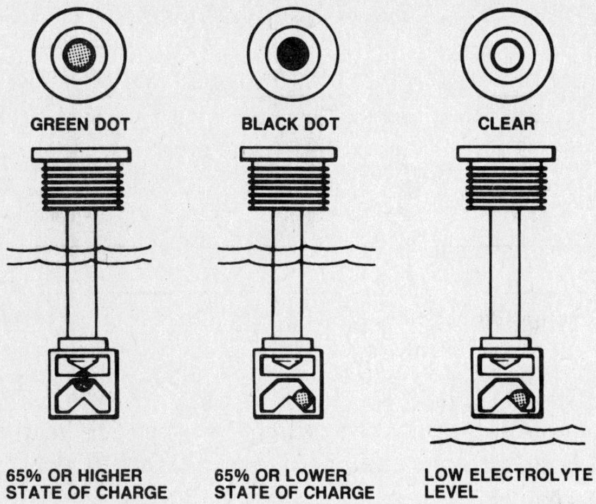

Figure 3-25. Many maintenance-free batteries have a state-of-charge indicator built into the top. Reading the color of the indicator gives you the state-of-charge.

OPEN CIRCUIT VOLTS	STATE OF CHARGE
12.6 or greater	100%
12.4 - 12.6	75 - 100%
12.2 - 12.4	50 - 75%
12.0 - 12.2	25 - 50%
11.7 - 12.0	0 - 25%
11.7 or less	0%

Figure 3-26. The results of an open circuit voltage test will indicate a battery's state-of-charge.

or blue) is suspended in a cage in the battery case. Depending on the specific gravity of the electrolyte, the ball will float or sink in its cage, changing the appearance of the indicator. If the colored dot is visible, the battery is at or above a 65-percent state of charge. A dark or black dot means the battery is below a 65-percent state of charge and must be charged. If the indicator is clear or light yellow, the battery is bad and must be replaced. Do not recharge the battery.

Some maintenance-free batteries have a two-ball state-of-charge indicator that contains both a red and a green ball. On these batteries, a visible green dot means that the battery is at or above a 75-percent state-of-charge and is ready for testing. A dark or black dot means the battery is between a 50- and 75-percent state of charge. A red dot means the battery is below a 50-percent state of charge and must be charged before testing. A clear or light yellow indicator means the battery is bad and must be replaced.

Finally, you can check a battery's state-of-charge using the open circuit voltage test. Battery temperature should be between 60° and 100°F (16° and 38°C) while you are conducting the test. If you charge the battery, allow it to stabilize for at least ten minutes before testing it. Use a DVOM to measure battery voltage to the nearest tenth of a volt. Then use the chart in figure 3-26 to determine the battery's state-of-charge.

Capacity (load) test

Perform a capacity (load) test only after determining that the battery's state-of-charge is within acceptable limits. The capacity test is the most important and meaningful method used to determine battery condition. This test measures the ability of a battery to provide cranking current while still maintaining sufficient voltage for ignition system operation. For this test, the battery electrolyte temperature should be approximately 80°F (27°C).

The best way to perform a capacity test is with a volt-amp tester or a charging-starting-battery (CSB) analyzer designed for this purpose. Always follow the equipment manufacturer's specific instructions for hooking up the tester leads and conducting the test. The following is a general procedure for performing a battery capacity test:

1. With the tester controls off, connect the tester leads to the battery. Observe correct polarity and make sure the test leads make good contact with the battery terminals.
2. If the tester you are using has an adjustment for battery temperature, turn it to the proper setting. Use a thermometer in one of the cells to determine the temperature of a conventional or low-maintenance battery. On maintenance-free units, estimate the temperature as closely as possible.
3. Refer to battery specifications to determine the correct cold-cranking or ampere-hour rating.
4. Turn the load control knob to draw battery current at a rate equal to one-half of the battery cold-cranking rating or three times the battery ampere-hour rating, as determined in Step 3. For example, if the battery is rated at 480 cold-cranking amperes, adjust the current draw to 240 amperes.
5. Maintain the load for 15 seconds while observing the voltmeter. Turn the control knob off immediately after 15 seconds of current draw.
6. The battery voltage at the end of the 15 seconds should not fall below 9.6 volts.

Battery service

The following sections cover basic battery service, including inspection, cleaning, replacement, charging, and jump-starting.

OPEN CIRCUIT VOLTAGE	BATTERY SPECIFIC GRAVITY*	STATE OF CHARGE	CHARGING TIME TO FULL CHARGE AT 80°F**					
			at 60 amps	at 50 amps	at 40 amps	at 30 amps	at 20 amps	at 10 amps
12.6	1.265	100%	FULL CHARGE					
12.4	1.225	75%	15 min.	20 min.	27 min.	35 min.	48 min.	90 min.
12.2	1.190	50%	35 min.	45 min.	55 min.	75 min.	95 min.	180 min.
12.0	1.155	25%	50 min.	65 min.	85 min.	115 min.	145 min.	280 min.
11.8	1.120	0%	65 min.	85 min.	110 min.	150 min.	195 min.	370 min.

*Correct for temperature. **If colder, it will take longer.

Figure 3-27. Battery charging times vary with state-of-charge, temperature, and charging rate.

Inspection

Inspect the battery for:

- Cracks, loose terminal posts, or other physical damage; replace any battery with such damage.
- Loose or damaged battery tray and holddown, or missing parts.
- Proper electrolyte level in each cell. If the caps are removable, add distilled or mineral-free water as needed to correct the electrolyte level; replace missing or damaged caps.
- Dirt and grease on the battery case and cover that can cause voltage leaks to ground; clean the battery as needed.
- Corrosion at the terminals, cable connectors, battery tray, and holddown; clean these parts as needed.
- Broken or corroded cables, wires, and insulation as well as loose, dirty, or damaged connectors; repair, replace, clean, and tighten the cables and connectors as needed.
- Proper positioning of the heat shield.

Cleaning

Before cleaning the battery, neutralize any corrosion on the terminals, connectors, and other metal parts with a solution of baking soda and water. Keep corrosion off painted surfaces and rubber parts, and do not allow the soda solution to enter the battery. Remove heavy corrosion with a stiff-bristled, non-metallic brush. Remove dirt and grease with a detergent solution or solvent. Rinse the battery and other parts with fresh water. Dry with a clean cloth or low-pressure compressed air. If the battery terminals and cable connectors are badly corroded, remove the cables from the battery (ground cable first) for cleaning. A special wire brush with internal and external bristles is a handy tool for cleaning battery posts and the inside of cable connectors. The posts and connectors must be completely clean for good electrical contact.

Replacement

Before removing the old battery, be sure you know which is the positive and which is the negative terminal. On most batteries, the positive post has a larger diameter than the negative post. Remove the ground (negative) cable from the battery first. Then remove the insulated (positive) cable. Loosen and remove the battery holddown and any heat shields that might interfere with battery removal. Lift the battery from the tray with a carrier or strap. If new cables are needed, install them at this time.

Install the new battery in the tray and secure the holddown and any heat shield parts. Connect the insulated (positive) cable to the battery first, then connect the ground (negative) cable. To help prevent corrosion on battery terminals, install anti-corrosion washers under the cable connectors, or apply a light coat of anti-corrosion compound to the terminals and connectors after installing the battery.

Battery charging

Battery charging is the process of applying current to a battery in order to bring it to a full state of charge. Charging input is measured in amperes over a period of hours. For example, a 20-ampere charging rate for three hours equals a 60-ampere-hour charging input.

Batteries are usually charged at rates from 3 to 50 amperes. Charging rates at the low end of this scale are called slow charging. Charging rates at the high end of the scale are called fast charging. Fast chargers are the most widely used type of charging equipment in service stations and garages.

Generally, any battery can be charged at any current rate, provided electrolyte gassing and spewing does not occur and electrolyte temperature stays below 125°F (52°C). When you have the time, however, it is best to charge at a slow rate of 5 to 15 amperes. Figure 3-27 gives typical charging times for various states of charge and charging rates.

Before charging a battery, check and adjust the electrolyte level. Leave the cell caps in

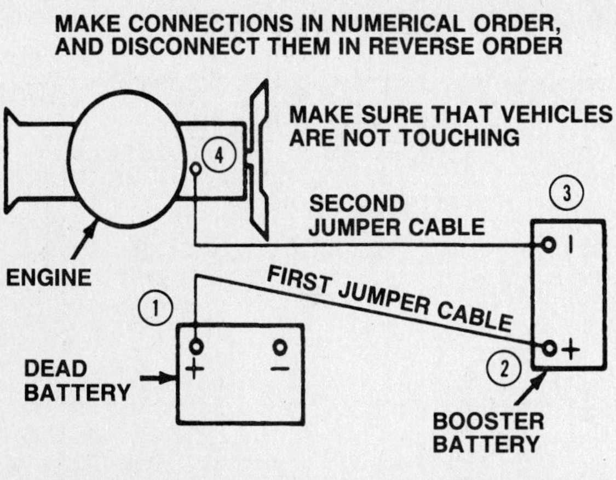

Figure 3-28. Connect jumper cables in the order shown when jump-starting a vehicle with a discharged battery.

place during charging, but be sure the vent holes are open. Whether charging the battery quickly or slowly, remember:

- Charge only in a well-ventilated area, away from sparks and open flame.
- Make sure the charger is off before connecting or disconnecting cables at the battery.
- Never attempt to charge a frozen battery.
- Always wear eye protection.

To charge a battery:

1. Disconnect both battery cables to avoid damage to the alternator or other electrical parts. If desired, the battery may be removed from the vehicle.
2. Connect the charger cables to the battery, checking for correct polarity.
3. Turn on the charger and set it to the desired charging rate.
4. Where possible, periodically check the temperature of the electrolyte. Stop charging and allow the battery to cool if the electrolyte temperature rises above 125°F (52°C).
5. Periodically check the voltage across the battery terminals. If the voltage rises above 15.5 volts at any time, lower the charging rate until the voltage drops below this maximum value.
6. The battery is fully charged when all cells are gassing freely and specific gravity does not increase for three hours. The fully charged specific gravity should be 1.260 to 1.280 (corrected for electrolyte temperature), unless otherwise specified by the battery manufacturer.
7. When charging a sealed maintenance-free battery, follow the manufacturer's specifications for charging rate and time. If the battery has a state-of-charge indicator, charge until the indicator shows a full charge. In any case, do not exceed the manufacturer's rate or time specifications.
8. After charging, wash and dry the top of the battery to remove any acid condensation due to gassing. Check the electrolyte level, and add mineral-free water as needed to correct the electrolyte level.
9. Install the battery if you removed it from the vehicle. Reconnect the insulated, positive cable first and then the ground cable.

Jump-starting

Refer to figure 3-28, and follow the procedure below to jump-start a vehicle safely and without damaging the charging system or producing sparks:

1. Set the parking brake, turn off the ignition and accessories, and place the shifter in neutral or park.
2. Connect one end of the first jumper cable to the positive terminal of the dead battery.
3. Connect the other end of the first cable to the positive terminal of the good battery.
4. Connect one end of the second cable to the ground terminal of the good battery.
5. Connect the other end of the second cable to a good ground on the engine block of the disabled car, as far from the battery as possible. The alternator mounting bracket is often a convenient place for this.
6. Start the car with the good battery. Then attempt to start the car that has the discharged battery. If the car does not start, wait a few minutes to allow the battery to charge.
7. After the disabled car starts, disconnect the second cable from the engine block first, and then from the good battery.
8. Disconnect the first cable from the good battery's positive terminal and from the dead battery's positive terminal.

Charging System Inspections and Tests

Alternators convert mechanical energy from the engine to electrical energy. If the alternator is defective, it may generate insufficient current to run the vehicle's circuits, or it may generate too much current, which can damage circuits. Inspect and test the charging system if you suspect that the alternator is defective.

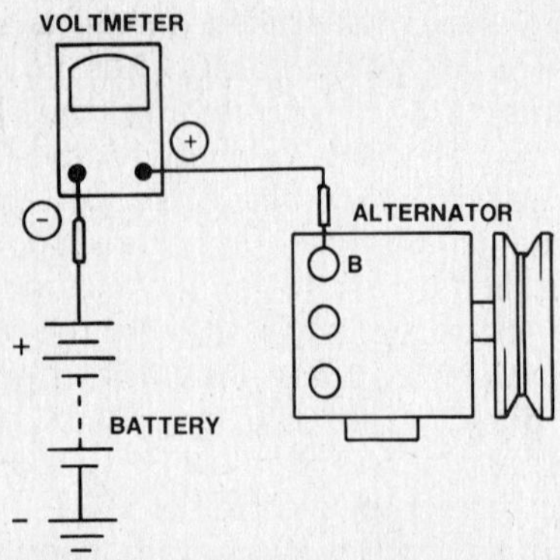

Figure 3-29. This is a common test meter hook-up for the circuit resistance test in a charging system.

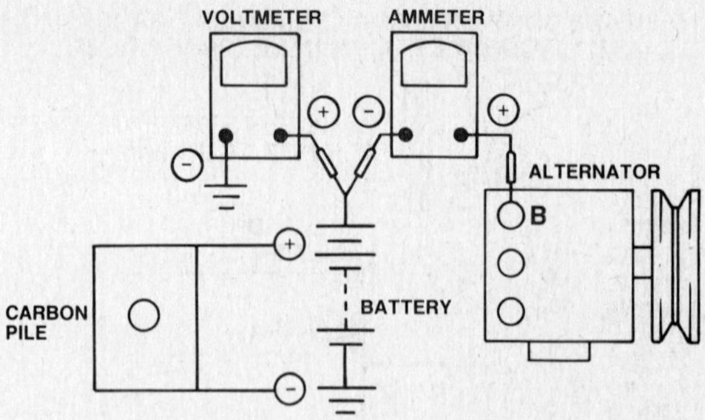

Figure 3-30. This is a common test meter hook-up for the current output test in a charging system.

Inspection
Many charging system problems are easy to repair. A visual inspection will reveal some problems:

- Check the battery electrolyte level, state of charge, and capacity as explained earlier. If the battery is bad, it will affect the performance of the charging system.
- Inspect the alternator drive belt. A loose or damaged drive belt will cause charging problems. Replace or tension the belt, if needed.
- Inspect system wiring and connections. Check fusible links for opens, and make sure electrical connectors are attached.
- Check the alternator and regulator mounts for loose or missing bolts. Replace or tighten them, as needed.

Charging system meter tests
Before beginning charging system tests, obtain the manufacturer's service manual for meter hook-up points and specifications. The following charging system tests are described in this section:

- Circuit resistance test
- Current output test
- Voltage output test.

If the results for these tests are not within the manufacturer's recommendations, refer to a manual on charging system service to repair or replace the alternator.

Circuit resistance test
Charging system circuit resistance is usually determined by the voltage drop test. Even though you might think to use an ohmmeter to measure resistance, in this case, it is easier to check resistance by measuring voltage drop.

Start the vehicle, and run the engine at 1,500 to 1,800 rpm. Use a voltmeter or DVOM to measure voltage between the battery positive terminal and battery output terminal at the alternator, figure 3-29. The meter should read about 12 volts. If there is a voltage drop of more than 0.1 volts (if the meter reads 11.90 volts or less), then there is excessive resistance in the circuit. Check for loose or corroded connections or damaged wires, and make repairs as needed.

Current output tests
A current output test checks the alternator's production of current. Alternators are rated according to the amount of current they produce. A vehicle manufacturer will choose an alternator that meets a vehicle's current demand. If the alternator is worn or damaged, it will not produce enough current to run the various electrical systems. Perform the current output test as follows:

1. Connect a carbon pile across the battery, connect a voltmeter to the battery positive terminal and ground, and connect an ammeter between the battery positive terminal and the battery terminal on the alternator, figure 3-30.
2. Run the engine at the manufacturer's recommended test speed, and adjust the carbon pile to maintain a steady 15-volt reading on the voltmeter, or to get the greatest possible current reading that is specified by the vehicle manufacturer.
3. Compare the ammeter reading to the vehicle manufacturer's specifications:
   - If it is within specifications, the charging system is working properly.
   - If it is below specifications, perform the voltage output test, as described in the next section.

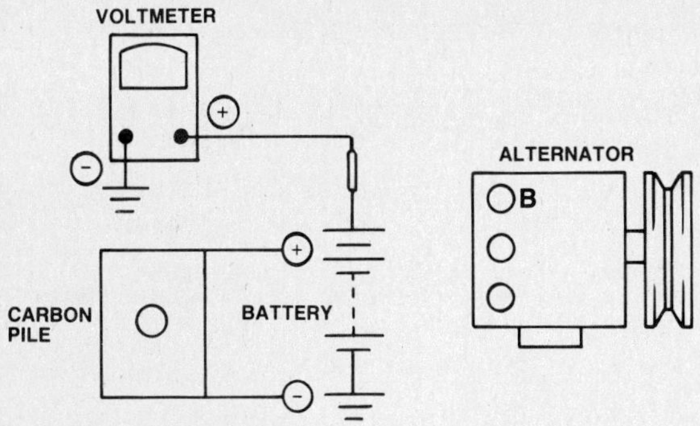

Figure 3-31. This is a common test meter hook-up for the voltage output test in a charging system.

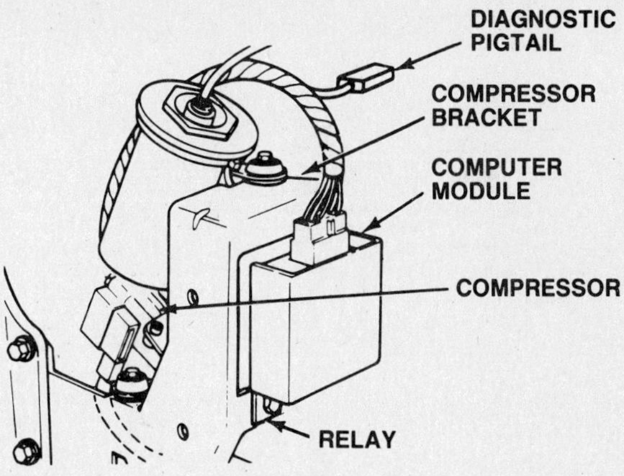

Figure 3-32. Ground the diagnostic pigtail located near the suspension computer to start the self-diagnostic test on this Ford automatic load-leveling suspension system.

Voltage output test

The voltage output test is recommended by some manufacturers to check alternator operation. To perform the voltage output test:

1. Connect a carbon pile across the battery, and connect a voltmeter to the battery positive terminal and ground, figure 3-31.
2. Run the engine at the manufacturer's recommended test speed, and use the carbon pile to load the battery. Do not allow system voltage to exceed 16 volts during this test.
3. If battery voltage drops below 13 volts, bypass the voltage regulator. This can be done by using a jumper wire, one of the special "full-fielding" tools that are available, or by shorting a special tab that is built into the alternator for this purpose.
4. Repeat the test.
   - If the battery voltage rises to about 16 volts, the voltage regulator may be faulty.
   - If the battery voltage remains low, remove the alternator for further testing.

EXAMPLE OF TROUBLESHOOTING AN ELECTRONICALLY CONTROLLED SUSPENSION SYSTEM

Some manufacturers use electronic controls for their steering and suspension systems, as discussed in Chapters Seven and Twelve of the *Classroom Manual*. Chapters Seven and Eleven of this *Shop Manual* have basic service procedures for these systems. For complete diagnostic and repair information, always refer to the vehicle manufacturer's service manual. The example in this section is to illustrate how to troubleshoot the electronic controls of a computerized suspension system.

The automatic leveling system used on some Ford, Lincoln, and Mercury models provides electronic control of the rear suspension. Height sensors at the rear axle send information to a suspension computer, which in turn signals the air compressor to fill air springs when the vehicle ride height is too low. If the ride height is too high, the computer signals a solenoid to release air from the air springs.

In this system, the height sensor controls two basic circuits:

- The compressor relay coil ground, which when activated, will start the compressor to fill the air springs.
- The exhaust solenoid coil ground, which when activated, will open the exhaust valve to vent air from the air springs.

Before performing electrical tests, check air lines for leaks and pressure loss. If you find a damaged or leaking nylon air line, fix it before starting the diagnostic procedure.

To perform the self-diagnostic test on this system, ground the diagnostic pigtail located in the wiring harness near the computer's connector on the compressor bracket, figure 3-32. The trouble codes will tell you which circuits are defective.

Check voltages at components along the circuit, or trace and repair any opens, shorts, or grounds in the circuitry.

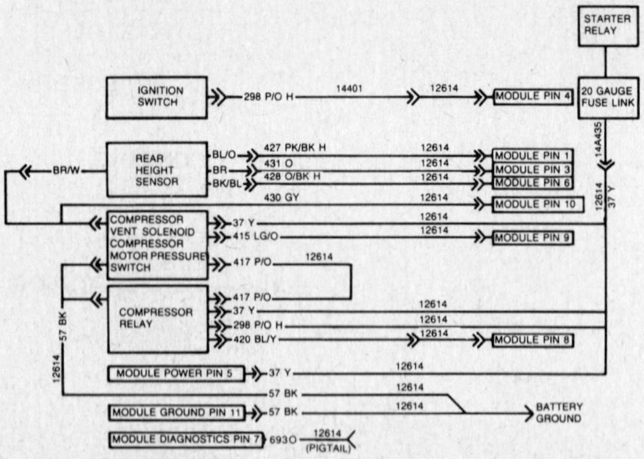

Figure 3-33. Use the vehicle manufacturer's circuit diagram and the trouble codes from the self-test to trace defective components or wires.

Figure 3-33 is a circuit diagram of the system. It tells you that the electrical power source is distributed from the main body wiring harness through the following circuits:

- Circuit 14A435 connects directly to circuit 12614 and supplies battery power to the system.
- Circuit 12614 transmits the battery voltage directly to the height sensor, compressor motor, relay, vent solenoid and computer module.
- Circuit 14401 is an extension of ignition switch circuit 298 (sends the ignition RUN position input to the computer).

From the trouble codes, you will be able to determine which circuits are defective. Then, using a DVOM and the various tests described earlier in this chapter, test the defective circuits. Compare meter readings with the vehicle manufacturer's specifications to identify defective parts.

PART TWO

Steering System Service

Chapter

4

Steering Wheel, Steering Column, and Steering Gear Service

If you determine that handling problems are due to a faulty steering wheel, steering column, steering shaft, or steering gear, inspect these components. This chapter contains procedures for removing, repairing, and replacing steering wheels, columns, and shafts, and procedures for removing, adjusting, and replacing steering gears. You can find diagnostic procedures for steering and suspension system problems in Chapter Two of this *Shop Manual*.

After working on the steering system, it is good practice to check and adjust wheel alignment, as described in Chapter Fourteen of this *Shop Manual*.

STEERING WHEEL AND COLUMN

The steering wheel and column are the vital link between the driver and the vehicle's wheels. Problems with the steering system reduce the driver's ability to control the vehicle, and if steering problems are severe, the car will be too dangerous to drive. Service procedures for the steering wheel and column include:

- Preliminary checks
- Removing the steering wheel
- Removing steering column accessories
- Removing the steering column
- Inspecting the steering column and shaft
- Installing the steering column and wheel.

Always carefully diagnose, repair, and then *reinspect* the steering system. If you are working in a shop, be sure to test drive the vehicle before returning it to the customer.

Preliminary Checks

When a customer comes into the shop, ask him or her to describe the vehicle's handling problems. Research the possible causes, and then inspect the vehicle. Preliminary checks will help you locate problem areas. Some common inspections recommended by many vehicle manufacturers include the:

- Steering wheel freeplay test
- Steering wheel neutral position check
- Visual inspection
- Test drive, as described in Chapter Two of this *Shop Manual*.

Steering wheel freeplay test

The steering wheel freeplay test measures steering response. Freeplay is the distance that you can turn the steering wheel before the vehicle's wheels actually move. If freeplay exceeds the manufacturer's recommended limits,

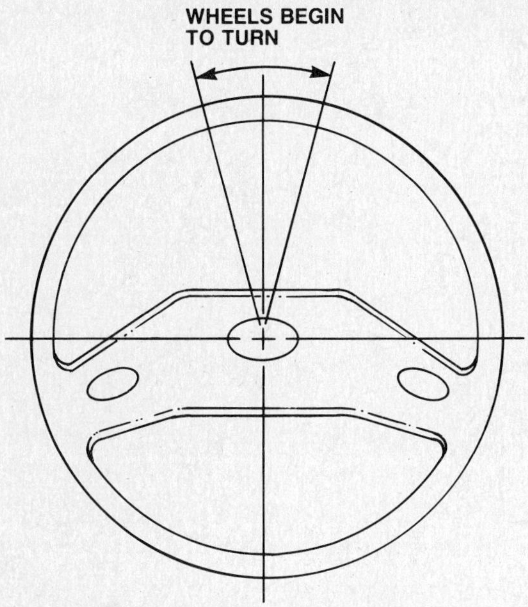

Figure 4-1. The steering wheel freeplay test measures the distance that you can turn the steering wheel before the vehicle's front wheels respond.

parts are likely worn in the steering column, gear, or linkage. Perform the test as follows:

1. Turn the steering wheel until the vehicle's front wheels are positioned straight ahead.
2. Ask another technician to watch the front wheels while you slowly turn the steering wheel.
3. When the front wheels move, measure the distance that the steering wheel has turned, figure 4-1.
4. Straighten the wheels again, and perform the test while turning the steering wheel in the other direction.
5. Compare your measurements with the manufacturer's steering wheel freeplay allowance, which can range from 0 to 1.5 inches (35 mm). If freeplay exceeds the manufacturer's recommendation, inspect steering components for wear.

Steering wheel neutral position check

When the vehicle's front wheels are positioned straight ahead, the steering wheel and steering gear should be centered. The steering wheel neutral position check will help you determine if the steering wheel and gear are properly positioned with respect to front wheel placement. Perform the test as follows:

1. Turn the steering wheel until the front wheels are straight. Ask another technician to help you do this if necessary.

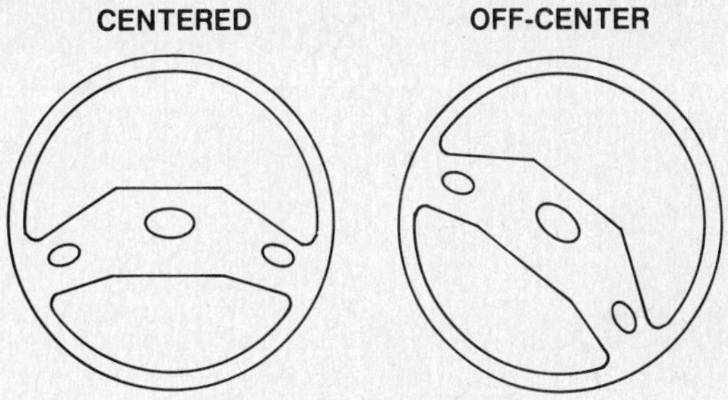

Figure 4-2. When the front tires and wheels are positioned straight ahead, the steering wheel should be centered, like the one on the left.

2. Check the position of the steering wheel. It should be centered, not turned to one side or the other, figure 4-2.
3. If the steering wheel is not centered, check front wheel alignment, as described in Chapter Fourteen of this *Shop Manual*. If the wheels are properly aligned and the steering gear is centered, remove and reposition the steering wheel according to the procedure later in this chapter.

Visual inspection

After checking the steering wheel for freeplay and position, visually inspect the steering column and accessories for any problems that may affect steering, figure 4-3. This inspection will help you determine if the vehicle is safe to test drive.

Are lock housing components, such as the ignition switch, turn signal, wiper, and light switch levers positioned correctly on the column? Ask the customer if there are any problems with these switches. If you have to remove column accessories while correcting a steering problem, you might as well replace defective switches as well.

If there is a gear shifter on the column, is it securely mounted?

If the steering column has a tilt mechanism, does it hold the steering wheel in position, does it in any way hamper steering wheel movement, and does it operate properly? If not, you should inspect and replace faulty components.

Does the horn work? If it does not, and the horn fuse is intact, check and repair electrical connections under the horn pad and in the steering column.

If the vehicle is equipped with an air bag, is the module intact? The module should be flat against the steering wheel, and the air bag itself should not be visible.

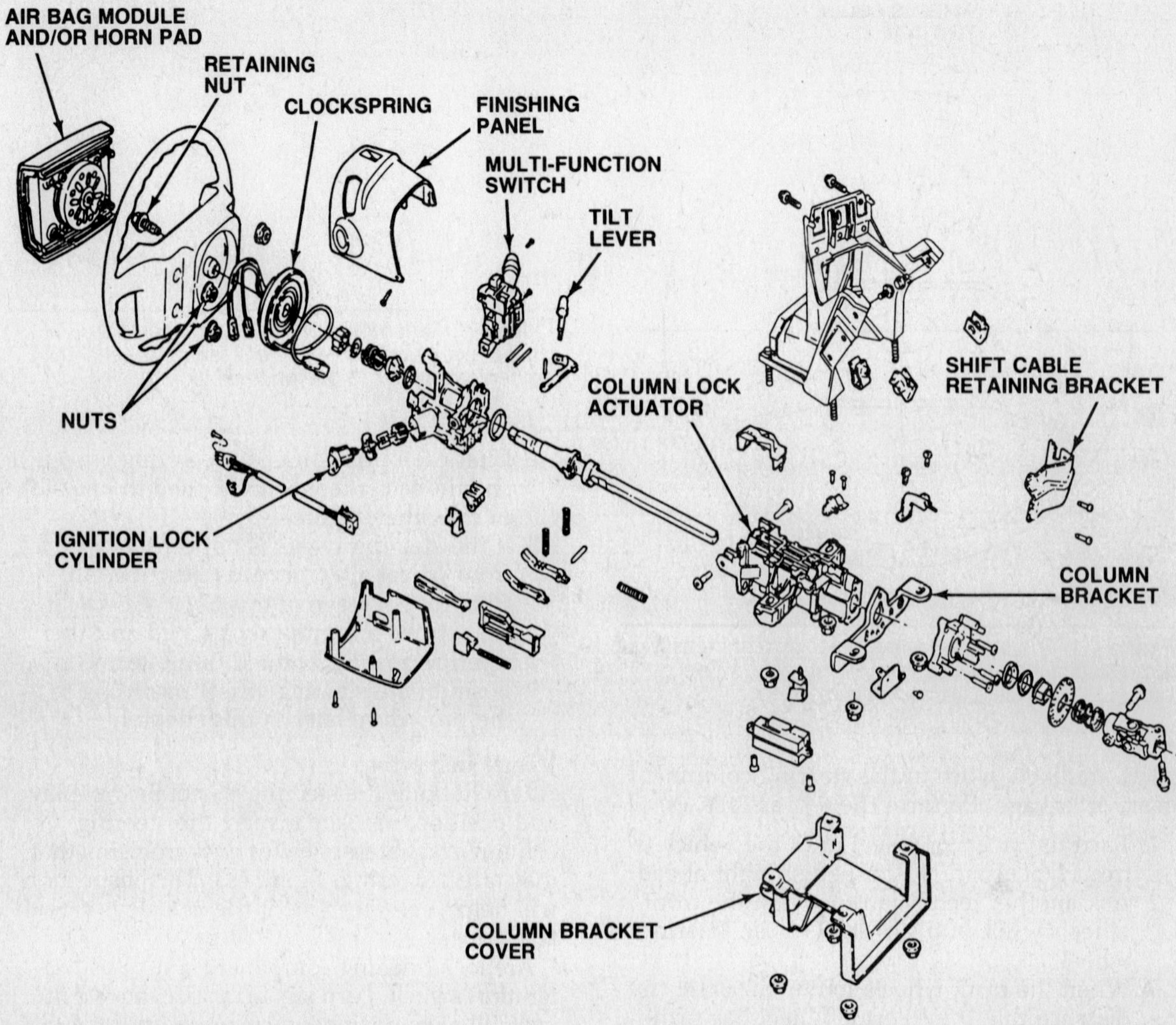

Figure 4-3. If column accessories are not secure, they may rub on the steering shaft and wheel, which increases steering effort. Check column accessories as part of a routine steering system inspection.

WARNING: Be careful when working on a vehicle that is equipped with an air bag. If the air bag discharges, you can be injured.

Do the shroud and dashboard finishing panels contact the steering wheel or column? Heavy steering, poor steering wheel return, and squeaking may be caused by panels rubbing on the steering wheel and column.

After inspecting the steering wheel and column accessories, test drive the vehicle, noting any handling problems.

Test drive
Begin the test by driving the vehicle slowly. This gives you a chance to check the severity of the steering problem. When test driving a car, check the steering feel as well as the overall handling. Careful observation will help you pinpoint areas for further inspection. See Chapter Two of this *Shop Manual* for information on locating steering system problems during the test drive.

Removing the Steering Wheel

Many late-model cars have an air bag installed in the center of the steering wheel. Be sure to check for and disable an air bag before removing the steering wheel.

WARNING: Follow the manufacturer's procedures for disabling and removing air bags, or you risk serious injury to yourself and damage to the car. Replacing a discharged air bag is costly.

After removing the air bag module, remove the steering wheel. If the vehicle does not use

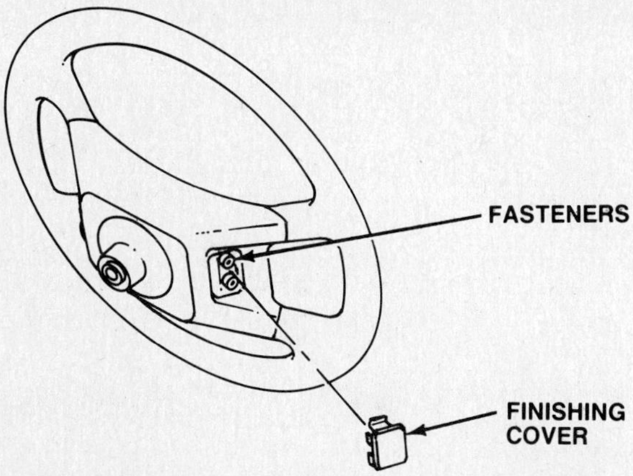

Figure 4-4. Pry off plastic finishing covers with a screwdriver to reveal fasteners that secure the air bag module to the steering wheel.

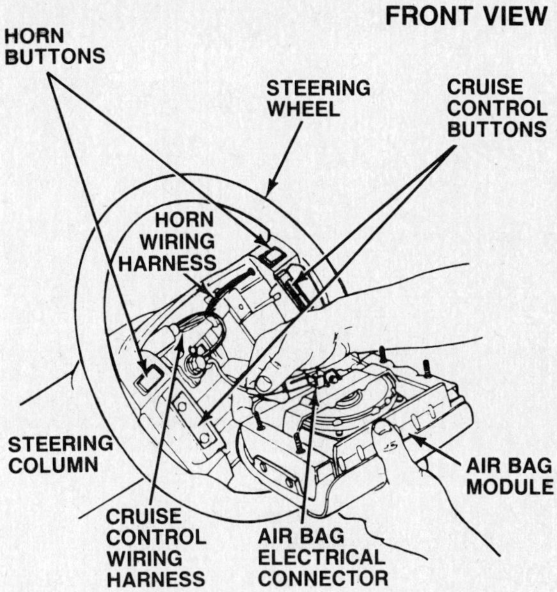

Figure 4-5. The air bag module is located in the center of the steering wheel. Remove it only after disconnecting the battery ground cable and the electrical connector.

an air bag, skip the following procedure and continue with the next one.

Air bag module removal
This procedure reviews the basic steps for removing an air bag module. Do not substitute these general instructions for the manufacturer's recommended procedure:

1. Turn the steering wheel until the vehicle's wheels are positioned straight ahead. Some components on the steering column can be removed only when the front wheels are straight.
2. Turn the ignition off, and disconnect the battery ground cable, which cuts power to the air bag ignitor.
3. *Wait at least 10 minutes before removing the air bag*. Some air bag systems have electronic components that will hold an electrical charge for a short period of time. If you do not wait long enough, there may be enough charge in the system to trigger the air bag.
4. Loosen and remove the nuts or screws that hold the air bag module in place. On some vehicles, these nuts or screws are located on the back side of the steering wheel, figure 4-3. On others, they are located on each side of the steering wheel. Nuts or screws may be concealed with plastic finishing covers that you must pry off with a screwdriver, figure 4-4. Use a socket wrench, Torx bit, or screwdriver as required, to remove the fasteners. Discard the old fasteners and replace them with new ones during reassembly.
5. Lift the air bag module from the steering wheel, and disconnect the electrical connector, which may be located under the module, figure 4-5, or below the steering wheel under a plastic cover.
6. Store the module (pad side up) on a shelf where it will not be damaged while you are working on the vehicle. *Do not* attempt to disassemble the air bag module. If the air bag is defective, replace the entire component.
7. Proceed to step 2 of the next procedure to remove the steering wheel.

Steering wheel removal
Remove the steering wheel using the following procedure as a guide:

1. Remove the horn cover by removing the nuts or screws that secure it. They are usually located on the back side of the steering wheel, figure 4-6. Some vehicles have horn buttons that are press-fit in place. Pry or twist off this type of horn cover.
2. Disconnect any electrical connectors under the horn cover. To make reassembly easier, code the connectors and wires with tape or tags.
3. Loosen and remove the steering wheel retaining nut with a socket wrench, figure 4-7.
4. Check the steering wheel and shaft for alignment marks, figure 4-8. These marks are placed by the manufacturer to help you align the steering wheel during assembly. If you do not see alignment marks and you

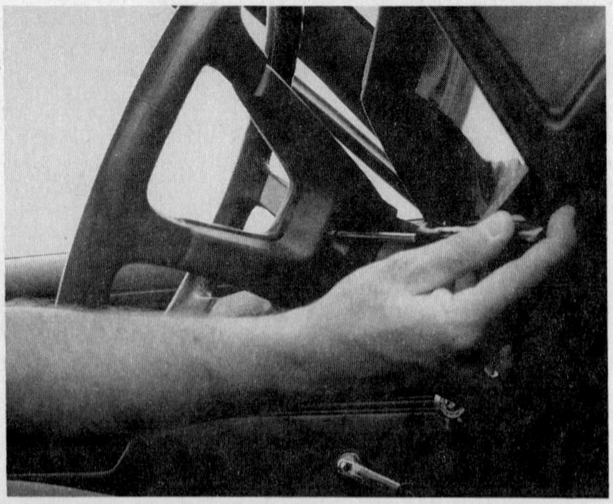

Figure 4-6. To remove a horn cover, loosen the screws that hold it in place.

Figure 4-7. After removing the horn pad or air bag module from the steering wheel, loosen and remove the retaining nut at the center.

are sure that the steering wheel is centered, use a felt pen to make alignment marks on the steering wheel and shaft before removing the steering wheel.

5. Remove the steering wheel. On some vehicles, you can rock the steering wheel back and forth, working it up the steering shaft, until it is loose enough to remove. On most late-model vehicles, however, you must use a special puller. Never use a hammer.
6. Locate the puller bolt holes in the steering wheel. Install the puller bolts, figure 4-9. Tighten the center bolt on the puller until the steering wheel lifts from the steering shaft. Do not use a hammer to loosen the steering wheel because you may damage or collapse the steering column and shaft.

Figure 4-8. Look for alignment marks on the steering wheel and shaft. These help you correctly position the steering wheel during reassembly. Make your own alignment marks if you do not find any.

Figure 4-9. To use this type of steering wheel puller, thread the center bolt into the steering shaft at the center of the steering column, and thread the two outer bolts into the steering wheel. Turning the center bolt on the puller forces the steering wheel up and off the shaft.

Removing Steering Column Accessories

When installing new steering column accessories, such as the turn signal mechanism or headlight switch, you do not have to remove the entire steering column from the car. Instead, you may have to remove the steering wheel, as described earlier, and some finishing panels.

Finishing panels, such as the driver's knee bolster and steering column cover, are held in

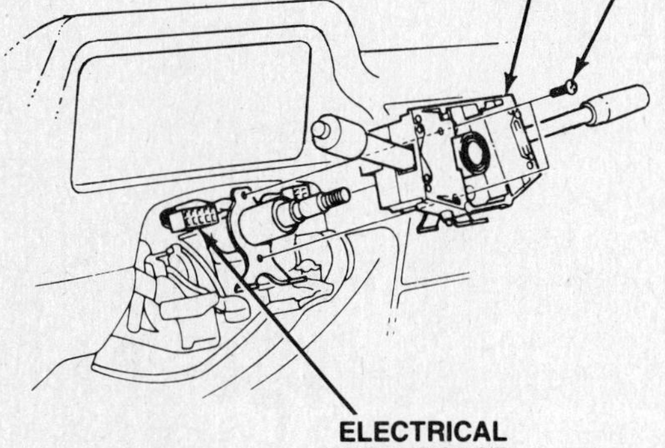

Figure 4-10. In many cars, steering column accessories are easy to remove and replace because they are combined into a single unit, called a combination or multi-function switch.

place with screws, which are sometimes concealed with plastic covers. Other panels can be removed by simply prying with a thin-blade screwdriver.

The procedures for removing steering column accessories vary depending on vehicle design, so you should use the manufacturer's service manual when you repair or replace column accessories. If the vehicle has a steering wheel lock, you may need a special tool available from the manufacturer to remove the lock before you can continue removing column accessories. On many vehicles, some of the accessories are combined into a single removable unit, called a combination or multi-function switch, figure 4-10.

This procedure for a 1990 Ford Taurus with steering column tilt is an example of how to remove column accessories, including a combination switch:

1. After removing the steering wheel, you can see the clockspring assembly or other electrical connection for the steering wheel wiring harness, figure 4-3. Disconnect the clockspring assembly or electrical connector, and remove the screws or retainers that hold it in place.
2. Remove the tilt lever from the column by turning it counterclockwise.
3. Disconnect the battery ground cable. Remove the ignition lock cylinder by rotating it to the RUN position. Then remove the retaining pin with a drift.
4. Loosen and remove the screws that hold the column shroud in place, and remove the shroud.
5. Remove the bolts that secure the instrument panel reinforcement, and remove the reinforcement.
6. If the car has a column shifter, disconnect the cable from the lock cylinder by loosening the fastening screw. Then unhook the cable from the shift tube hook. If the shifter is on the console, remove the cable retaining bracket and screws, and remove the cable.
7. Disconnect the electrical connectors for the combination or multi-function switch and the ignition switch.
8. Remove the screws that hold the combination switch to the steering wheel hub, figure 4-10, and remove the switch.

If an accessory in the combination switch is defective, such as the turn signal switch, replace the entire combination switch with a new or rebuilt one.

On many early model vehicles, steering column accessories are replaced individually. After determining if accessory failure is due to electrical or mechanical problems, disassemble and replace components as needed. This procedure is for a 1966 Chevrolet Chevelle, figure 4-11:

1. Disconnect the battery ground.
2. Remove the steering wheel as described earlier.
3. Lift the cancelling cam and spring from the column.
4. Loosen and remove the screw that fastens the turn signal lever, and remove the lever.
5. Disconnect the turn signal switch from the wiring harness at the instrument panel.
6. Loosen and remove the screws that hold the turn signal switch in place, and lift the switch from the housing.
7. Replace components that are damaged, and reassemble in the reverse order.

Removing the Steering Column

The steering column is rarely removed and serviced. However, if the vehicle was involved in a collision, the steering column and shaft may have collapsed, and you should replace them. The column and shaft are designed to collapse on impact, so that if the driver is thrown forward in a head-on collision, the steering column and shaft will give, lessening

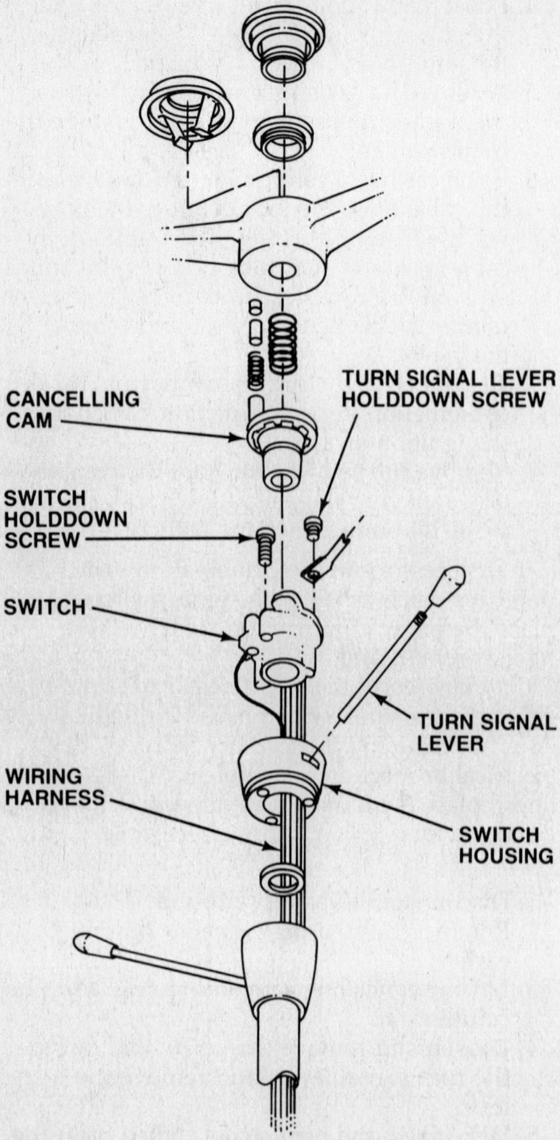

Figure 4-11. Older vehicles have fewer steering column accessories. If a column accessory is defective, it can usually be replaced as an individual component.

upper-body injury to the driver. Remove the steering column as follows:

1. Remove the steering wheel, and disconnect any accessories that would prevent you from removing the steering column, such as the gear shift cable and any electrical connections.
2. Remove the steering shaft joint cover, which is usually located at the base of the steering column on the floor of the passenger compartment, figure 4-12. Many manufacturers use a rubber boot or tube to cover

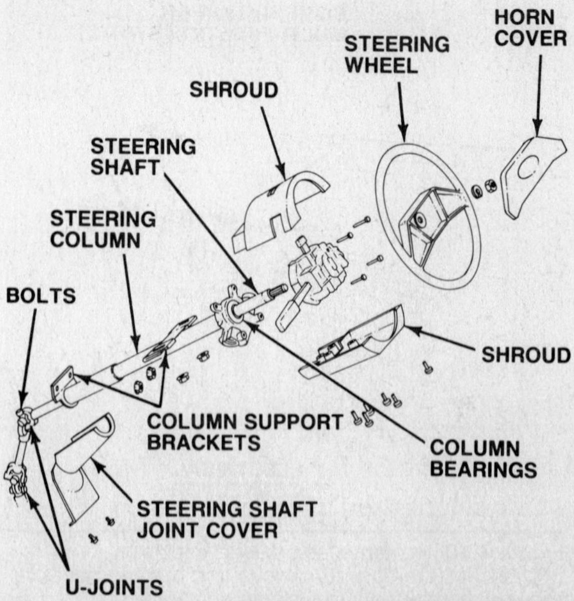

Figure 4-12. On some vehicles, a steering shaft joint is accessible from the passenger compartment. It is usually covered with a rubber boot or plastic shroud at the base of the steering column.

the joint, but some may use a bracket and rubber gasket that bolt to the floor. If the steering shaft does not disconnect at the floor of the passenger compartment, look for the steering shaft joint in the engine compartment, figure 4-13.

3. Loosen and remove the bolts at the U-joint or other coupling, figure 4-12.
4. Loosen and remove the nuts, bolts, or screws that secure the column support brackets to the floor or underside of the dash, figure 4-12. As you remove these brackets, support the steering column.
5. Remove the steering column from the car. You may have to work the steering shaft around slightly to disconnect the joint.

Inspecting the Steering Column and Shaft

After you have removed the column, inspecting the steering column and shaft is not difficult. You should check:

- Steering shaft joints, such as U-joints, flexible couplings, or pot couplings for wear
- Steering shaft bearings for wear or binding
- Steering column or shaft for collapse. If the vehicle was in a collision, the steering column and shaft may be damaged.

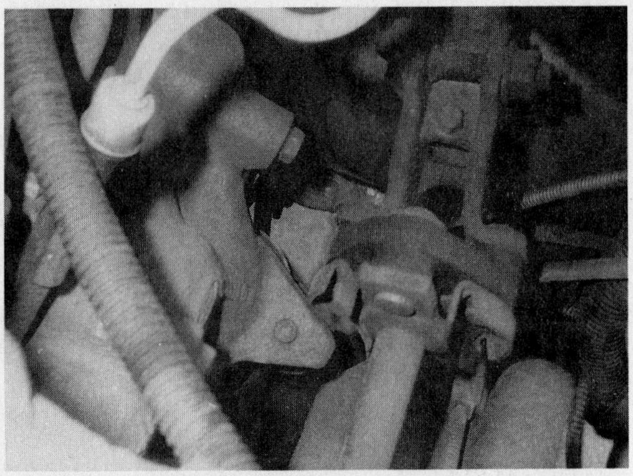

Figure 4-13. Steering shaft joints may be located in the engine compartment, as in this Ford truck.

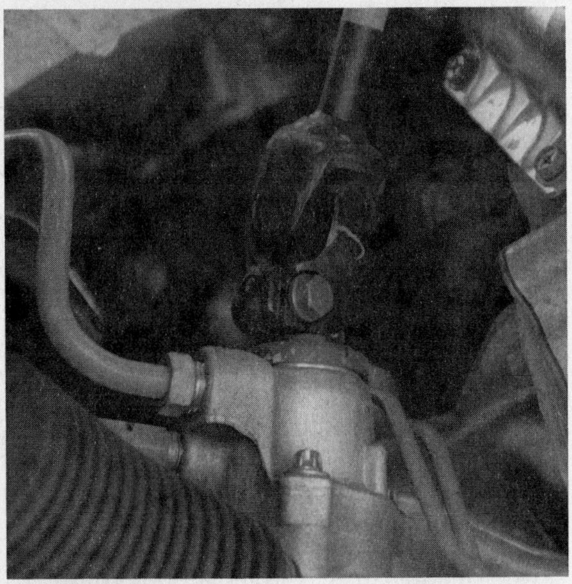

Figure 4-14. This U-joint connects the steering shaft and steering gear input shaft on a rack and pinion steering gear.

Checking and replacing steering shaft joints

Joints allow flexibility in the steering shaft while still transferring rotary motion from the steering wheel to the steering gear. When these joints wear, motion is not transferred smoothly, and the steering will feel loose or uneven.

The number of joints in the steering shaft varies. Some manufacturers use a single joint that joins the steering shaft to the steering gear, while others have a number of joints along the shaft. The three most common joints used in the steering shaft are U-joints, flexible couplings, and pot joints.

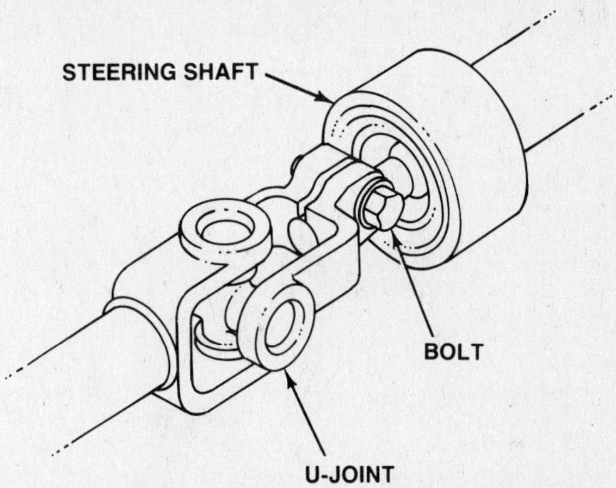

Figure 4-15. Loosen the bolt that clamps the U-joint assembly to the steering shaft, and slip the joint off the steering shaft.

Checking and replacing U-joints

Many manufacturers use a U-joint at the base of the steering shaft to join it to the steering gear input shaft, figure 4-14. Some have U-joints higher up on the steering shaft to allow for flexibility. If the U-joints are hard to get to, refer to the procedure in this chapter titled "Checking and replacing the steering shaft and bearings" to access these U-joints on the upper shaft. Then use the following procedures to check and replace them:

1. Remove the rubber boot that covers the U-joint at the base of the steering shaft in the passenger or engine compartment.
2. Use a screwdriver to lever the U-joint by applying pressure to the screwdriver handle. If there is play, replace the joint with a new one.
3. Check all steering shaft U-joints in this manner, replacing any that are defective.
4. If the U-joint is not worn, lubricate it with the manufacturer's recommended grease.

If U-joints are worn, replacing them is not difficult. Install new U-joints as follows:

1. Loosen the bolt that secures the U-joint to the steering shaft, figure 4-15.
2. Slip the U-joint off the splined shaft, and replace it with a new one.
3. Tighten the bolt to hold the new U-joint in place.
4. Replace the other half of the U-joint on the adjacent steering shaft section using the same method. If the other portion of the joint is an integral part of the shaft, remove and replace the shaft.

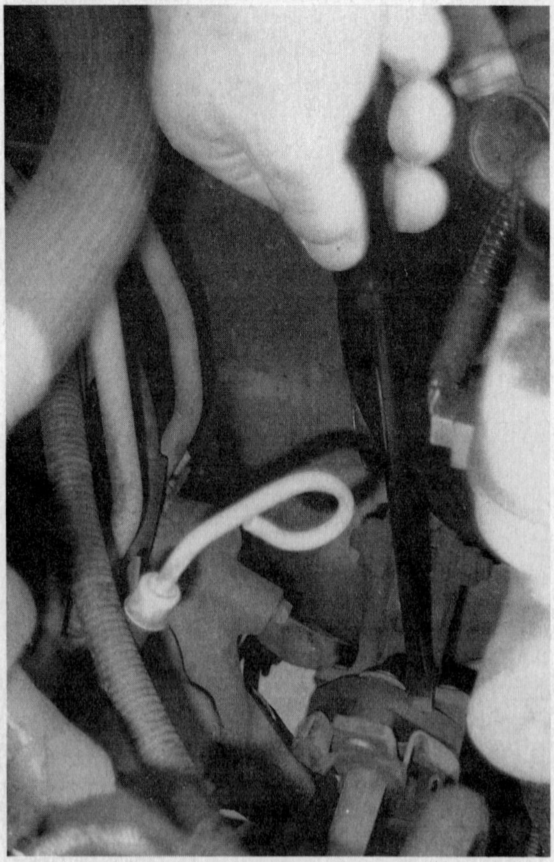

Figure 4-16. A rubber bushing in this flexible coupling damps vibration and road shock that would otherwise be transmitted through the steering shaft to the steering wheel. Check bushings for deterioration.

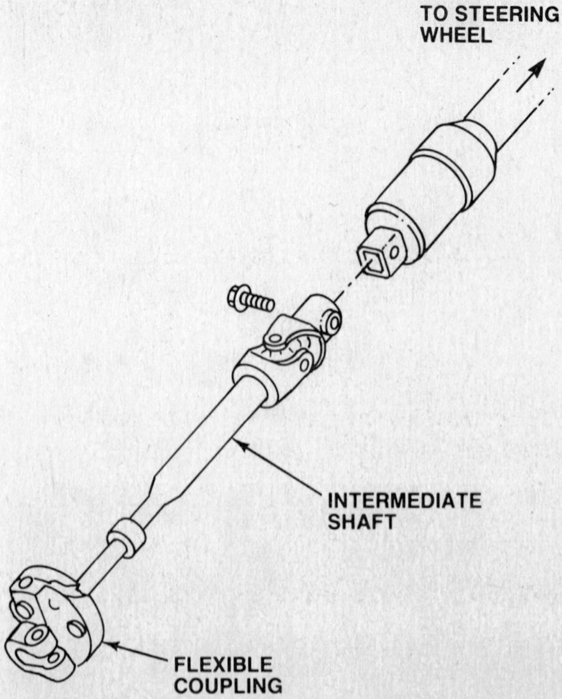

Figure 4-17. In this Ford Ranger, the flexible coupling is an integral part of the intermediate shaft. To replace the coupling, you must replace the intermediate shaft.

Checking and replacing flexible couplings

A flexible coupling joins the two ends of the steering shaft with flat, metal plates and a rubber bushing, figure 4-16. The bushing allows a small amount of flexibility in the steering shaft and absorbs vibration, but it is prone to wear. Check a flexible coupling as follows:

1. Check the rubber bushing for deterioration due to contamination or wear.
   - Prod the rubber with a screwdriver. If the rubber feels mushy, it is probably contaminated with oil.
   - Check bushing thickness. It should be about one-half inch (13 mm) thick.
2. Grasp the shaft on each end of the coupling and twist it. If there is play, the flexible coupling is worn and should be replaced.

If the rubber bushing is deteriorating or the joint is worn, replace the flexible coupling with a new one as follows:

1. Loosen and remove the bolts that secure the flexible coupling to the steering shaft.
2. On some vehicles, the flexible coupling is part of an intermediate shaft, and you must replace the intermediate shaft when you replace the flexible coupling, figure 4-17.
3. Slip each end of the flexible coupling off the steering shaft, and replace the coupling with a new one.
4. Tighten the bolts.

Checking and replacing pot joints

Pot joints, figure 4-18, are not common on late-model vehicles. However, some late-model Chrysler and Oldsmobile rear-wheel-drive (RWD) vehicles use them. Check and replace pot joints as follows:

1. Grasp the steering shaft on both sides of the joint, and twist it.
2. If you feel roughness, or if you cannot turn the shaft evenly, replace the pot joint bearings or the entire joint.
3. Remove the rubber boot covering the joint.
4. Pry off the joint retaining ring with a screwdriver, and pull the coupling cover from the steering shaft.
5. Remove the bearings and bearing clip from the pivot pin.
6. Check the bearings, pivot pin, bearing clip, and spring for wear.

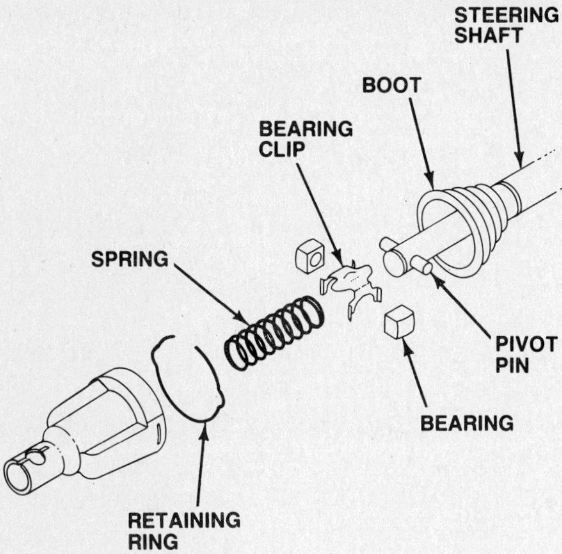

Figure 4-18. Pot joints, sometimes called ball and trunnion joints, are uncommon in late-model cars. If there is play in this type of joint, replace the bearings or the entire joint.

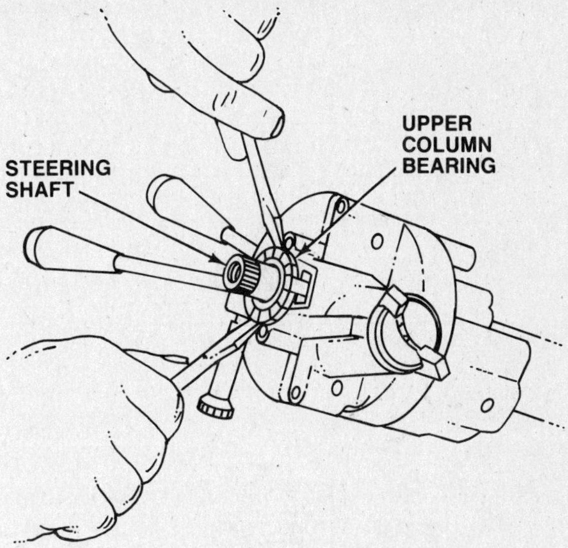

Figure 4-19. After removing the snapring that holds the upper column bearing in place, use two screwdrivers to pull the steering shaft up from inside the column.

If these parts are worn, replace them with new ones, or replace the entire pot joint.

1. Remove the steering shaft from the vehicle using the procedure described later in this chapter.
2. Support the steering shaft in a vise, and press the pivot pin from the shaft. Slide the boot off the shaft.
3. Clean dirt and grease from the steering shaft, and slide a new boot onto the shaft.
4. Press a new pivot pin into the shaft. Be sure the pin is centered, or the bearings will bind when they are installed.
5. Apply the manufacturer's recommended grease to the inside and outside of the bearings. Position the bearings and bearing clip on the pivot pin.
6. Slide the cover over the bearing assembly, and install the retaining ring. Twist the joint to check for binding.

Checking and replacing the steering shaft and bearings

Although the steering shaft and bearings will probably last the life of the car, sometimes the bearings wear due to inadequate lubrication. If the steering wheel rattles during turns, but all of the connections are tight, check column bearings. Most cars have only an upper or lower column bearing, but some have both.

Replace an upper bearing with the steering column in the car. Before removing this bearing, remove the steering wheel and any accessories that are blocking access to the upper bearing retainer plate or snapring. Before removing the lower bearing, remove the steering column from the vehicle, as described earlier in this chapter. Use the following steps as a guide to remove the steering shaft or to check and replace bearings:

1. Remove the steering wheel and any steering column accessories that block access to the bearing retainer plate, as described earlier in this chapter. Disconnect the U-joint or other coupling at the base of the steering shaft.
2. Remove the upper bearing retainer plate, which is fastened to the steering hub with screws or simply placed over the bearing.
3. Using snapring pliers, remove the snapring that holds the bearing onto the steering shaft.
4. Use two screwdrivers as levers to pull the steering shaft up in the column, exposing the bearing, figure 4-19.
5. Use screwdrivers to work the bearing up the shaft until it comes off.
6. If you are replacing the lower bearing or removing the steering shaft from the column, remove the steering column from the vehicle, as described earlier. Then remove the steering shaft by pulling it out at the lower end of the steering column.
7. If the lower column bearing is worn, remove it and press a new one in place. In some vehicles, you must replace a bearing

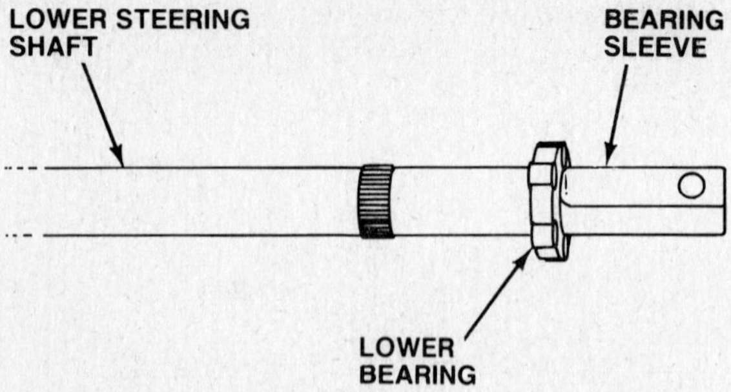

Figure 4-20. In the Ford Aerostar, the lower column bearing is part of a sleeve that slides onto the end of the steering shaft. To replace the bearing, you must replace the entire sleeve.

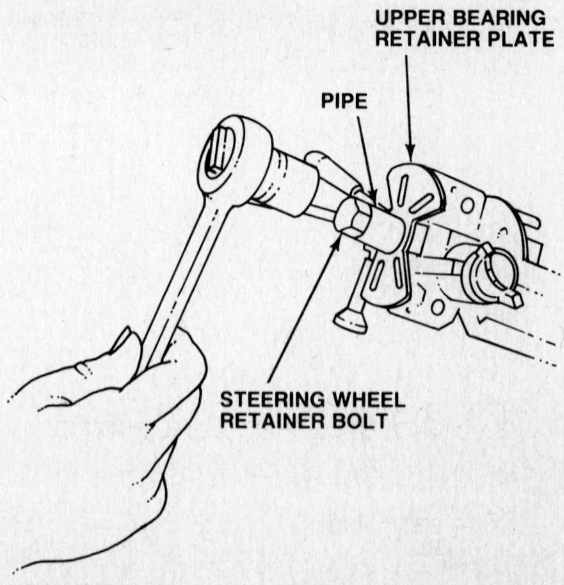

Figure 4-21. Use the manufacturer's recommended tool or a piece of pipe to press the upper steering shaft bearing in place.

sleeve assembly, which contains the lower column bearing, figure 4-20. Use the manufacturer's recommended grease to lubricate the bearing.

8. Install the steering shaft in the column.
9. Before installing the upper bearing, use a punch to roughen the surface of the steering shaft where the bearing will seat. This ensures that the bearing will fit snugly.
10. Use the manufacturer's recommended grease to lubricate the bearing. Push the bearing down the shaft as far as possible, and replace the retainer plate.
11. Place a piece of pipe over the bearing, and insert the steering wheel retainer bolt through the pipe and into the steering hub, figure 4-21. Tighten the bolt until the bearing seats. Then remove the retainer bolt and the pipe.
12. Install the snapring.
13. Replace steering column accessories, and install the column in the vehicle.

Checking for a collapsed steering column
If a customer's vehicle was involved in a collision, the steering column and shaft may have collapsed. Manufacturers use one or a combination of methods described below to construct collapsible steering columns and shafts, figure 4-22.

As part of a general steering column and shaft inspection, some manufacturers recommend that you measure the length of the steering shaft, figure 4-22A. If the length is not within specified limits, replace the shaft.

Many manufacturers use a breakaway bracket that yields to the forward force of the driver during a collision, figure 4-22B. The bracket breaks away upon impact, causing the steering column to drop. If the vehicle you are working on uses this method of collapsible steering, be sure that the breakaway bracket is secure and not partially broken. Replace it, if necessary.

Some manufacturers use shear pins and a telescoping tube, figure 4-22C. Upon impact, the pins break and the telescoping tube collapses. In vehicles equipped with this type of collapsible steering column, look for pieces of plastic as you disassemble the steering column. If you find pieces of plastic, replace the pins or the entire steering column according to the manufacturer's recommendations.

Some manufacturers construct the steering column jacket with a meshed section, figure 4-22C. Upon impact, the meshed area collapses. When disassembling this type of steering column, check the column and shaft for bends in the tubing, and replace any defective parts.

Because steering columns and shafts collapse, never use a hammer to remove or install them. Impact from hammer blows is likely to weaken the steering column and shaft.

Installing the Steering Column and Wheel

After inspecting and replacing worn or broken parts, reassemble and install the steering column and shaft, as described in the following sections.

Reassembling the steering column
Install the shaft and bearings in the steering column, and replace the column accessories,

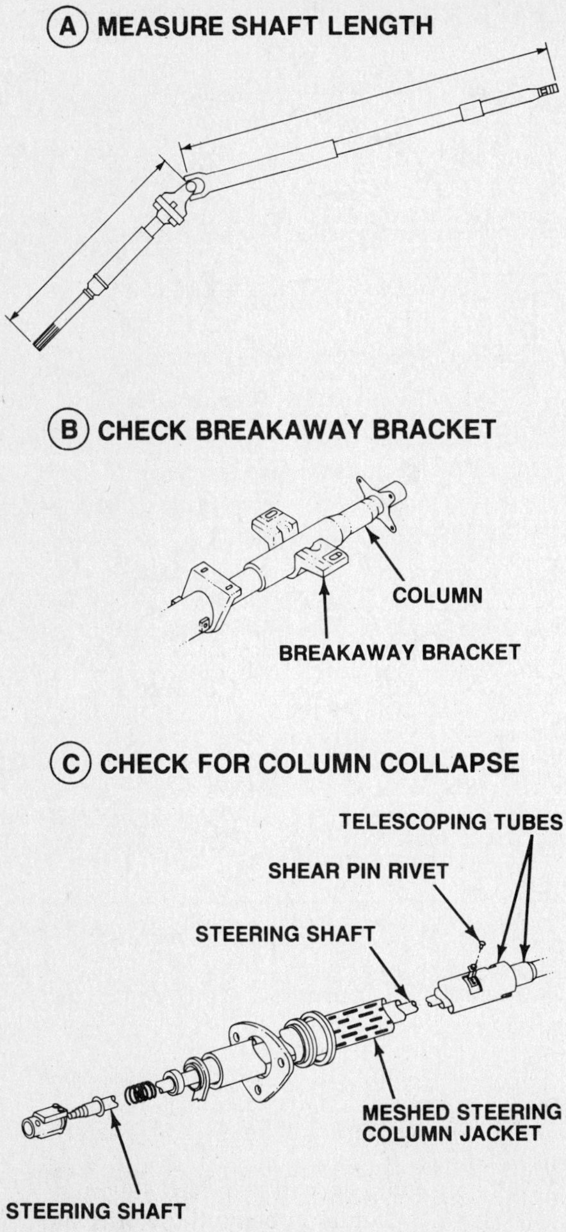

Figure 4-22. To check for steering column or shaft collapse, measure the length of the steering shaft (A), inspect breakaway brackets for cracks or looseness (B), and look for bends and breaks in the meshed area of the steering column jacket or for pieces of broken plastic shear pins (C).

such as the turn signal and light switch levers. Then install the column in the vehicle. Many manufacturers recommend that you install the steering wheel and air bag module after installing the steering column.

Installing the steering column
Be sure to install the column right-side-up so that when you install the steering wheel, the alignment marks on it will line up with those on the hub.

1. Position the car's front wheels straight ahead, and check that the steering gear is centered. To center the steering gear, see the procedures later in this chapter.
2. With the steering column in the vehicle, connect the U-joint or coupling at the end of the steering shaft to the steering gear or intermediate shaft. Tighten the bolts at the joint.
3. Replace the brackets that support the steering column. Replace the boot, if there is one, that covers the joint in the passenger compartment.
4. Install the steering wheel, being sure to align the marks on the steering wheel and hub. Torque the steering wheel retaining nut to the manufacturer's specifications to seat the steering wheel on the shaft.
5. Connect wiring harnesses and the clockspring electrical connector.
6. Replace the air bag module, if the car is so equipped. Reconnect the air bag electrical connector and the battery ground cable.
7. Replace any finishing panels.
8. Test drive the vehicle before returning it to the customer.

STEERING GEARS

Standard steering gears, figure 4-23, and rack and pinion steering gears, figure 4-24, convert the rotation of the steering wheel and steering shaft to linear motion, which moves the steering linkage and turns the vehicle's wheels. The rest of this chapter describes removal, inspection, and replacement of standard steering gears and rack and pinion steering gears.

Power steering systems are covered in Chapter Six of this *Shop Manual*. Power steering gears are similar to the steering gears discussed in this chapter, the primary difference being the incorporation of a hydraulic-assist system, figure 4-25. When you repair *power steering systems*, use the information in this chapter to repair or replace the steering gears, but also use the additional information on hydraulic systems in Chapter Six of this *Shop Manual*.

Troubleshooting Steering Gears

You can use the preliminary checks at the beginning of this chapter to diagnose problems

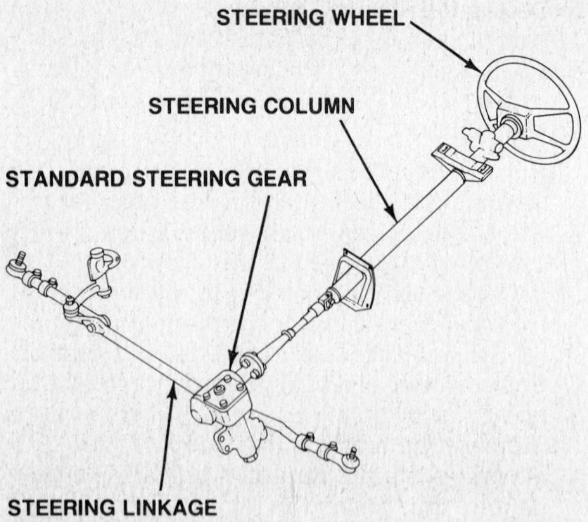

Figure 4-23. Standard steering gears are used mainly on trucks, vans, and some large cars.

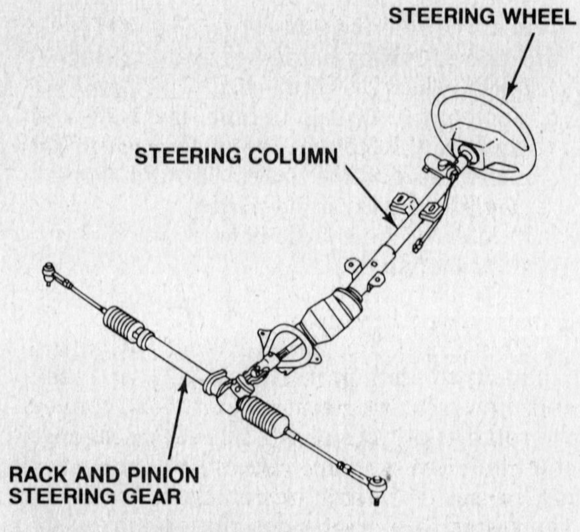

Figure 4-24. Rack and pinion steering gears have replaced standard steering gears in most late-model cars and some light trucks.

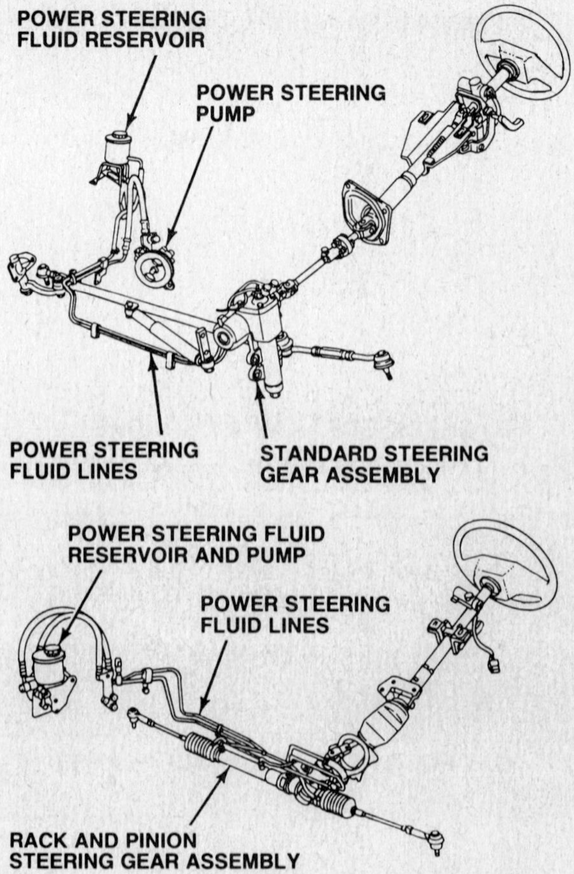

Figure 4-25. Standard steering gears and rack and pinion steering gears may be equipped with hydraulic assist, called power steering. Power steering reduces the steering effort required by the driver.

with the steering gear. For example, the steering wheel freeplay test indicates looseness in the steering system. If you do not find the source of the problem after inspecting the steering shaft and column, check the steering gear. Information from the test drive, as described in Chapter Two of this *Shop Manual*, will also help you pinpoint steering gear problems.

STANDARD STEERING GEARS

Standard steering gears are used on most early model vehicles and some larger late-model cars, trucks, and vans. Rack and pinion steering systems are used on most smaller, late-model vehicles, and will be discussed later in this chapter.

There are four types of standard steering gear, the most common being the worm and roller and the recirculating ball types. The two other types, the cam and lever and the worm and sector steering gears, are uncommon, and you will rarely encounter them in your work. Refer to Chapter Four of the *Classroom Manual* for more information on these steering gears. This chapter contains the following procedures for the worm and roller and the recirculating ball type steering gears:

- Visual inspection
- Adjustment
- Removal
- Teardown and inspection
- Installation.

Figure 4-26. Try to move the steering gear with your hands. If the bolts are loose or the rubber bushings are worn, the steering gear will feel loose.

Figure 4-27. This U-joint is located between the steering gear and steering shaft. If the joint is worn, the driver will feel play when turning the steering wheel.

Visual Inspection

Verify your findings from the test drive by inspecting the steering gear. Use the following general inspection procedure:

1. Position the lift supports under the vehicle chassis using the manufacturer's recommended lift points, and raise the vehicle. See Chapter One of this *Shop Manual* for more information on lifting a vehicle.
2. Locate the steering gear housing. It should be mounted securely to the vehicle's crossmember or frame, figure 4-26. If it is not, replace missing washers or bushings, tighten the bolts, and check it again.

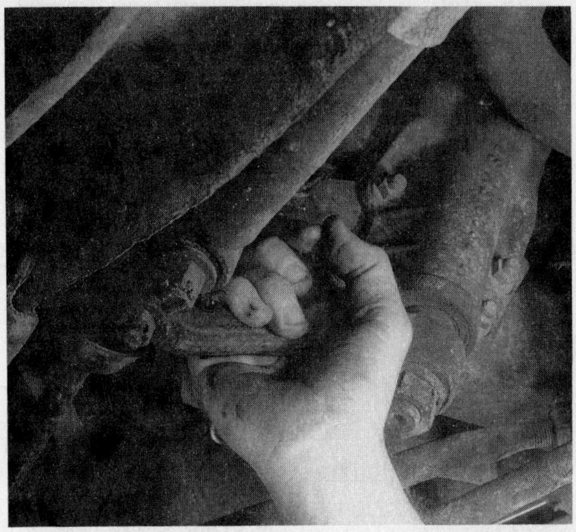

Figure 4-28. The pitman arm transfers movement from the steering gear to the steering linkage. If there is play between the pitman arm and sector shaft, steering will feel loose.

3. Grasp the steering shaft in one hand and the intermediate shaft or gear housing in the other, and try to twist them in opposite directions, checking the joint for play. If the joint is loose, figure 4-27, tighten the bolts that secure it, and check for play again. If there is still play, the joint is worn. Replace it, as described earlier in this chapter.
4. Check the connection between the steering gear sector shaft and the pitman arm, figure 4-28. Replace the securing nut or cotter pin if they are missing.
5. Check the steering gear for leaks. If you find leaks, disassemble and inspect the gear for wear, and add more lubricant as described later in this chapter. Replace the gear if necessary.

Always use the manufacturer's recommended steering gear lubricant. Using a lubricant with the wrong viscosity (thickness) may hinder gear operation. Low temperatures will thicken some lubricants, which in turn causes a noticeable increase in steering effort. For cold climates, some manufacturers recommend light-weight or temperature-resistant lubricants, and some recommend increasing the gear lash slightly. If you are servicing a vehicle that will be operated in extremely cold temperatures, check the vehicle manufacturer's specifications for lubricant and gear lash.

If you cannot correct steering gear problems after tightening connections and checking the lubricant, adjust the gears according to the procedures in the following sections.

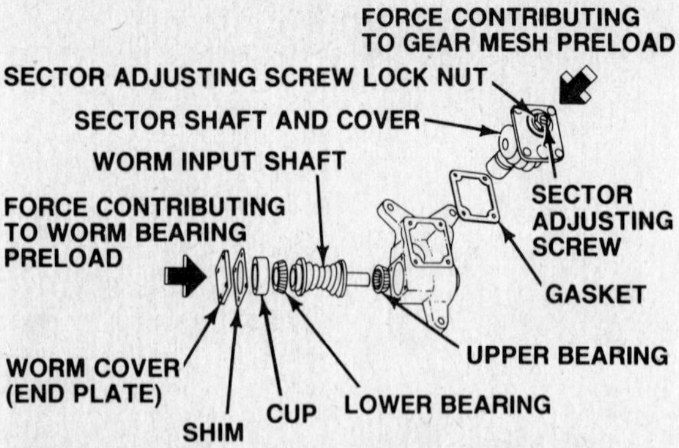

Figure 4-29. Two common standard steering gear adjustments are worm bearing preload and gear mesh preload. The force that the bearings and adjusters place on the worm shaft and sector shaft determine these preload measurements.

Steering Gear Adjustments

Adjustments take up slack in the steering gear due to wear. Before adjusting the gear, be sure that steering shaft joints are tight and that the gear is securely mounted to the vehicle's frame or crossmember. Any play in the steering shaft or linkage will affect measurements and adjustments.

You will need a low-calibration torque wrench (inch-pounds) to measure gear tolerances. Adjustments for standard steering gears include:

- Worm bearing preload
- Gear mesh preload
- Sector shaft endplay.

As discussed in Chapter Four of the *Classroom Manual*, worm bearing preload and worm endplay measurements are related. When checking worm bearing preload, you are measuring the torque needed to overcome the force that the bearings place on the worm shaft, figure 4-29.

Gear mesh preload and gear lash are also related. Gear mesh preload is a measurement of how much torque is required to overcome the force that the gears are placing on each other. When gear mesh preload is high, the steering wheel is difficult to turn because the worm and sector or ball nut and sector are meshed tightly together, figure 4-29. Gear lash is a measurement of how much play is allowed before the gears turn.

Sector shaft endplay is not adjustable in all steering gears. It is a shim adjustment that is made between the sector shaft and its adjusting screw. If the distance between the sector shaft and adjusting screw is too great, the sector shaft may bind near the end of its range of travel.

If you are rebuilding a steering gear, adjust sector shaft endplay first (if it is adjustable on the steering gear that you are working on). Then adjust worm bearing preload and gear mesh preload. If you are not disassembling the steering gear, then simply adjust worm bearing preload and gear mesh preload.

Checking and adjusting worm bearing preload

Worm bearing preload and endplay are measurements and adjustments that are made simultaneously because when you change one, you change the other. Most manufacturers give specifications for worm bearing preload.

You can check worm bearing preload with the steering gear in the vehicle. Some manufacturers recommend that you remove the steering gear before adjusting preload, especially if the adjustment is made with shims.

1. Before checking worm bearing preload, disconnect the pitman arm and sector shaft, using a puller if needed. See Chapter Five of this *Shop Manual* for removing the pitman arm from the sector shaft.
2. Remove the horn pad and any other accessories, as described earlier in this chapter, to access the steering wheel retaining nut.
3. Turn the steering wheel in either direction until it reaches the end of its travel. Do not try to force the steering wheel past the stop, or you may damage the steering gear.
4. Turn the steering wheel in the opposite direction until you feel a slight resistance (about one-quarter turn). This point of resistance is where the gears begin to mesh, and this is where the worm bearing preload measurement is taken.
5. Place a torque wrench on the steering wheel retaining nut, and turn the wheel about one and one-half turns, figure 4-30. Note the torque reading, which indicates worm bearing preload. Compare worm bearing preload with manufacturer's specifications:
   a. If worm bearing preload is within specifications, proceed to the next adjustment, gear mesh preload.
   b. If worm bearing preload is not within specifications, continue with the adjustment procedure.
   c. If gear movement feels rough, the bearings or gear may be damaged.

Figure 4-30. Before checking worm bearing preload, position the steering gears so they are just beginning to mesh. Then turn the steering wheel with a torque wrench. Compare the torque reading with the manufacturer's specifications.

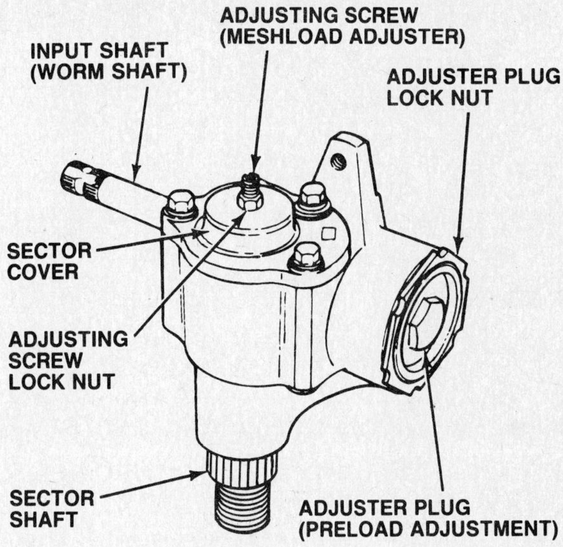

Figure 4-31. If the steering gear has an adjuster plug and lock nut at the worm shaft, turn the plug in (clockwise) to increase preload and out (counterclockwise) to decrease preload on the worm bearing.

Tear down and inspect or replace the steering gear according to the procedures given later in this chapter.

6. If the steering gear uses an adjuster plug, adjust worm bearing preload as follows, figure 4-31:
   a. Loosen the adjuster plug lock nut so that you can turn the adjuster plug. Turning the plug in (clockwise) increases worm bearing preload, and turning the plug out (counterclockwise) decreases worm bearing preload.
   b. Check the worm bearing preload again after making the adjustment.
   c. When worm bearing preload meets the manufacturer's specifications, tighten the lock nut.
   d. Check and adjust gear mesh preload, as described in the next section.
7. If the steering gear uses shims to adjust worm bearing preload:
   a. Remove the steering gear from the vehicle using one of the procedures described later in this chapter.
   b. Remove the drain plug, if there is one, and drain lubricant from the steering gear. If you do not find a plug, drain the lubricant after removing the worm cover (end plate).
   c. Remove the bolts, and remove the worm cover, figure 4-29.
   d. Adjust endplay by adding or removing shims under the worm cover. Removing shims reduces the clearance between the worm cover and bearing, which applies more pressure to the bearing when the cover is torqued into place. This increases the worm bearing preload. Remove shims to increase worm bearing preload, and add them to decrease bearing preload.
   e. Replace the worm cover and gasket. Torque the cover in place according to manufacturer's specifications. Recheck the worm bearing preload, and readjust if needed.
   f. Check gear mesh preload as described in the next section before refilling the steering gear with lubricant and installing it in the vehicle.

Checking and adjusting gear mesh preload
After adjusting worm bearing preload, check and adjust gear mesh preload. Gear mesh preload adjustments can be made with the steering gear in the vehicle. If you removed the steering gear from the vehicle to make worm bearing preload adjustments, do not reinstall it. You can check gear mesh preload with the steering gear out of the vehicle by checking preload at the worm shaft instead of the steering wheel. Use the following procedure:

1. Turn the steering wheel or worm shaft in either direction until it reaches the end of its travel. Then turn it in the opposite direction, counting the number of turns it

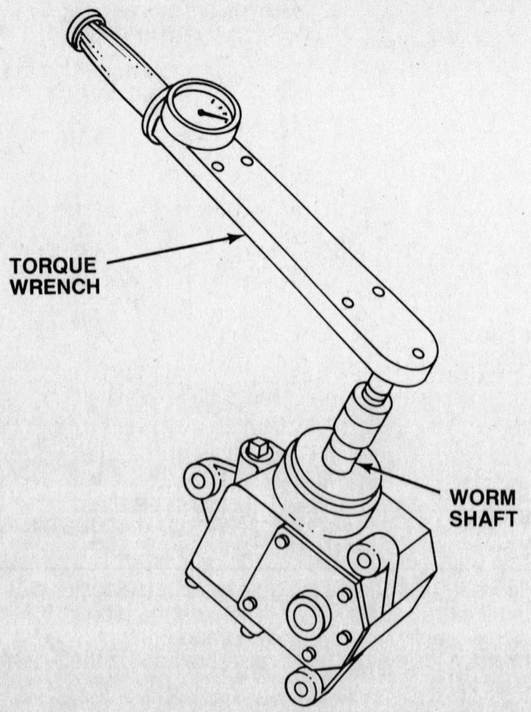

Figure 4-32. Center the steering gear, and check gear mesh preload by turning the worm shaft with a torque wrench. Compare the results with the manufacturer's specifications.

takes to reach the other stop. Center the steering gear by turning the steering wheel or worm shaft back one-half the number of turns you counted.

2. Place a torque wrench on the steering wheel nut or worm shaft, figure 4-32, and turn it one-half turn. Compare the reading with manufacturer's specifications for gear mesh preload.
3. To adjust gear mesh preload, loosen the lock nut on the sector adjusting screw, figure 4-29. To increase gear mesh preload, turn the adjusting screw clockwise. Turn it counterclockwise to decrease gear mesh preload.
4. Hold the adjusting screw in place with a screwdriver, and tighten the lock nut.
5. Center the steering gear again, and recheck gear mesh preload.
6. If you drained the lubricant from the steering gear, replace the drain plug, and refill the gear housing with the recommended lubricant. Reinstall the steering gear according to the procedure described later in this chapter.

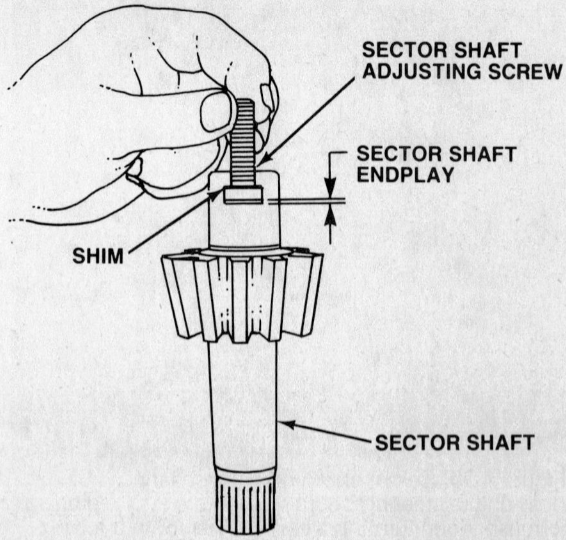

Figure 4-33. Adjust sector shaft endplay by inserting a shim between the adjuster screw and the sector shaft. Check the adjustment with a feeler gauge.

7. Reconnect the pitman arm to the sector shaft, and reinstall the horn cover and any steering column accessories that you removed.

Checking and adjusting sector shaft endplay

Sector shaft endplay is not adjustable in all steering gears, and is an adjustment made only when the gear is disassembled. Use the following procedure as a guide to adjust sector shaft endplay before reassembling the gear:

1. Before installing the sector shaft in the steering gear, check sector shaft endplay, which is the distance between the sector shaft adjusting screw and the sector shaft, figure 4-33.
2. Use a feeler gauge to check endplay.
3. If endplay exceeds the manufacturer's specifications, install a shim to reduce play.
4. Hold the adjuster screw between your fingers, and turn the sector shaft. The sector shaft should turn freely. If it does not, you have used a shim that is too thick. Replace the shim with another that will still increase endplay slightly, but will allow the shaft to turn freely. Be sure that endplay does not exceed the manufacturer's recommendations.
5. After adjusting sector shaft endplay, assemble the steering gear, as described later in this chapter.

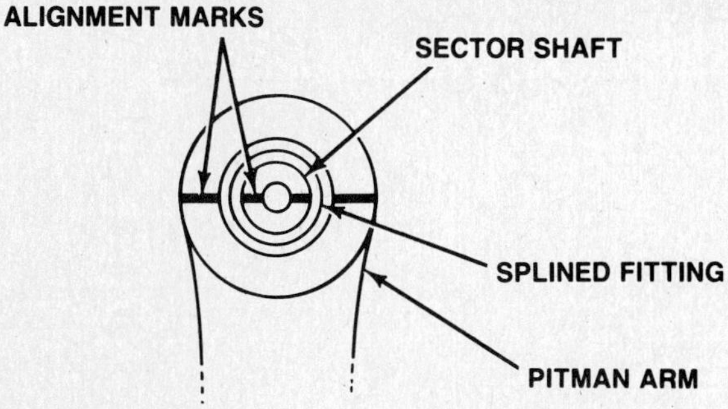

Figure 4-34. Reference marks help you align the pitman arm and sector shaft during reassembly.

Removing the Steering Gear

If you cannot adjust the steering gear to the manufacturer's recommended tolerances, or if the adjustments do not remedy steering problems, then check the gear for worn parts. For most late-model vehicles, you must replace the entire steering gear if a component in the gear is worn. Remove a standard steering gear as follows:

1. Position the lift supports under the vehicle chassis using the manufacturer's recommended lift points, and raise the vehicle. See Chapter One of this *Shop Manual* for more information on lifting a vehicle.
2. Remove the lock nut and cotter pin, if there is one, that secures the pitman arm to the sector shaft. Mark the pitman arm and sector shaft before you separate them, figure 4-34. You will use these marks to align the sector shaft and pitman arm during reassembly.
3. Use a puller to separate the pitman arm and sector shaft, figure 4-35. Do not use a hammer to loosen the pitman arm, as this can damage the steering gear. If they are difficult to separate, apply penetrating oil and try again.
4. Disconnect the steering shaft from the worm shaft (steering gear input shaft) by removing the bolts at the joint.
5. Loosen and remove the bolts, washers, and bushings that secure the steering gear housing to the vehicle. Remove the steering gear.

Steering Gear Teardown and Inspection

Although it is uncommon to rebuild steering gears (mainly because individual replacement

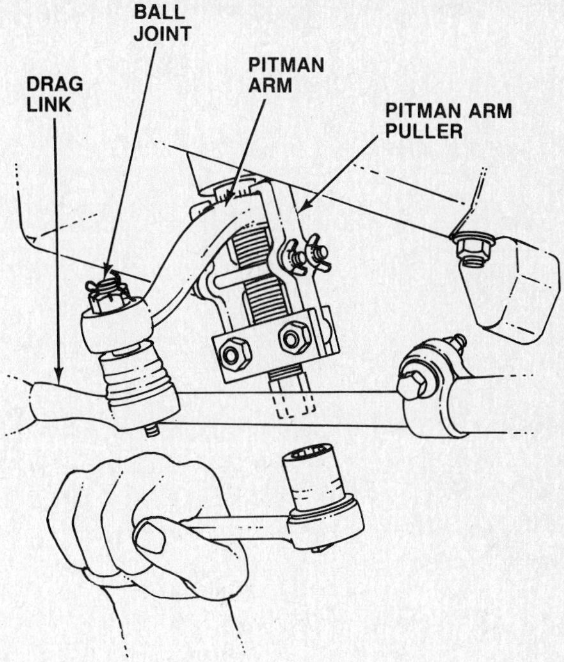

Figure 4-35. Use a puller to separate the pitman arm and sector shaft. Do not use a hammer, as this may damage the steering gear.

parts are not readily available), you can disassemble the steering gear to inspect components. This section includes procedures for disassembling, inspecting, and reassembling worm and roller and recirculating ball type steering gears.

Worm and roller steering gear

If you suspect that parts are worn and a replacement gear is not available, you can disassemble, inspect, and replace components, as needed.

Disassembly

After removing the worm and roller steering gear from the vehicle, clean the gear housing before disassembling it on a clean work bench. Any dirt that enters the steering gear will accelerate wear.

1. Loosen the sector shaft adjusting screw.
2. Remove the drain and fill plugs, and drain the lubricant from the gear.
3. Clamp the steering gear in a vise.
4. Loosen and remove the screws or bolts that hold the sector shaft cover in place. Lightly tap the end of the sector shaft with a soft-face hammer to loosen the cover. Then remove the cover and sector shaft as

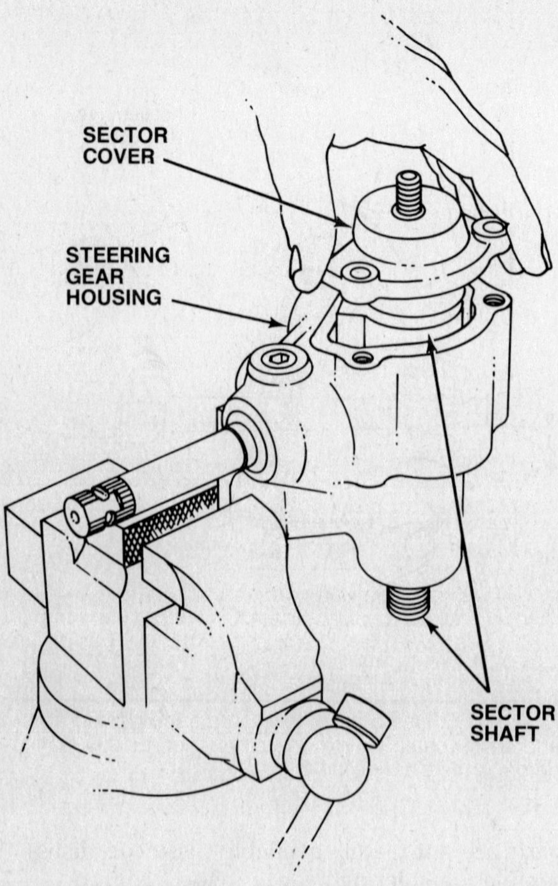

Figure 4-36. Clamp the steering gear in a vise, and remove the cover and sector. You may have to tap the sector cover with a soft-face hammer to loosen it.

a single piece, figure 4-36. Remove the sector cover gasket, if there is one.

5. Remove the adjusting screw lock nut from the sector shaft and cover, and turn the adjusting screw in (clockwise) to separate the sector shaft and cover, figure 4-37.
6. If the sector shaft bearing in the cover is worn, remove the bearing with a puller.
7. Loosen the screws or bolts that hold the worm shaft cover (end plate) in place. Remove the cover, gasket, bearing cup, and lower bearing.
8. Remove the worm shaft and upper bearing. You may have to use a puller or hammer and socket to remove the upper bearing cup from the gear housing, figure 4-38.
9. If they are worn, remove the sector shaft bearing and oil seal from the gear housing with a drift and hammer.
10. Use a puller to remove the worm shaft upper oil seal.

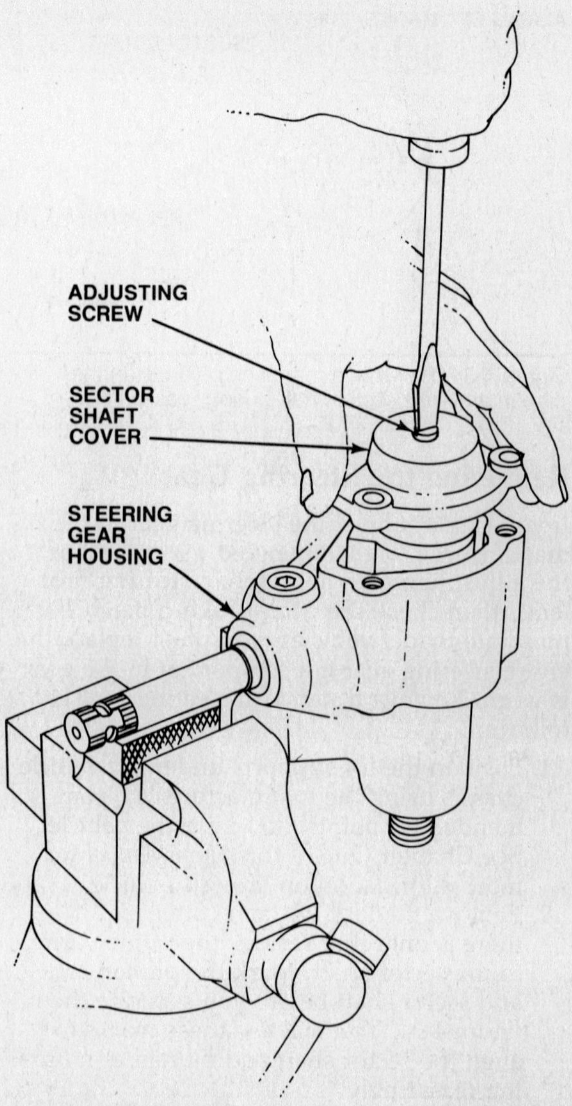

Figure 4-37. The sector shaft is secured to the cover with an adjusting screw. Separate the sector shaft from the cover by turning the screw. Then inspect the bearing.

Inspection

Inspect parts carefully for signs of wear or other damage before reassembling the worm and roller steering gear.

1. Clean steering gear parts in solvent, and allow them to dry.
2. Inspect the bearings and bearing cups for indentations, chipping, or other indications of breakdown, figure 4-39. Replace worn bearings.
3. Check the worm and sector teeth for wear or chipping, and replace parts as needed.
4. Inspect for contamination, such as metal flakes or chips in the housing, which may

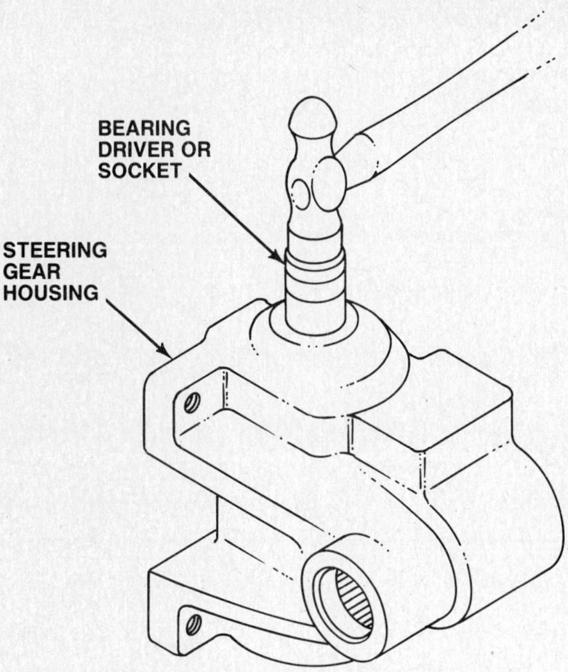

Figure 4-38. Remove bearings from the gear housing using the manufacturer's recommended puller. Some manufacturers suggest using a socket and hammer to drive the bearing from the housing.

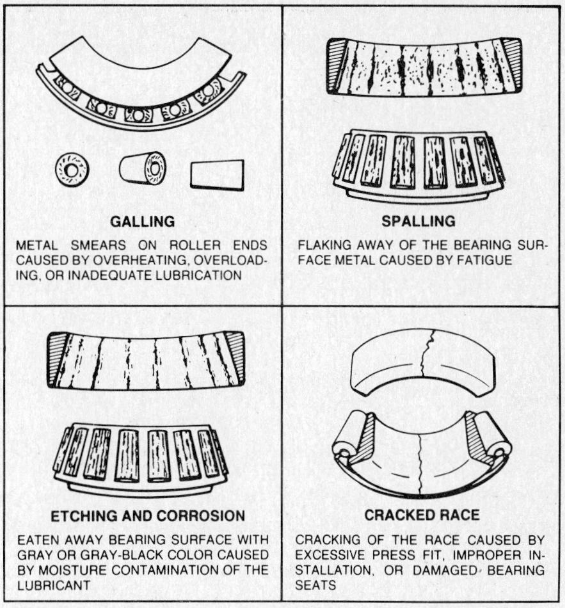

Figure 4-39. Bearing wear can be caused by a number of factors. Inspect the bearings to determine why they are worn, and correct the problem before assembling the gear.

cause binding. Clean or replace parts as needed.

5. Inspect the housing for cracks or damage, and replace the housing if you find damage.

If replacement parts are not available, replace the entire steering gear.

Reassembly

Reassemble the worm and roller steering gear on a clean workbench. Be sure parts are dirt-free and in good condition.

1. Install a new sector shaft bearing in the sector shaft cover, if the old one is worn.
2. Place the sector shaft in the cover, and install the adjusting screw and lock nut. Do not tighten the lock nut.
3. Install a new sector shaft bearing in the gear housing if the old one is worn. Also install the worm shaft bearing cup in the housing if you removed it during disassembly.
4. Install the worm shaft and its upper bearing in the housing. Place the lower bearing and cup on the worm shaft.
5. Install the worm shaft cover (end plate). Be sure to replace the shims or gaskets under the cover. These are used to adjust worm bearing preload. Torque the worm shaft cover bolts to the manufacturer's specifications.
6. Check worm bearing preload, and adjust it as described earlier in this chapter.
7. Lubricate the worm gear with the manufacturer's recommended lubricant, and install the worm shaft seal.
8. Install the sector shaft, sector cover, and gasket. Tighten the bolts to specifications.
9. Center the worm shaft, and check the gear mesh preload, as described earlier in this chapter.
10. Fill the gear housing with the manufacturer's recommended lubricant, and replace the filler plug. Install the steering gear.

Recirculating ball steering gear

The recirculating ball steering gear is the most common standard steering gear used in vehicles today. If you suspect that parts are worn and a replacement gear is not available, you can disassemble, inspect, and replace components, as needed.

Disassembly

After removing the recirculating ball steering gear from the vehicle, clean the gear housing before disassembling it on a clean work bench. Any dirt that enters the steering gear will accelerate wear.

1. Clamp the steering gear in a vise, and center the worm shaft, as described in the gear adjustment section.

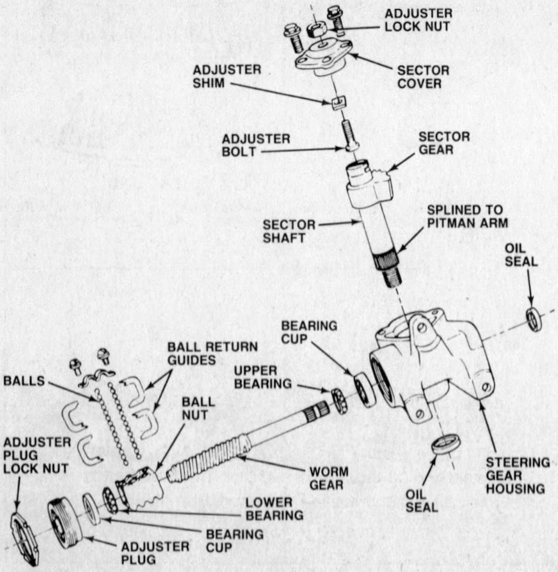

Figure 4-40. The recirculating ball steering gear is much like the worm and roller type, except that it uses a ball nut instead of a worm shaft.

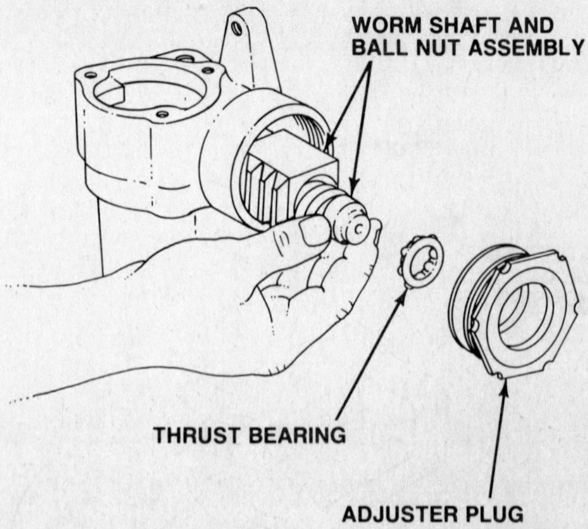

Figure 4-41. The worm shaft and ball nut assembly, along with the bearing, are under an adjuster plug in the recirculating ball steering gear.

2. Loosen and remove the screws or bolts that hold the sector shaft cover in place, figure 4-40. Lightly tap the end of the sector shaft with a soft-face hammer to loosen the cover. Then remove the sector cover and sector shaft as a single piece. Remove the sector cover gasket, if there is one.
3. Separate the sector shaft and cover by turning the adjusting screw clockwise. Be sure to keep the shim with the adjusting screw, because this shim is used to adjust sector shaft endplay.
4. Loosen the worm bearing adjuster lock nut, and remove the adjuster plug, figure 4-41. Remove the worm shaft thrust bearing.
5. Remove the worm shaft and ball nut as an assembly. Do not let the ball nut run down to either end of the worm shaft while you are removing the assembly because this will damage the ends of the ball guides.
6. Remove the sector and worm shaft seals with a screwdriver, and throw them away.
7. Remove the bearing cup from the adjuster plug with a puller. Remove the other bearing cup from the housing with a drift.
8. Check the bearing in the sector cover. If the bearing is damaged and is removable, remove it with a puller. If it is not removable, then replace the entire sector cover.
9. Check the needle bearing in the gear housing. If the bearing is damaged, remove it with a puller. If it is not removable, replace the housing.

Inspection

Carefully inspect the steering gear components for wear or damage, and replace any parts as needed. If replacement parts are not available, then replace the entire steering gear.

1. Clean steering gear parts in solvent and allow them to dry. Be sure that old gasket material is removed from the mating surfaces at the sector shaft cover and gear housing.
2. Inspect the bearings and bearing cups for indentations, chipping or other indications of breakdown, figure 4-39. Replace worn bearings.
3. Inspect the ball nut gear teeth for chipping or wear. Rotate the worm shaft in the ball nut. If the worm shaft does not turn smoothly, or if you feel resistance, replace the ball nut and worm shaft as an assembly.
4. Inspect the fit of the sector shaft in the sector cover bearing, and replace parts if needed. If you must replace the sector shaft or cover, be sure to use the correct one. Sector shaft and cover construction varies depending on whether the gear has a constant or variable gear ratio.
5. Inspect for contamination, such as metal flakes or chips in the housing, which may cause gear binding. Clean or replace parts as needed.

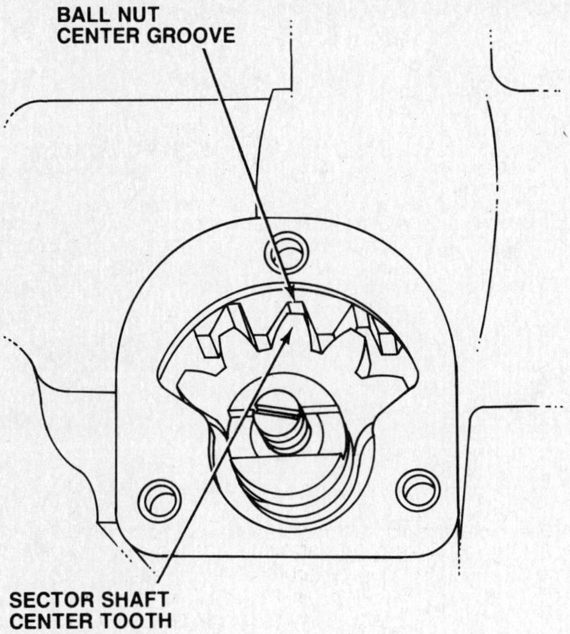

Figure 4-42. The sector shaft and ball nut should be centered during reassembly to ensure that the steering gear has a full range of motion.

6. Inspect the housing for cracks or damage, and replace it if needed.
7. Check the sector shaft endplay between the adjuster screw and the bottom of the adjusting slot with a feeler gauge, figure 4-33, as described earlier in this chapter. If clearance exceeds manufacturer's recommendations, add shims to correct sector shaft endplay.

Reassembly

Like the worm and roller, the recirculating ball steering gear must be assembled at a clean work bench.

1. Install the worm shaft bearing cup in the gear housing and the adjuster plug bearing cup in the adjuster plug, if you removed them during disassembly. You may need a special installer to do this.
2. Install the sector shaft seal and worm shaft seal in the housing if you removed them.
3. Clamp the steering gear housing in a vise.
4. Apply the manufacturer's recommended grease to the worm shaft bearings, the sector shaft needle bearing in the gear housing, and the sector shaft bearing in the sector cover.
5. Place a bearing on the worm shaft, and slide the worm shaft and ball nut assembly into the gear housing.
6. Place the other worm shaft bearing in the adjuster plug bearing cup. Install the adjuster plug and lock nut in the gear housing. Screw the plug into the housing until there is no play. Do not fully tighten the lock nut.
7. Insert the sector shaft adjusting screw and shim into the end of the sector shaft, and check sector shaft endplay if you have not already done so. Use shims to adjust sector shaft endplay, as described earlier in this chapter.
8. Pack the steering gear housing with the manufacturer's recommended grease.
9. Rotate the worm shaft until the ball nut is centered on the worm shaft.
10. Install the sector shaft and adjusting screw. The center tooth of the sector shaft should rest in the center groove of the ball nut, figure 4-42.
11. Add additional grease to the housing, and apply the proper sealant or gasket to the sector shaft cover.
12. Place the sector shaft cover over the opening, and turn the adjuster screw to seat the cover. Install the sector cover bolts, and tighten them to the manufacturer's specifications.
13. Adjust the worm bearing preload and gear mesh preload, as described earlier, before installing the gear in the vehicle.

Installing a Standard Steering Gear

Use the following procedure to install a standard steering gear:

1. Center the steering gear by counting the number of turns needed to turn the worm shaft from stop to stop. Then center the steering gear by turning the worm shaft one-half that number of turns from either stop position.
2. Use new bolts, washers, and bushings (if any) to attach the steering gear to the frame or crossmember. Torque the gear housing in place according to the manufacturer's specifications.
3. Position the steering wheel so that it is in the neutral position. Connect the steering shaft and worm shaft at the U-joint or other coupling, and tighten the bolts to manufacturer's specifications.
4. Align the splines and marks on the sector shaft and pitman arm.
5. Put the lock nut on the sector shaft threads. Torque the nut to the manufacturer's specifications. This will push the pitman arm onto the sector shaft. Install a new cotter pin, if one is required.

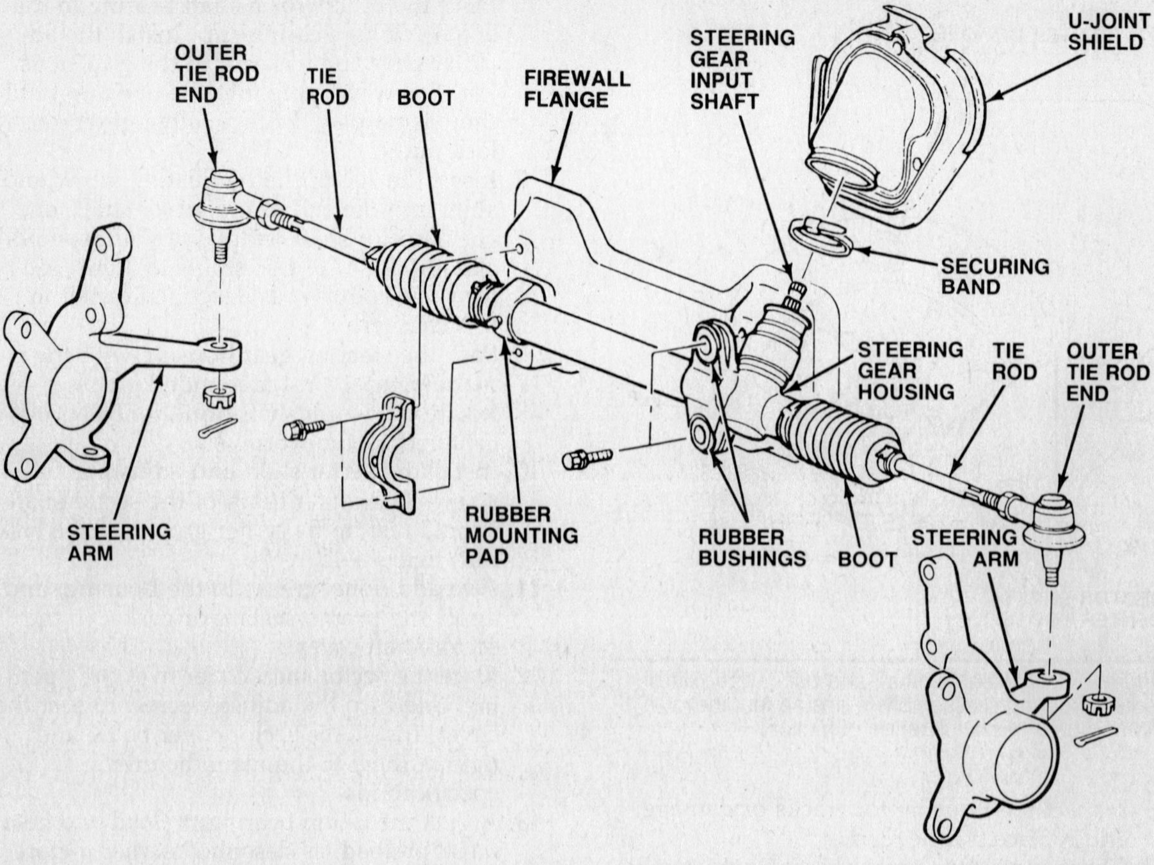

Figure 4-43. Many rack and pinion steering gear components can be inspected while the rack is installed in the vehicle.

6. Check that the steering gear is centered with respect to the steering wheel. The number of turns from center to lock should be equal on both sides.

RACK AND PINION STEERING GEARS

Rack and pinion steering gears are used on nearly all late-model FWD passenger cars. Most rack and pinion steering systems have hydraulic assist (power steering), although some smaller cars have manual rack and pinion steering. You can find additional information on hydraulic assist in Chapter Six of this *Shop Manual*.

If you are having trouble with a rack and pinion steering system:

- Perform a visual inspection
- Check and adjust the rack and pinion
- Remove, disassemble, and inspect or replace parts, as needed.

Visual Inspection

There are a number of components that you can inspect without removing the rack assembly from the vehicle. Use the following procedure as a guide, figure 4-43:

1. Position the lift supports under the vehicle chassis using the manufacturer's recommended lift points, and raise the vehicle. See Chapter One of this *Shop Manual* for more information on lifting a vehicle.
2. Locate the rack assembly, which is connected to the vehicle frame or crossmember. Sometimes the rack is positioned high in the engine compartment, behind the transaxle. If you do not see the rack, follow the tie rods that extend from the steering knuckles to the rack.
3. Check tie rods and steering knuckles for bends, which will affect steering and wheel alignment. If tie rods are bent, remove and replace them as described in Chapter Five of this *Shop Manual*. If the

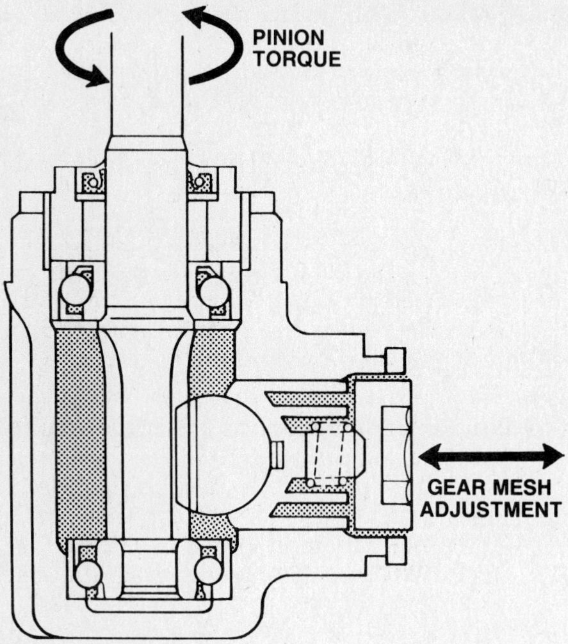

Figure 4-44. Tighten the rack support cover or add shims underneath it to increase pinion torque. Pinion bearing preload can be adjusted in some vehicles by placing shims under the pinion gear end cover.

steering knuckles or spindles are bent, replace them as described in Chapter Nine of this *Shop Manual*.

4. Inspect the rubber boots (sometimes called "bellows"), which cover the inner tie rod ends. If the boots are torn, check the inner tie rod end joints for wear, as described in Chapter Five of this *Shop Manual*. Tears in the boot allow dust and water into the rack assembly, which results in accelerated wear at the joints. If the joints are still in good condition, replace the boots as described in the reassembly procedure later in this chapter.
5. Inspect the outer tie rod end ball joints for play, as described in Chapter Five of this *Shop Manual*. If outer tie rod end ball joints are worn, replace them, as described in Chapter Five of this *Shop Manual*.
6. Check that the rack is firmly mounted to the vehicle frame or crossmember by grasping the rack and trying to move it. If you find loose mounting bolts, tighten them. If you find bent mounting brackets, replace them. The rack should rest solidly against the frame or crossmember.
7. Check the rubber bushings that are sometimes used between the rack housing and the vehicle frame or crossmember. If these bushings are deteriorating, the rack will feel loose. Leaking fluids in the engine compartment such as oil, brake fluid, and power steering fluid, contaminate rubber bushings and accelerate wear. Repair leaks if you find them. Check the bushings by prodding them with a screwdriver. If they are dry, cracked, and brittle, or soft and spongy, replace them with new ones. Use the rack removal procedure later in this chapter as a guide when replacing the bushings or brackets.

Adjusting Rack and Pinion Gear Tolerances

After checking the steering column and linkage for play, if the steering still feels loose, adjust the steering gear. The two adjustments for rack and pinion steering gears are:

- Pinion torque, which is adjustable in all rack and pinion steering gears
- Pinion bearing preload, which is adjustable in some rack and pinion gears.

In most vehicles, you must remove the rack and pinion assembly from the vehicle before making gear adjustments.

Pinion torque

Pinion torque is a measure, in foot-pounds or Newton-meters, of the effort required to turn the pinion, figure 4-44. It is similar to gear mesh preload in the standard steering gear. Pinion torque is changed by adjusting the position of the rack support cover or by adding or removing shims to adjust how closely the rack and pinion gears mesh.

1. Remove the rack and pinion assembly from the vehicle according to the procedure described later in this chapter.
2. Put the rack and pinion assembly in a vise, and turn the pinion shaft with a torque wrench. Note the torque reading, and compare it with the manufacturer's recommended pinion torque.
3. If pinion torque does not meet specifications, adjust it in one of the following ways:
   - Some rack and pinion steering gears have a screw or cover that you can loosen or tighten to adjust the pinion torque. Loosen the lock nut and turn the cover or screw clockwise to increase pinion torque and counterclockwise to decrease pinion torque, figure 4-45.

Figure 4-45. This rack and pinion gear, which is used on a power steering system, has an adjuster plug on the side that you turn to adjust pinion torque. The plug at the base of the gear is for adjusting pinion bearing preload.

- On other rack and pinion steering gears, you must add or remove shims that are under the rack support cover to adjust pinion torque, figure 4-46.

4. After making adjustments, check the pinion torque again. Tighten the lock nut or rack cover bolts if the measurement meets the manufacturer's specifications.

Pinion bearing preload
Pinion bearing preload is a measure of how much force the bearings place on the pinion shaft. Very few rack and pinion steering gears allow for a pinion bearing preload adjustment. If the system you are working on has shims under the pinion end cover or an adjuster at the base of the gear, use the following procedure to adjust pinion bearing preload:

1. Remove the rack and pinion assembly from the vehicle according to the procedure described later in this chapter.
2. Put the rack and pinion assembly in a vise.
3. Loosen the rack support cover so that its pressure does not affect pinion movement.
4. Turn the pinion shaft with a torque wrench. Note the torque reading, and compare it with the manufacturer's recommended pinion bearing preload.
5. Add or remove shims from under the pinion end cover, figure 4-46, or turn the adjuster at the base of the gear, figure 4-45, to increase or decrease pinion bearing preload.
6. Check pinion bearing preload again after making adjustments.
7. Then adjust pinion torque, as described in the previous section.

Removing and Installing the Rack and Pinion Assembly

Use the following procedure as a guide to remove and replace a rack and pinion steering gear. If you are working on a power rack and pinion system, be sure to refer to Chapter Six of this *Shop Manual* for information on servicing a hydraulic system.

1. Position the lift supports under the vehicle chassis using the manufacturer's recommended lift points, and raise the vehicle. See Chapter One of this *Shop Manual* for more information on lifting a vehicle.
2. Disconnect the steering shaft from the pinion shaft by loosening and removing the lower U-joint or pinch bolt at the pinion shaft.
3. On some vehicles, it is a good idea to remove the front wheels for easier access to steering linkage components.
4. Remove the cotter pins and castle nuts from the outer tie rod end ball joints. Use the procedure in Chapter Five of this *Shop Manual* to remove the ball joints from the steering knuckles or spindles.
5. Loosen and remove the bolts and brackets that hold the rack housing in place. If there are rubber bushings under the brackets, remove them as well.
6. Remove the rack and pinion assembly, and inspect it as described later in this chapter. Replace parts as needed, or replace the entire rack and pinion assembly.
7. To install the rack, replace worn rubber bushings or bent mounting brackets. Bolt the rack housing with the bushings and brackets to the frame or crossmember. Torque the bolts to manufacturer's specifications.
8. Count the number of turns needed to turn the pinion shaft from stop to stop. Then center the rack by turning the pinion shaft one-half the number of turns from stop to stop.
9. Center the vehicle's steering wheel in the neutral position.
10. Join the pinion shaft and steering shaft at the coupling or U-joint. Tighten the bolts to the manufacturer's recommended torque.

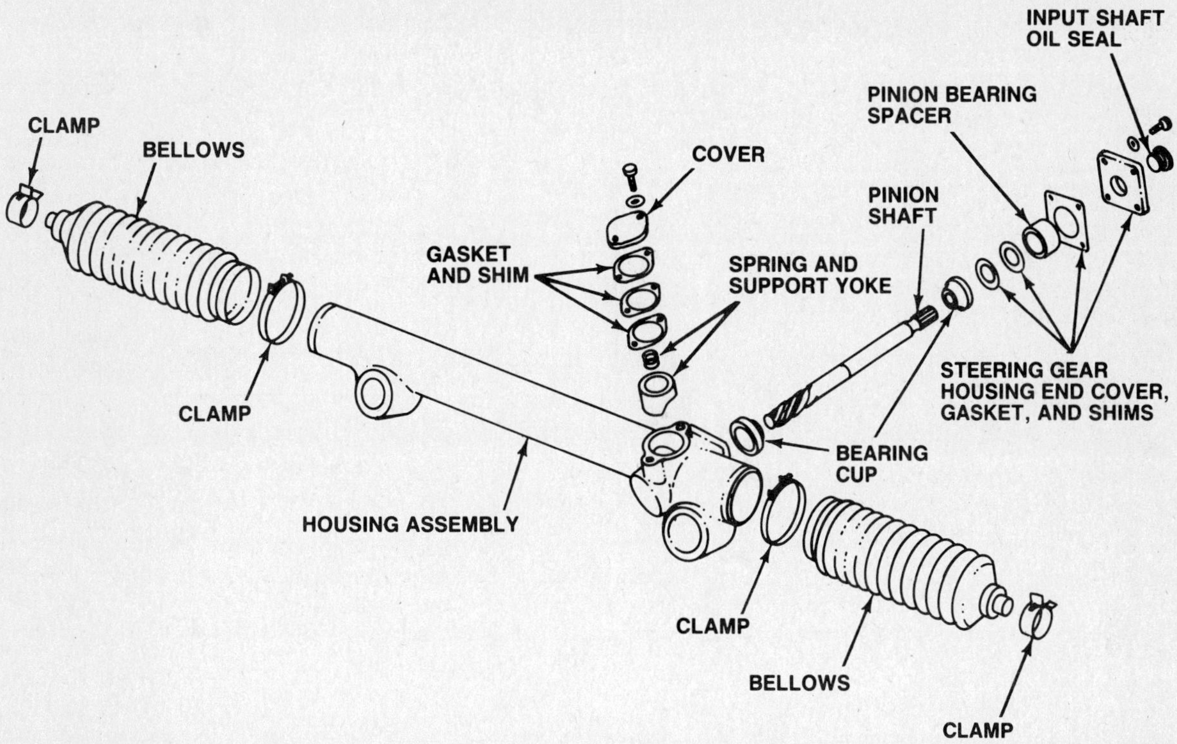

Figure 4-46. Notice that the shims in this rack and pinion steering gear are used to adjust pinion torque and pinion bearing preload.

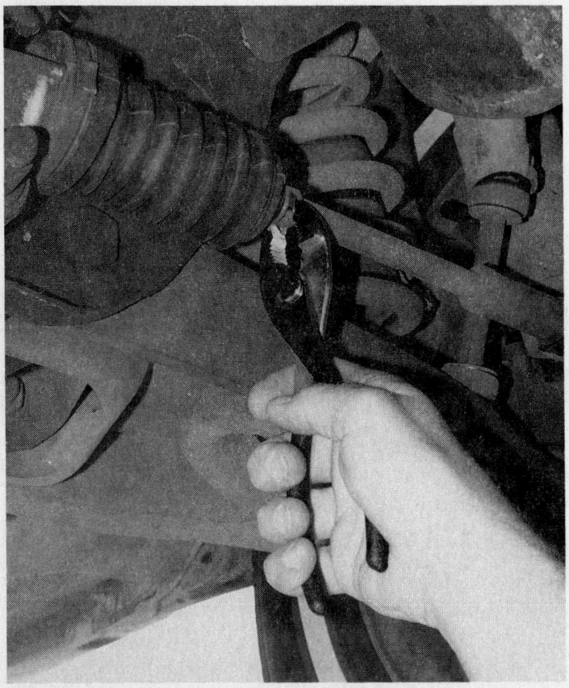

Figure 4-47. The inner tie rod ends on a rack and pinion steering gear are covered with rubber boots, which are secured with clips, clamps, or wires.

11. Install the tie rod end ball joints in the sockets on the steering knuckles or spindles. Tighten the castle nut on the ball joint to the manufacturer's recommended torque, which will seat the ball joint. Install new cotter pins.
12. Put the wheels on the car, and tighten the wheel nuts. Lower the vehicle. Check and adjust wheel alignment as described in Chapter Fourteen of this *Shop Manual*.

Disassembling and Inspecting a Rack and Pinion Steering Gear

Most shops will replace a rack and pinion steering assembly rather than repair it. Some shops rebuild worn rack and pinion gears and resell them. If replacement parts are available, use the following procedure to disassemble and replace defective parts on a rack and pinion steering gear after removing it from the vehicle:

1. Secure the rack and pinion assembly in a vise.
2. Remove the clip, clamp, or wire that secures the rubber boot covering the inner tie rod end, figure 4-47. If the fastener

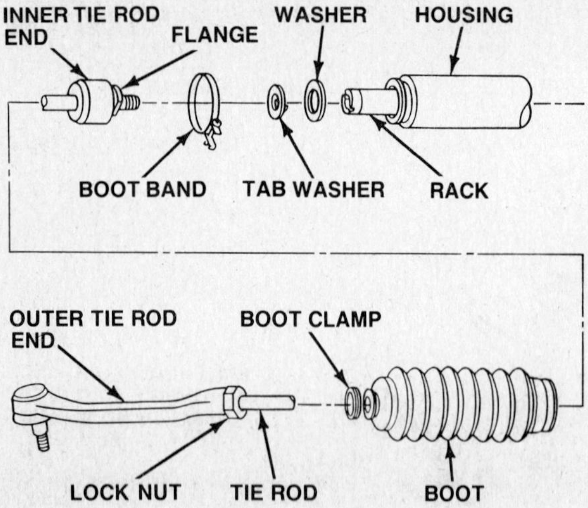

Figure 4-48. A tab washer or other fastener holds the tie rod to the rack at the inner tie rod end. Clips, clamps, or wires hold the boots in place.

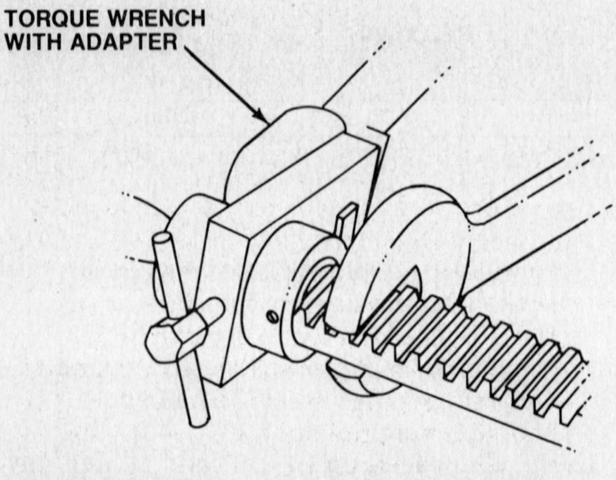

Figure 4-49. On some vehicles, you need a wrench and special adapter to turn the inner tie rod end housing when separating the inner tie rod end from the rack.

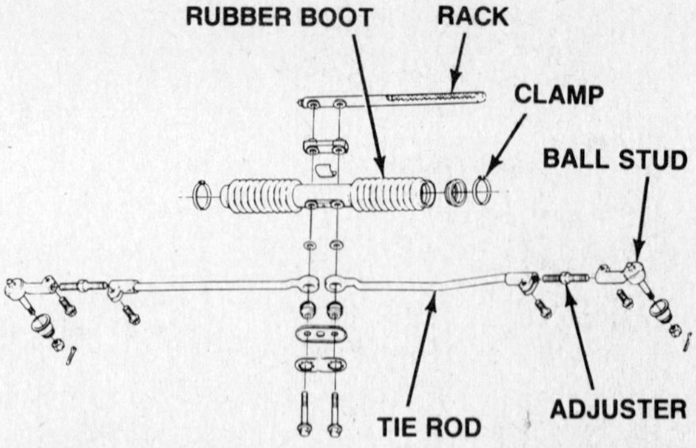

Figure 4-50. On this rack and pinion steering gear, the tie rod ends bolt to the center of the rack, rather than at the ends.

does not hold the rubber boot in place securely, throw the old fastener away, and replace it with a new one later.

3. Drain any lubricant from the boot. Slide the boot toward the outer tie rod end. Check the boot for tears, and replace it if needed.
4. Separate the rack and inner tie rod ends as follows:
   - Locate the lock pin or tab washer at the inner end of the tie rod, figure 4-48. Remove the lock pin with a punch and hammer or an Allen wrench, or straighten the tab on a tab washer with pliers or a hammer. Remove the tie rod.
   - On some vehicles, disconnect the inner tie rod by turning the ball housing with a wrench and special adapter, figure 4-49, while holding the rack in place with another wrench. Then pull the tie rod off the rack.
   - Some rack and pinion steering gears use inner tie rod ends that attach at the center of the rack rather than at the ends, figure 4-50. When you remove and replace tie rods on this type system, it is a good idea to work with one tie rod at a time rather than removing and replacing both tie rods simultaneously. Simply remove the securing bolt and remove the tie rod.
5. Inspect the tie rod as follows:
   a. Check the joint where the inner tie rod end meets the rack. Replace the rack or tie rod if the joint is worn.
   b. Inspect the outer tie rod end ball joints, as described in Chapter Five of this *Shop Manual*. Replace outer tie rod ends if the ball joints are worn.
   c. If the tie rods are bent, replace them.
6. Remove the rack support lock nut and cover, figure 4-51. On some vehicles, you simply loosen and remove the bolts that hold the rack support cover in place, figure 4-46. Then remove shims, springs, washers, and the rack support. Inspect these parts as follows:
   - Inspect the rack support, or guide, for wear or damage.
   - Check the spring for deterioration.

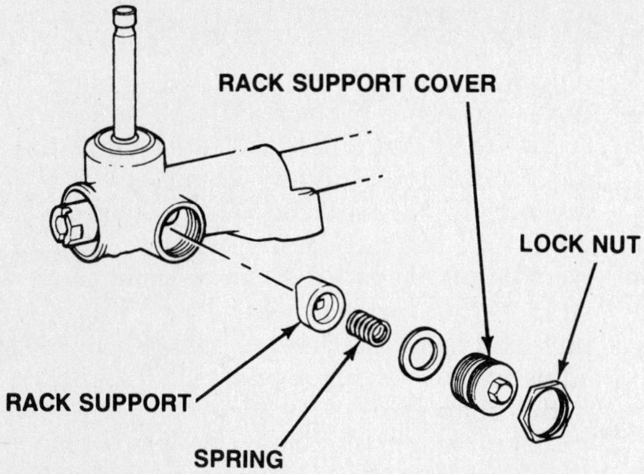

Figure 4-51. The rack support and spring are located under the rack support cover. Be sure the spring is intact, as it holds tension between the pinion and rack.

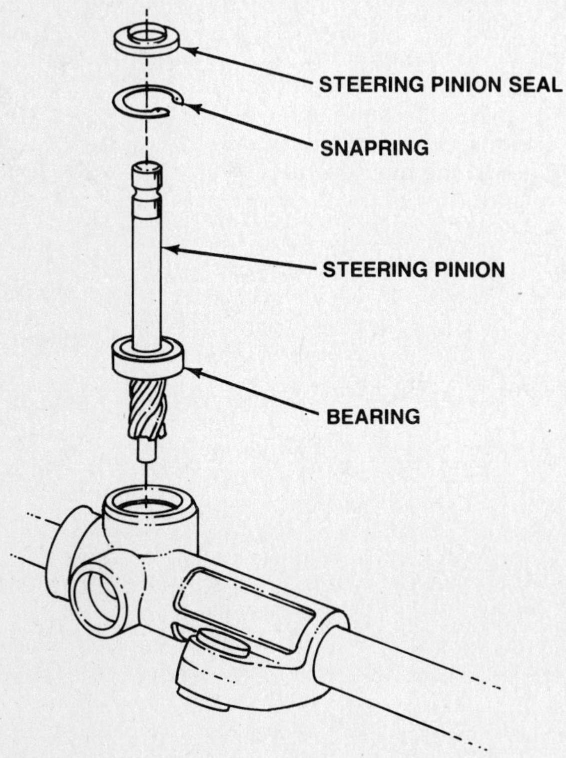

Figure 4-52. Replace the felt seal or oil seal when reassembling the rack and pinion gear. This ensures that the steering gear will remain adequately lubricated and dust free.

7. If the vehicle has an oil seal or felt dust cover at the pinion cover, remove the seal or cover and discard it, figure 4-52. Replace it with a new one during assembly.

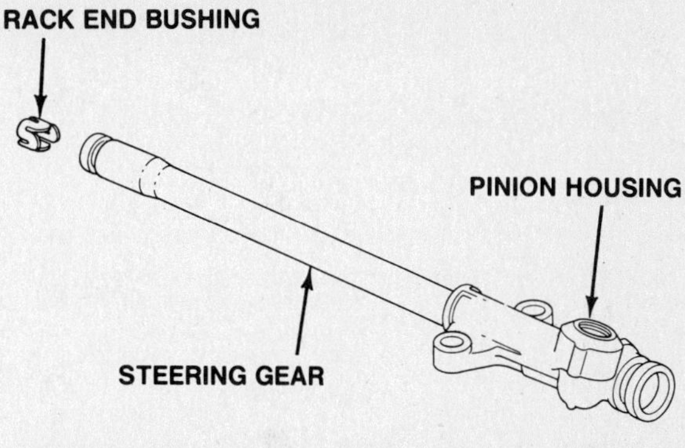

Figure 4-53. Check the rack end bushing for wear, and replace it if needed before reassembling the gear.

8. Some vehicles have a cover that bolts over the pinion shaft, figure 4-46, and others simply use a snapring to hold the pinion and its bearing in place, figure 4-52. Remove the pinion cover bolts, the pinion cover, and any shims, or use pliers to remove the snapring that holds the pinion bearing in place.
9. Hold the pinion, and tap on the gear housing with a soft-face hammer. This will separate the pinion from the housing. Remove the pinion assembly, and inspect the pinion teeth for wear or chipping.
10. Remove the rack from the housing by grasping the housing in one hand, and sliding the rack out. On most vehicles, it is easier to pull the rack from the pinion housing end, figure 4-53, and not from the narrow end of the rack housing. Inspect the rack for damage:
    - Check the teeth for wear or chips.
    - Check the back surface of the rack for wear or damage.
11. If possible, remove the lower pinion bearing from the gear housing once the rack is removed. Also if possible, remove the bearing from the pinion shaft. You will need a special puller to do this. Check the condition of the pinion bearings, and replace them if they are worn.
12. Finally, use pliers to remove the rack end bushing in the housing, figure 4-53. On some vehicles, you must remove a snapring, and then use a special puller to remove the end bushing. Check the end bushing for wear, and replace if needed.

Reassembling the Rack and Pinion Steering Gear

Following the inspection, replace any worn parts, and reassemble the rack and pinion steering gear as follows:

1. Grease the rack bushing, and install it in the housing, figure 4-53. Be sure to use the manufacturer's recommended grease. Secure the rack bushing with a snapring, if applicable.
2. If you removed the lower pinion bearing, apply grease to a new bearing, and press-fit it into the housing.
3. Apply grease to the rack, working the grease into the gear teeth. Grease the back of the rack as well.
4. Slide the rack into the housing.
5. If you removed the pinion bearing from the pinion, install a new one using a bearing installer. Apply grease to the pinion teeth, bearing, and seal.
6. Install the pinion in the gear housing.
7. Install the adjustment shims, the pinion cover, and the bolts. Torque the cover in place, and adjust pinion bearing preload if it is adjustable on the rack and pinion assembly you are working on. Use the procedure described earlier in this chapter. If the pinion simply uses a snapring, install the snapring over the pinion bearing to hold it in place. Install the felt dust cover.
8. Apply grease to the rack plunger or support at the point where it slides against the rack. Install the plunger, spring, any adjustment shims, and the rack support cover. Torque the rack support cover bolts in place. If the vehicle uses an adjuster plug and lock nut instead, install the adjuster plug and lock nut. Be sure to check and adjust pinion torque, as described earlier in this chapter.
9. Install new tie rods using a new lock pin or tab washer. Hammer or screw in the lock pin, or bend the tab washer with pliers or a hammer to hold the assembly together. For tie rods with ball housings, use the wrench and special adapter to tighten the tie rod to the rack.
10. Replace the boot over the inner tie rod end, and clip, clamp, or wire it in place.
11. Place the outer tie rod end ball stud in the steering arm socket. Torque the castle nut that secures the outer tie rod end in place. Replace the cotter pin.
12. Add the manufacturer's recommended lubricant to the rack and pinion steering gear.

Chapter

5

Steering Linkage Service

Compared to other systems in late-model vehicles, the steering linkage appears simple because it has few moving parts. But built into this simple design are precise angles and lengths that must be within the manufacturer's specifications so the car will steer properly. Slight variations in lengths and angles, or looseness in the system will prevent the steering linkage from smoothly transferring steering gear movement to the wheels. This chapter contains information to help you identify, inspect, adjust, and replace steering linkage components.

Chapter Two of this *Shop Manual* has diagnostic information for problems, such as accelerated tire wear and steering difficulties, that may be caused by defective steering linkage components. Because some wheel alignment problems are caused by worn or damaged steering linkage parts, check the linkage before performing a wheel alignment, as described in Chapter Fourteen of this *Shop Manual*.

IDENTIFYING THE STEERING LINKAGE

When a vehicle with steering problems comes into the shop, determine the type of steering gear and linkage it has before the inspection and test drive. It is important to identify the steering linkage *before* test driving the car because different linkages have different driving characteristics.

Chapter Five of the *Classroom Manual* contains detailed descriptions of the steering linkage designs in use today and how they operate. As a review, these linkages are:

- Parallelogram, figures 5-1 and 5-2
- Rack and pinion, figures 5-3 and 5-4
- Parallelogram variants, figure 5-5
- Single-tie-rod, figures 5-6 and 5-7
- Center point (equal arm), figures 5-8 and 5-9
- Long-and-short-arm (compensated pitman), figure 5-10
- Haltenberger, figures 5-11 and 5-12.

In your day-to-day work, you will most often see parallelogram and rack-and-pinion steering linkages. The other linkages are less common, but you should be familiar with them. For example, you will see the Haltenberger steering linkage only on Ford trucks.

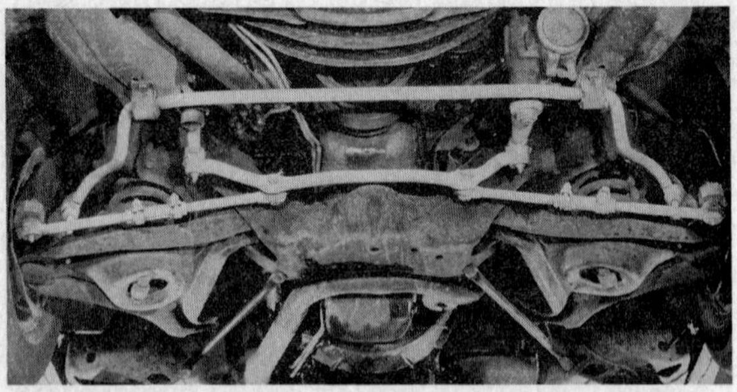

Figure 5-1. You will find parallelogram steering linkages on many large cars and trucks with standard (non-rack and pinion) steering gears.

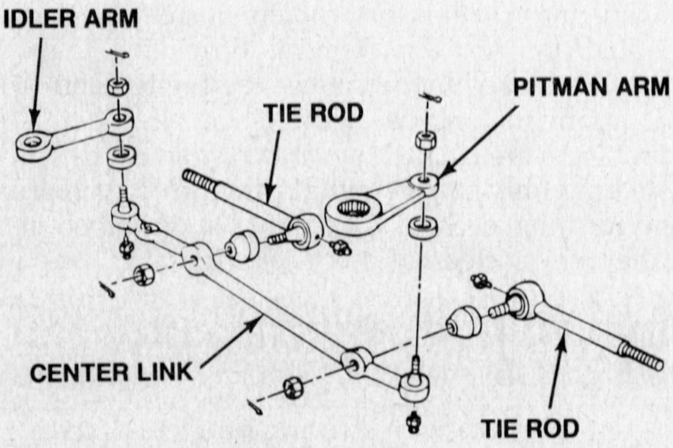

Figure 5-2. The parallelogram steering linkage is made up of one pitman arm, one idler arm, one center link, and two tie rods. The linkage components are connected with ball joints.

INSPECTING AND REPLACING COMPONENTS

Perform the test drive, using Chapter Two of this *Shop Manual* as a guide. Then use the following procedures to inspect and replace the following steering linkage components, figure 5-13:

- Ball joints
- Pitman arm
- Center link, relay rod, or drag link
- Idler arm
- Tie rods
- Steering damper
- Steering arm.

Use new cotter pins when you replace or reassemble the steering linkage to ensure that steering linkage components are secure.

Inspecting the Overall Condition of the Car

The "dry-park" test is a good overall steering linkage inspection. During this test, you place

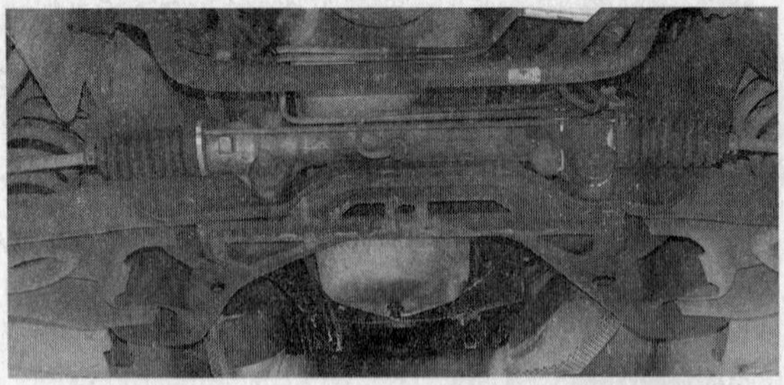

Figure 5-3. Rack and pinion steering gears are used on most light-weight, late-model cars. To locate the rack, follow the tie rods from the steering knuckle to the rack.

stress on steering linkage components by turning the steering wheel back and forth while the car is standing still. This makes any slack in the steering system easier to detect than when the car is moving or when it is on a lift. For both power and non-power steering systems, conduct the dry-park test as follows:

1. Park the car on an alignment rack or on level ground. If you use an alignment rack, be sure to lock the turntables before driving the car onto the rack, and keep the turntables locked during this test.
2. Ask an assistant to sit in the driver's seat. With the ignition off, have the assistant turn the steering wheel back and forth while you inspect the steering linkage.
3. Check for slack as the wheels move back and forth, and note any parts that are worn or need adjustment. Pay close attention to the following inspection points, figure 5-13:
   - Outer tie rod ends
   - Ball joints or bushings between the tie rods and center link, drag link, or rack
   - Threads and clamps on the tie rod adjusting sleeves
   - Ball joint or pivot between the pitman arm and center link
   - Splines between the pitman arm and steering gear sector shaft
   - Ball joint between the idler arm and center link
   - Idler arm mount at the frame or crossmember
   - Bushing on the steering gear or rack mount.

Besides normal wear, the most common cause of damage to the steering linkage is collision. Even a minor driving mishap, such as driving one wheel over a curb, may bend a tie rod or center link. Ask your customer about previous accidents, even if the vehicle does not show obvious signs of damage.

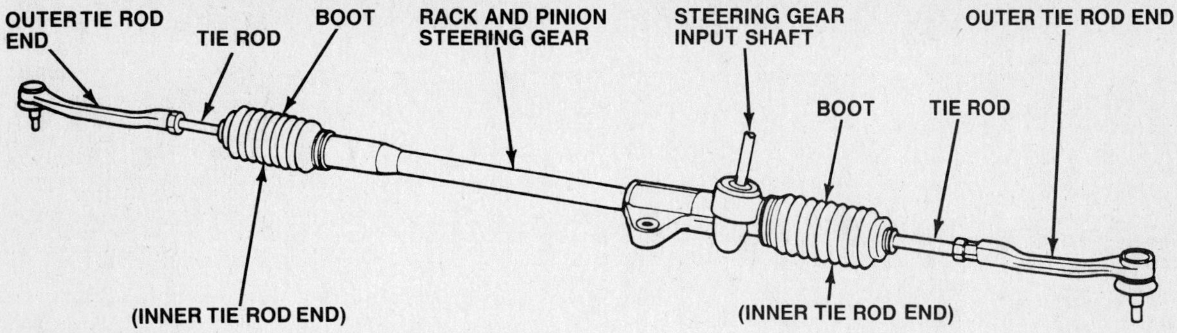

Figure 5-4. The only steering linkage components on a rack and pinion steering gear are the tie rods, which extend from the rack and connect to the steering knuckle.

VARIANTS OF PARALLELOGRAM STEERING LINKAGE

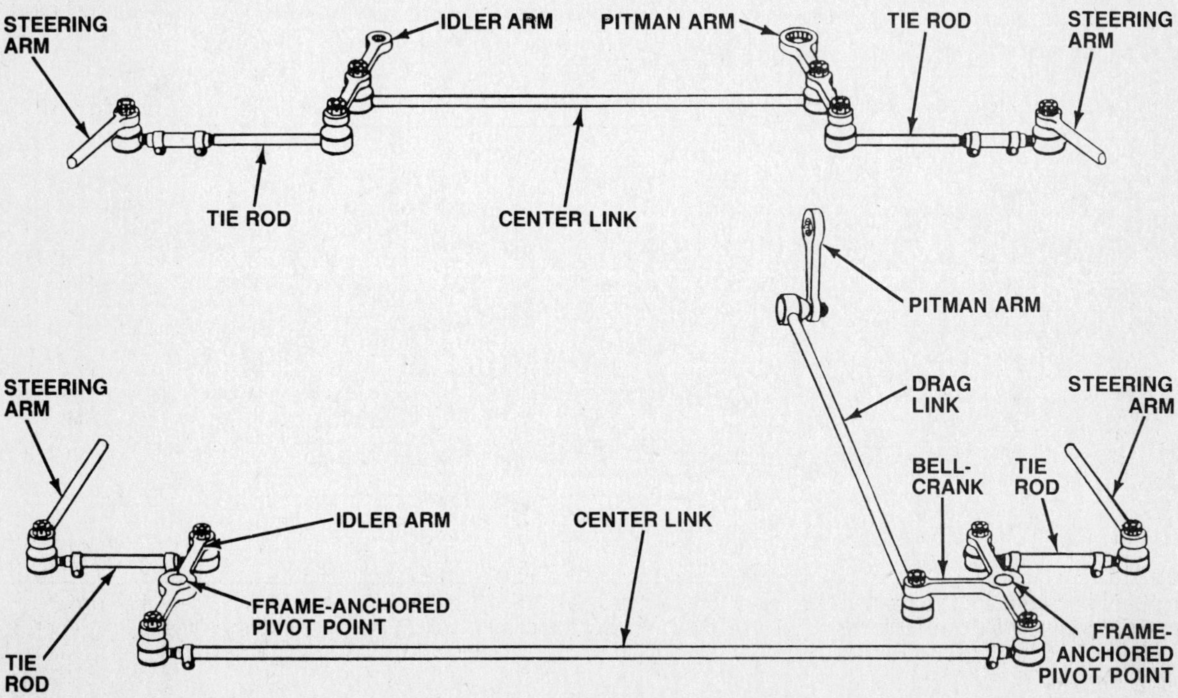

Figure 5-5. These two steering linkages are variants of the parallelogram linkage because they use additional parts that are not standard on the parallelogram steering linkage design.

Walk around the car, and check for paint that looks new. New paint is a sign of recent body work. Look down the side of the car, checking body panel alignment. If body panels are misaligned, the frame may be bent. When the car is up on a lift, look for wrinkles or welds on the frame and body panels, which indicate that the car has had extensive body work. Other signs of undercarriage damage include missing or cracked undercoating, cracked or flaking paint, and fresh rust on components. Finally, use standard wheel alignment measurements, such as ride height and thrust angle, which are discussed in Chapter Fourteen of this *Shop Manual*, for additional information on the condition of the frame.

Figure 5-6. The single-tie-rod linkage is different from a parallelogram linkage in that a tie rod runs from one steering arm to the other, and a drag link connects one of the steering arms to the pitman arm.

Figure 5-8. The center point steering linkage is so named because the tie rods are joined at the center of the vehicle with an intermediate steering arm.

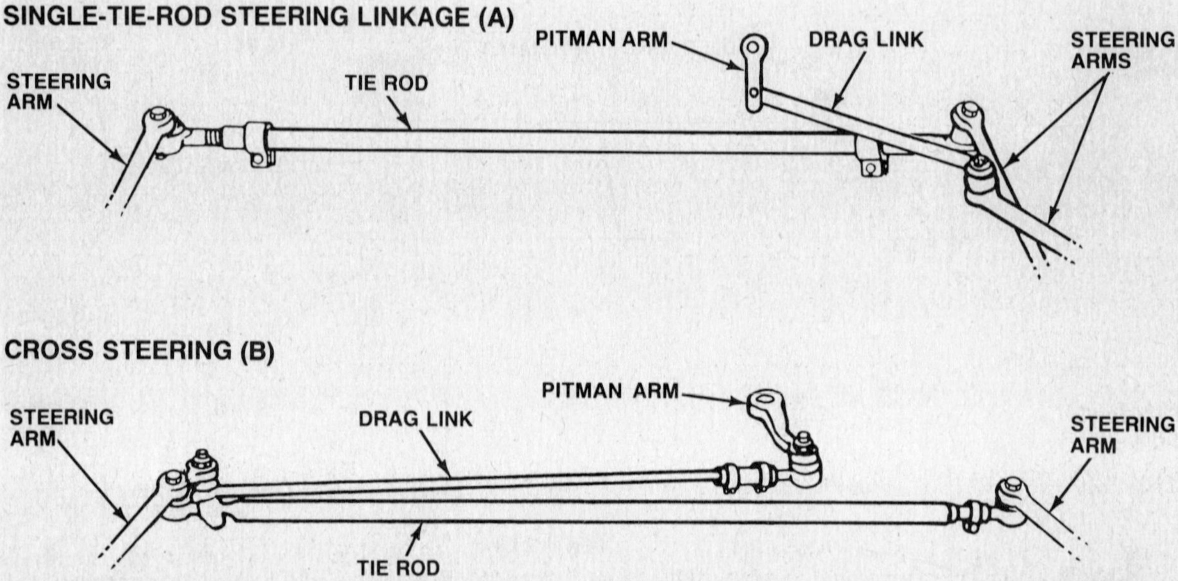

Figure 5-7. There are two variations of the single-tie-rod steering linkage. The drag link may connect the pitman arm to the driver's side steering arm (A), or it may connect the pitman arm to the passenger's side steering arm (B).

It is important to remember that if the frame is bent, steering linkage and suspension components cannot be bolted squarely to the frame. This makes correct linkage adjustment and wheel alignment difficult. If the frame is severely bent, have it professionally straightened before you replace and adjust steering linkage components or align the wheels.

When you repair the steering linkage, replace cotter pins with new ones. Be sure to position fasteners and clamps according to the automaker's instructions. Some manufacturers have diagrams that show you where to position the clamps, figure 5-14. If a clamp is not in its proper place, it may hit other parts or the vehicle frame during suspension travel.

Finally, after servicing the steering linkage, it is good practice to check wheel alignment before releasing the car to the customer.

Servicing Steering System Ball Joints

Looseness in the steering linkage is often due to worn ball joints. A steering linkage ball joint, figure 5-15, has a socket, tapered stud, wear surface or bearing, preload spring, and dust cover or seal. Steering system ball joints must withstand the lateral, or side-to-side, motion placed on them when the driver turns the steering wheel.

CENTERPOINT STEERING LINKAGES

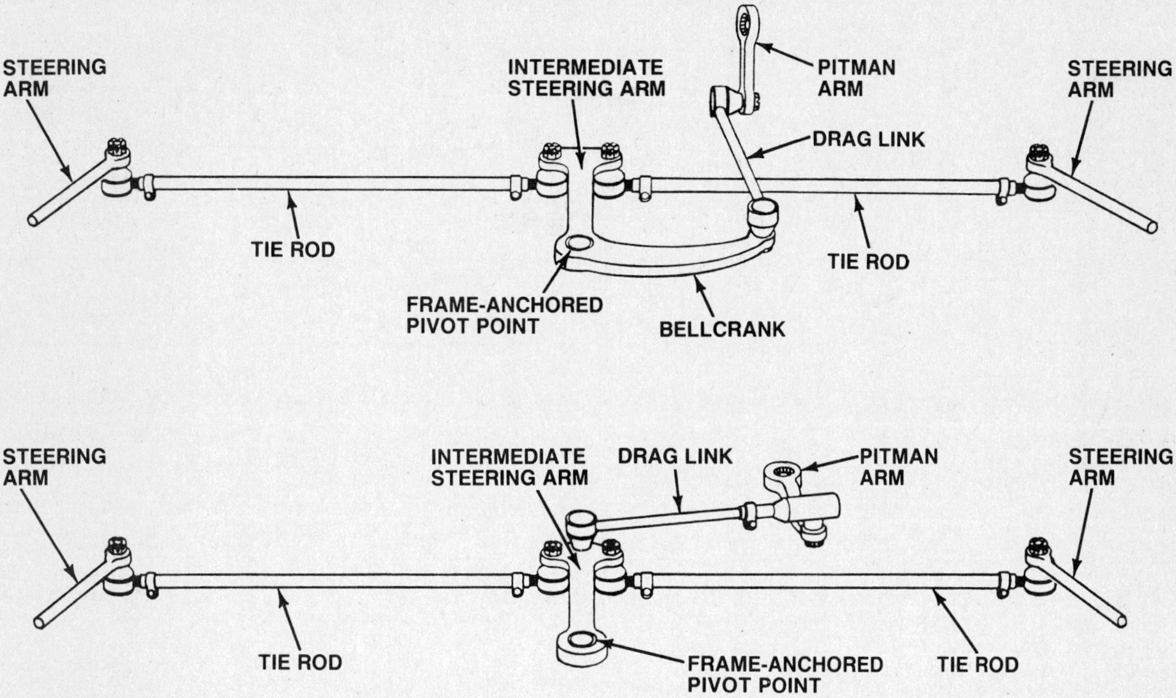

Figure 5-9. The center point steering linkage uses two tie rods of equal length that are joined at the center of the vehicle by an intermediate steering arm. A drag link, or a drag link and bell crank, join the intermediate steering arm to the pitman arm and steering gear.

LONG-AND-SHORT-ARM STEERING LINKAGE

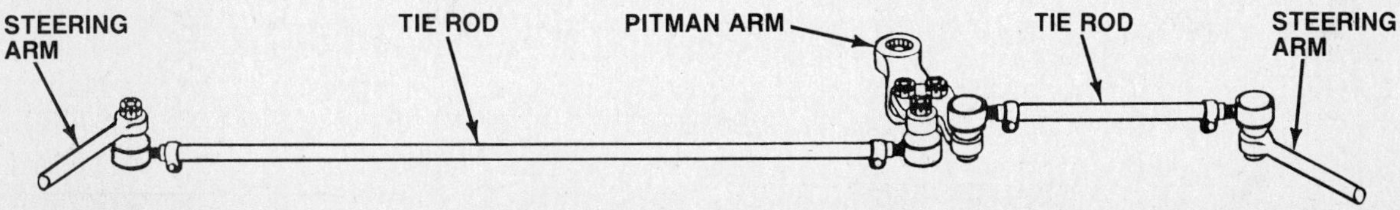

Figure 5-10. The long-and-short-arm steering linkage simply consists of two tie rods and a pitman arm. There is no center link or idler arm.

Figure 5-11. The Haltenberger steering linkage has an unusual design, and is used on some Ford trucks.

Ball joints are located at the outer tie rod ends of parallelogram and rack and pinion steering linkages, and between the inner tie rod ends and the center link in parallelogram steering systems, figure 5-16. Use the section titled "Identifying the Steering Linkage" to help locate ball joints in other types of steering linkages.

Checking ball joints
While performing the dry-park test, watch the ball joints. If they allow sloppy steering action,

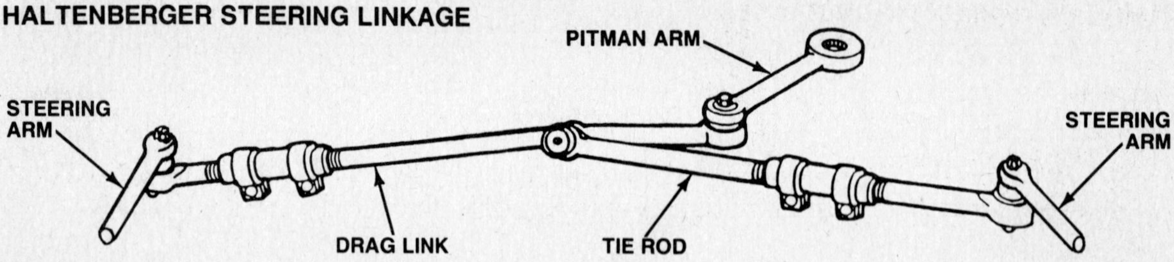

Figure 5-12. The Haltenberger steering linkage uses a pitman arm, drag link, and tie rod to transfer motion from the steering gear to the wheels.

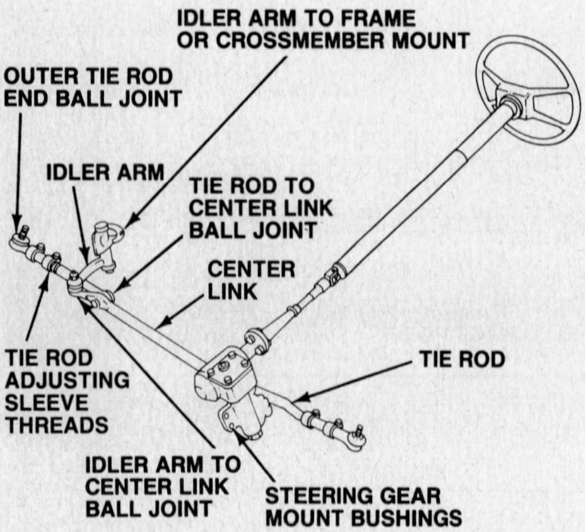

Figure 5-13. When conducting the dry park test on a steering system that uses a standard steering gear, look for play in these areas.

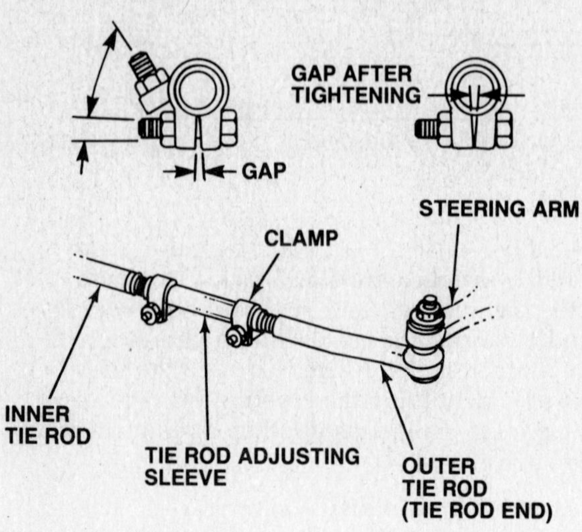

Figure 5-14. Many car manufacturers provide diagrams to help you position steering linkage clamps and bolts. This ensures that the clamps and bolts do not interfere with other undercarriage components.

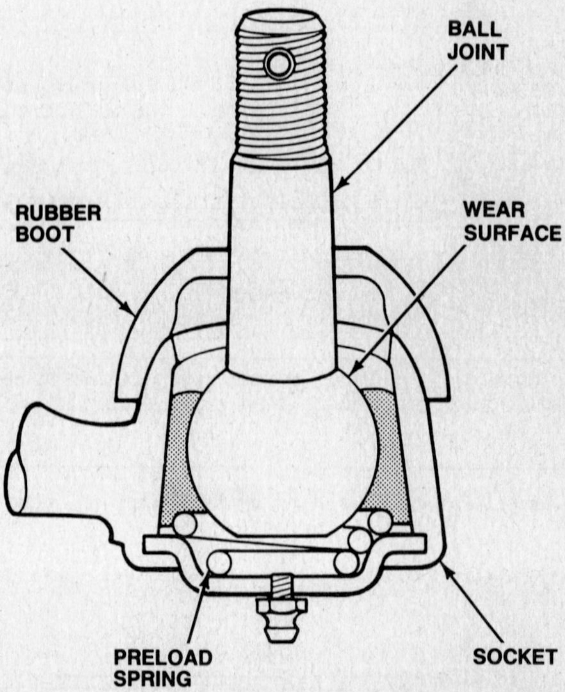

Figure 5-15. The construction of steering system ball joints is similar to that of suspension system ball joints. Steering system ball joints are smaller, however.

then they are likely worn. Use the following steps to further investigate ball joint condition:

1. Use your finger to feel for tears in the rubber boot covering the ball joint. Tears allow lubricant to escape from the ball joint and allow dirt in. You should replace the joint if you find this type of damage, figure 5-17.
2. Pinch the rubber boot. It will feel firm if the ball joint is adequately lubricated. Some ball joints have grease fittings that allow you to add lubricant, if needed.
3. To check for ball joint wear, push up and down on the tie rod, watching for movement at the joint. If you find play, the ball joint is worn and should be replaced.

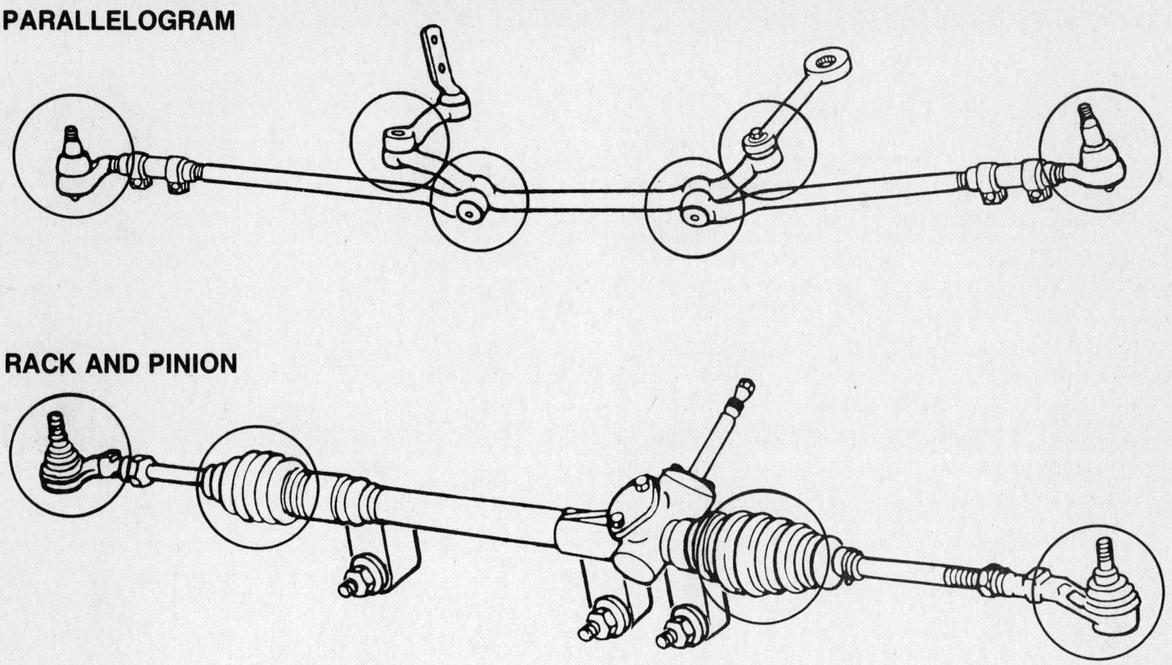

Figure 5-16. If steering feels loose during the dry park test or test drive, check all ball joints in a parallelogram or rack and pinion steering linkage.

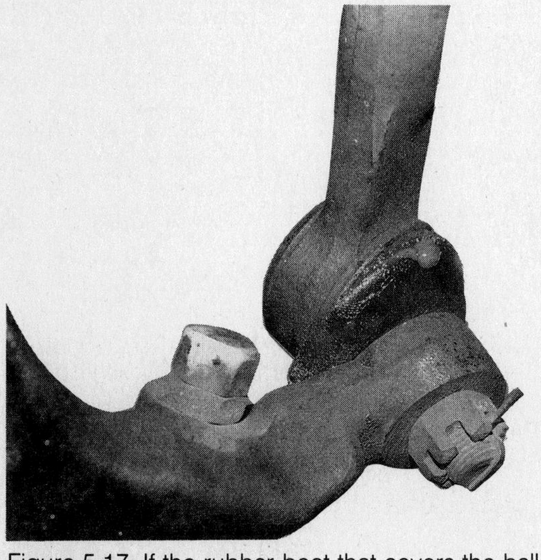

Figure 5-17. If the rubber boot that covers the ball joint is torn, replace the joint.

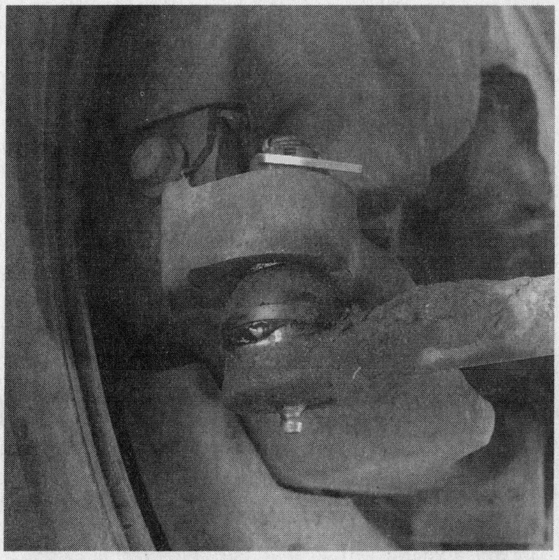

Figure 5-18. The grease fitting on this ball joint allows you to add grease at the service intervals recommended by the vehicle manufacturer.

Lubricating ball joints

On many late-model vehicles, ball joints are permanently sealed. These joints cannot be lubricated, and when they wear, you must replace them. Older vehicles and some aftermarket ball joints, however, have grease fittings, figure 5-18, which allow you to add grease.

Lubricate ball joints, as described in the following procedure, at the vehicle manufacturer's recommended intervals:

1. Wipe away dirt from the ball joint and grease fitting. Some manufacturers use a grease plug in the ball joint rather than a fitting. If this is the case, remove the

Figure 5-19. The adapter on a grease gun allows you to lubricate ball joints that have a grease fitting.

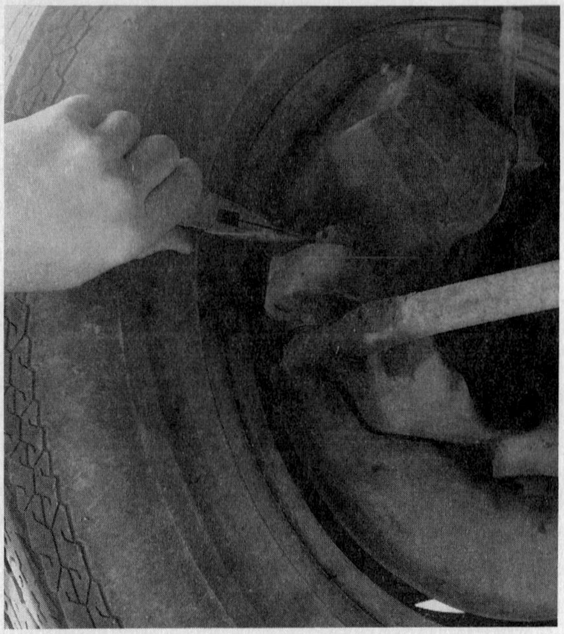

Figure 5-20. Most steering and suspension components use cotter pins to secure castle nuts. Straighten the pin, and then pull it out with pliers.

grease plug, and replace it with a fitting that matches your grease gun.

2. Add grease to the joint until the boot feels firm, figure 5-19. Do not overfill the joint, as the boot may break from excessive pressure.
3. Wipe excess grease from the ball joint fitting, and install the grease plug, if there is

Figure 5-21. Turn the bolt on the tie rod puller to force the ball joint from the socket.

one. Some shops leave the grease fittings in the ball joints for easier lubrication service next time.

Separating and installing ball joints
When ball joints wear, steering feels sloppy and the vehicle may wander during the test drive. Because both conditions are dangerous, you should replace components, such as tie rod ends or the center link, if their ball joints are worn. Use the following procedure as a guide to separate and install ball joints. If needed, refer to sections later in this chapter for removing and replacing steering linkage components.

1. Position the lift supports under the vehicle chassis using the manufacturer's recommended lift points, and raise the vehicle. See Chapter One of this *Shop Manual* for more information on lifting a vehicle.
2. Remove the cotter pin that secures the castle nut on the ball joint, figure 5-20. Remove the nut, using penetrating oil, if needed, to loosen it.
3. Separate the ball joint from its socket using one of the three methods described below:
   - Place the forked end of the tie rod puller between the ball joint and socket, figure 5-21. Be sure that you do not catch the rubber boot between the puller and the ball joint, or you will rip the boot. Position the other end of the puller on the end of the stud, and tighten the bolt until the ball joint is pressed from the socket.
   - Place a ball joint separator, more commonly known as a ''pickle fork'', between the ball joint and the socket,

Figure 5-22. This wedge-shaped pickle fork is used to lever the ball joint from the socket.

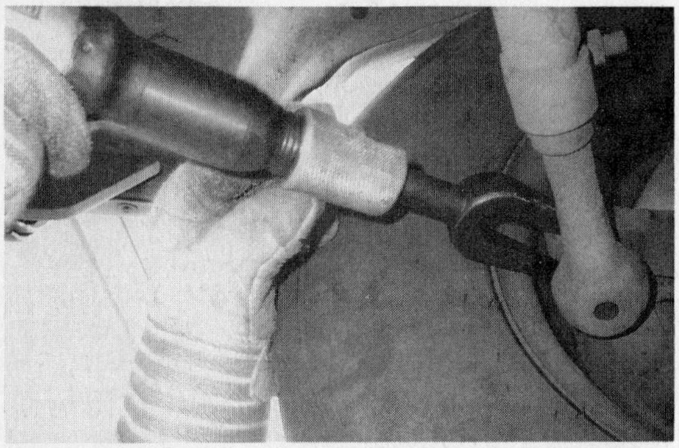

Figure 5-23. Using a pickle fork attachment on an impact wrench makes separating ball studs a quick and easy task.

figure 5-22. Drive the pickle fork with a hammer until the ball joint separates from the socket. Using this removal procedure will likely damage the rubber boot, so be sure to have a new ball joint on hand.

- Use an impact wrench with a pickle fork attachment, figure 5-23, to break the ball joint loose from the socket.

When installing new ball joints, always use new castle nuts and cotter pins. Install ball joints as follows:

1. Inspect and clean the ball joint and taper so that the ball joint will fit tightly against the inner surface of the socket.
2. Before installing the new joint, check the rubber boot for tears.
3. Place the ball joint in the socket, and install the castle nut. Tighten the nut to the manufacturer's recommended torque, figure 5-24, which will cause the ball joint to seat in the socket.
4. Install a new cotter pin to secure the nut.
5. After working on the steering system, check wheel alignment and adjust if needed.

Figure 5-24. Torquing the castle nut to the specified torque value ensures that the ball joint is seated in the socket.

RBS ball joints

Some Ford vehicles use RBS (rubber bonded socket) ball joints, figure 5-25, which have inspection and replacement procedures that differ from those of standard ball joints. Instead of using a metal socket, ball joint, and preload spring like standard ball joints, RBS joints use a rubber-covered ball stud that is pressed into, or "bonded" to, the metal socket. When the ball joint moves, the rubber pulls it back to its original position. To identify RBS joints, check the bottom of the socket. You will find a small hole, which allows air to escape from the socket when the ball joint moves. Also, RBS joints do not have rubber boots that cover the entire joint.

RBS joints have "memory steer". Be sure that the vehicle's wheels and steering linkage are positioned straight ahead before tightening the joints. RBS joints will "remember" the position in which they were tightened, and return to that position.

The main advantage of RBS joints is that they do not require lubrication. RBS joints are virtually maintenance free because there is no metal-to-metal contact, so they take a long

RBS JOINT

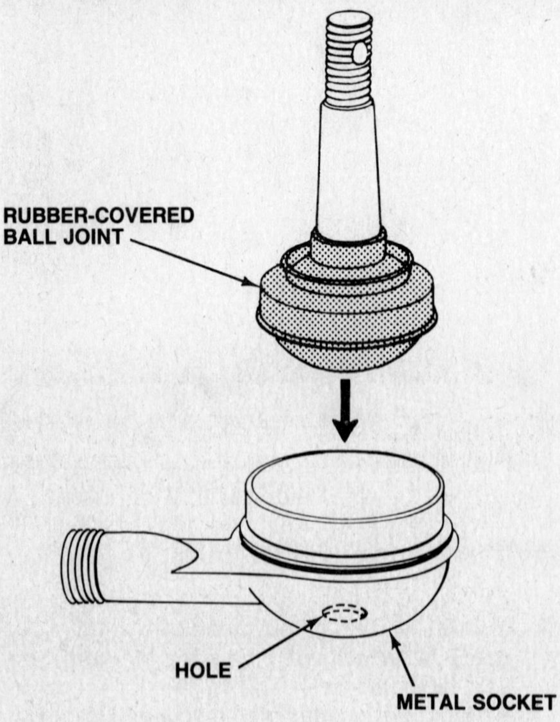

Figure 5-25. You can identify RBS joints by a hole drilled in the base, which allows air to escape when the stud is pressed in place by the manufacturer. Also, RBS joints do not have rubber dust boots.

time to wear. Servicing these joints is simply a matter of inspecting for play and replacing the joints if needed.

RBS joints do not allow any free-play unless they are worn. Pull up and down and twist the tie rod. You should feel resistance to both up-and-down and rotary motion. If there is play, replace the joint because the rubber is likely torn somewhere in the metal socket. RBS tie rod ends are removed and replaced like standard tie rod ends. In fact, Ford recommends that you replace worn RBS joints *with* standard tie rod ends. If only *one* RBS joint is worn, replace *both* joints with standard tie rod ends.

Pitman Arm

The pitman arm connects the steering gear sector shaft to the rest of the steering linkage, figure 5-26. Any looseness between the sector shaft and pitman arm or between the pitman arm and steering linkage will cause noticeable driveability problems. Watch the pitman arm carefully while conducting the ''dry-park'' test.

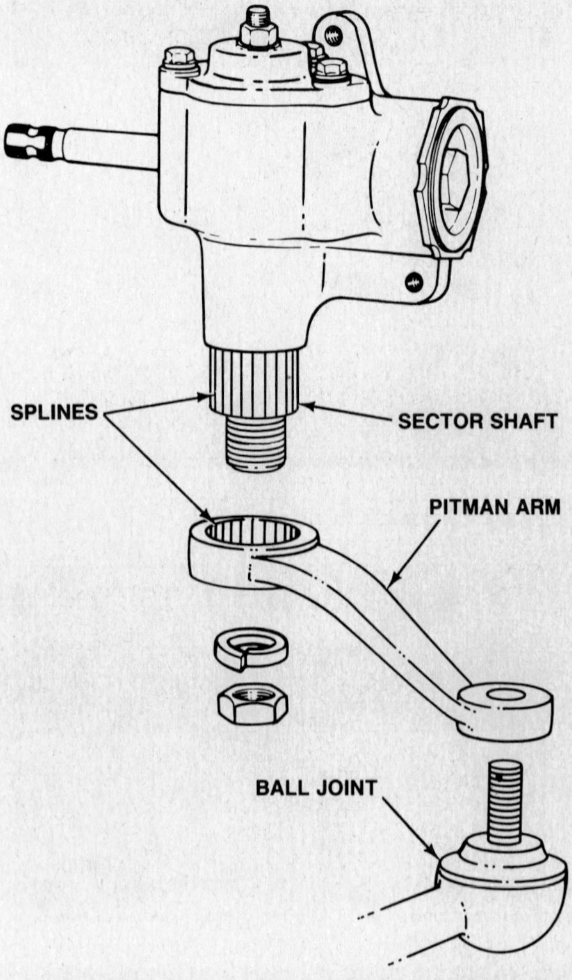

Figure 5-26. Be sure that the splines on the pitman arm and steering gear sector shaft are not stripped. If they are, the pitman arm may not transfer motion smoothly to the steering linkage.

Inspection

The pitman arm and steering gear sector shaft have an extremely tight splined fitting. If you notice looseness, remove the pitman arm from the sector shaft, and check the splines, figure 5-26. If the splines are stripped, replace the pitman arm.

The other end of the pitman arm connects to the center link, drag link, or tie rod with a ball joint, figure 5-26. Check the ball joint for wear, as described earlier.

If the pitman arm is bent, do not attempt to straighten it. Instead, replace it.

Replacement

The pitman arm will normally last the life of the car. But if it is bent, if it has a worn-out integral ball joint, or if you must remove it to

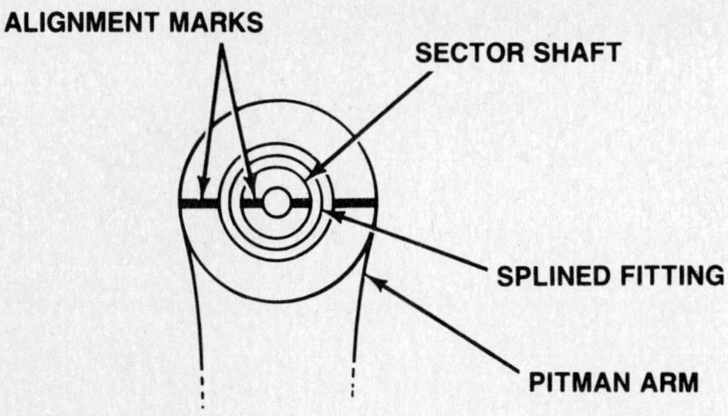

Figure 5-27. Make alignment marks on the pitman arm and sector shaft with chalk or paint, or scribe marks into the metal.

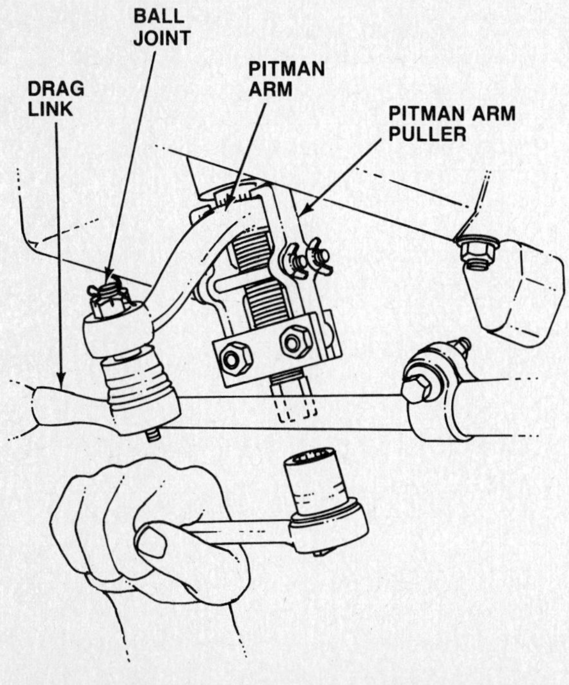

Figure 5-28. The splined fitting between the pitman arm and steering gear sector shaft is sometimes difficult to separate. Use a pitman arm puller for best results.

repair another component, then remove and replace the pitman arm as follows:

1. Position the lift supports under the vehicle chassis using the manufacturer's recommended lift points, and raise the vehicle. See Chapter One of this *Shop Manual* for more information on lifting a vehicle.
2. Remove the cotter pin and castle nut that secure the pitman arm to the center link, drag link, or tie rod. Discard the cotter pin and castle nut. You must replace them with new ones during reassembly.

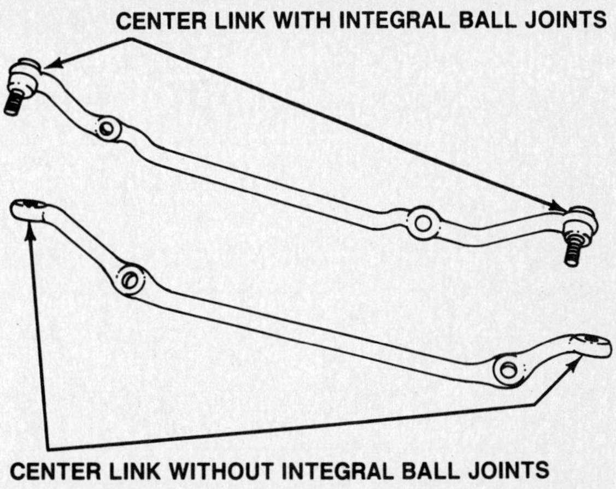

Figure 5-29. Some center links have integral ball joints, while in other steering systems, the center link has sockets to accept ball joints from the pitman and idler arms.

3. Using a puller or pickle fork, as described earlier in this chapter, separate the ball joint between the pitman arm and center link, drag link, or tie rod.
4. Remove the castle nut and cotter pin that secure the pitman arm to the sector shaft. If you are removing the pitman arm to repair another component and plan to reinstall the old pitman arm, mark the pitman arm and sector shaft before you separate them, figure 5-27. You will use these marks to align the sector shaft and pitman arm during reassembly.
5. Use a puller to separate the pitman arm and steering gear sector shaft, figure 5-28. If they are difficult to separate, apply penetrating oil and try again.

CAUTION: Manufacturers do not recommend heating parts with a torch to remove them because heat will weaken the metal.

6. To install the pitman arm, put the pitman arm ball joint into the socket on the center link, drag link, or tie rod. On some steering linkages, the ball joint is in the center link rather than the pitman arm, figure 5-29. Install a new castle nut, and torque it to the manufacturer's specifications to press the joint in place. Install a new cotter pin. Do not use a hammer to install steering linkage ball joints, as you may damage them.
7. Position the front wheels straight ahead. Check the steering gear to be sure that it is centered, as described in Chapter Four of this *Shop Manual*. Center the steering gear, if necessary.

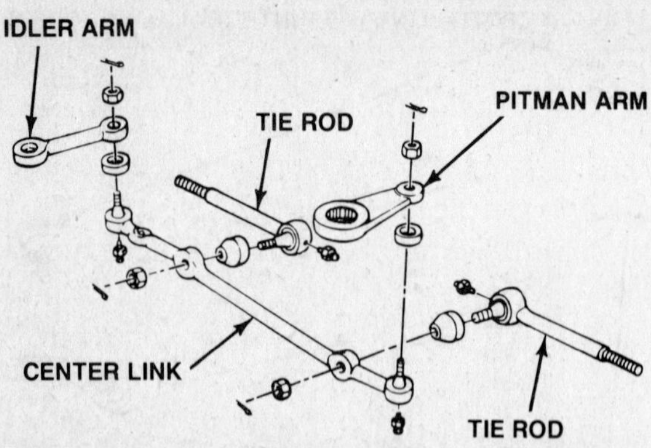

Figure 5-30. To ensure that steering linkage components do not come apart while the vehicle is moving, use new cotter pins when you reassemble the linkage.

8. Center the steering wheel and shaft, as described in Chapter Four of this *Shop Manual*.
9. Align the marks on the sector shaft and pitman arm. Do not move the steering gear from its center position.
10. Put the castle nut on the sector shaft threads. Torque the nut to the manufacturer's specifications. This will push the pitman arm onto the shaft. Install a cotter pin, if there is one.
11. Check parallelism, as described later in this chapter. Finally, check wheel alignment.

Center Link, Relay Rod, and Drag Link

Not all steering linkages have a center link. Some have a relay rod or drag link. If you are working on a vehicle that has a drag link, some of the information in this section is applicable, but you should also refer to the section on tie rods for additional pertinent information.

Inspection
There are two types of center link — those with integral ball joints, and those without, figure 5-29. If the ball joints are not built into the center link, then they are incorporated into the parts that connect to the center link, such as the pitman and idler arms. When inspecting the center link, check the ball joints as described earlier in this chapter. If they are worn and the center link has integral ball joints, replace the center link. If the ball joints are not integrated into the center link, then replace the pitman arm, idler arm, or tie rods, as needed.

Check the center link, drag link, or relay rod for bends. Replace a bent center link. Never attempt to straighten it.

Replacement
Before removing the center link, note its position carefully. It is easy to mistakenly install the center link up-side-down or backward.

As a reminder, if the vehicle uses RBS joints, be sure to straighten the wheels before installing the center link and tightening the joints. If the RBS joints are worn, replace them with standard tie rod ends, as recommended by Ford. To remove the center link:

1. Position the lift supports under the vehicle chassis using the manufacturer's recommended lift points, and raise the vehicle. See Chapter One of this *Shop Manual* for more information on lifting a vehicle.
2. Remove and discard the castle nuts and cotter pins that secure the tie rods, idler arm, and pitman arm to the center link, relay rod, or drag link, figure 5-30.
3. Using a puller or pickle fork, as described earlier in this chapter, separate the idler arm, pitman arm, and tie rod ball joints from the center link. Remove the center link.
4. Before installing a new center link, check it for bends or stress cracks.
5. Position the new center link in the vehicle. Place the idler and pitman arm ball joints into the sockets on the center link. Thread the castle nuts onto the studs.
6. Place the tie rod ball joints into the sockets on the center link, and thread the castle nuts onto the studs.
7. Torque the nuts to the manufacturer's specifications, which will press the ball joints in place, figure 5-31.
8. Install new cotter pins.
9. Check and adjust parallelism as described later in this chapter. Finally, check and adjust wheel alignment.

Idler Arm

In a parallelogram steering linkage, the idler arm supports the center link on the end opposite the pitman arm. If the idler arm is bent or loose, it will not hold the center link in position. This will allow toe settings to change while the car is moving, adversely affecting steering and wheel alignment.

Inspection
While performing the "dry-park" test, check for movement at the idler arm mount. You

Figure 5-31. Torque the castle nuts on the center link ball joints to the manufacturer's specifications.

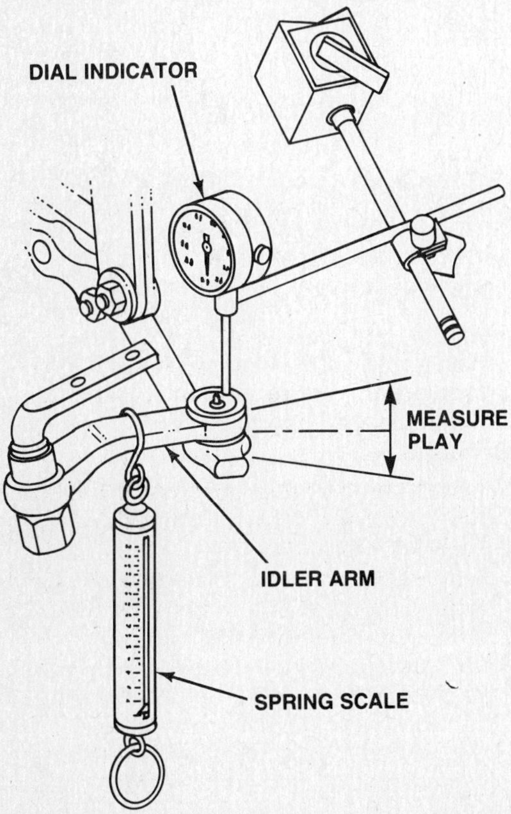

Figure 5-32. Use a dial indicator to measure play and a spring scale to measure the pulling force applied to the idler arm.

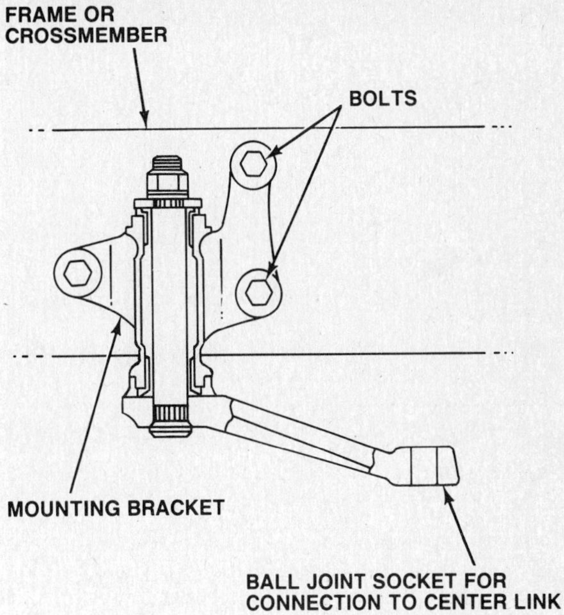

Figure 5-33. The idler arm should pivot smoothly, and the bolts securing it to the frame should be tight.

must replace the idler arm or secure the mount if there is movement. Inspect the idler arm as follows:

1. Place a dial indicator against the idler arm dust cap to measure movement when force is applied, figure 5-32.
2. Pull down on the idler arm with 25 pounds (11 kg) of force, using a spring scale to measure the force applied. If the idler arm deflects more than 0.125 inch (3 mm), replace it.
3. Check these following points, figure 5-33:
   - Be sure that the bolts that secure the idler arm to the frame or crossmember are tight.
   - Check for cracks in the frame or crossmember at the idler arm mount. If the frame supporting the idler arm is cracked or bent, repair it before replacing the idler arm.
   - Check the idler arm and its mounting bracket for bends or damage. Be sure the idler arm pivot points operate smoothly.
   - Check the ball joint between the idler arm and center link. If it is worn, you must replace the center link or idler arm, as required.

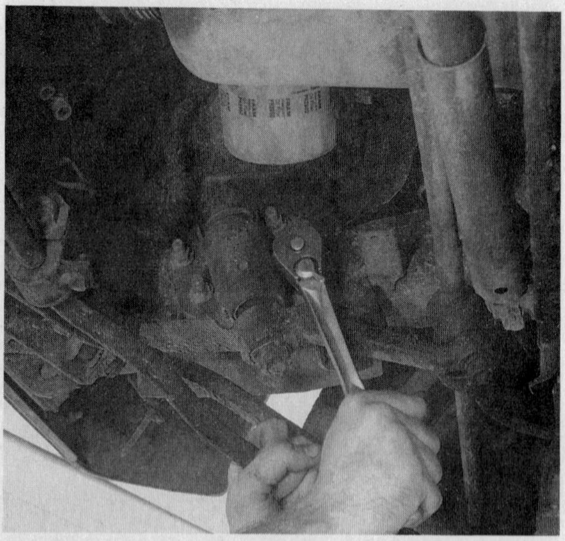

Figure 5-34. Remove the bolts that hold the idler arm to the frame or crossmember. There may be a rubber bushing under the mount that damps vibration. Be sure to remove the bushing, if there is one.

Removal

If the idler arm is worn or damaged, use the following procedure to replace it. Afterwards, check and adjust parallelism as described later in this chapter.

1. Position the lift supports under the vehicle chassis using the manufacturer's recommended lift points, and raise the vehicle. See Chapter One of this *Shop Manual* for more information on lifting a vehicle.
2. Loosen and remove the castle nut and cotter pin from the ball joint that connects the idler arm and center link.
3. Using a puller or pickle fork, as described earlier in this chapter, separate the ball joint between the idler arm and center link.
4. Remove the bolts that hold the idler arm mounting bracket to the vehicle's frame or crossmember, figure 5-34. Remove the idler arm, bracket, and rubber bushing, if any.
5. Check the frame or crossmember where the idler arm was mounted. This area should not be cracked and the bolt holes should be intact. Note that in some vehicles, the frame or crossmember may have elongated holes for adjusting parallelism at the idler arm, as described later in this chapter. Repair any problems before replacing the idler arm, or you may not be able to properly adjust parallelism and align the wheels.

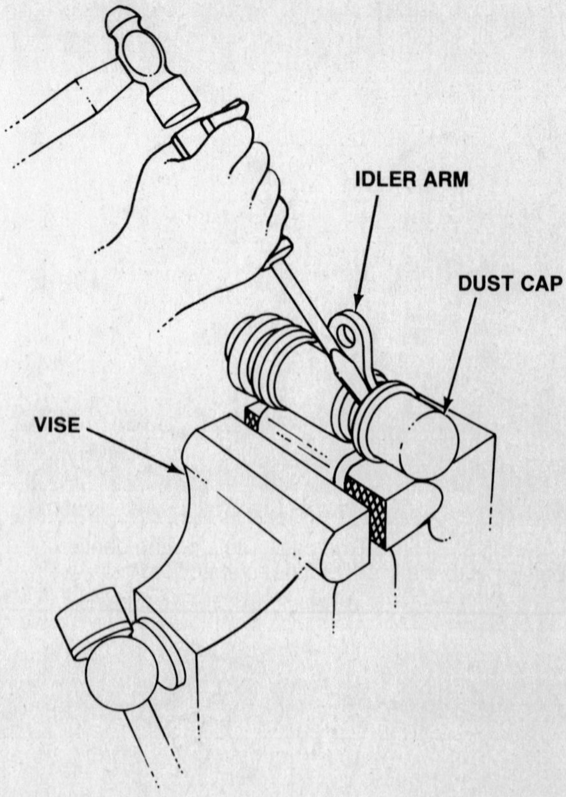

Figure 5-35. On some vehicles, you can disassemble and refurbish the idler arm. First remove the dust cap with a screwdriver and hammer.

Repair

On most vehicles, if the idler arm is worn, you must replace the entire assembly. If the idler arm has replaceable parts, however, you can refurbish it:

1. Secure the idler arm in a vise. If there is a dust cap, pry it off with a screwdriver, figure 5-35.
2. Remove the nut under the dust cap that secures the idler arm to the mounting bracket, and remove the idler arm.
3. Use a torque wrench to turn the idler arm, figure 5-36. If it does not turn smoothly, or if it does not meet the manufacturer's recommended torque, replace the bushings.
4. Remove the oil seal and bushings with a screwdriver. Press or drive new bushings in place. You may need the manufacturer's factory tool to do this, figure 5-37. Replace the oil seal.
5. Install the idler arm in the mounting bracket, and replace the securing nut and dust cap.

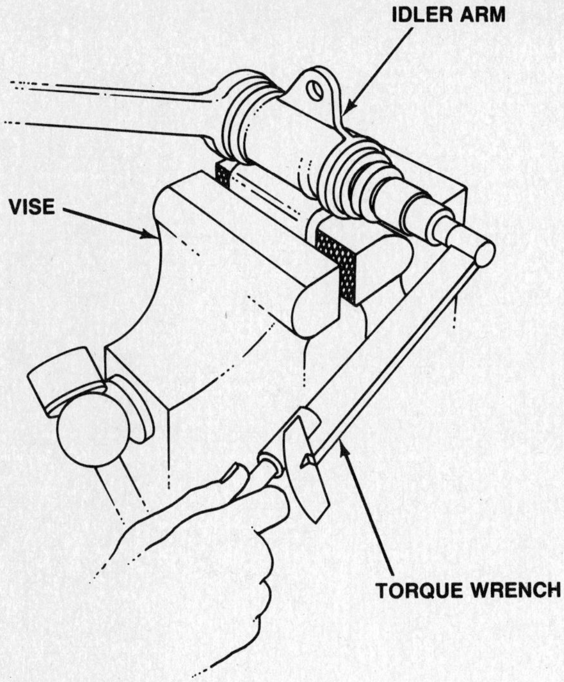

Figure 5-36. Some manufacturers suggest that you check the idler arm bushings with a torque wrench.

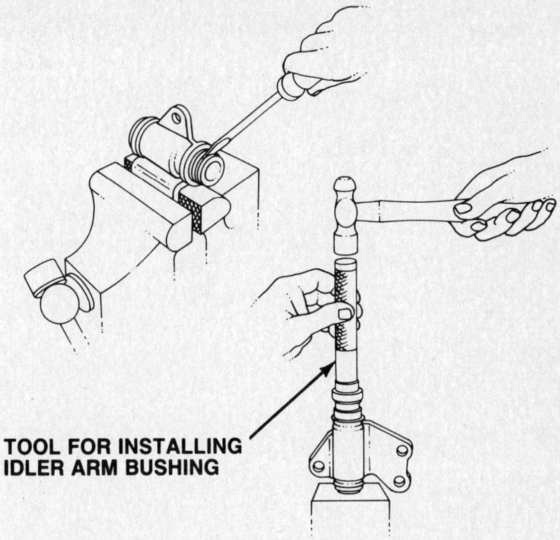

Figure 5-37. On most vehicles, when the idler arm is worn, you must replace the entire assembly. On others, you can replace the bushings and oil seals if needed.

Installation
The molded rubber bushing used on some idler arms aids steering. You must center the steering wheel and gear, and straighten the vehicle's wheels before installing the idler arm.

1. Place the ball joint in the socket to join the idler arm and center link. Install the castle nut, torquing it to specifications. Install the cotter pin.
2. Position the idler arm mounting bracket and rubber bushing, if any, against the vehicle's frame or crossmember, and tighten the bolts as specified by the manufacturer.
3. Check and adjust parallelism, as described later in this chapter. Also check and adjust wheel alignment, as described in Chapter Fourteen of this *Shop Manual*.

Tie Rods

Tie rods join the steering linkage or rack to the steering knuckles at the vehicle's wheels. Defective tie rods can cause a variety of problems, ranging from loose steering to abnormal tire wear and bump steer. Fortunately, it is not difficult to inspect, adjust, and replace tie rods.

Some trucks with solid front axles and single-tie-rod steering linkages use a drag link that is adjustable like a tie rod, figure 5-38. Refer to the manufacturer's instructions if you must adjust or replace a drag link. If the drag link in the vehicle that you are working on is similar to a tie rod, you can inspect it according to the procedures described in this section.

Inspection
Watch for play at the tie rod ball joints during the dry-park test. Use the following procedure to inspect tie rods.

1. Tie rod ball joints are located at the outer tie rod ends, and between the inner tie rod ends and the center link on parallelogram steering linkages. Rack and pinion steering systems have ball joints at the outer tie rod ends, and ball-and-socket joints at the inner tie rod ends.
2. Check the condition of the ball joint boots, figure 5-39, and check for ball joint wear using the dry-park test, as described earlier.
3. Rack and pinion steering systems do not have standard ball joints at the inner tie rod ends. Instead, they have ball-and-socket joints that are covered with rubber boots or bellows, figure 5-40, as described in Chapter Four of the *Classroom Manual*. Check rack and pinion inner tie rod ends for play by pinching the boots at the joint while conducting the dry-park test. You will feel the joints move if there is play. If possible, remove the boots or bellows and

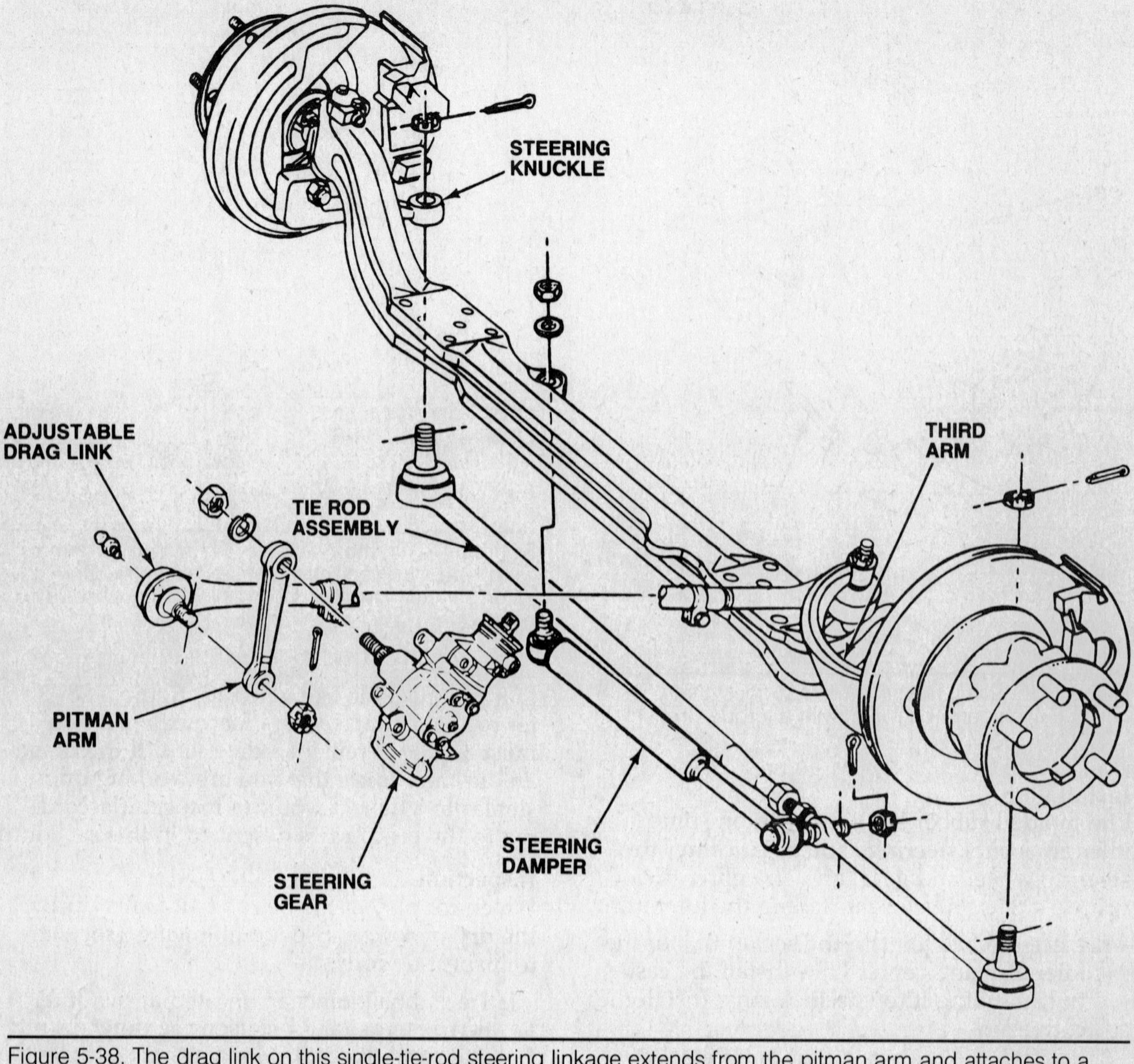

Figure 5-38. The drag link on this single-tie-rod steering linkage extends from the pitman arm and attaches to a "third arm", which is very much like a steering arm.

visually inspect the joint, as described in Chapter Four of this *Shop Manual*. If there is play, replace the tie rod.

4. Check tie rods for cracks, excessive rust, or other signs of wear. Check for bends in the rod, which will affect tie rod length and steering geometry. Replace damaged tie rods.
5. Use a wrench or pliers to loosen the lock nuts or clamps on the tie rod ends. Then use a torque wrench to turn the adjuster nut or sleeve. If it takes more than 9 Nm (80 in-lb) of force to turn the adjuster nuts or adjusting sleeve, the threads are probably damaged or rusted. Clean the threads. If the tie rod end is still difficult to adjust, replace the tie rod.

Replacement

Use the following procedure to remove and replace tie rods.

1. Remove the cotter pin and castle nut that secure the outer tie rod end ball joint in its socket at the steering knuckle.
2. Using a puller or pickle fork, as described earlier in this chapter, separate the ball joint from the steering arm socket.
3. If you are replacing only the tie rod end, use the following procedure. If you are replacing the entire tie rod, proceed to step 4 or 5, as applicable.
   a. Before loosening the tie rod end, make a mark on the threads, figure 5-41. You can align this mark with the new

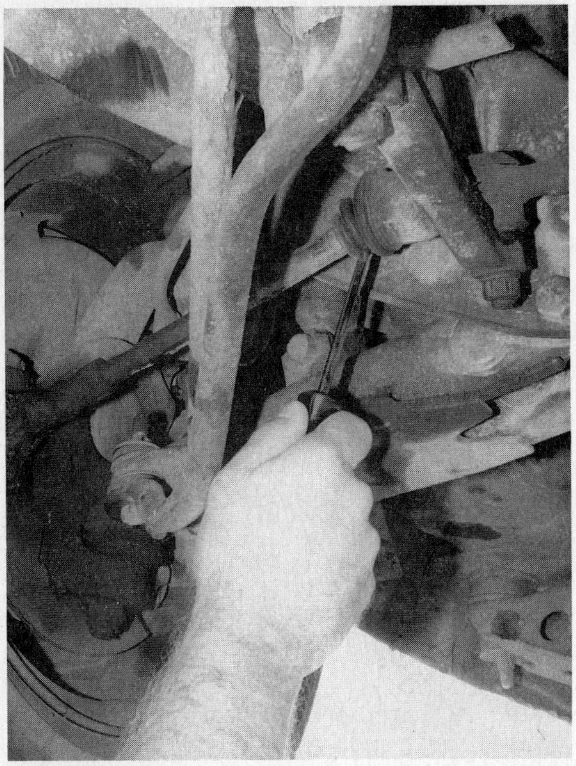

Figure 5-39. Be sure to check the entire tie rod, which includes the ball joints and their boots, during an inspection.

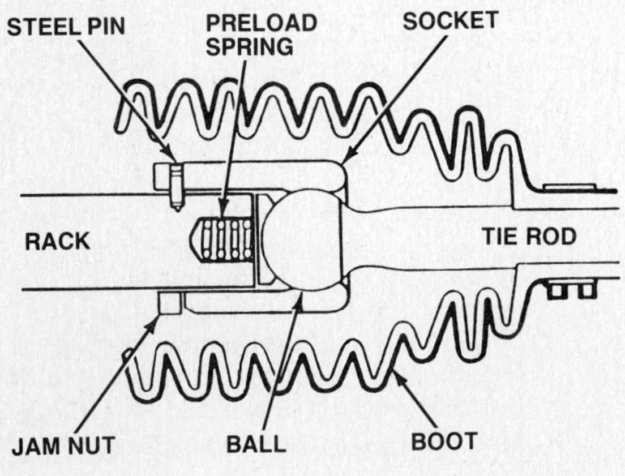

Figure 5-40. The inner tie rod end on a rack and pinion steering system is covered with a large rubber boot. If you pinch the boot at the location of the ball joint, you can check the joint for play.

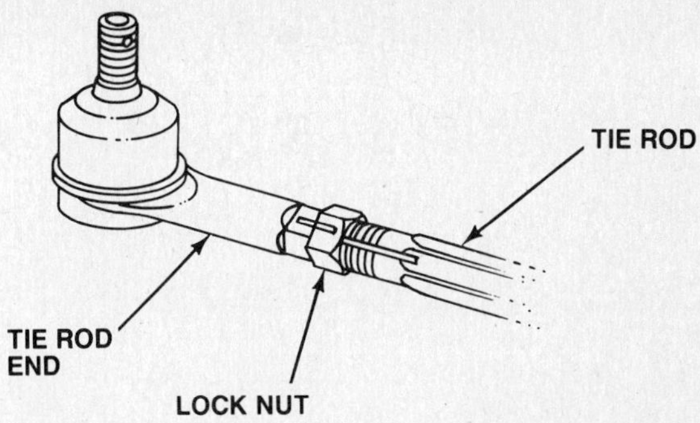

Figure 5-41. Mark the threads on the tie rod before removing the tie rod end. This will help you install the new tie rod end in the proper position.

Figure 5-42. Removing the tie rod end is simply a matter of unscrewing it from the tie rod.

tie rod end to ensure that the new tie rod is the same length as the old one. Some manufacturers recommend measuring the exposed threads on the tie rod before removing the tie rod end.

b. Loosen the lock nut or adjusting sleeve on the tie rod, and unscrew the tie rod end, figure 5-42.
c. Put the new tie rod end in place, ensuring that it lines up with the marks that you made earlier, or that the exposed threads are the same length as before.
d. Place the ball joint in the steering arm socket, replace the castle nut, and torque the nut in place. Replace the cotter pin. Check and adjust toe.

4. If you are replacing the entire tie rod on a standard steering gear system:
   a. Remove the cotter pin and castle nut that secure the inner tie rod end ball joint to the center link, drag link, or relay rod.
   b. Using a puller or pickle fork, as described earlier in this chapter, separate the inner tie rod end ball joint from its socket, figure 5-43.

Figure 5-43. If you have to remove the entire tie rod, you must separate the inner tie rod end from the center link.

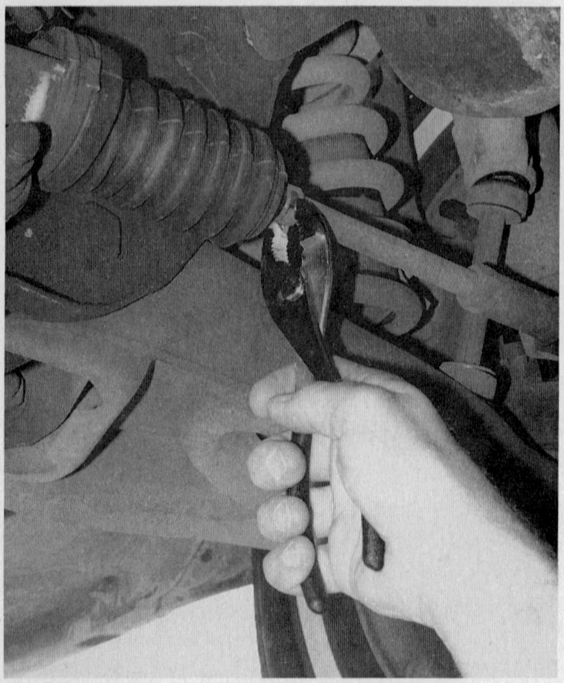

Figure 5-44. The inner tie rod end on a rack and pinion steering system is covered with a rubber boot. Remove the clip, clamp, or wire that secures the boot.

c. To install a new tie rod, place the ball joint at the inner end of the tie rod into the socket on the center link, drag link, or relay rod.
d. Replace the castle nut and torque it in place. Install the cotter pin.
e. Put the outer tie rod end ball joint into the steering arm socket. Be sure that the new tie rod is the same length as the old one.
f. Torque the castle nut that secures the outer tie rod end in place. Replace the cotter pin, and check and adjust toe.

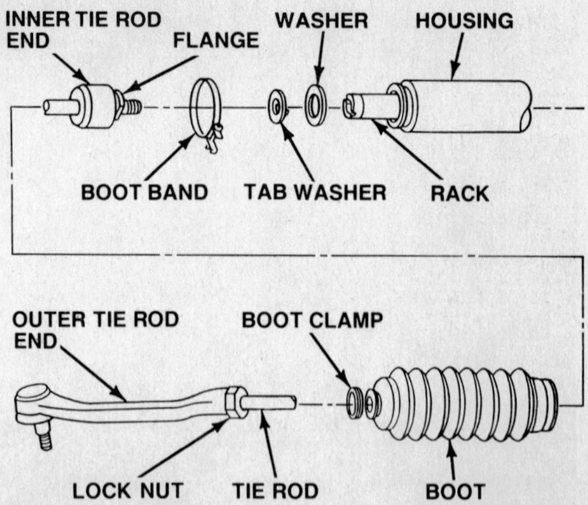

Figure 5-45. On some rack and pinion steering systems, the inner tie rod end is locked in place with a tab washer. Other systems use a lock pin to hold the inner tie rod end in place.

5. If you are replacing the entire tie rod on a rack and pinion steering gear system:
   a. Remove the clip, clamp, or wire that secures the rubber boot covering the inner tie rod end, figure 5-44. Remove the boot by slipping it down and off the tie rod.
   b. Locate the lock pin or tab washer at the inner end of the tie rod, figure 5-45. Remove the lock pin with a punch and hammer or an Allen wrench, or straighten the washer with pliers. Remove the tie rod. On some vehicles, disconnect the inner tie rod by turning the ball housing with a wrench and special adapter, figure 5-46, while holding the rack in place with another wrench. Then pull the tie rod off the rack. Never use an impact wrench to remove the inner tie rod end.
   c. Install a new tie rod using a new lock pin or tab washer. Push in the lock pin, or bend the tab washer with pliers or a hammer to hold the assembly together. For tie rods with ball housings, use the wrench and special adapter to tighten the tie rod to the rack.
   d. Replace the boot over the inner tie rod end, and clip, clamp, or wire it in place.
   e. Place the outer tie rod end ball joint in the steering arm socket. Be sure that the new tie rod is the same length as the old one.

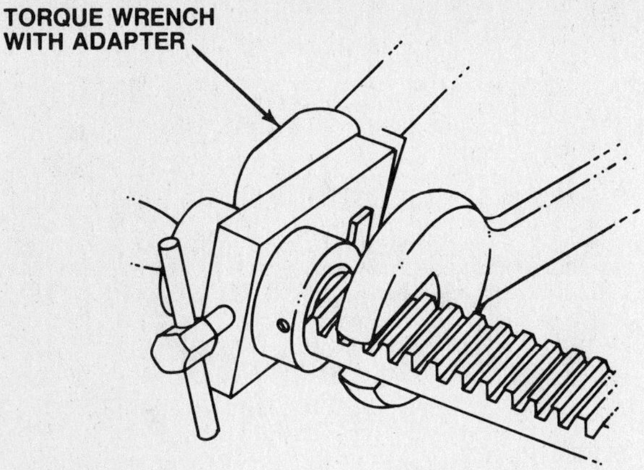

Figure 5-46. On some rack and pinion steering systems, a wrench with a special adapter is used to turn the inner tie rod end ball housing to separate the joint.

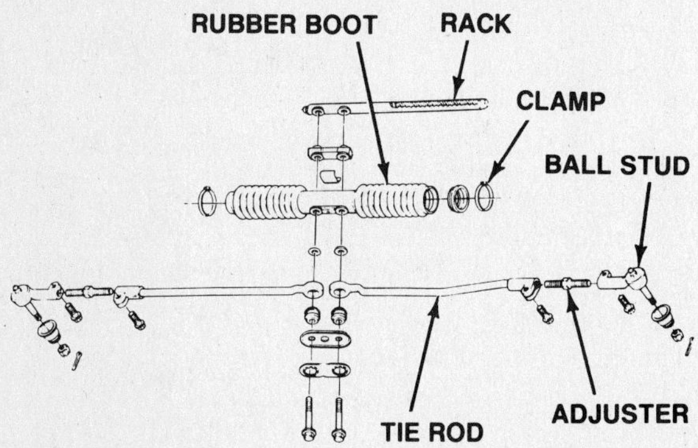

Figure 5-47. On this particular rack and pinion steering gear, the tie rods connect at one end of the rack with bolts and bushings, rather than at each end of the rack with ball joints.

f. Torque the castle nut that secures the outer tie rod end in place. Replace the cotter pin, and check wheel alignment.

Some rack and pinion steering gears use inner tie rod ends that attach at one end of the rack rather than at each end, figure 5-47. The manufacturer recommends that when you remove and replace tie rods on this type of system, it is a good idea to work with one tie rod at a time rather than removing and replacing both tie rods simultaneously.

Check tie rod lengths after replacing tie rods or tie rod ends. If the tie rods are not adjusted to the manufacturer's recommended lengths, bump steer may result. Bump steer occurs when the outer tie rod end cannot travel in the same arc as the suspension parts when the wheel goes over a bump. The tie rod resists suspension movement, and pushes or pulls the steering arm either inward or outward, causing the vehicle to turn abruptly. Checking and adjusting wheel alignment after replacing the tie rods will prevent this problem.

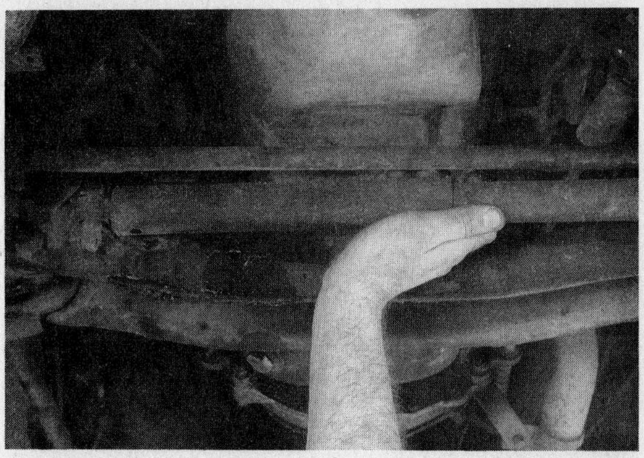

Figure 5-48. The steering damper is an oil-filled shock absorber that cushions the steering linkage from road shock.

Steering Damper

You will find steering dampers on some non-power standard and rack and pinion steering gear systems. Steering dampers are not used on power steering systems because the power piston acts like a steering damper. For more information on power steering, see Chapter Six of this *Shop Manual*.

A steering damper is a shock absorber that is located between the vehicle frame or crossmember and the steering linkage, figure 5-48. If you feel kickback or shimmy in the steering wheel, check the condition of the damper. Steering dampers are sealed, and cannot be repaired. If the damper is damaged, replace it.

Inspection

Although the steering damper is a relatively trouble-free part, check it occasionally, as follows:

1. Check the steering damper for leaks. If there is excessive oil or grease around the shaft seal, figure 5-49, the seal is worn and you must replace the damper. A slight oily film is acceptable. Be sure that the oil or grease on the seal is from the damper and not from a leak elsewhere in the vehicle.
2. Check the ball joint or bushing between the damper shaft and steering linkage, figure 5-50. If it is worn, replace the damper.
3. Remove the bolt that secures the damper to the crossmember or frame, figure 5-51. If there is a rubber bushing between the

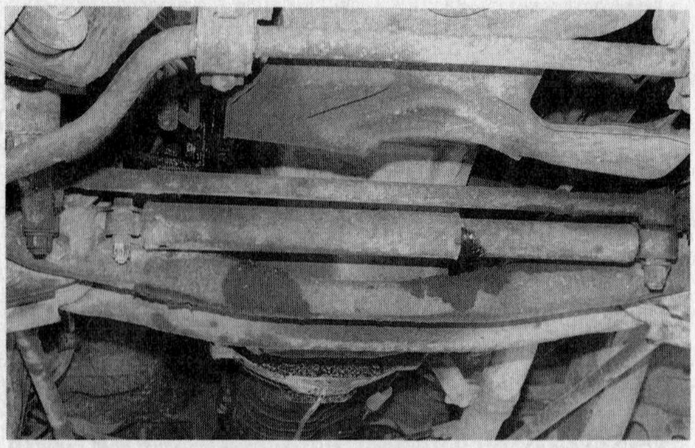

Figure 5-49. If oil is leaking past the steering damper oil seal, replace the damper.

Figure 5-50. Check the bushings or ball joints at each end of the steering damper for wear.

Figure 5-51. One or two bolts secure the steering damper to the vehicle frame or crossmember. Notice that the other end of this steering damper is bolted to the pitman arm.

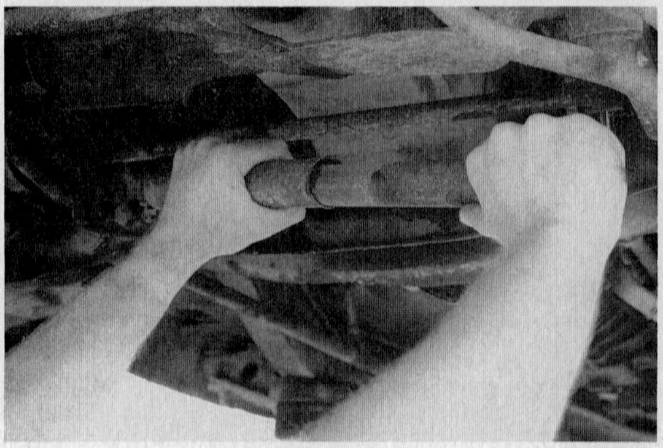

Figure 5-52. Push and pull the plunger on the steering damper. Replace the damper if the plunger does not travel smoothly.

damper and the crossmember or frame, check it for deterioration, and replace it if needed.

4. Push and pull the damper through its full range of travel, figure 5-52. It should move smoothly without binding. Replace the damper if there is no resistance to movement, or if movement is uneven.

Replacement

The steering damper is not difficult to replace because you do not have to remove other steering linkage components.

1. Position the lift supports under the vehicle chassis using the manufacturer's recommended lift points, and raise the vehicle. See Chapter One of this *Shop Manual* for more information on lifting a vehicle.
2. Remove the cotter pin from the nut that secures the damper to the crossmember or frame. Also remove the cotter pin from the nut securing the damper to the steering linkage.
3. Remove the bolt holding the steering damper to the frame or crossmember, figure 5-51. Be sure to remove the washer and bushing, if any. Discard the old bushing, and use a new one when you install the new damper.
4. Use a puller or pickle fork, as described earlier in this chapter, to separate the ball joint that connects the steering damper to the steering linkage. On some vehicles, the steering damper simply *bolts* to the steering linkage, figure 5-53. If this is the case, remove the nut or bolt holding the steering damper to the steering linkage. Be sure to remove the washer and bushing, if any. Discard the old bushing and use a new one when you install the new damper.
5. Bolt the new damper to the frame or crossmember, replacing the washer and bushing. Torque the bolt to the manufacturer's specifications, and install the cotter pin.
6. Place the steering damper ball joint in the socket on the steering linkage, and torque the castle nut to the manufacturer's specifications, or simply bolt the damper to the linkage. Install the cotter pin. Check and adjust toe.

Figure 5-53. The damper may connect to the steering linkage with a ball joint, or it may simply bolt in place.

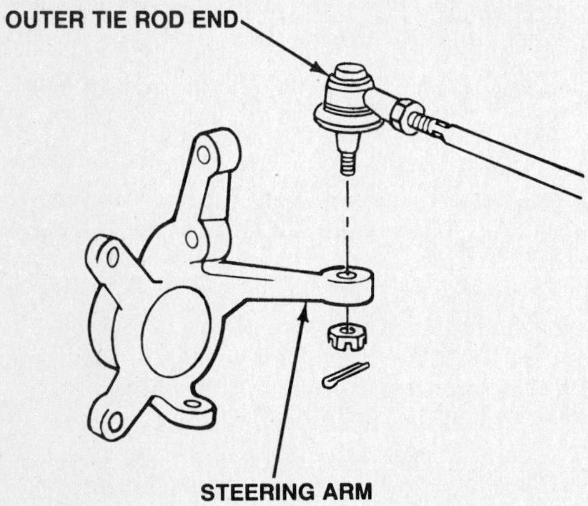

Figure 5-54. In many vehicles, the steering arm is a cast part of the steering knuckle.

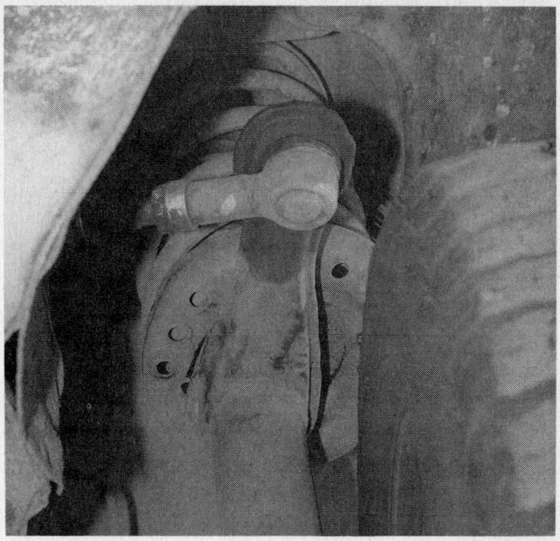

Figure 5-55. On this vehicle, the steering arm is on the MacPherson strut.

Steering Arm

The steering arm may be an integral part of the steering knuckle, figure 5-54, or it may bolt onto the knuckle. It can also be a part of the MacPherson strut, figure 5-55. Tie rods from the steering linkage attach to the steering arms to turn the vehicle's wheels.

Inspection

Steering arm position determines one of the non-adjustable alignment angles known as ''toe-out on turns''. If this alignment angle is not within the manufacturer's recommended specifications, then the steering arm may be bent. Toe-out on turns is discussed in detail in Chapter Fourteen of the *Classroom Manual*.

Inspect the steering arm for bends and cracks. If you find a bend, do not attempt to straighten it. Instead, replace the steering arm, steering knuckle, or strut bracket.

Some trucks with solid front axles use a drag link from the pitman arm to a ''third arm''. The third arm extends from the steering knuckle and is similar to a steering arm. Check the ''third arm'' for bends or damage if the vehicle you are working on has one.

Replacement

If the vehicle has removable steering arms, use the procedure that follows to remove them. If the steering arms are not removable, use the procedure in Chapter Nine of this *Shop Manual* to remove the steering knuckle, or the procedure in Chapter Eight of this *Shop Manual* to remove the strut bracket.

1. Position the lift supports under the vehicle chassis using the manufacturer's recommended lift points, and raise the vehicle. See Chapter One of this *Shop Manual* for more information on lifting a vehicle.
2. Remove the wheel and brake system accessories, if necessary, to gain access to the nuts or bolts that hold the steering arm in place. Remove the cotter pins or wires that secure the steering arm nuts or bolts.
3. Remove the tie rod end ball joint from the steering arm using a puller or pickle fork, as described earlier in this chapter.
4. Loosen and remove the steering arm nuts or bolts, and remove the steering arm, figure 5-56. On some vehicles, the steering arm and knuckle may have a splined fitting, and you may have to use penetrating oil to separate them.

Figure 5-56. In this example, the steering arm can be unbolted from the knuckle.

5. Install the new steering arm in reverse order. Carefully examine the steering arm to be sure that you have the correct part. Steering arms for the right and left sides of the vehicle look similar, but if you accidentally switch them, the vehicle's steering geometry will be incorrect.
6. Use new nuts or bolts and cotter pins or wires when installing the new steering arm.

MEASUREMENTS AND ADJUSTMENTS

Some manufacturers recommend that you measure parallelism to check the steering linkage after replacing parts. Parallelism is an attribute of parallelogram steering systems. However, you can make a similar measurement on rack and pinion systems that checks the horizontal position of the rack.

Parallelism

Manufacturers design parallelogram steering linkages so that when the center link is parallel to the ground, the car steers smoothly. This condition in which the center link is parallel to the ground is called "parallelism", figure 5-57.

If the center link is not parallel, which may be caused by a bent or incorrectly positioned idler or pitman arm, parallelism is affected. This, in turn, affects tie rod position, resulting in a variety of steering problems.

Two problems caused by improper tie rod position are bump steer and unequal turning radii for left and right turns. If the vehicle that you are working on has these problems, check parallelism.

Checking parallelism

Check and adjust parallelism after you replace parts in the steering linkage. Remember, you can only check parallelism on vehicles with parallelogram steering linkages.

PARALLELISM

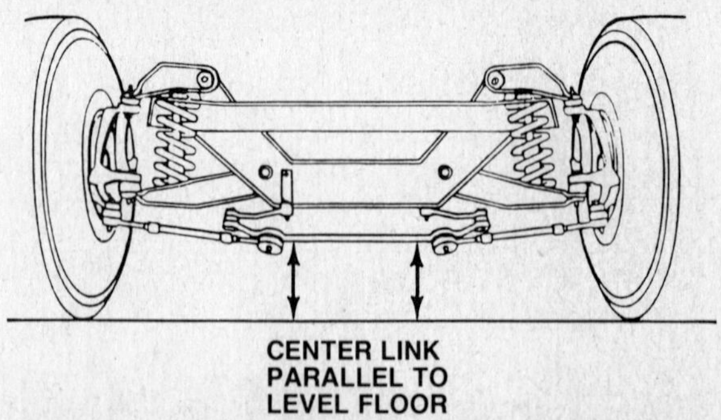

Figure 5-57. In a parallelogram steering system, the center link must be parallel to the ground.

1. Park the car on an alignment rack or on level ground. If you use an alignment rack, be sure to lock the turntables before driving the car onto the rack, and keep them locked during this procedure.
2. Position the vehicle's wheels straight ahead, and center the steering gear.
3. Check vehicle ride height, as described in Chapter Fourteen of this *Shop Manual*. Be sure the ride height is equal from side to side. Also check and correct tire pressures, if needed.
4. Locate the points where the pitman and idler arms connect to the center link. Use a tape measure to measure the distance from these points to the ground, figure 5-58. Some center links have flat machined surfaces for making these measurements.
5. The distance from the pitman arm side of the center link and the idler arm side of the center link to the ground should differ by no more than $^1/_{16}$ inch (1.5 mm). If the difference is greater than $^1/_{16}$ inch, then adjust parallelism.

The general procedure described above will work for most parallelogram steering linkages, although if the vehicle ride height is not equal on each side of the car, if tire pressures are incorrect, or if the surface on which the vehicle is parked is not level, then parallelism measurements will be off. Some manufacturers recommend more elaborate procedures for checking parallelism. The following procedure, used for some General Motors vehicles, is an example:

1. Park the car on an alignment rack or on level ground. If you use an alignment rack, be sure to lock the turntables before driving the car onto the rack, and keep them locked during this procedure.
2. Position the vehicle's wheels straight ahead, and center the steering gear.

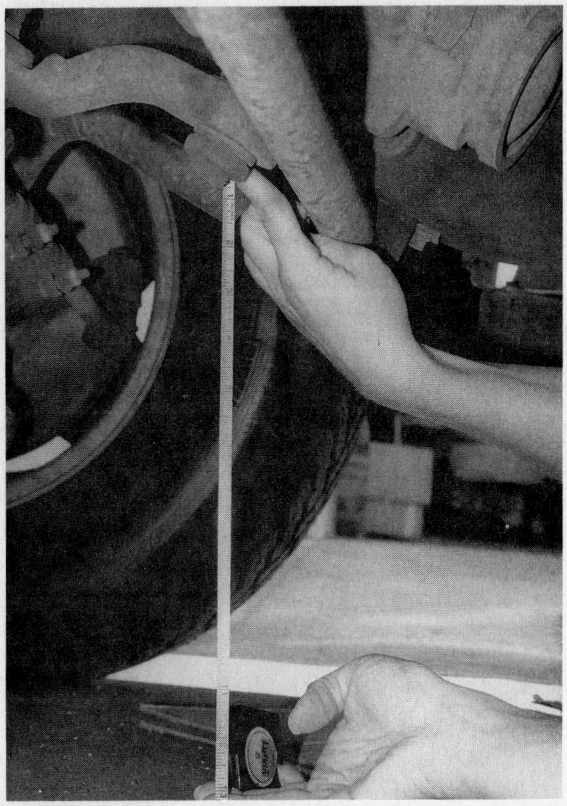

Figure 5-58. Use a tape measure to check the distance from the pitman and idler arms to the ground. If these distances are unequal, the center link is not parallel, and you must make adjustments to the steering linkage.

3. Check the vehicle ride height, as described in Chapter Fourteen of this *Shop Manual*. Be sure the ride height is equal from side to side. Also check and correct tire pressures, if needed.
4. Turn the steering wheel until the ball joint on the tie rod end is directly under the upper control arm bolt, figure 5-59.
5. Place a socket on the control arm bolt. The socket should be long enough so that you can measure from the tie rod end ball joint straight up to the socket, as in the illustration.
6. Measure the distance from the tie rod end to the socket. Then take the same measurement on the other side of the vehicle.
7. If the difference between these measurements is more than about 1/25 inch (1 mm), then adjust parallelism.

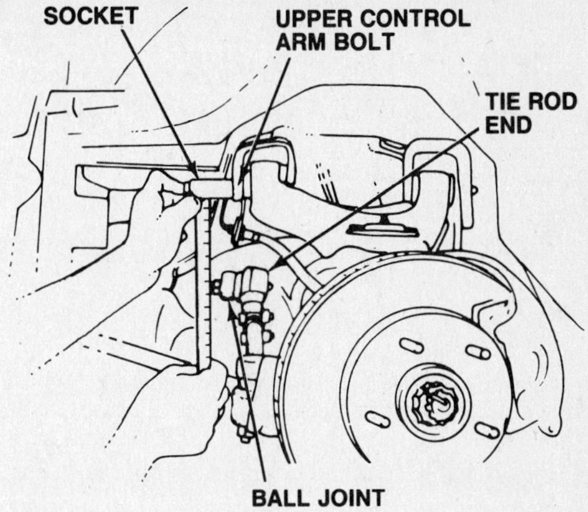

Figure 5-59. To check parallelism on some GM vehicles, use a measuring tape to check the distance from a socket on the upper control arm bolt to the tie rod end. Then do the same on the other side of the vehicle.

Adjusting parallelism

Parallelism is adjusted at the idler arm. The most common method for adjusting parallelism is described below:

1. Position the lift supports under the vehicle chassis using the manufacturer's recommended lift points, and raise the vehicle. See Chapter One of this *Shop Manual* for more information on lifting a vehicle.
2. Adjust parallelism at the idler arm. Use either of the following methods to move the idler arm until the difference between the two measurements falls within the manufacturer's recommended tolerance for parallelism:
   a. On some vehicles, you must loosen the mounting bolts that secure the idler arm to the frame, and reposition the idler arm. The frame has slots that allow you to reposition the idler arm mounting bracket higher or lower on the frame, figure 5-60. Then tighten the bolts.
   b. On other vehicles, you must turn a threaded bushing at the base of the idler arm mounting bracket, which changes the position of the idler arm, figure 5-61.
3. Make parallelism measurements again to check your work.

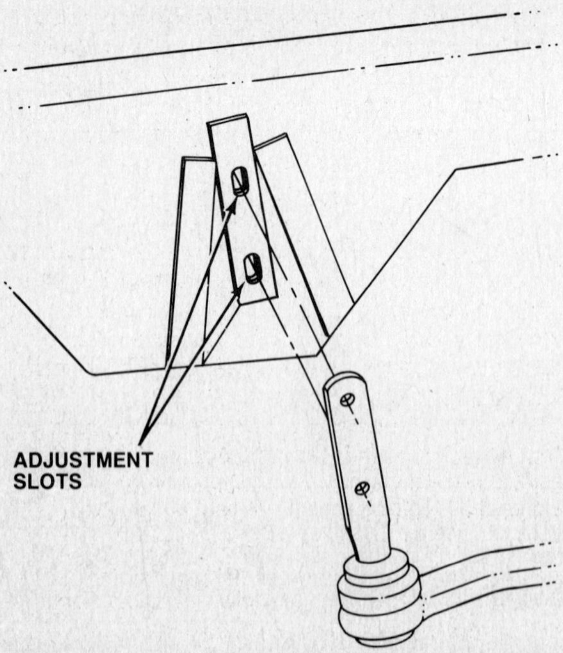

Figure 5-60. Some steering systems have adjustable idler arms. On this system, the bolt holes in the frame or crossmember are elongated so that you can change the idler arm's position.

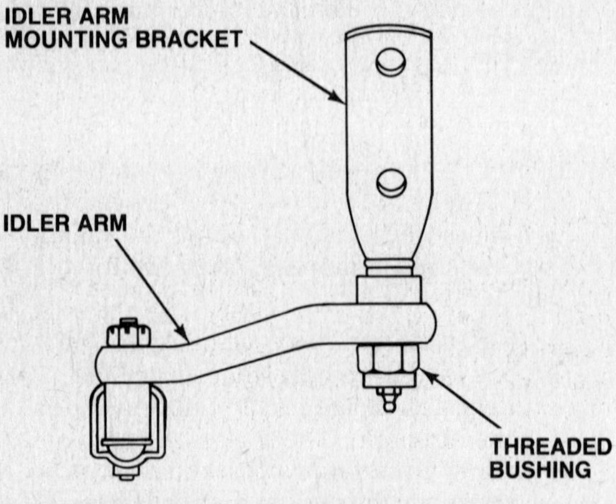

Figure 5-61. This type of idler arm is adjusted by turning the threaded bushing at the base of the idler arm mount.

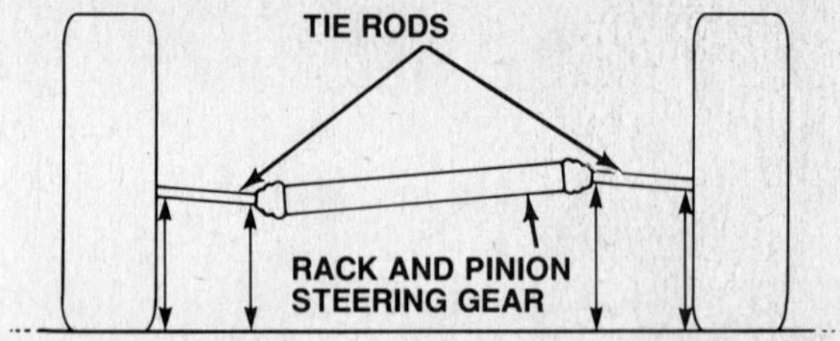

Figure 5-62. In a rack and pinion steering system, the rack should be parallel to the ground. If it is not, you can readjust the rack to correct this problem.

Rack Horizontal Position

Just as the center link in a parallelogram steering system must be parallel with the ground, so must the rack in a rack and pinion system. If the rack is not parallel with the ground, then the tie rods are also not positioned properly, figure 5-62.

Check the rack horizontal position as follows:

1. Park the car on an alignment rack or on level ground. If you use an alignment rack, be sure to lock the turntables before driving the car onto the rack, and keep them locked during this procedure.
2. Position the vehicle's wheels straight ahead, and center the steering gear.
3. Check the vehicle ride height, as described in Chapter Fourteen of this *Shop Manual*. Be sure the ride height is equal from side to side. Also check and correct tire pressures, if needed.
4. Use a tape measure to measure the distance from the outer tie rod end to the ground and from the inner tie rod end to the ground on one side of the car, figure 5-62. Calculate the difference between these numbers. Then do the same for the other side of the vehicle.
5. Compare the two differences. If there is more than 1/2 inch (13 mm) difference, then the rack is not parallel to the ground, and the car likely has driveability problems. Adjust the rack position using the procedures described in Chapter Four of this *Shop Manual*.

Chapter

6

Power Steering System Service

Power steering systems are similar to manual systems, except that they have additional parts to allow for hydraulic assist, figure 6-1. These parts are the power steering pump and drive belt, hydraulic lines and fittings, and the control valve and power piston or cylinder.

As mentioned in Chapter Six of the *Classroom Manual*, there are two types of power steering systems. Integral power steering uses a power piston located inside the steering gear to help turn the vehicle's wheels. Linkage power steering, which is uncommon in late-model cars, uses a power cylinder that is outside the steering gear.

When investigating a power steering problem, check the fluid level and drive belt tension first, then check for leaks. If you do not find any problems, conduct pump pressure tests to help pinpoint defective components. Finally, repair hoses and fittings, the power steering pump, or the steering gear, as needed, using the procedures in this chapter.

POWER STEERING FLUID

Power steering fluid, figure 6-2, is specially designed to work in the high-pressure, high-heat conditions of the power steering hydraulic system, and is prone to chemical breakdown by oxidation. To reduce fluid temperature and breakdown, some vehicle manufacturers equip the power steering system with special radiator-like coolers at the pump return hose.

Never substitute any other type of lubricant for power steering fluid unless specified by the vehicle manufacturer. The wrong fluid may not be able to withstand the temperature and pressure extremes in the system. Also, if the fluid is chemically incompatible with the rubber in the power steering system, hoses and seals will swell or deteriorate.

Power steering fluid is a clear, yellow- to amber-colored liquid. Some vehicle manufacturers specify ATF (automatic transmission fluid) for the power steering system. Two types of ATF used in power steering systems are DEXRON II, which is a clear, red liquid, and Type F, which is a clear, amber-colored liquid. Use the following procedures to check fluid level and condition. If the system is low on fluid or if the fluid is contaminated, inspect for and repair leaks or other sources of contamination. Then flush the system to remove debris, which will prevent further damage. Finally, refill the system with the manufacturer's recommended power steering fluid or ATF, and bleed air from the system.

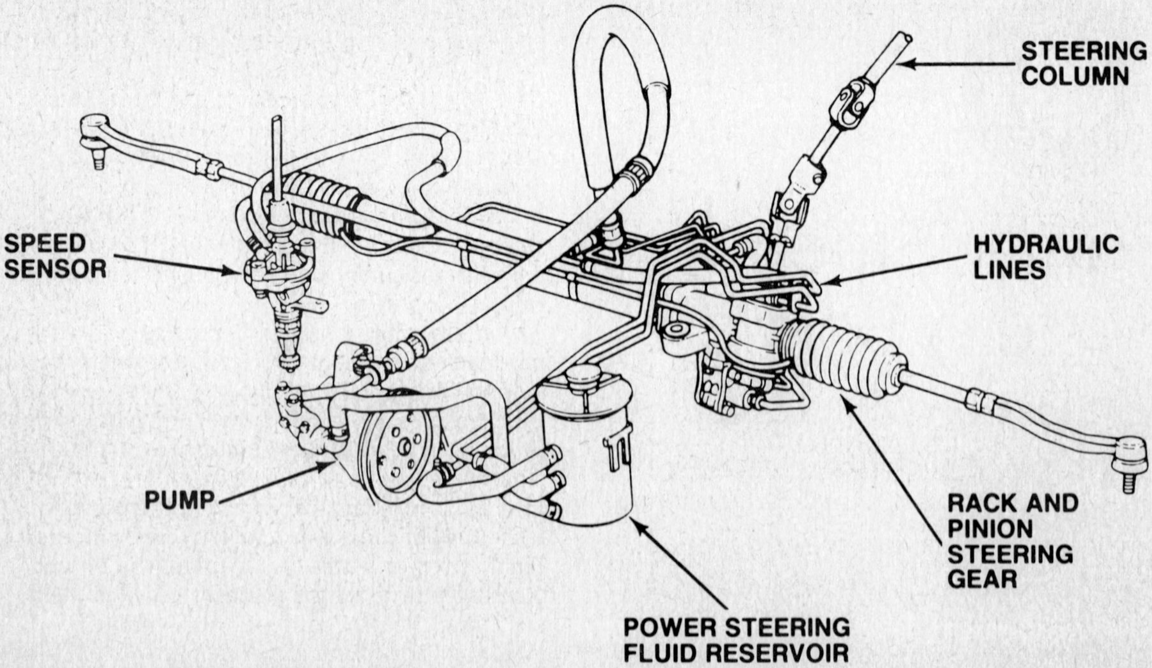

Figure 6-1. If the steering gear has hydraulic lines leading to it, the vehicle has power steering. Otherwise, the steering gear and linkage look much like those used on a manual steering system.

Figure 6-2. Use the power steering fluid recommended by the vehicle manufacturer. Some manufacturers recommend fluid specially formulated for the power steering system, and others recommend using ATF in the power steering system.

Checking and Filling Power Steering Fluid

Check the power steering fluid level at the manufacturer's recommended service interval, which is normally every six to twelve months, or whenever the vehicle's power steering system is not operating properly. Signs of low fluid level include a humming or buzzing sound during turns, as well as increased steering effort. If the fluid drops to extremely low levels, the power steering hydraulic system will overheat, making the car dangerous to drive. Power steering systems will function even when the hydraulics fail, but turning the steering wheel will be very difficult. This sudden loss of steering control due to fluid loss is quite surprising and can be dangerous if the driver is accustomed to hydraulic assist.

Check the power steering fluid level as follows:

1. Park the car on level ground, and locate the power steering fluid reservoir, which may be integral with or remote from the power steering pump. An integral reservoir is part of the power steering pump, figure 6-3, and is found on vehicles that have large engine compartments where space to fit the pump is not restricted. A remote reservoir is separate from the pump, but is connected to it with a suction hose, figure 6-1. Remote reservoirs are used in engine compartments that have space restrictions, such as in FWD (front-wheel-drive) vehicles.
2. Wipe dirt from the cap and from around the reservoir opening with a shop towel. This will keep dirt from entering the system. If dirt enters the steering gear and pump, it will act as an abrasive and accelerate wear.

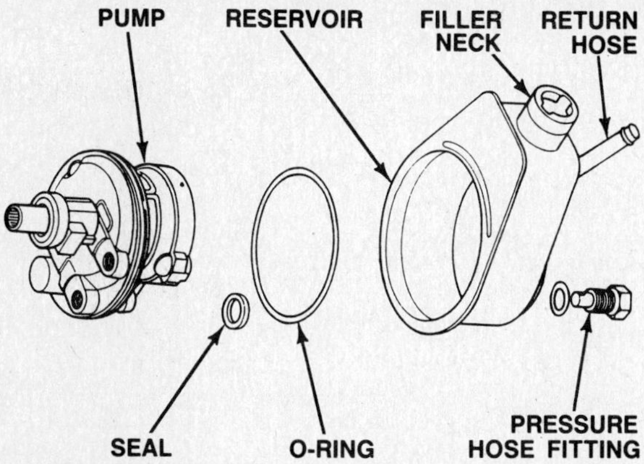

Figure 6-3. On this type of power steering pump, the fluid reservoir is integral with the pump. When you fill the reservoir, you are actually filling the pump body.

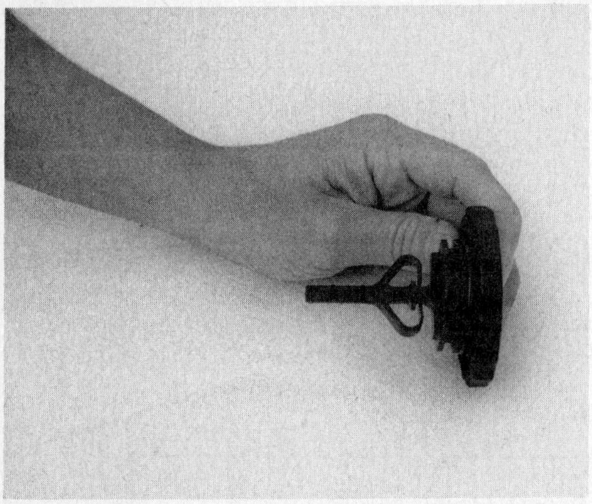

Figure 6-4. Read the power steering fluid level on the dipstick, which is attached to the power steering fluid reservoir cap.

3. Check for any fluid that leaked past the reservoir cap. If you find fluid around the cap, then the reservoir may be too full.
4. Loosen and remove the cap. The cap may simply twist off, or it may be held in place with a wing nut.
5. Locate the fluid level indicator marks. Most power steering fluid reservoirs have graduations on the dipstick, figure 6-4, which is attached to the reservoir cap. Remove the cap, wipe the dipstick with a clean towel, and insert the dipstick in the fluid again. Then remove the dipstick to read the fluid level. Others systems have graduations on the reservoir body, figure 6-5. Still others do not have any graduations. On these systems, the fluid level must be about 1/2 inch (13 mm) from the top of the reservoir.

Figure 6-5. On some power steering systems, the fluid level indicators are stamped into the metal or plastic fluid reservoir.

6. Determine if the fluid level is correct. Most vehicles have two level indicators, one to check the fluid level at normal operating temperature, and another to check it when cold. Most vehicle manufacturers suggest that you check the fluid at normal operating temperature, which is 120° to 180°F (50° to 80°C). If you are not sure, refer to the vehicle's service manual, or check the top and sides of the fluid reservoir. Sometimes, instructions are stamped on the reservoir or cap.
7. Check the fluid condition by dabbing a little on a piece of white paper. Use the information in the next section of this chapter to interpret fluid condition. If the fluid is contaminated, flush the system before filling it with fresh fluid.
8. Add or remove fluid as needed until the level is correct. On many late-model cars, the type of power steering fluid needed is stamped on the reservoir or cap. If you are refilling a system that is empty, or if the fluid level is extremely low, raise the front end of the vehicle before adding power steering fluid. Adding fluid when the front wheels are elevated reduces air pocket formation in the gear and pump.
9. Wipe any fluid that may have spilled on the outside of the reservoir, and replace the cap.
10. Bleed the hydraulic system if needed, as described later in this chapter.

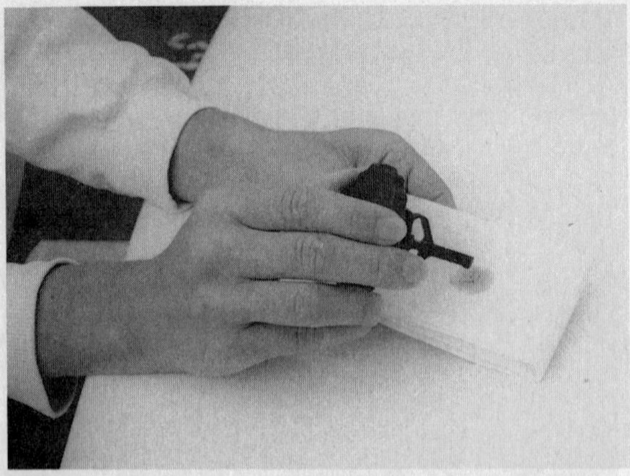

Figure 6-6. When you check the power steering fluid level, inspect its condition as well. If the fluid is thick and dark in color, or if it has particles in it, flush the hydraulic system and refill with fresh fluid.

11. After driving the vehicle, check the power steering fluid level again. If the level is low, check for leaks, as described later in this chapter.

Interpreting Power Steering Fluid Condition

While checking the power steering fluid level, you should also check its condition. Dab some fluid on a piece of white paper, figure 6-6, and inspect it. The color should be clear and light. Oxidized fluid is brown, is slightly thicker than fresh fluid, and smells like varnish.

Check the fluid for metal or black particles. Metal particles indicate worn pump bearings, or binding steering gear teeth. Black particles indicate deteriorating hoses and seals. If the fluid is foamy or cloudy, air is in the system, which reduces hydraulic assist.

Repair any problems that you find, and flush the power steering system. If the fluid has air in it, bleed the system, as described later in this chapter.

Checking the Hydraulic System for Leaks

To check for leaks, position the lift supports under the vehicle chassis using the manufacturer's recommended lift points, and raise the vehicle. See Chapter One of this *Shop Manual* for more information on lifting a vehicle. Check the engine compartment for a yellow-, amber-, or red-colored liquid. If the vehicle manufacturer uses fluorescent dye in the power steering fluid, check the engine compartment for leaks using a black light.

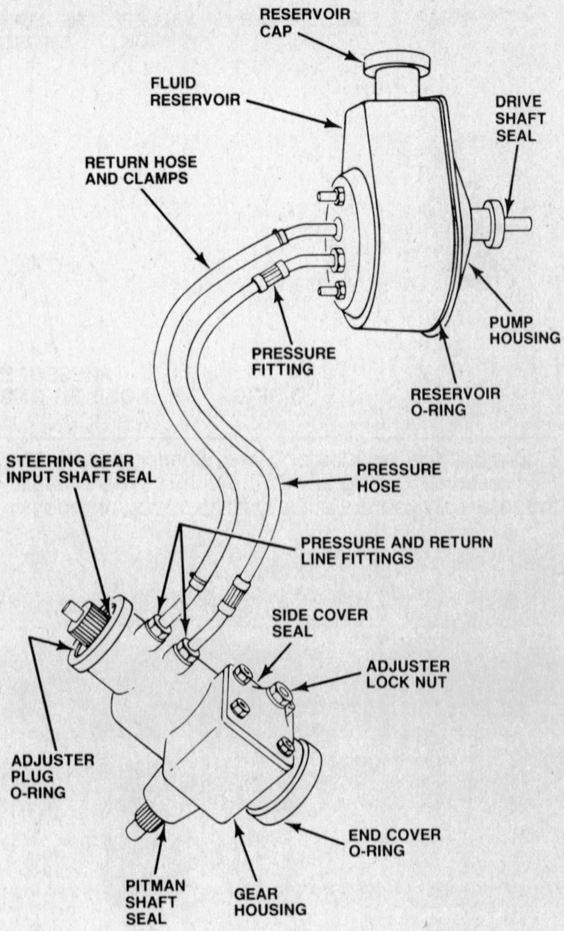

Figure 6-7. When you inspect the power steering system for leaks, pay extra attention to these areas, where seals and O-rings may need replacement.

If you find power steering fluid, follow the leak upward and forward in the engine compartment until you locate the faulty component. Power steering components tend to leak at the seals and O-rings when the vehicle gets older, figure 6-7. If the engine compartment is extremely dirty or if there are numerous fluid leaks, you can steam clean the engine, drive it for a few minutes, and then check again for leaks.

Check the power steering fluid reservoir. If the cap does not make a tight seal, fluid may spill from the reservoir. Check along hoses and fittings, and inspect the pump and steering gear for fluid seepage at gaskets and seals.

Rack and pinion type steering gears can lose fluid without any evidence of leakage. A worn pinion seal or inner and outer rack seals, will allow fluid to seep from the hydraulic system, figure 6-8, but the fluid will not leak into the

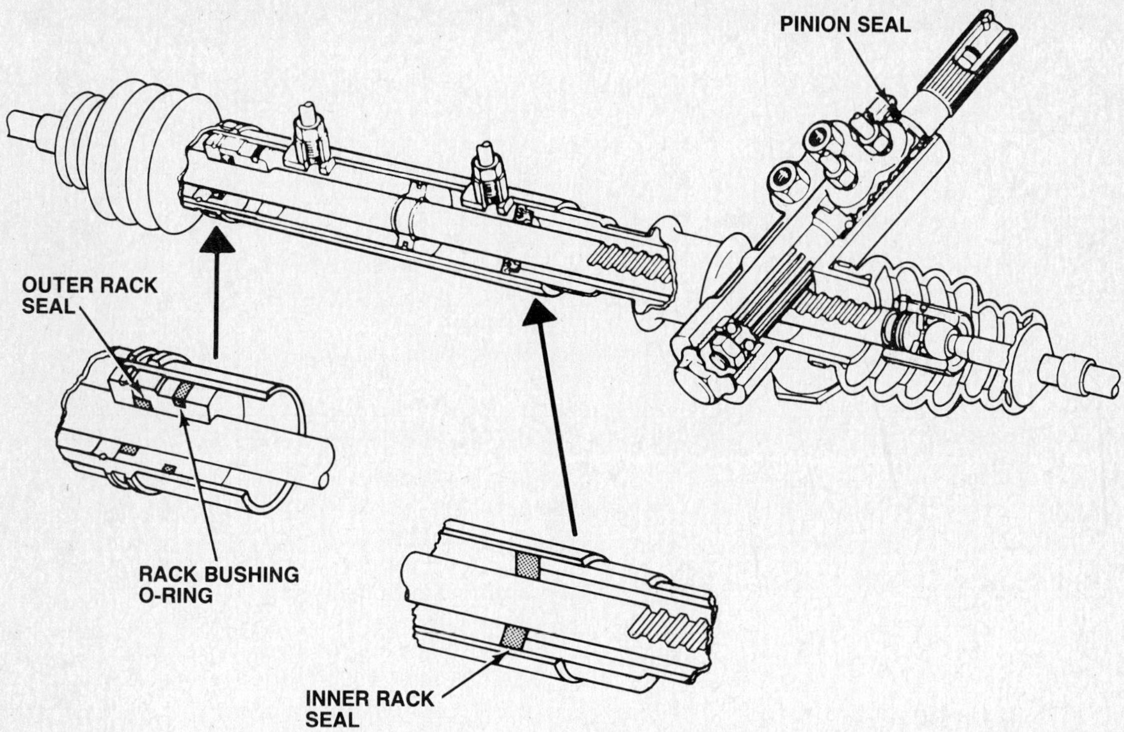

Figure 6-8. When the rack or pinion seals are worn, fluid leaks past them and accumulates in the rubber boots instead of leaking onto the ground.

engine compartment. Instead, it will accumulate in the rubber boots. Refer to the section in this chapter on rack and pinion steering gears for more information on detecting leaks of this type.

When you locate the source of a leak, use the procedures later in this chapter to inspect, repair, or replace defective components.

Flushing the System

If power steering fluid is contaminated, make any needed repairs, and then flush the hydraulic system. Some manufacturers suggest that you flush the hydraulic system after installing a new component to remove any debris that may be inside the new part. Have an assistant help you flush the power steering hydraulic system using the following procedure:

1. Position the lift supports under the vehicle chassis using the manufacturer's recommended lift points, and raise the vehicle. See Chapter One of this *Shop Manual* for more information on lifting a vehicle.
2. Place a drain pan under the pump. Remove the low-pressure return hose from the pump, and place the hose in the drain pan. Plug the return hose fitting at the pump.
3. Have an assistant fill the reservoir with fluid while the engine is running. Turn the steering wheel from stop to stop to allow the fluid to circulate through the entire steering gear. Pour about one quart (one liter) of power steering fluid through the system.
4. Repeat steps two and three until the power steering fluid runs clear.
5. Stop the engine, reconnect the return hose, and fill the hydraulic system. Run the engine for about fifteen minutes, turning the steering wheel from stop to stop.
6. Bleed the system, as described in the following section.

Bleeding the System

Pressurized fluid in the power steering system moves the power piston in the steering gear or the power cylinder at the steering linkage to turn the vehicle's wheels. If there are air bubbles in the fluid, the fluid will place uneven pressure against the power piston or cylinder, and steering will feel spongy or jerky. As fluid circulates through the hydraulic system, most air bubbles are released at the fluid reservoir. However, if pockets of air get trapped in the

FORD'S CII POWER STEERING PUMP

Figure 6-9. If you are bleeding the power steering system on a Ford vehicle that uses this CII pump, remember that you must use special procedures, which include pulling vacuum on the hydraulic system, to remove air from the fluid.

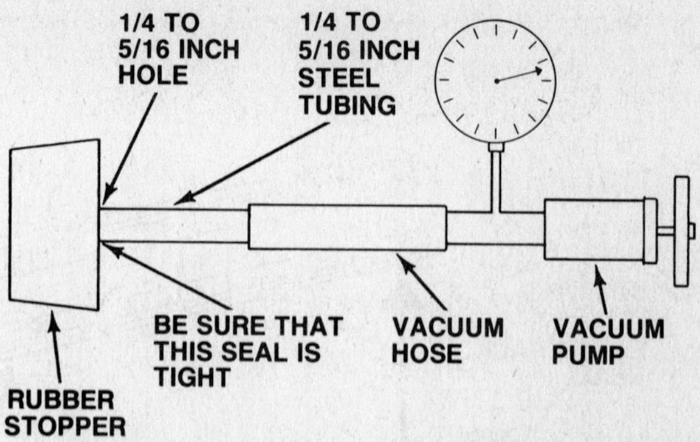

Figure 6-10. For power steering systems that use Ford's CII pump, pull vacuum at the fluid reservoir using this special vacuum pump setup. This bleeds air from the system.

steering pump or gear, the pump and gear will make groaning sounds during operation, and the steering system will not give full power assist.

Bleed the hydraulic system to release trapped air. You do not have to bleed the system after a routine reservoir check and refill. But after draining the system to replace parts, or after refilling a fluid reservoir that was extremely low, bleed the power steering hydraulic system as follows:

1. If the manufacturer recommends it, drive the vehicle up on ramps so that the front end of the car is elevated.
2. Run the engine for about ten minutes to allow the power steering fluid to reach normal operating temperature, which is about 120° to 180°F (50° to 80°C).
3. Turn the steering wheel *slowly* from lock to lock several times, which helps release air bubbles that are trapped in the steering gear. Do not hold the steering wheel in the lock position for more than a few seconds. Have an assistant watch the fluid reservoir, topping it up whenever it runs low.
4. Stop the engine, and check the fluid condition. If the fluid is foamy or milky, it still has air in it. Wait a few minutes, and bleed the system again, repeating this procedure.

The Ford model CII pump, figure 6-9, has an aluminum housing, a plastic reservoir, and a reservoir cap with an O-ring seal. You cannot bleed this system using the procedure described earlier. Instead, you must pull vacuum on the system, as follows:

1. Obtain a rubber stopper that fits tightly in the reservoir opening, and drill a hole in the top. Push a piece of steel tubing through the hole in the rubber stopper, figure 6-10. Then use a vacuum hose to connect the rubber stopper assembly to a vacuum pump.
2. Fill the reservoir to the full mark. Place the rubber stopper in the reservoir opening. Raise the front end of the vehicle, and start the engine.
3. Pull 15 in-Hg (380 mm-Hg) of vacuum on the system for three minutes, while turning the steering wheel from lock to lock. Be sure to maintain vacuum while doing this.
4. Remove the vacuum pump and rubber stopper assembly, and check the condition of the power steering fluid. If it is clear, top up the reservoir if needed, and replace the reservoir cap. If it is foamy or milky, repeat this procedure.

DRIVE BELTS

The power steering pump is driven by a belt, which is turned by the engine crankshaft pulley. For the power steering pump to operate properly, this belt must be in good condition, properly tensioned, and aligned. As drive belts age, they wear and stretch, so you must check them periodically. Manufacturers recommend checking and adjusting drive belts about every six to twelve months. Replace the belt, if needed.

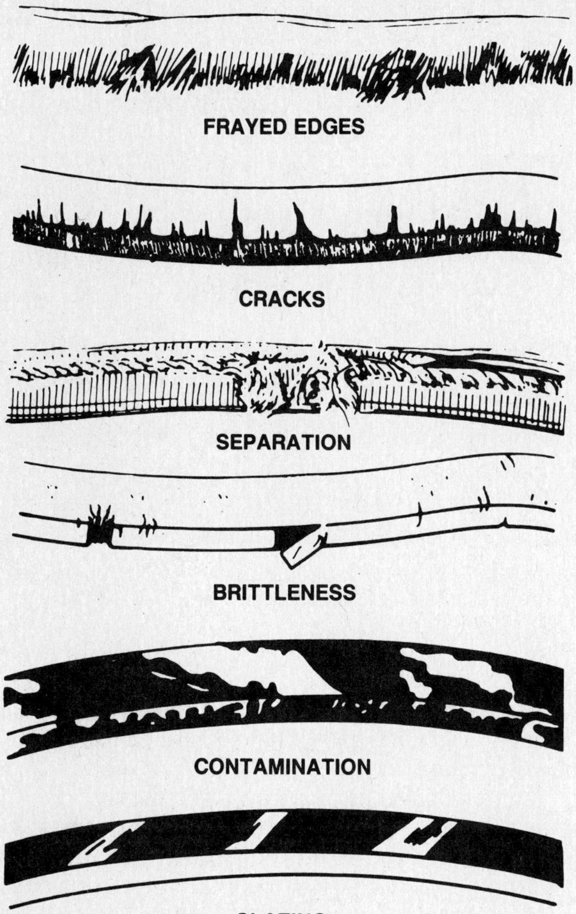

Figure 6-11. Check the belt that drives the power steering pump. If the belt has any of the defects shown, replace it.

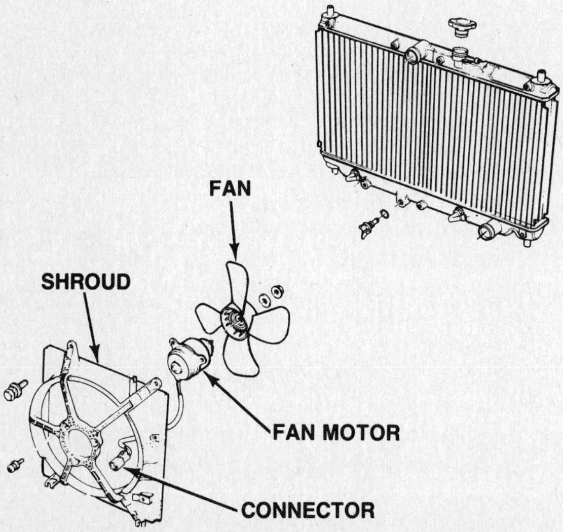

Figure 6-12. Disconnect the electrical connector for the fan before you inspect or change drive belts.

Figure 6-13. Flat, ribbed serpentine belts are common on late-model vehicles, although some vehicles still use V-belts to drive accessories.

Checking Belt Condition

A damaged drive belt will keep the power steering pump from operating at full capacity. Reduced steering assist and noises under the hood may indicate a damaged belt. Some common types of belt damage include, figure 6-11:

- Frayed edges
- Cracks
- Separation
- Brittleness
- Oil contamination
- Glazing.

Before inspecting drive belts, disconnect the battery ground cable. If the vehicle that you are working on uses an electric fan, as many FWD vehicles with transverse-mounted engines do, then disconnect the electrical connector for the fan before checking the belts, figure 6-12.

Determine which type of belt is driving the power steering pump. There are two types — V-belts and serpentine belts. V-belts are thick and have a V-shaped cross section. Serpentine belts are flat with V-shaped ribs on the underside, figure 6-13. About 90 percent of late-model vehicles have serpentine belts.

When checking a V-belt, inspect the underside and sides by twisting the belt, figure 6-14, as the top will not exhibit wear. You may have to turn the engine over a few times to examine the full length of the belt. If the sides are shiny and hard, the belt is glazed, which results in belt slippage on the pulley. A glazed belt will eventually crack.

Check both V-belts and serpentine belts for oil contamination, which breaks down rubber, causing it to crumble. Replace oily belts, and fix the source of the oil leak.

When checking a serpentine belt for wear, do not twist it more than 90 degrees, as this can damage the belt. Inspect a serpentine belt as you would a V-belt, plus check the ribs for

Figure 6-14. Wear and damage will occur on the sides and underside of a drive belt, so you must twist it when inspecting the belt.

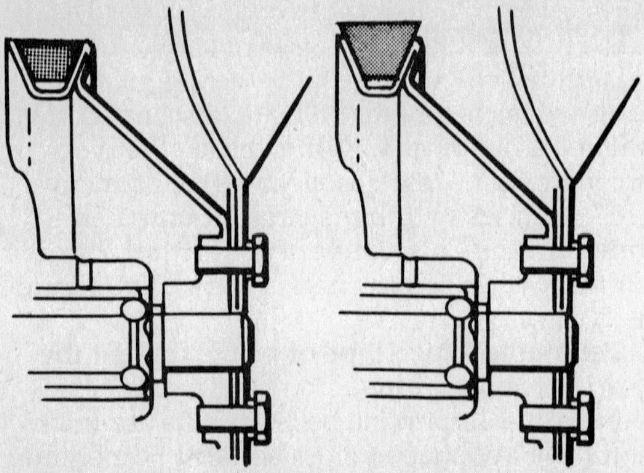

Figure 6-15. A V-belt should be flush with or about $^1/_{16}$ of an inch (2 mm) above the top of the pulley.

cracking. A few random cracks across the belt are normal, but if you find cracks spaced every $^1/_2$ to $^3/_4$ of an inch (13 to 19 mm), replace the belt.

Check the pump and crankshaft pulley. If they are worn, the belt will ride too low in the pulley groove, and it may wear prematurely. Tops of V-belts should be even with the top of the pulley or ride about $^1/_{16}$ of an inch (2 mm) above the pulley, figure 6-15. Check pulleys for burrs that may accelerate belt wear.

Checking and Adjusting Belt Tension

A loose pump drive belt will slip under load, and a tight belt will wear out quickly and damage accessory bearings. Decreased power assist and a high-pitched squeal from the engine compartment are signs of a slipping power steering pump drive belt.

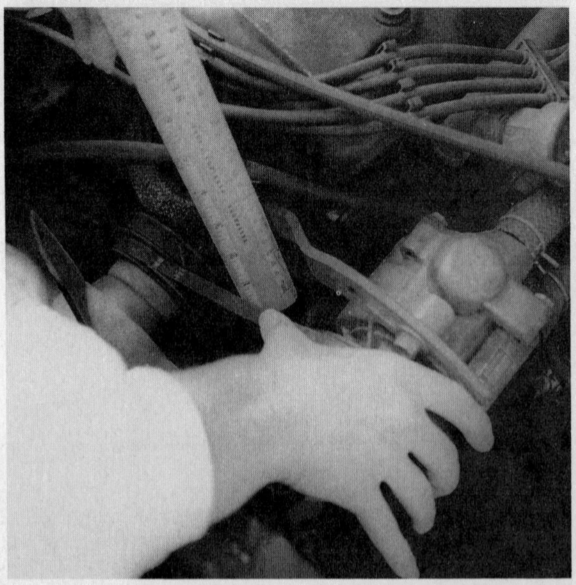

Figure 6-16. You can measure belt deflection with a ruler to estimate belt tension.

As part of a general inspection and whenever you replace drive belts, check and adjust belt tension. Tension specifications vary depending on whether the belts are new or used. Vehicle manufacturers often give two specifications, one for new belts, and one for used belts. Be sure to use the correct specification.

Checking belt tension

Manufacturers suggest that you use the deflection method or a tension gauge to check belt tension. The tension gauge is more accurate. After checking tension, adjust the belt if needed, as described later in this chapter. Check belt tension using the deflection method as follows:

1. Find the longest length of drive belt between two pulleys, and locate the midpoint of this length.
2. With your thumb or forefinger, either push the belt away from you or pull it toward you using moderate pressure, figure 6-16.
3. Measure the amount of deflection from the belt's normal position, and compare it with the manufacturer's specifications. Most belts should deflect from $^1/_2$ to $^3/_4$ inch (13 to 19 mm).
4. Adjust belt tension if needed, using one of the methods described later in this chapter.

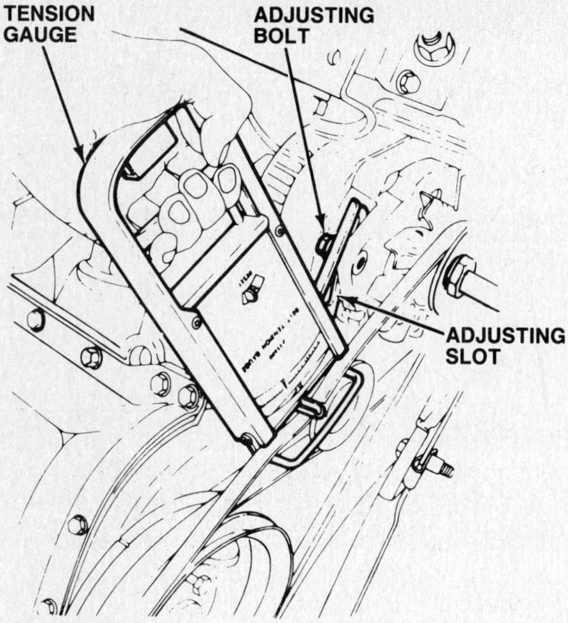

Figure 6-17. A tension gauge is an accurate method for checking drive belt tension. Most vehicle manufacturers recommend this method.

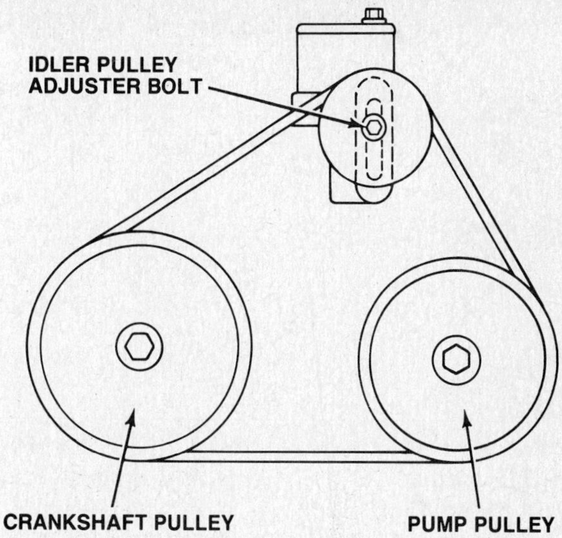

Figure 6-18. Move the idler pulley on its bracket to tension the accessory belt, and tighten the securing bolt.

Check belt tension using a tension gauge as follows:

1. Find the longest length of drive belt between two pulleys, and locate the midpoint of this length.
2. Position the tension gauge on the drive belt, and press the gauge handle, figure 6-17.
3. Measure the pounds of force needed to deflect the belt, and compare this reading with the manufacturer's specifications. New belts deflect at about 120 to 150 pounds (530 to 670 N), and used belts deflect at about 90 to 120 pounds (400 to 530 N). Belts are considered used after only a few minutes of use. Serpentine belts normally require more force to deflect.
4. Adjust belt tension, if needed, using one of the methods described in the next section.

Adjusting belt tension
V-belts are tensioned in three ways:

- With an adjustable idler pulley
- By repositioning the pump or another accessory on a pivoting bracket
- By using shims between the pulley halves to change belt tension.

Serpentine belts usually have a self-adjusting tensioning pulley, but you should check tension anyway to be sure that the pulley is operating properly.

To tension a belt with an adjustable idler pulley, figure 6-18, loosen the pulley adjuster bolt, and use a screwdriver to move the idler pulley until it tensions the belt. Tighten the idler pulley bolt, and check belt tension with a gauge or by using the deflection method. Readjust the idler pulley, if needed.

For systems that use a pivoting bracket to tension the belt, figure 6-19, loosen the adjustment bolt on the pivoting component, and turn the adjustment screw to move the bracket. If the bracket does not have an adjustment screw, lever it with a screwdriver or piece of wood. Never pry against the component itself, as the housing may bend. Tighten the bolt that holds the bracket in place, and recheck belt tension.

Some Volkswagens use shims to adjust belt tension, figure 6-20. Place shims between the pulley halves to open the pulley groove wider. This causes the belt to ride lower in the pulley groove, and belt tension is reduced. Remove shims to make the pulley groove narrower. This causes the belt to ride higher in the groove, and belt tension is increased. Recheck the tension after making adjustments.

For a vehicle with an automatic belt tensioner, figure 6-21, as with most serpentine belt systems, lever the tensioner up before putting

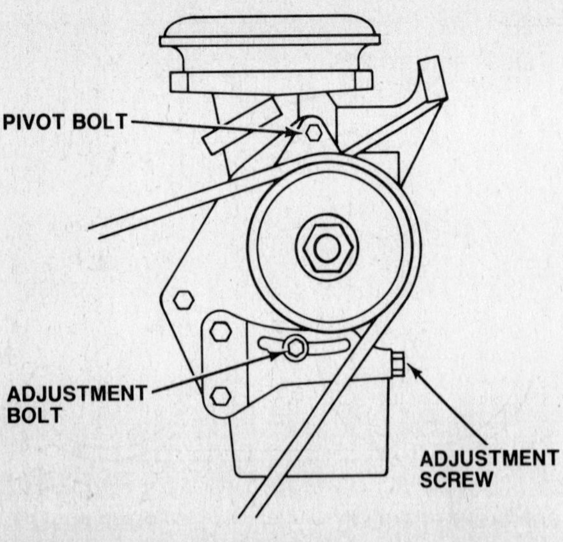

Figure 6-19. To adjust belt tension on this vehicle, loosen the adjustment bolt, turn the adjustment screw, and pivot the accessory and its bracket to tension the belt.

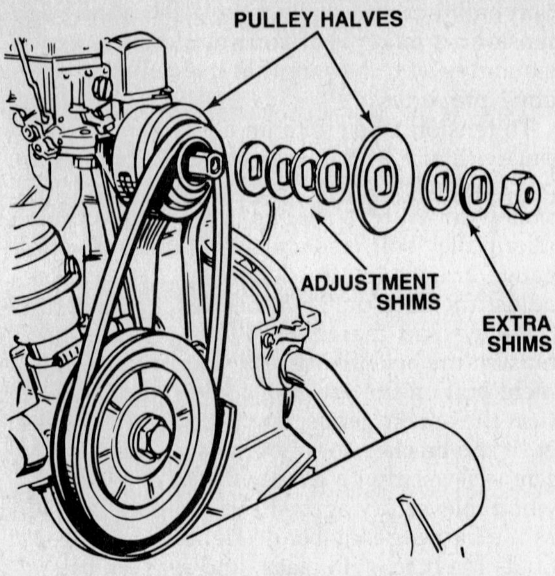

Figure 6-20. This Volkswagen uses a split pulley on the generator to adjust belt tension. Add or remove shims from between the pulley halves to change belt tension.

the belt on the pulleys. Once it is on, release the automatic tensioner, and check with a tension gauge.

After adjusting belt tension, run the engine for fifteen to twenty minutes. Then recheck tension using the manufacturer's specifications for a used drive belt.

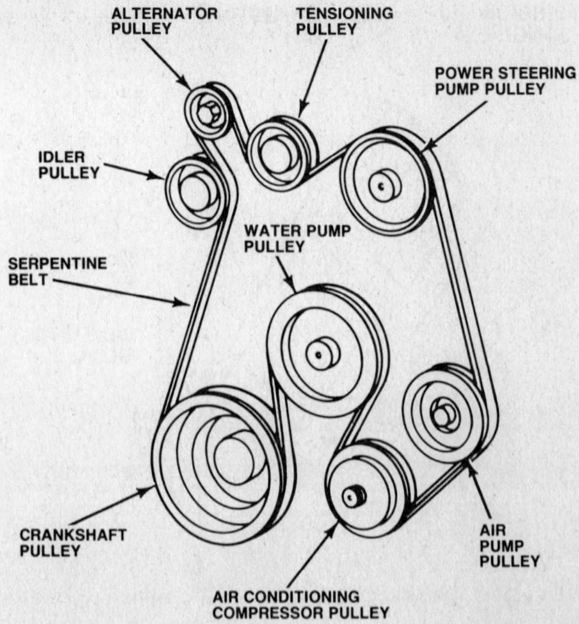

Figure 6-21. This serpentine belt system uses an automatic belt tensioner. After installing the belt, release the tensioning pulley. Spring tension on the pulley puts the required amount of force on the belt to ensure correct tension.

Removing and Replacing the Drive Belt

Drive belts are fairly inexpensive and easy to replace. However, in some vehicles, space in the engine compartment is limited, which makes removing and replacing belts a little more difficult. Use the following procedures to remove a V-belt or serpentine belt, inspect the pulley system for damage, and replace the belt.

Removing a V-belt

If you find damage when inspecting the belts, remove a V-belt as follows:

1. Disconnect the battery ground cable. Although this is not absolutely necessary, it is a safety precaution, especially if the car has an electric fan. You can also disconnect the electrical connector for the fan.
2. Loosen the bolts on the idler pulley or pivoting accessory bracket, or turn a draw bolt on the accessory bracket, figure 6-22.
3. Ease the tension on the belt by pivoting the accessory or levering the idler pulley. If you use a pry bar to pivot the accessory, apply pressure against the bracket, not the accessory housing, as some housings are made of aluminum and will bend.
4. Slip the belt off the pulleys.

Figure 6-22. Before removing a belt, loosen the bolts on the idler pulley or bracket. Then pivot the accessory until the tension on the belt loosens.

Figure 6-23. Lever the self-tensioning pulley away from the belt, so that you can remove the belt from the pulleys.

Removing a serpentine belt
Before you remove a damaged serpentine belt, locate a diagram of how it is routed around the accessory pulleys, which you can find in the manufacturer's service manual. If you do not have access to the manual, sketch your own diagram. In some vehicles, routing a serpentine belt can be confusing if you do not know the belt path.

1. Disconnect the battery ground cable. Although this is not absolutely necessary, it is a safety precaution, especially if the car has an electric fan.
2. Most vehicles with serpentine belts use a self-tensioning pulley. Loosen belt tension by levering the pulley away from the belt, figure 6-23.
3. Slip the belt off the pulleys.

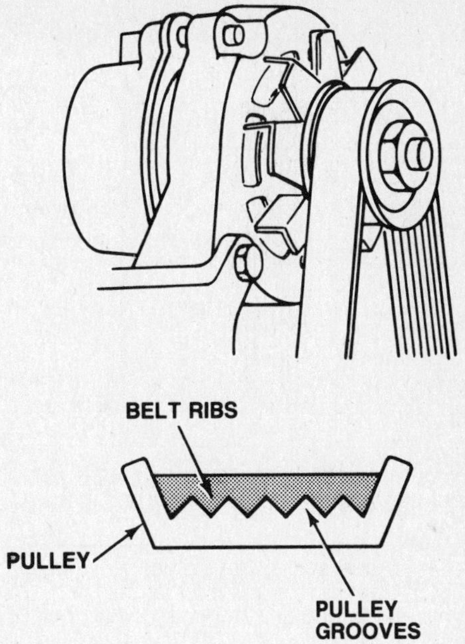

Figure 6-24. On a serpentine belt system, the pulleys must be free of dirt and grease, which can become embedded in the pulley grooves. The belt ribs must ride in these grooves.

Checking for causes of belt wear
Inspect the pulleys for defects that may contribute to belt wear:

- Turn the pulleys with your hands, listening for noise, which may indicate worn bearings. Replace bearings as needed.
- Check and repair oil leaks that may be contaminating belts.
- Check the pulleys for dirt and grease embedded in the groves. This is especially important for serpentine belt systems where belt ribs ride in the grooves, figure 6-24. Use a wire brush to clean the pulleys if needed.
- Check the pulleys for chips, which will damage the belt, and replace the pulley if needed.
- Check pulley alignment. The pulleys should not appear out-of-round, and they should mount squarely on the drive shafts.

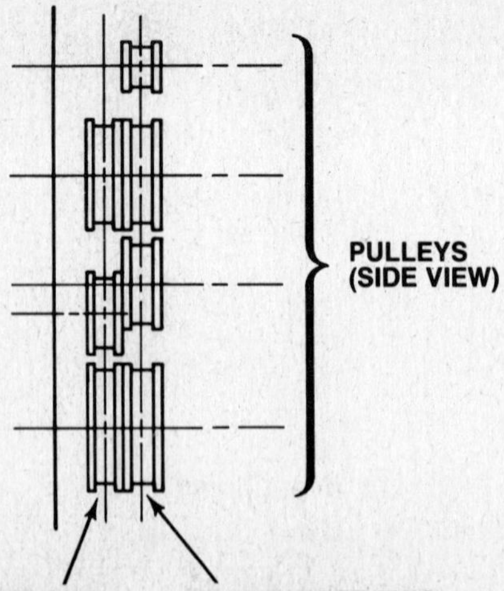

Figure 6-25. Accessory pulleys must be installed correctly on their drive shafts so that they line up with each other. If the pulleys are uneven, the drive belts may twist.

The pulleys should be lined up, figure 6-25, so that the belts lay flat as they run along the pulleys.

Replacing drive belts
Use the correct size and type of belt, as recommended by the vehicle manufacturer, when replacing a damaged drive belt.

1. Route the belt over the accessory pulleys. If you need help with belt routing, use the manufacturer's diagram or the sketch that you made before you removed the belt.
2. If the belt system uses a self-tensioning pulley, lever the pulley forward. Route the belt around the pulleys, placing the belt on the tensioning pulley last. Then release the self-tensioning pulley.
3. If the belt system uses an idler pulley or pivoting accessory bracket, loosen the securing bolts, and move the idler pulley or accessory bracket forward so you can slip the belt over the pulleys. Then lever the idler pulley or accessory bracket to tension the belt. Remember that you must not lever against accessory housings. Tighten the securing bolts.
4. Reconnect the battery ground cable if you disconnected it earlier.
5. Check and adjust belt tension.

POWER STEERING PRESSURE TESTS

If fluid level and belt tension are not causing the power steering problem, pressure test the hydraulic system. Power steering pressure tests help pinpoint leaks or blockage in defective steering system components.

There are three pressure tests. Begin by performing the system pressure test, which gives you information about the overall condition of the hydraulic system. Then test the pump and steering gear individually to check the cause of pressure drops or fluctuations detected during the system test.

System Pressure Test

The system pressure test checks the overall condition of the power steering hydraulic system.

1. Place a drain pan under the power steering pump and steering gear to catch fluid that may drip while you are connecting the pressure gauge.
2. Check the power steering fluid reservoir. If it is not full, top it up with fluid before conducting the test.
3. Disconnect the high-pressure fluid hose at the power steering pump using the proper size flare nut wrench.
4. Install the pressure gauge in the fluid line that runs between the pump outlet and the steering gear inlet, figure 6-26. Be sure to position the pressure gauge near the pump and the shutoff valve near the steering gear.
5. Tighten the pressure gauge fittings. To avoid distorting or rounding-off the fittings, use a flare nut wrench, figure 6-27.
6. Open the valve on the pressure gauge completely by turning it counterclockwise.
7. Start the engine, and run it to normal operating temperature. When the engine is at normal operating temperature, the power steering fluid will be about 120° to 180°F (50° to 80°C), which is the temperature at which you should test the system. Use a thermometer to check the temperature of the fluid, and check the fluid level again before you begin the test.
8. Read the pressure on the gauge, and compare it to the manufacturer's specifications. System pressures can range between 80 and 200 psi (550 to 1380 kPa). Some manufacturers give system pressure specifications at a specific engine RPM, so check the specifications carefully:
   - If system pressure is above specifications, check the hoses and connections for blockage. Then pressure test the system again.

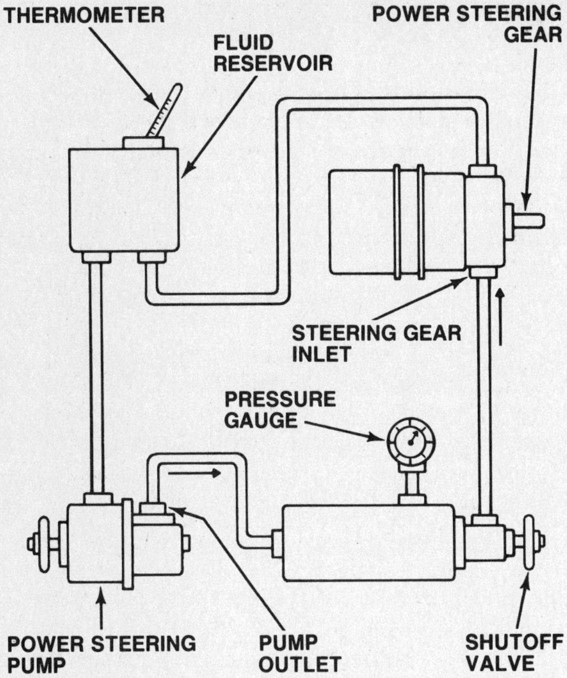

Figure 6-26. Install a pressure gauge with a shutoff valve between the power steering gear and pump. The shutoff valve on the pressure gauge should be toward the steering gear, as shown.

- If system pressure is below specifications, the pump or steering gear may be leaking. Pressure test them to help find the source of the leak.
- If system pressure is within specifications, then the hydraulic system is likely in good condition. Many manufacturers suggest that you check pump and steering gear pressures as a precautionary measure. Testing pump pressure will ensure that the system is operating properly, and that the flow control valve in the pump is not compensating for pressure problems elsewhere in the hydraulic system.

Pump Pressure Test

The pump pressure test checks flow control valve operation and maximum pump output. While conducting this test, do not close the valve on the pressure gauge for more than five seconds. Pressure build-up can damage valves and seals in the pump.

1. Connect the pressure gauge in the fluid line that runs between the pump outlet and the steering gear inlet, as described in the system pressure test procedure.

Figure 6-27. Use a flare nut wrench to remove and install high-pressure fittings. A regular open-end wrench will strip flare fittings.

2. With the engine running, open and close the shutoff valve several times. Do not keep the valve closed for more than five seconds.
3. Record the pressure each time you close the valve.
4. Compare the recorded pressure readings with the manufacturer's specifications:
   - If pressures are within the manufacturer's specifications and the difference between pressures is less than 50 psi (350 kPa), the pump is operating normally. Conduct the steering gear pressure test.
   - If pressures are high, or if they are within the manufacturer recommended specifications but the difference between pressures is more than 50 psi (350 kPa), the flow control valve is faulty. Remove the pump. Replace the flow control valve, or replace the entire pump.
   - If pressures are lower than the manufacturer's recommended specifications, but are within 50 psi (350 kPa) of each other, the pump is faulty. Remove the pump and replace seals, as needed, or replace the entire pump.

Steering Gear Pressure Test

Conduct the steering gear pressure test if system and pump tests are within specifications:

1. Connect the pressure gauge in the fluid line that runs between the pump outlet

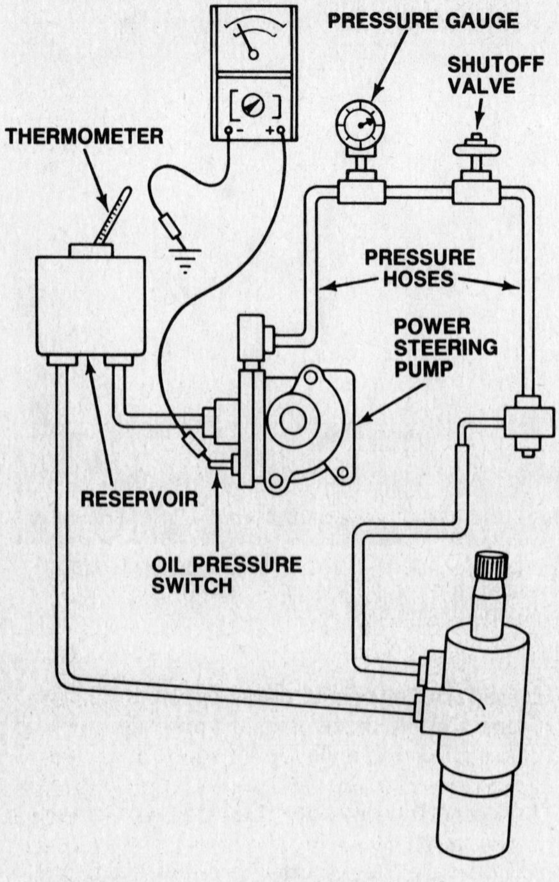

Figure 6-28. Check oil pressure switch continuity with an ohmmeter when system pressure reaches the manufacturer's recommended specification.

and the steering gear inlet, as described in the system pressure test procedure.

2. With the engine running, turn the steering wheel to the full left lock position and full right lock position. Do not leave the steering wheel in the lock position for more than a few seconds. Check and record pressure at each position.
3. Compare these readings with the manufacturer's recommended specifications:
   - Compare the highest steering gear pressure with the pressure reading obtained during the pump pressure test. If these readings are nearly the same, the steering gear is operating normally.
   - If gear pressure is lower when the steering wheel is in the lock position, then the steering gear is leaking internally. Disassemble and repair the steering gear, or replace the gear using the procedures given later in this chapter.

Oil Pressure Switch Test

When hydraulic system pressure increases, such as when the steering wheel is turned against a stop, an extra burden is placed on the power steering pump because it has to operate against this excess pressure. This burden is passed along to the engine, which must turn the pump pulley.

In many vehicles with electronic engine controls, engine RPM will increase when pressure rises in the power steering hydraulic system. An oil pressure switch senses power steering system pressure and signals the engine computer to increase idle speed. This keeps the engine from stalling. If the vehicle that you are working on uses an oil pressure switch, check its operation with an ohmmeter as follows:

1. Use steps one through seven of the system pressure test procedure, described earlier in this chapter, to prepare for the oil pressure switch test.
2. Disconnect the oil pressure switch connector at the power steering pump. Connect one ohmmeter lead to the switch and the other lead to ground, figure 6-28.
3. Gradually close the valve on the pressure gauge, which increases pressure in the system.
4. When the ohmmeter indicates that the switch is activated (continuity), check the pressure on the gauge. Compare the pressure reading with the manufacturer's specification for the pressure at which the switch should activate.
5. Open the valve on the pressure gauge to decrease system pressure. When the ohmmeter indicates that the switch is deactivated (no continuity), check the pressure gauge again. Compare the pressure reading with the manufacturer's specifications for the pressure at which the switch should deactivate.
6. If the pressures do not meet manufacturer's specifications, replace the oil pressure switch and conduct the test again.

HOSES AND FITTINGS

Special reinforced hoses and flare fittings are used in the high-pressure fluid lines that connect the pump and steering gear. Hoses and fittings on the low-pressure return lines from the steering gear to the reservoir and pump are not reinforced, figure 6-29.

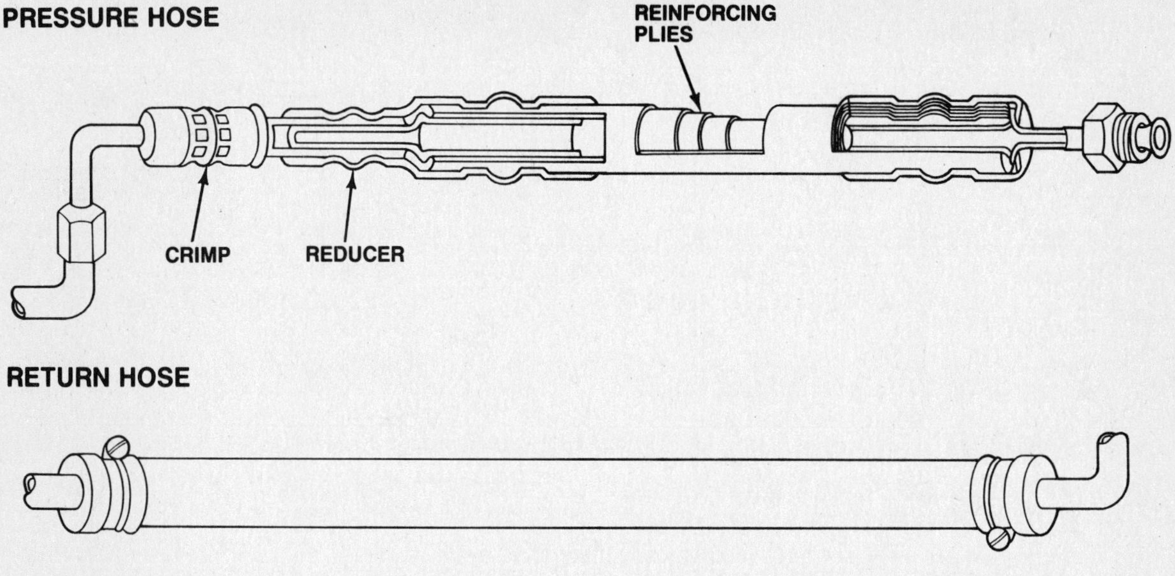

Figure 6-29. High-pressure power steering hoses are made of reinforced rubber and have special flare fittings. Low-pressure return hoses are not reinforced, and they use clamps instead of high-pressure fittings.

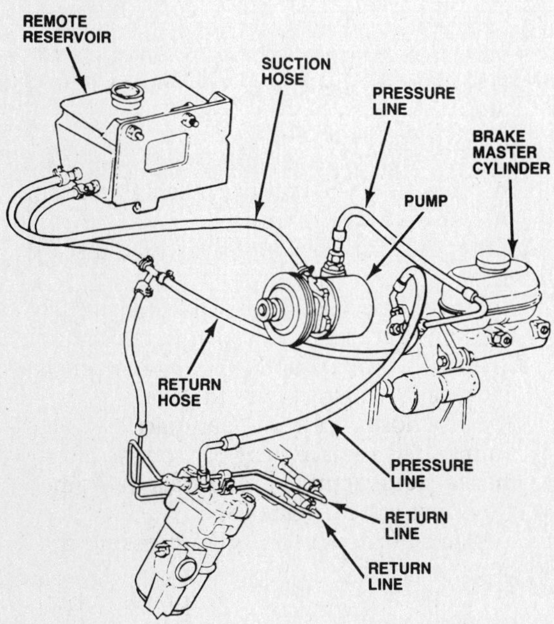

Figure 6-30. On this Ford power steering system, the power brakes use pressurized fluid from the power steering pump.

Periodically, you should inspect the power steering system hoses and fittings for leaks, using the following guide lines:

- Check the engine compartment for oil leaks. If power steering hoses are covered with oil, replace them. Oil will deteriorate rubber hoses.
- Check hoses for obvious external damage such as swelling, kinks, cracks, or abrasions. Check for leaks at the fittings. Replace any damaged hoses and fittings. If high-pressure fittings are damaged, replace the entire hose. If a high-pressure flare fitting is leaking, loosen the fitting, rotate it slightly, and retighten the fitting. Sometimes reseating the fitting will stop leaks.
- Replace brittle hoses that have been damaged by excess heat in the engine compartment. In FWD cars, components are packed tightly together, and damage to power steering hoses by high heat is common. Inspect these hoses carefully.
- When hoses deteriorate on the inside, they may feel spongy. Check hoses for sponginess by pinching them with your fingers. Replace damaged hoses. Remember that low-pressure return hoses are softer than high-pressure hoses because they are not reinforced.
- Check the power steering fluid for small bits of rubber, which also indicate that hoses are deteriorating internally. If you find rubber particles in the power steering fluid, replace the hoses and flush the system. Particles will clog fluid passages and make the pump work harder. This condition will eventually wear out the pump.
- On some vehicles, such as the 1990 Ford Econoline, the power brake and power steering systems are integrated, figure 6-30. When you check for leaks in these

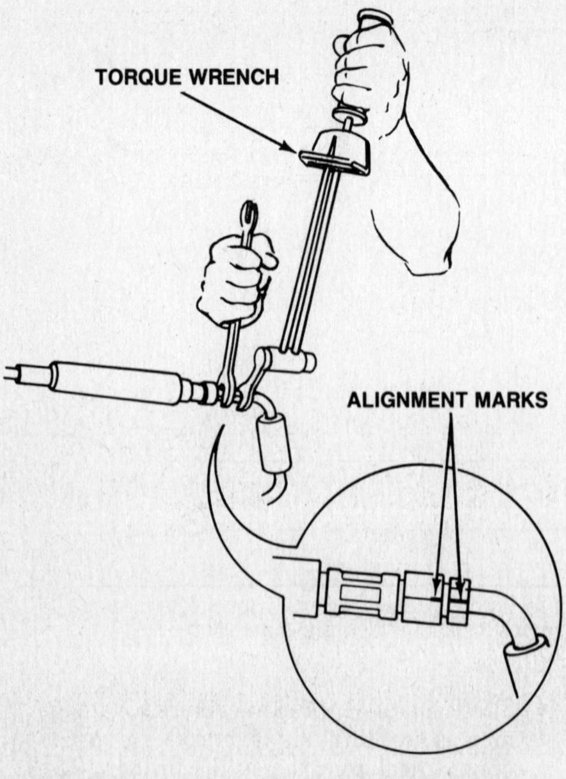

Figure 6-31. Align the marks on the high-pressure hose and pump outlet fitting before torquing the fitting in place.

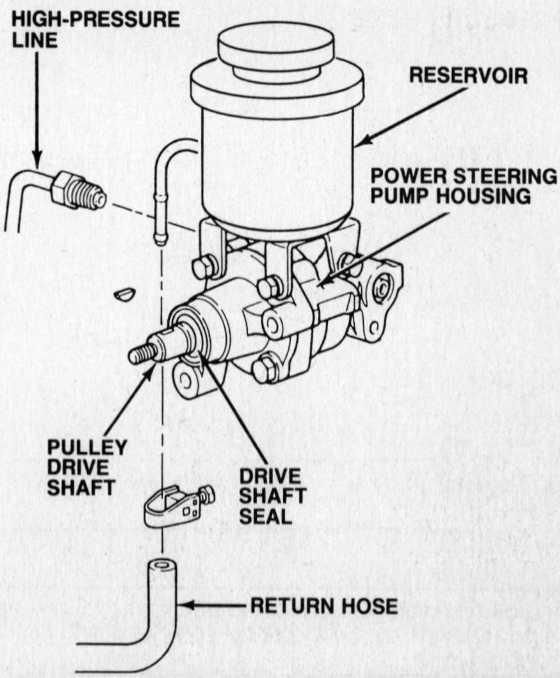

Figure 6-32. You do not have to remove the pump from the vehicle to inspect it for leaks. Simply check the pump housing and fittings for oil seepage.

power steering systems, check power brakes lines as well.

- Some systems have a power steering fluid cooler located in the return circuit of the hydraulic system. This cooler may look like a small radiator, or it may just be an extra length of hose located in a well-ventilated area of the engine compartment. Check the fluid cooler for leaks. If you remove the cooler, be sure to replace it with the proper part so the power steering fluid will not overheat.

When you replace hoses and fittings, be sure to use the parts recommended by the manufacturer. Never use vacuum hose, brake line, or other types of hose in place of power steering hoses. Replace hoses and fittings using the following procedure:

1. If the power steering system is integrated with the power brake system, loosen the brake reservoir cover to release pressure in the system.
2. Disconnect the power steering low-pressure hose at the pump or reservoir. Low-pressure hoses simply press or clamp in place, and you can use a screwdriver or pliers to remove them.
3. Catch the fluid in a container, and discard it.
4. Remove damaged hoses. Disconnect high-pressure flare fittings with a flare nut wrench. Do not use an open-end wrench because it will strip flare fittings.
5. Replace hoses with the manufacturer's recommended replacement part or the aftermarket equivalent. Route hoses the same way they were originally.
6. Replace O-rings on the high-pressure fittings, if required.
7. On some vehicles, the high-pressure hose fitting and the pump outlet fitting have alignment marks, figure 6-31. Align them before tightening the fitting. Tighten the fittings to the manufacturer's recommended torque.
8. If you suspect that power steering system components are contaminated with bits of hose, flush the system, as described earlier in this chapter. Then refill the system with power steering fluid. Start the engine, bleed the system, and check for leaks. Recheck the fluid level, and top up if needed.

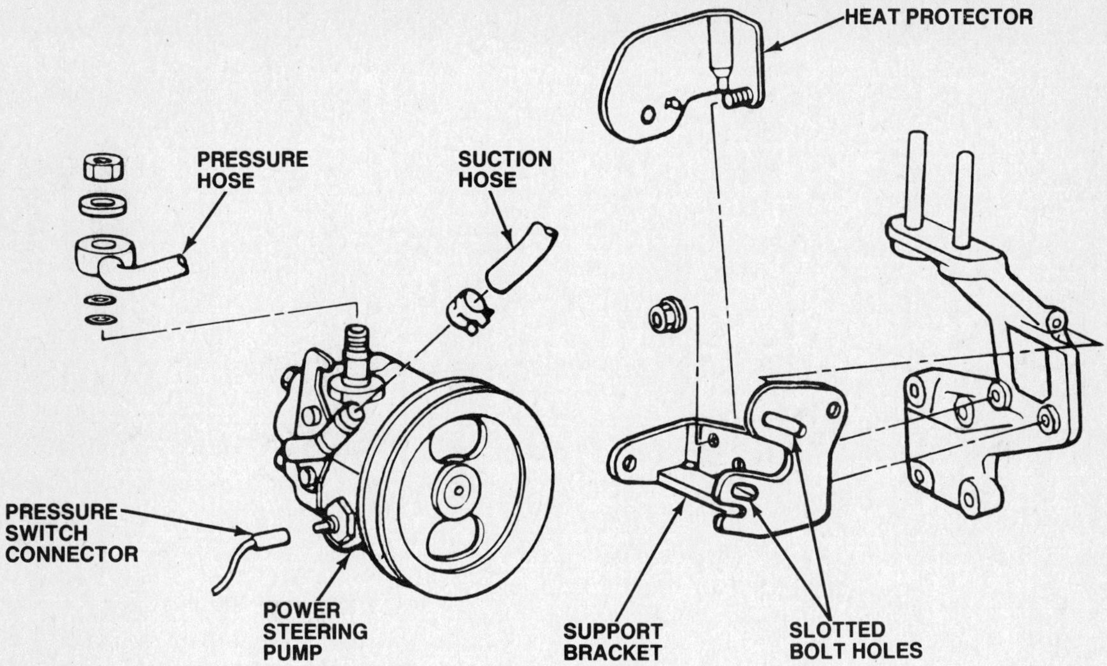

Figure 6-33. Metal brackets hold the power steering pump in place in some vehicles. If the mounts are bent or damaged, the pump pulley will not align with other accessory pulleys.

POWER STEERING PUMP

The power steering pump circulates high-pressure fluid to the steering gear. If the pump drive belt is broken or loose or if the fluid level is low, the pump will not work properly, if at all. Check drive belt tension and power steering fluid level as described earlier in this chapter before inspecting the pump for damage.

Inspecting the Pump, Mounts, and Pulley

Inspect the pump, mounts, and pulley for damage, and determine if they should be replaced. You can correct some power steering pump problems by replacing parts rather than the entire pump.

Pump inspection

Remove the pump drive belt as described earlier in this chapter. Then inspect the pump, figure 6-32, using the following notes as a guide:

- Check the pulley drive shaft bearings for wear. Grasp the pulley with your hands, and try to move it back and forth on the shaft. If you can move the pulley, the bearings are worn. Replace the pump.
- Check the pulley drive shaft seal. If you find fluid around the seal, replace the pump.
- If the fluid reservoir is attached to but is not integral with the pump, check for leaks between the pump and reservoir. Replace the O-ring between the pump and reservoir if needed. Check the reservoir cap for leaks as well.
- Check pump fittings for leaks. If you find leaks, and the hose fittings are intact, then the pump fittings are worn. Replace the pump.
- Check the pump housing for damage. Prying against the pump to adjust belt tension can damage the housing. If the pump is dented, it may leak at the seams, and must be replaced.

Mount inspection

When you adjust belt tension or remove the pump for service, check the mounting brackets, figure 6-33. If bolt holes in the mount are out-of-round or if the mounts are bent, the pump will be loose, which can affect belt tension and cause pulley alignment problems. If the mounts are damaged, replace them. On some vehicles, the power steering pump bracket is shared with the alternator, figure 6-34.

Pulley inspection

While you are servicing the power steering system, check the pump pulley for warps. Remove the belt, and check the pulley grooves

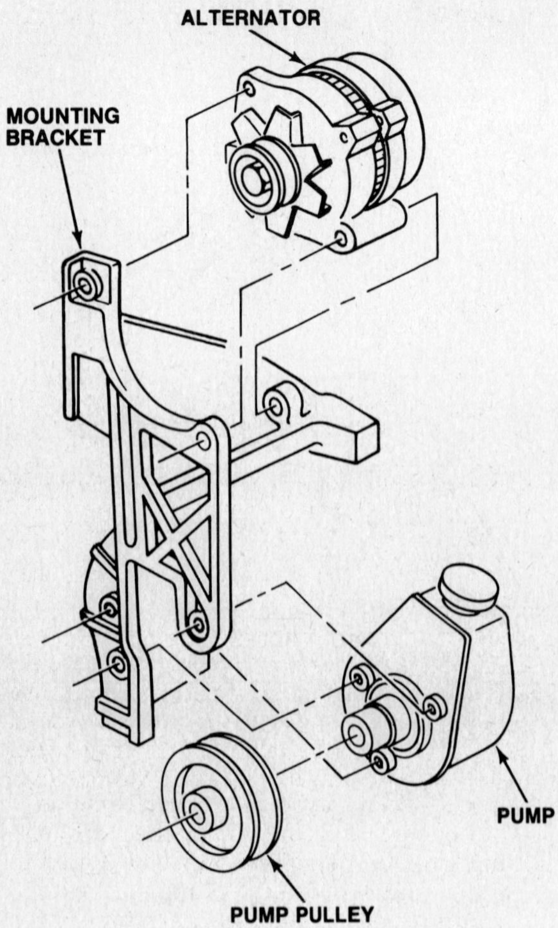

Figure 6-34. On some vehicles, the power steering pump will share a mounting bracket with another accessory. In this example, the alternator and power steering pump share a mounting bracket.

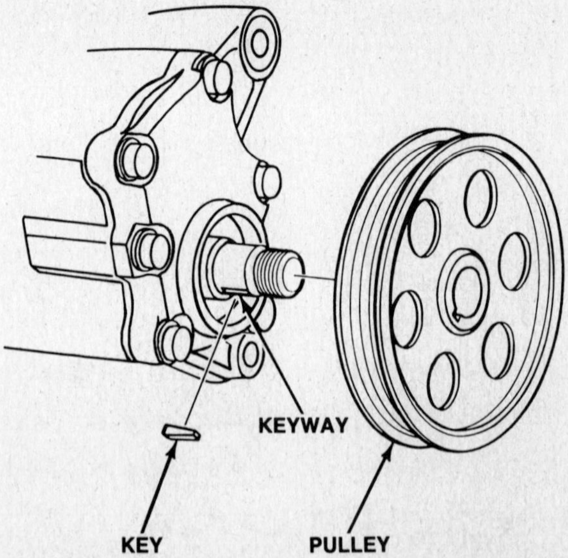

Figure 6-35. Some power steering pulleys are held to the pump drive shaft with a key that fits in a keyway. If the key or keyway are worn, the pulley will feel loose on the shaft.

for gouges or burrs, which will damage the belt. If the pulley is damaged, replace it.

Be sure that the pulley is secured to the drive shaft. For example, if the pulley is held in place with a key, figure 6-35, be sure that the keyway and key are not worn.

Removing and Replacing the Pump

Most shops will replace a power steering pump, rather than rebuild it, if it leaks fluid or is otherwise not operating properly. Remove and replace a pump as follows:

1. Disconnect the battery ground cable before working near belts and pulleys.
2. Clean dirt from the pump high-pressure and return hose fittings so that when you disconnect the fittings, dirt will not enter the system.
3. Use plastic to cover engine components that are below the pump so that when you remove hoses, power steering fluid does not get all over the engine.
4. Remove the low-pressure fluid return hose. Allow fluid to drain into a container, and discard the fluid according to local regulations.
5. Plug hose ends or cover them with plastic bags to keep dirt from entering the system.
6. Remove the pump drive belt as described earlier in this chapter.
7. Remove the bolts that secure the pump to the vehicle, figure 6-36. Be sure to support the pump while you are removing the bolts. Then lift the pump from the engine compartment.
8. Remove the pulley from the old pump. Because many new power steering pumps do not come with a pulley, use the old pulley if it is in good condition. See the procedure later in this chapter for removing and installing pulleys.
9. Replace pump mounts at this time if they are damaged.
10. Place the new pump in position in the engine compartment. Install the bolts.
11. Route the drive belt around the pump pulley and other accessory pulleys. Use the procedures described earlier in this chapter to install the belt and adjust tension.

Figure 6-36. Support the power steering pump while you remove the bolts securing the pump to the bracket.

12. Remove the plastic bags or plugs from the hoses and fittings on the pump. Connect the high-pressure and return hoses as described earlier in this chapter. Tighten fittings to the manufacturer's recommended torque specifications.
13. Fill the power steering fluid reservoir, bleed the system, and then top up the reservoir as described earlier in this chapter. Check the fittings for leaks after driving the vehicle.

Removing and Replacing the Mounts and Pulley

If the pump mounts or pulley are damaged, replace them to ensure proper pump operation. Refer to the previous procedure to remove and install the pump.

Pump mount replacement
Inspect the pump mounts as described earlier in this chapter, and replace any that are damaged.

1. Remove the power steering pump from the vehicle, using the previous procedure as a guide.
2. Loosen and remove the bolts that hold the mounts in place, figures 6-33 and 6-34.
3. Discard the old mounts, and replace them with new ones. On some vehicles, the power steering pump and alternator share a mounting bracket. Remove the alternator, if needed, to replace the power steering pump mount.
4. Install the power steering pump, using the previous procedure as a guide.

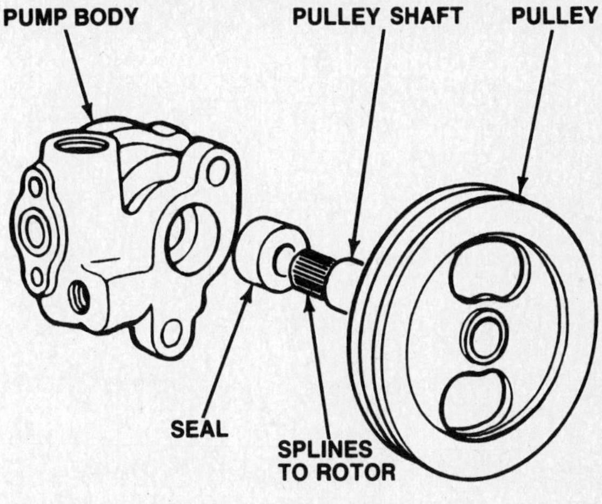

Figure 6-37. On some steering pumps, the pulley and drive shaft are a single unit. If the pulley is damaged, you must replace the assembly.

Pump pulley replacement
Often, new power steering pumps do not include a pulley, so you must remove the pulley from the old pump and put it on the new one. If the pulley is damaged, purchase a new one.

On some vehicles, the pulley and drive shaft are a single unit, figure 6-37. If the pulley is damaged, you must replace it during a pump rebuild, as described later in this chapter. If the pulley and shaft are not a single component, remove and replace the pulley as follows:

1. Remove the power steering pump from the vehicle as described earlier in this chapter.
2. On some pumps, you do not have to remove the mounting brackets for a rebuild. Check the pump, and remove the mounting brackets, if needed.
3. Clamp the pump in a soft-jaw vise. If the bracket is still attached to the pump, clamp the bracket in the vise to avoid damaging the pump.
4. Remove the nut that secures the pulley on the pump drive shaft. Hold the pulley while you remove the nut, figure 6-38. Some manufacturers have a tool that bolts to the pulley so that you can hold the pulley still while loosening the nut. On some pumps, the securing nut may have left-hand threads. Other pumps are not secured with a bolt, but are simply pressed onto the shaft. For these pumps, you need a special puller and installer to service the pulley.

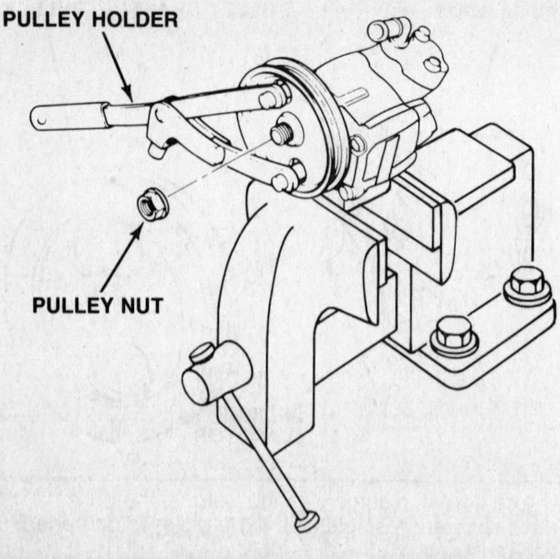

Figure 6-38. The pump pulley may be press-fit in place, or it may be held on the driveshaft with a nut. Use a pulley holder to keep the pulley from turning while you remove the nut.

5. Use a special pulley puller or a three-jaw puller to remove the pulley, figure 6-39. Never use a hammer or an impact wrench, as you can damage the drive shaft or internal pump components.
6. Remove the key, if there is one.
7. Clean dirt or rust from the pulley shaft.
8. Install a new pulley on the pump drive shaft, replacing the key, if there is one. Many manufacturers recommend that you use a special installer so that you do not damage the pump, figure 6-39.
9. Replace the nut on the end of the drive shaft. Torque the nut to the manufacturer's recommendations.

Rebuilding the Pump

Most shops replace the entire pump with a new one if it is damaged or leaking. Replacing pump internal components, such as the rotor and cam ring or drive gears and plungers, is time consuming, and the repair is not always successful. However, you can replace some seals and gaskets to stop leakage without much difficulty. Some manufacturers recommend removing and cleaning the flow control or pressure relief valves as part of a pump rebuild.

Flow control and pressure relief valves keep pump pressure from exceeding a maximum setting. If these valves malfunction, pressure will build up in the pump and damage seals.

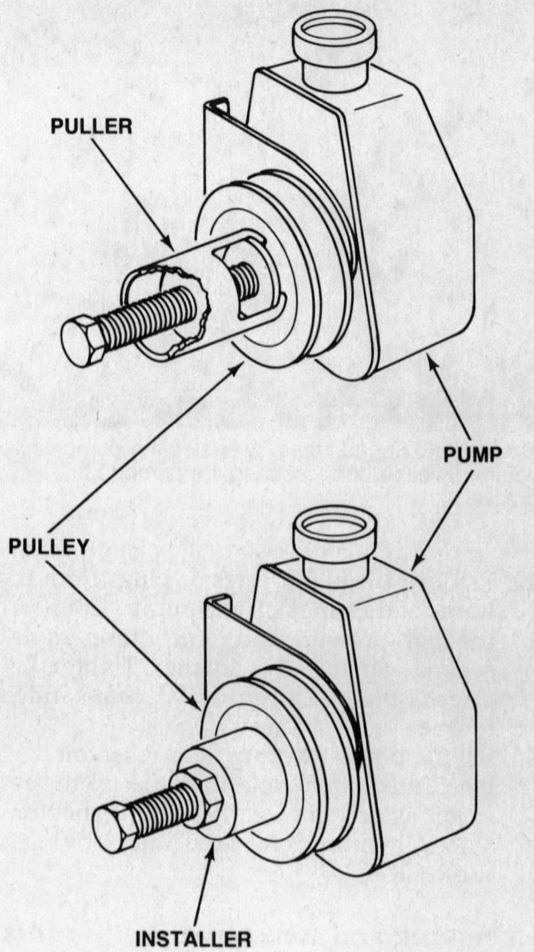

Figure 6-39. To avoid damaging the pump on some vehicles, you must use special puller and installer tools when changing the pulley. Never use a hammer.

On vehicles with electronic engine controls, a power steering pressure switch signals the on-board computer to increase engine speed when needed. This keeps the engine from stalling when the power steering pump must load the engine during heavy power assist.

Because there are slight variations in pump design, use the vehicle manufacturer's recommended procedure to rebuild a power steering pump. However, use the following procedure as a guide:

1. Remove the power steering pump from the engine compartment, and remove the pump pulley, as described earlier in this chapter.
2. Check the drive shaft bearing for wear. For many vehicles, if the bearing is worn, you must replace the entire pump. For others, the bearing is mounted on the drive shaft,

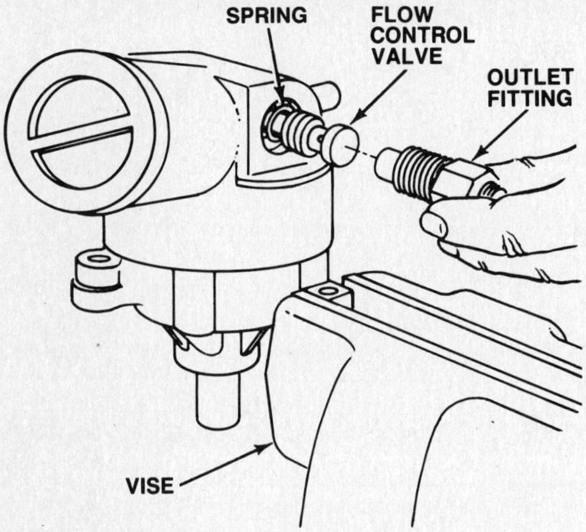

Figure 6-40. Unscrew the pump outlet fitting, and remove the flow control valve and spring that are inside the port.

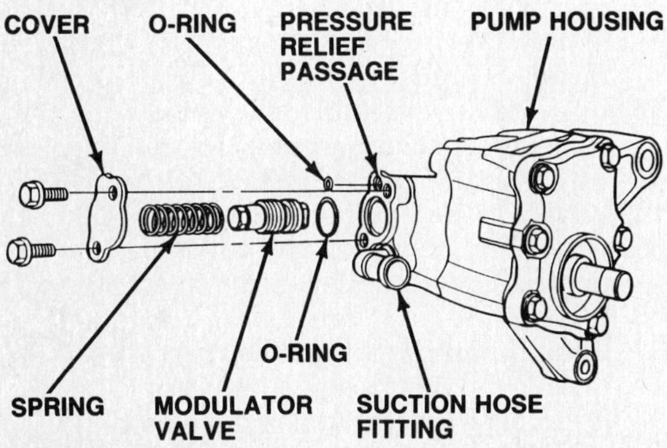

Figure 6-41. On this power steering pump, you must remove a cover plate on the pump body to withdraw the control valve and spring from the pump housing.

and you can remove and replace the drive shaft. On a few vehicles, you can use a puller to remove the bearing.

3. Clamp the pump in a soft-jaw vise. Do not bend the pump housing.
4. Remove the high-pressure outlet fitting from the pump body by unscrewing it with a wrench, figure 6-40. Remove any O-rings from under the outlet fitting with a screwdriver.
5. Use a magnet to lift the flow control valve and spring from the pump housing port. In some vehicles, you must remove a cover plate on the pump housing to expose the control valve, figure 6-41. Then remove the control valve along with any O-rings and springs.

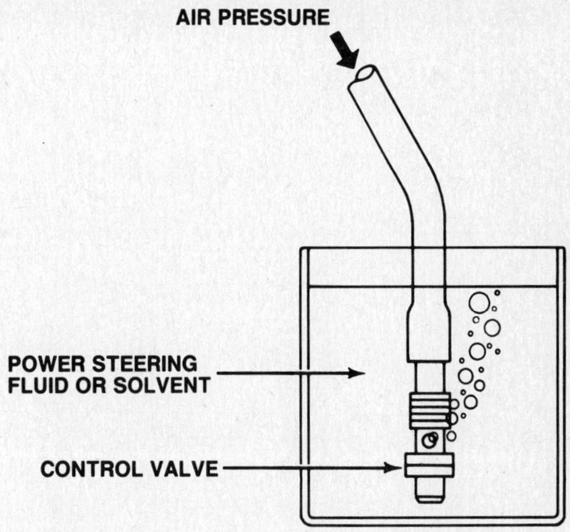

Figure 6-42. Submerge the control valve in solvent or power steering fluid, and blow air through the valve to remove debris. Check the manufacturer's specifications for the maximum pressure allowed through the valve.

6. Clean the valve in solvent to remove debris that may be clogging the valve opening. Some manufacturers recommend blowing air through the control valve to remove debris, figure 6-42. Set the valve aside, and continue disassembling the pump. If the valve is defective, discard it and replace it with a new one.
7. If the fluid reservoir is attached to the pump but is removable, loosen the bolts that hold it in place, if there are any. Then remove the reservoir by twisting and pulling up. Remove and discard the O-ring that seals the reservoir and pump mating surfaces.
8. Remove the pump end cover by prying out the retaining ring that secures the cover to the pump, figure 6-43. Other pumps use bolts to secure the front cover. Remove the bolts, front cover, seals, and dowel pins, figure 6-44.
9. Remove the shaft seal from the pump housing or housing cover by prying it out with a screwdriver, figure 6-45.
10. Press a new shaft seal into the pump housing or pump cover using a ball peen hammer and seal installer. Set the cover aside until later.
11. Remove the rotating assembly, which may include the rotor, cam ring, and slippers, figure 6-46, or gears, figure 6-47, depending on the type of pump you are working

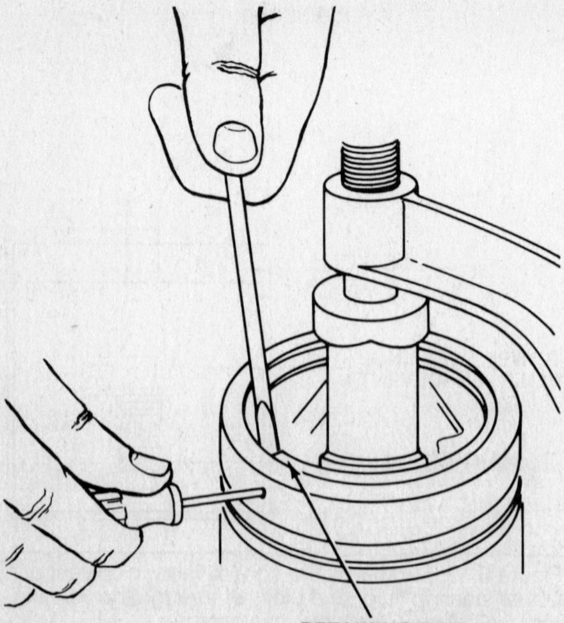

Figure 6-43. On this power steering pump, remove the retaining ring from the pump cover. Then loosen and remove the cover.

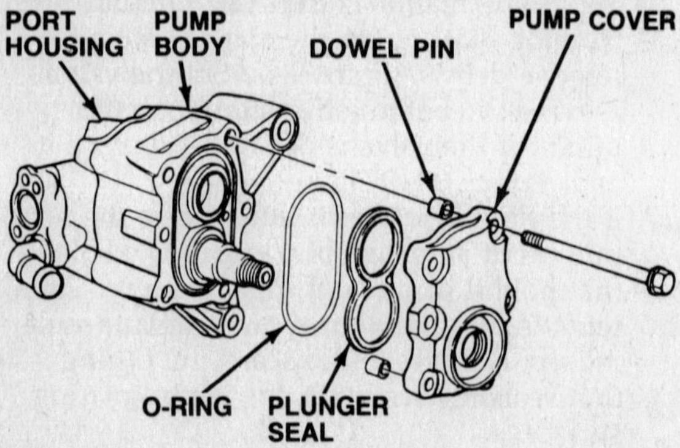

Figure 6-44. This power steering pump cover simply bolts to the housing. After removing the cover bolts, tap the pump cover with a soft-face hammer to loosen it.

on. You may have to use a soft-face hammer to loosen the rotating assembly from the pump housing.

12. Remove and replace the shaft seal if you did not do so earlier.
13. Clean parts with solvent and air dry them. Do not use solvent on O-rings or rubber seals.
14. Inspect internal parts for wear, and replace the pump, if needed. If the rotating assembly is in good condition, install it in the pump housing.

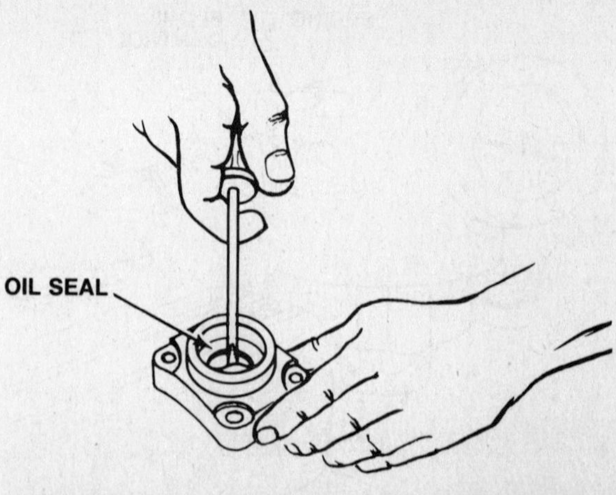

Figure 6-45. In most power steering pumps, the shaft oil seal can be removed with a screwdriver. Press a new one in place.

15. Lubricate the O-rings with power steering fluid, and replace the pump cover. If the cover bolts in place, install the bolts and tighten them to the manufacturer's recommended torque.
16. Coat the flow control valve and spring with power steering fluid. Coat the O-ring with steering grease. Install the control valve, spring, O-ring, and housing port cover. Torque the control valve cover bolts in place. Replace the pump outlet fitting if there is one.
17. If the pump has an attached reservoir, lubricate and install new O-rings to keep the reservoir from leaking, and install the reservoir.
18. Install the pump pulley, as described earlier in this chapter. Install the pump and drive belt in the vehicle. Adjust belt tension, add fluid, bleed the system, and top up the reservoir, if needed. Check for leaks after driving the vehicle.

POWER-ASSISTED STEERING GEARS

Power assisted steering gears are much like manual gears in both construction and operation. Both rack and pinion steering gears and standard steering gears, such as recirculating ball types, can have power assist.

Power steering gears have a hydraulic circuit that supplies pressurized fluid through a control valve, as described in Chapter Six of the *Classroom Manual*. When the driver is not turning the steering wheel, the control valve is centered by a centering mechanism (usually springs). When the driver turns the steering

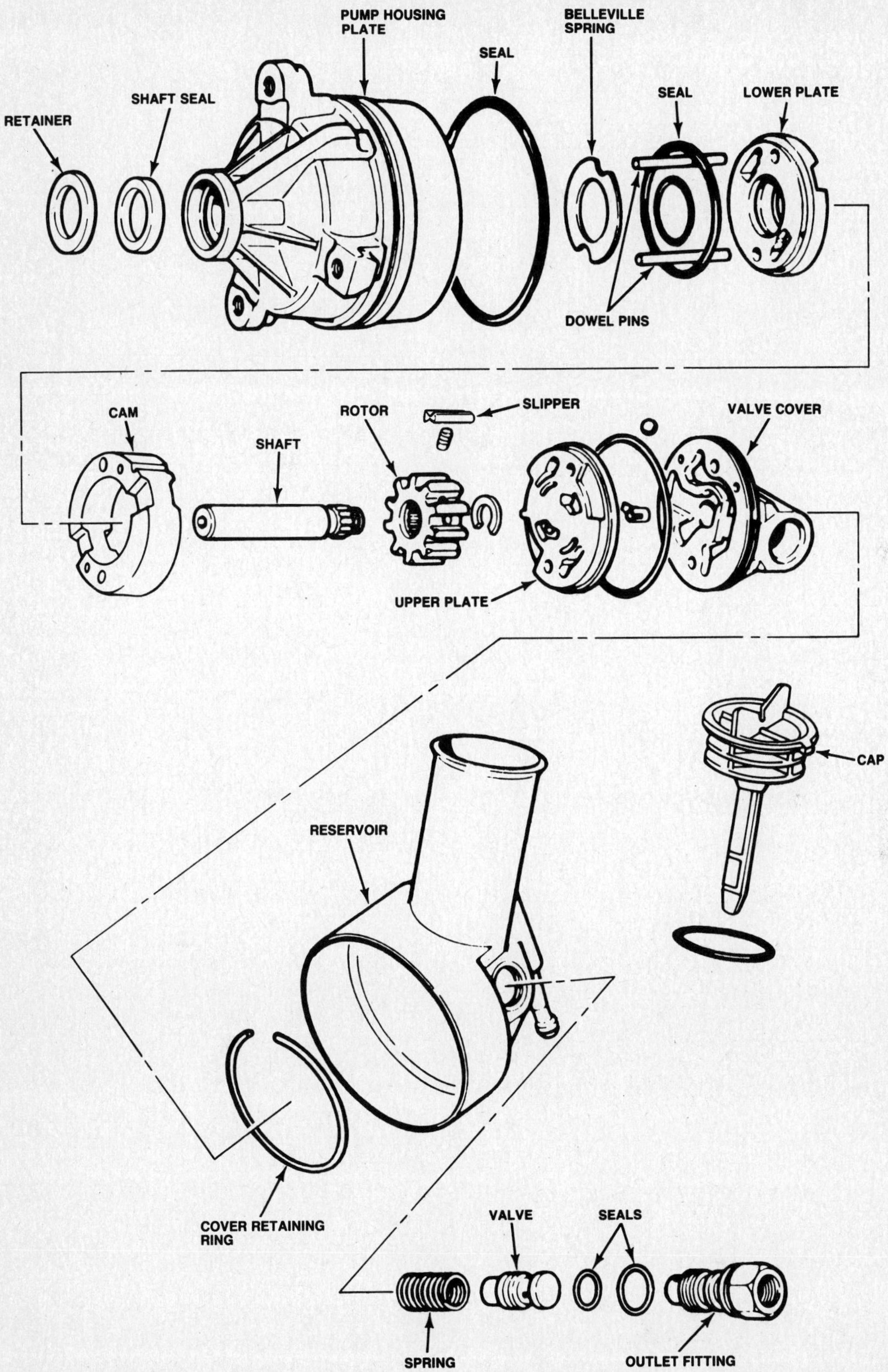

Figure 6-46. This Ford model CII pump is a slipper type pump. Replace seals and O-rings when rebuilding it.

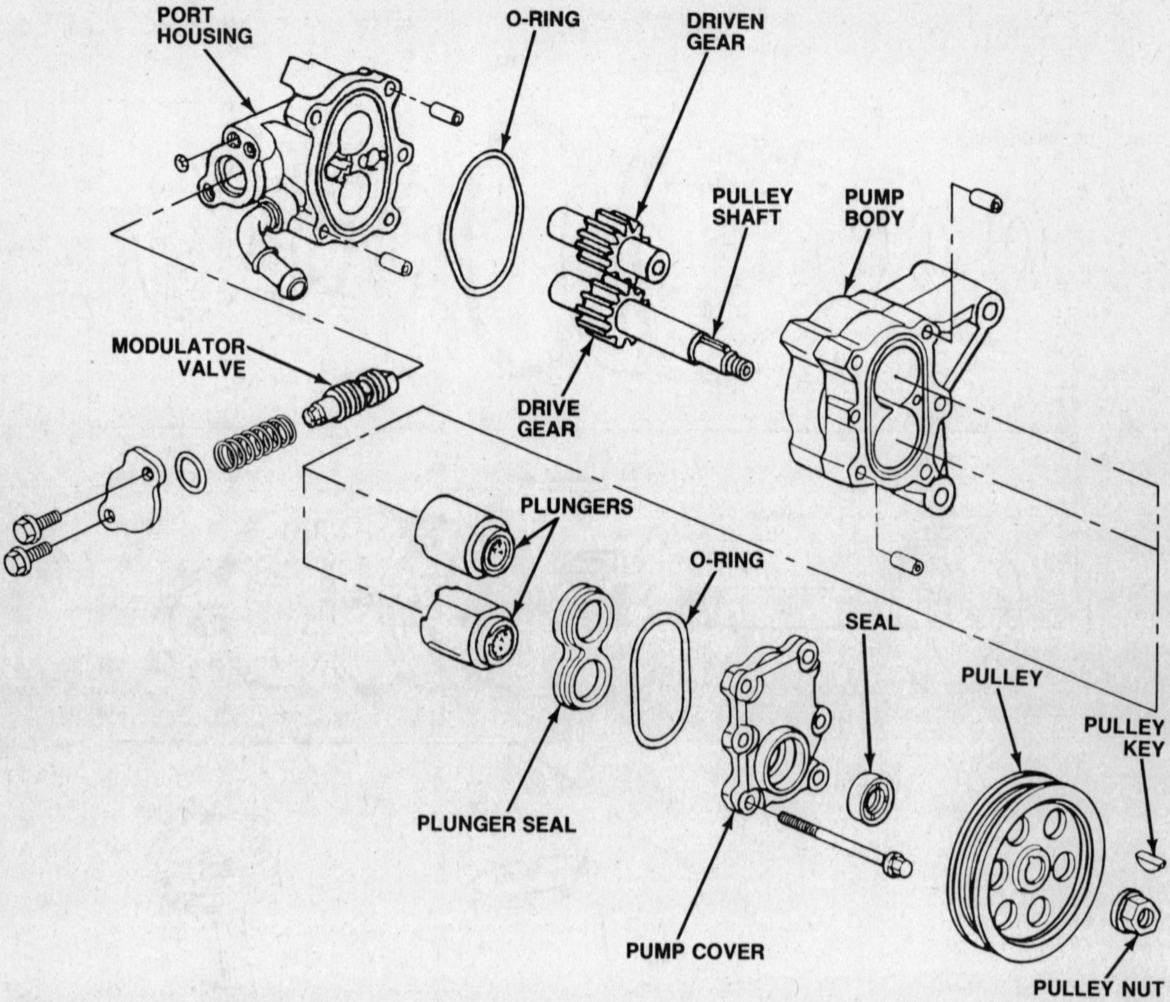

Figure 6-47. This Honda power steering pump, uses gears and plungers to increase pressure in the hydraulic system. When the plungers, gears, and housing wear, replace the pump.

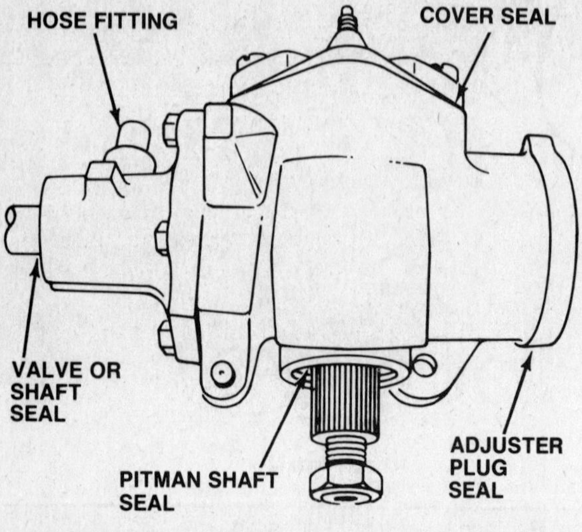

Figure 6-48. Check a standard power steering gear for leaks at the points indicated in the illustration.

wheel, the control valve guides fluid to the appropriate side of the power piston or cylinder to move the steering linkage.

The procedures for inspecting a steering gear vary depending on the type of steering gear used (standard or rack and pinion) and the type of power steering system used (linkage, integral, or variable assist).

Inspecting Power Steering Gears

Check power steering gears for leaks. For a standard steering gear, figure 6-48, inspect the following areas:

- The seal between the gear cover and the gear housing
- The adjuster plug seal or O-ring
- The pitman shaft seal
- The valve or shaft seal

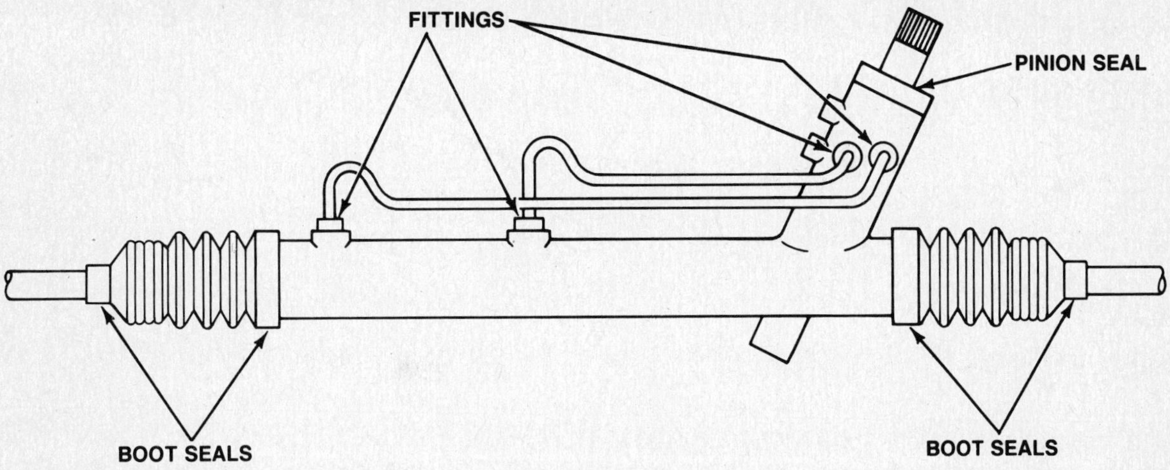

Figure 6-49. Check fittings, boot seals, and the pinion seal for seepage. Also check inside the boots, as leaks from the internal rack seals can cause fluid to accumulate in the boots.

- The hose fitting
- The gear housing (cracks).

For a rack and pinion steering gear, figure 6-49, inspect the following areas for leaks:

- The pinion seal
- The boot seals
- The hydraulic line fittings.

On rack and pinion steering gears, leaks may not be immediately obvious because power steering fluid can leak into the boots and be contained there without leaking on the ground. Check inside the boots for power steering fluid.

If you find fluid in the boots, you can determine if the inner or outer rack seal is leaking, figure 6-50. Remove the boots, and have an assistant turn the steering wheel while you check for leaks. If fluid leaks from the passenger side, the outer rack seal is leaking. If fluid leaks from the driver's side, the inner rack seal is leaking. You cannot determine the source of the leak by seeing which boot contains fluid because when air is transferred from boot to boot during turns, so is fluid.

If the power steering gear is leaking, replace it. Most repair shops do not rebuild power steering gears for a number of reasons. If a steering gear is leaking, then internal parts and the gear housing are likely worn. Also, rebuilding a power steering gear can be time consuming, rather complicated, and not always successful. Shop owners find that it is cost effective to replace the steering gear rather than spend the time rebuilding it.

Removing and Installing Power Steering Gears

Chapter Four of this *Shop Manual* contains the procedures for removing and replacing both standard and rack and pinion steering gears. Use this additional information to remove and replace power steering gears:

1. Some manufacturers suggest that you steam clean the gear housing, hoses, fittings, and fluid reservoir before removing the gear. Dry them with compressed air. This will prevent dirt and water from entering the hydraulic system when you disconnect hoses.
2. Position the lift supports under the vehicle chassis using the manufacturer's recommended lift points, and raise the vehicle. See Chapter One of this *Shop Manual* for more information on lifting a vehicle.
3. Place a drain pan under the gear to catch fluid. Disconnect the return hose at the steering gear, and secure the hose at a point higher than the fluid reservoir. Then disconnect and secure the high-pressure hose. If the power steering system has additional hoses that lead to a fluid cooler, disconnect and plug these hoses as well. Cover the hose ends with plastic.
4. Plug the fittings on the steering gear so that dirt will not enter.
5. Use the procedure in Chapter Four of this *Shop Manual* to remove and install the steering gear. After disconnecting hydraulic lines, procedures for removing and installing a power steering gear are the same as those for a manual steering gear.

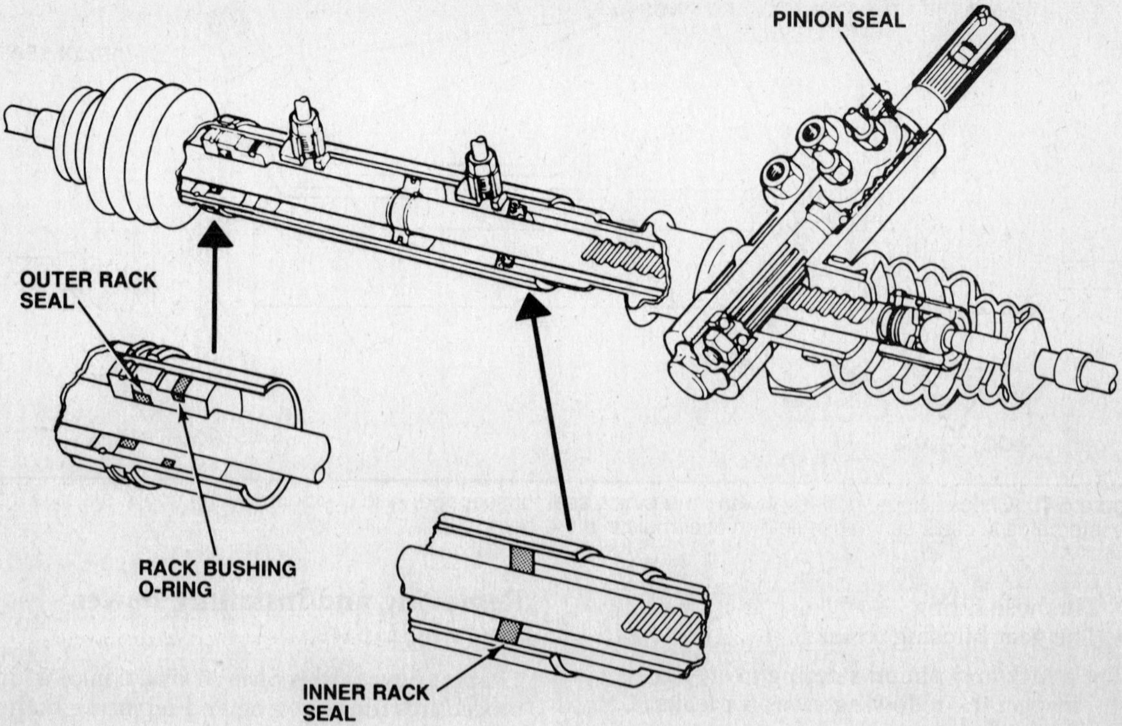

Figure 6-50. When fluid leaks past the inner or outer rack seals, it accumulates in the boots instead of leaking into the engine compartment.

6. Replace O-rings, if any, on the pressure and return hose fittings.
7. Connect the pressure and return hoses to the steering gear. Tighten the connections with a flare nut wrench.
8. Add power steering fluid to the reservoir and bleed the system as described earlier in this chapter. Some manufacturers recommend that you flush the system after installing new components. Drive the vehicle and check the gear for fluid leaks.

Steering Gear Adjustments

Adjustments for power steering gears are similar to those for manual steering gears. Power steering gears sometimes have additional adjustments to ensure that hydraulic fluid flows smoothly and evenly.

Standard steering gear
Adjustments for the standard power steering gear include worm bearing preload, gear mesh preload, and control valve centering. Not all standard power steering gears have all three adjustments, and adjustment procedures may differ slightly. Adjust worm bearing preload and gear mesh preload as described in Chapter Four of this *Shop Manual*. Then check the control valve to see if it is centered.

Checking and centering the control valve
If the control valve is not centered, the car may pull, lead, drift, and have poor returnability in one direction. Check that the control valve is centered by raising the vehicle on a lift with the front wheels positioned straight ahead. Then start the engine. If the front wheels turn when the engine starts, the control valve is not centered.

If the steering gear uses a rotary control valve, you cannot center the control valve. Instead, you must rebuild or replace the entire steering gear.

If the steering gear has a spool-type control valve, such as in the Chrysler Ram Van, figure 6-51, you can center the control valve by repositioning it in the valve bore.

1. Loosen the screws that secure the valve body to the steering gear.
2. Use a soft-face hammer to tap the end plug, which moves the valve body downward. To move the valve up, tap the screw that secures the valve body to the steering gear.
3. Check for valve centering again.
4. If the valve is centered, tighten the bolts that secure the valve body to the steering gear. If it is not centered, continue tapping

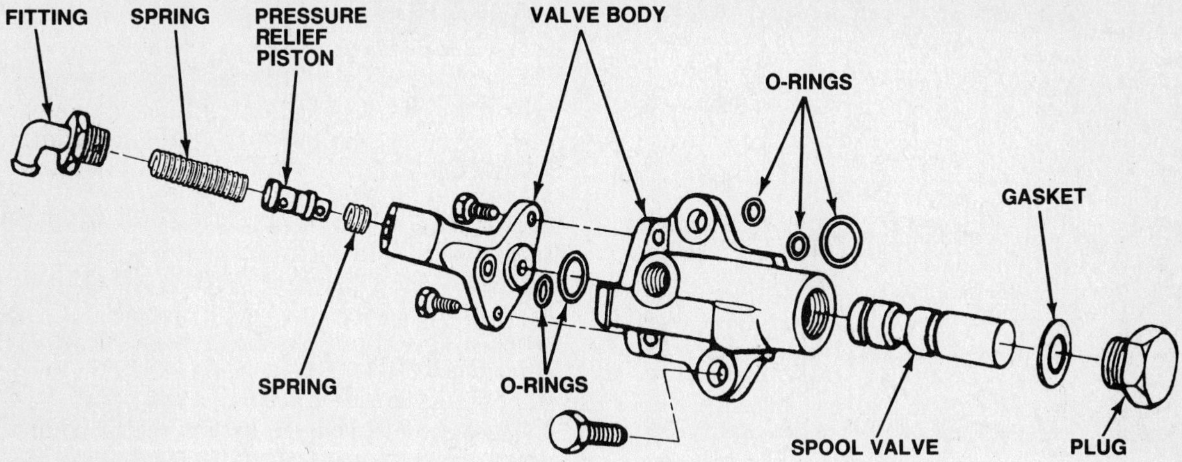

Figure 6-51. Reposition the valve body by tapping it with a soft-face hammer until the spool-type control valve is centered.

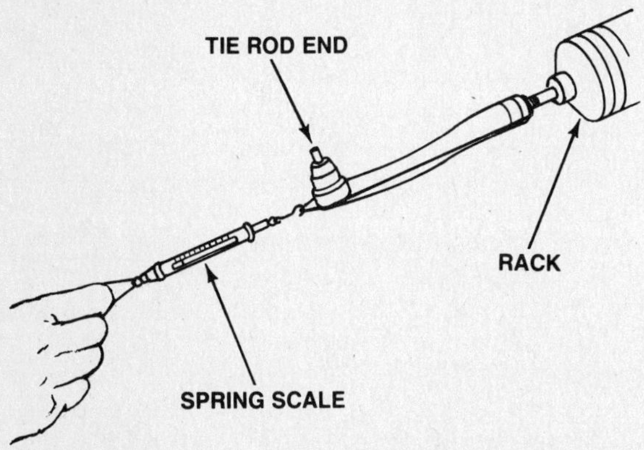

Figure 6-52. Use a spring scale to check the force required to slide the rack. If the force exceeds manufacturer's recommendations, readjust pinion torque.

the valve body until the control valve is properly positioned.

Rack and pinion steering gear
Adjustments for power rack and pinion gears are similar to those for manual gears. Use the procedures in Chapter Four of this *Shop Manual* to adjust pinion torque and pinion bearing preload, if adjustable. Then check rack sliding force:

1. Disconnect the tie rod ends from the steering knuckles, and disconnect the steering shaft from the pinion input shaft as described in Chapter Four of this *Shop Manual*.
2. Fill the fluid reservoir, and bleed the system.
3. Start the engine and run it to normal operating temperature. The power steering fluid should be about 122° to 176°F (50° to 80°C) before you begin testing the system. Check the fluid level again.
4. Center the steering rack as described in Chapter Four of this *Shop Manual*.
5. Pull one of the tie rods with a spring scale, figure 6-52. If the pulling force does not meet the manufacturer's recommendations, readjust pinion torque, as described in Chapter Four of this *Shop Manual*.
6. Recheck the rack sliding force with the spring scale. If it is still out of specifications, replace the steering gear.

Chapter

7

Four-Wheel-Steering System Service

Four-wheel steering improves a vehicle's handling because it turns the rear wheels in response to a variety of factors, such as steering wheel turning angle, steering velocity, and vehicle speed. Four-wheel steering is sometimes abbreviated "4WS" in manufacturer or aftermarket service literature.

Chapter Seven of the *Classroom Manual* describes several four-wheel-steering systems, figure 7-1, each differing in the type of input and output controls used to turn the rear wheels. The Honda system is mechanically controlled, the Mazda system uses a combination of mechanical, electronic, and hydraulic controls, and the Nissan system uses a combination of electronic and hydraulic controls, called Super HICAS (High Capacity Actively Controlled Suspension). Because four-wheel-steering systems are rather sophisticated, you should always refer to the manufacturer's service manual to diagnose and repair them.

Although there are not many vehicles with four-wheel steering, some are out on the road, so you should familiarize yourself with these systems. One thing that may help you understand four-wheel steering is that some inspection and service procedures are similar to those for two-wheel-steering systems.

In this chapter, you will learn methods for identifying a vehicle with four-wheel steering. You will also learn some of the basic service procedures for the rear steering gear, center or transfer shaft, power steering fluid pump, and rear steering linkage.

IDENTIFYING A FOUR-WHEEL-STEERING SYSTEM

While you are inspecting a vehicle's steering system, check for one or more of the following items, which may indicate that the vehicle has four-wheel steering, figure 7-1:

- A shaft that extends from the front steering gear to a rear steering gear
- A steering gear or power cylinder at the rear axle
- Stepper motors, hydraulic valves, and other electrical or hydraulic controls at the rear axle
- An over-size power steering pump, and hydraulic lines that lead to the rear of the vehicle
- Tie rods or other steering linkage that lead to rear wheel steering arms.

Use the following procedures as a guide to inspect, adjust, and replace some of the components in four-wheel-steering systems.

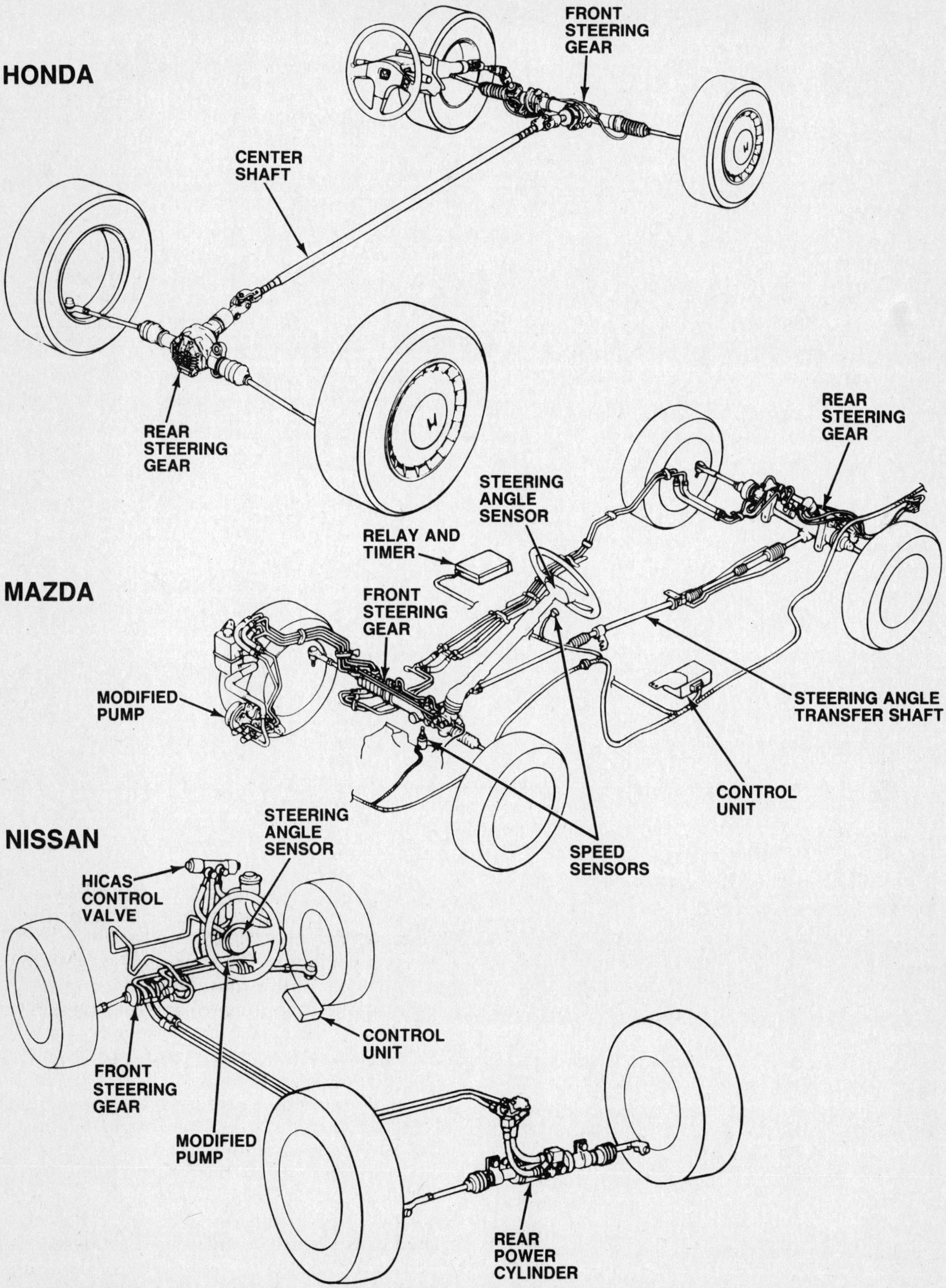

Figure 7-1. Honda, Mazda, and Nissan, as well as other companies, manufacture vehicles with four-wheel steering. These three examples represent the types of four-wheel-steering systems on the road today.

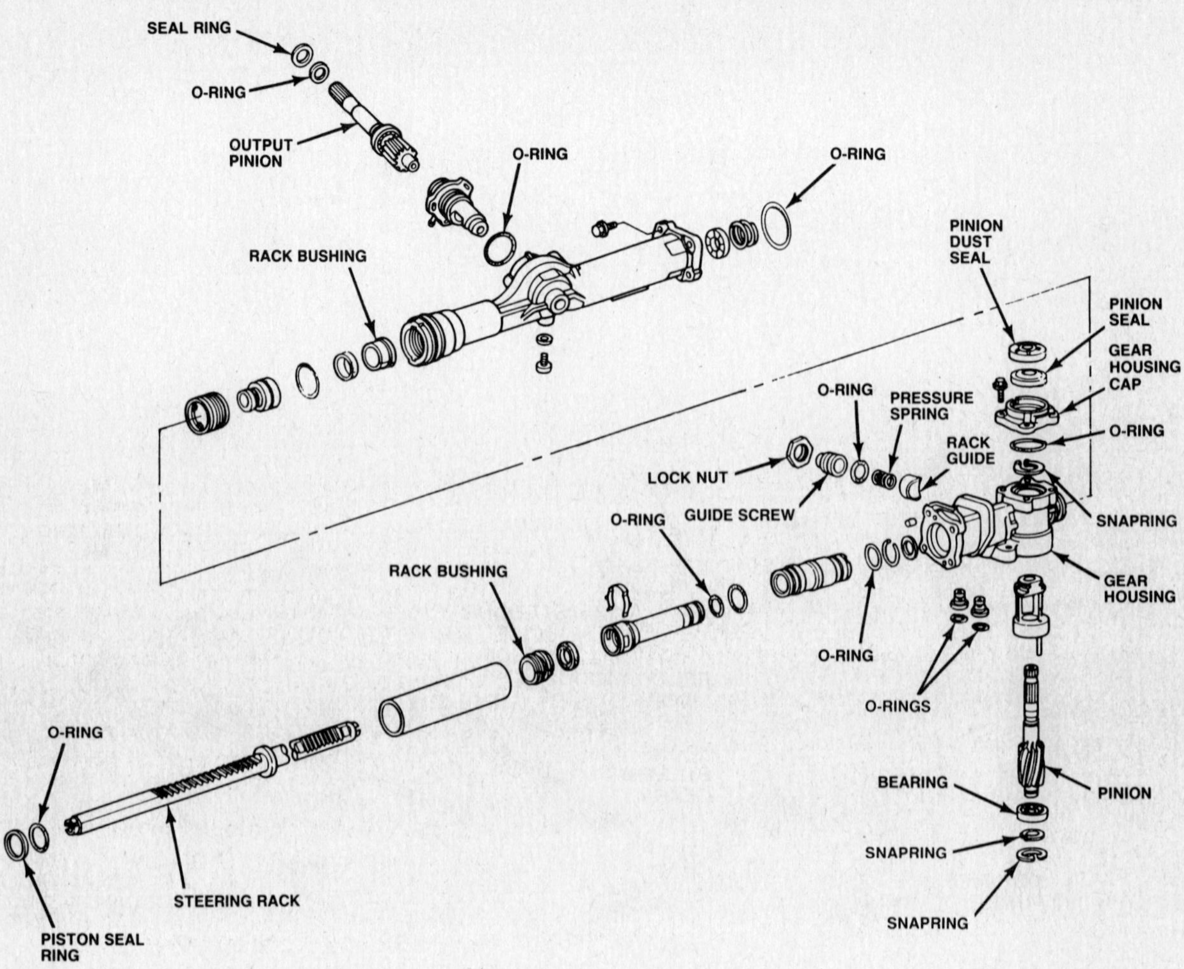

Figure 7-2. Service the front steering gear on a Honda four-wheel-steering system like any other rack and pinion steering gear. The only difference is that this gear has an extra output pinion that drives the center shaft.

REAR STEERING GEAR OR POWER CYLINDER

The Honda and Mazda four-wheel-steering systems use a steering gear to move the rear wheels, while the Nissan system uses a completely hydraulically actuated power cylinder.

Honda

Because a center shaft joins the front and rear steering gears, procedures for servicing the front steering gear on a Honda four-wheel-steering system are slightly different than those for a two-wheel steering system gear. Before servicing either the front or rear gears, you must remove or disconnect the center shaft, as described later in this chapter.

Servicing the front steering gear
The front steering gear, figure 7-2, is very much like the rack and pinion gear on a two-wheel-steering system, except that it has an additional output pinion shaft that turns the center shaft.

When you adjust the rack guide on the front steering gear of this four-wheel-steering system, use the procedure for a regular steering gear, as described in Chapter Four of this *Shop Manual*. Rack guide adjustment specifications for the four-wheel-steering system differ from those of the regular steering gear, so be sure to check the manufacturer's service manual for the correct specifications.

If you must remove the front steering gear, install the center lock pin in the rear steering gear, which will keep the rear gear centered. Then disconnect the center shaft, as described later in this chapter. Finally, remove the front steering gear, as described in Chapters Four and Six of this *Shop Manual*.

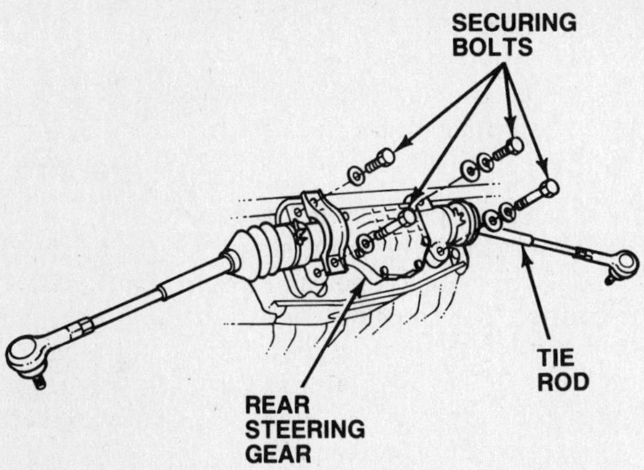

Figure 7-3. To remove the rear steering gear on this Honda, disconnect the tie rod ends from the steering arms, remove the bolt that secures the U-joint between the center shaft and rear steering gear, and remove the bolts that hold the gear to the vehicle.

Removing and replacing the rear steering gear

If you must remove the rear steering gear, figure 7-3, install the center lock pin and disconnect the center shaft, as described later in this section. Then use the following procedure to remove the rear steering gear:

1. Position the lift supports under the vehicle chassis using the manufacturer's recommended lift points, and raise the vehicle. See Chapter One of this *Shop Manual* for more information on lifting a vehicle. Remove the rear wheels.
2. Disconnect the tie rod ends from the rear steering arms using a pickle fork or one of the other methods described in Chapter Five of this *Shop Manual*.
3. After installing the center lock pin, disconnect the center shaft from the rear steering gear, as described later in this chapter.
4. Remove the bolts that secure the rear steering gear to the vehicle frame or crossmember, and remove the rear steering gear.
5. Install the rear steering gear in the reverse order. Be sure to center the front and rear steering gears before connecting the center shaft. The procedure for centering the rear steering gear is described later in this section. The procedure for centering the front steering gear is described in Chapter Four of this *Shop Manual*.
6. Check wheel alignment, as described in Chapter Fourteen of this *Shop Manual*.

Installing the center lock pin

The center lock pin holds the rear steering gear in place when you disconnect the center shaft

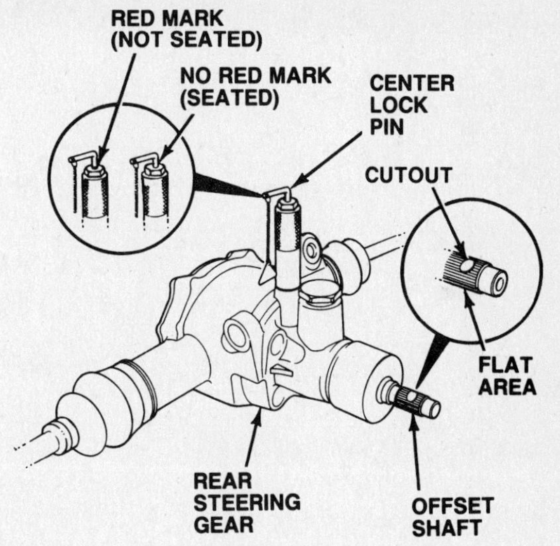

Figure 7-4. The center lock pin is a special manufacturer's tool that locks the rear steering gear in place when you remove the cap bolt and disconnect the center shaft. When the red mark is no longer visible, the center lock pin is seated.

or remove the rear steering gear from the vehicle, figure 7-4.

1. Center the steering wheel and front steering gear as described in Chapter Four of this *Shop Manual*.
2. Position the lift supports under the vehicle chassis using the manufacturer's recommended lift points, and raise the vehicle. See Chapter One of this *Shop Manual* for more information on lifting a vehicle.
3. Use a Torx bit to remove the cap bolt from the rear steering gear.
4. Install the center lock pin in the opening. Turn the steering wheel slightly to the right and left until the center lock pin is seated. It is seated when the red mark on the pin is no longer visible. You can now remove the rear steering gear, and it will be centered when you install it later. If you cannot seat the center lock pin, remove the rear steering gear from the vehicle and center the gear using the procedure in the next section.
5. When you install the rear steering gear, be sure to remove the center lock pin after reconnecting the center shaft. If you forget to do this, you will damage the gears.

Centering the rear steering gear

Use this procedure if you cannot seat the center lock pin, which indicates that the steering gear is not centered. You must remove the rear steering gear from the vehicle before centering it.

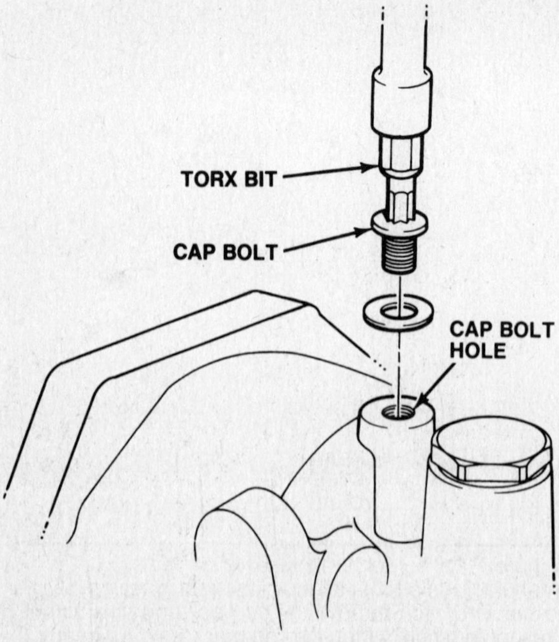

Figure 7-5. Look through the cap bolt hole in the rear steering gear, and center the notch on the offset shaft by turning the shaft.

1. Use a Torx bit to remove the cap bolt from the rear steering gear, figure 7-5.
2. While looking through the hole, turn the offset shaft until you can see the cutout in the shaft through the hole.
3. Then install and seat the center lock pin in the bolt hole, as described in the previous section.
4. Try to turn the offset shaft. If the steering gear is centered, the red mark on the center lock pin will not be visible, and the offset shaft will not turn. If you can still turn the shaft, remove the lock pin and repeat steps 2 and 3.
5. Slide the tie rod boots off the inner tie rod ends so that you can see the stroke rod. The length of stroke rod protruding from each side of the gear should be equal, indicating that the stroke rod is centered. If not, remove the center lock pin, and turn the offset shaft in the appropriate direction to center the stroke rod. Then repeat steps 2 through 4 to center the rear steering gear.
6. After installing the rear steering gear, as described earlier, remove the center lock pin to prevent damage to the gear.

Mazda

Service the front steering gear in the Mazda four-wheel-steering system as you would a power steering gear in a two-wheel system. The rear steering gear operates in response to input from the steering angle transfer shaft, electric stepper motor, and hydraulic control valve.

Removing the front steering gear
Before servicing the front steering gear on a Mazda four-wheel-steering system, you must disconnect the steering angle transfer shaft from the front steering gear output shaft, figure 7-6, as described later in this chapter. Then remove the rack as you would a regular power rack and pinion steering gear.

Removing the rear steering gear
Before removing the rear steering gear, disconnect the steering angle transfer shaft, as described later in this chapter. Use the following procedure to remove the rear steering gear, figure 7-7:

1. Position the lift supports under the vehicle chassis using the manufacturer's recommended lift points, and raise the vehicle. See Chapter One of this *Shop Manual* for more information on lifting a vehicle.
2. Remove the rear wheels.
3. Disconnect the tie rod ends from the rear steering arms using a pickle fork or one of the other methods described in Chapter Five of this *Shop Manual*.
4. Disconnect the hydraulic lines at the solenoid valve and steering gear with a flare nut wrench, using a container to catch any fluid that may leak from the lines. Be sure to plug the hydraulic lines and the openings on the solenoid valve and rear steering gear so that dirt does not enter the system.
5. Disconnect the solenoid valve electrical connectors.
6. Remove the bolts that hold the stepper motor to the rear steering gear, and disconnect the stepper motor.
7. Remove the bolts that secure the rear steering gear to the vehicle frame or crossmember, and remove the steering gear.
8. Before reinstalling the rear steering gear, replace the oil filter in the solenoid valve, as described later in this chapter.
9. Check the fluid level in the rear steering gear, as described later in this chapter, and add fluid if needed. If fluid is leaking from the oil pan, replace the gasket, as described later in this chapter.
10. Install the rear steering gear in the reverse order. Be sure to adjust the rear steering angle, as described later in this section, and center the front steering gear before

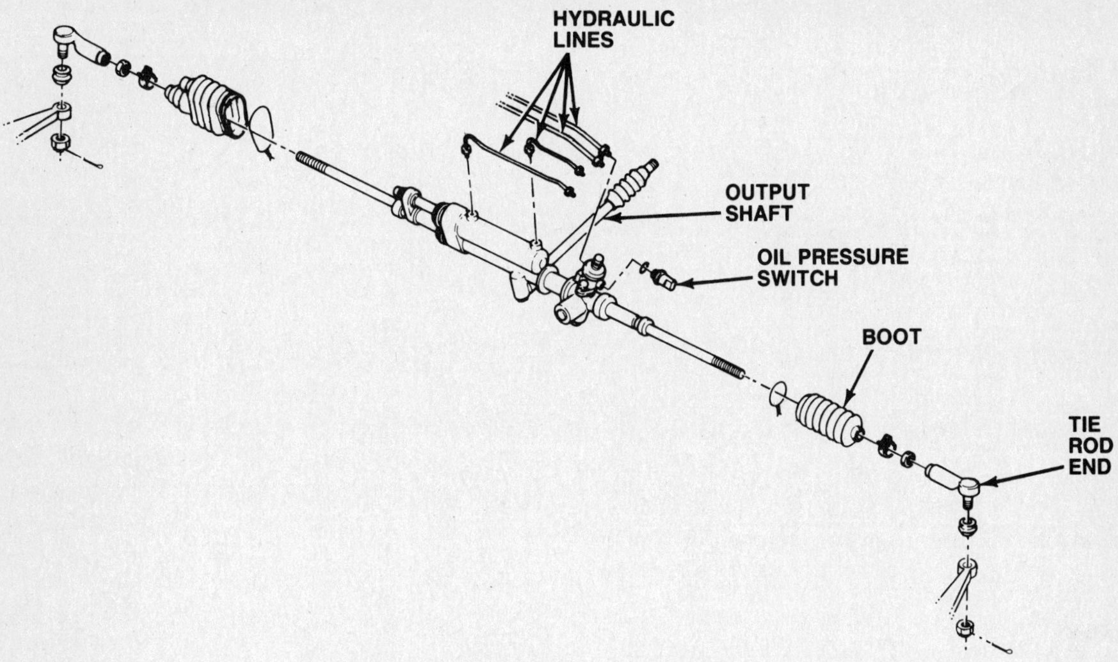

Figure 7-6. The front gear on the Mazda four-wheel-steering system looks like a fairly ordinary rack and pinion steering gear. One difference is an additional pinion called an output shaft that links to the steering angle transfer shaft.

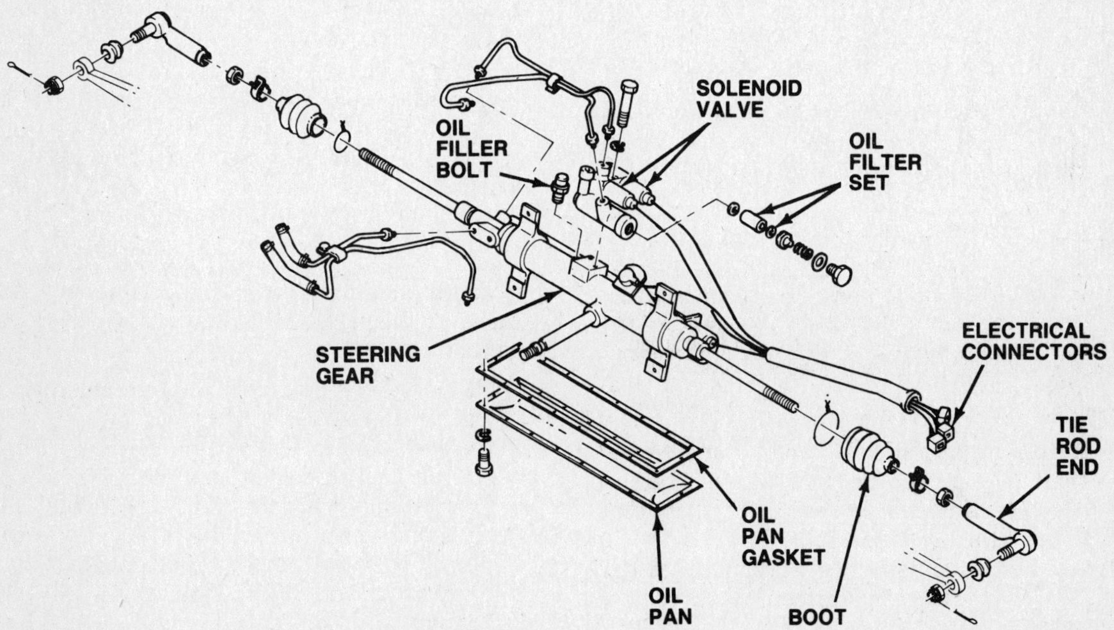

Figure 7-7. In addition to the tie rod ends, steering angle transfer shaft, and mounting bolts, you must disconnect the stepper motor electrical connectors and the solenoid valve hydraulic lines before removing the rear steering gear in a Mazda system.

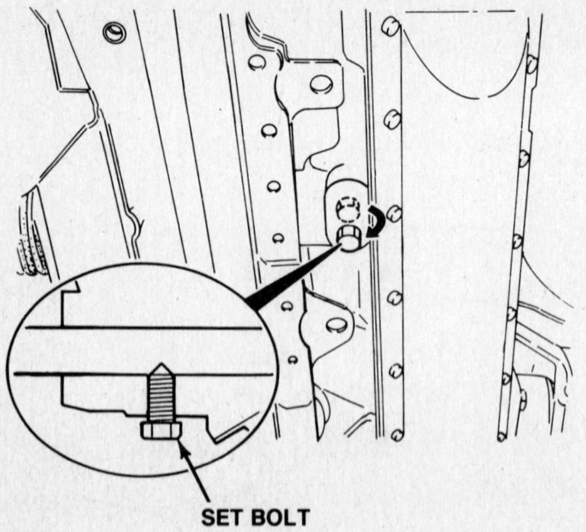

Figure 7-8. The set bolt used to adjust the Mazda rear steering gear is like the center lock pin used to center the rear steering gear in the Honda. Be sure to remove the set bolt before driving the vehicle.

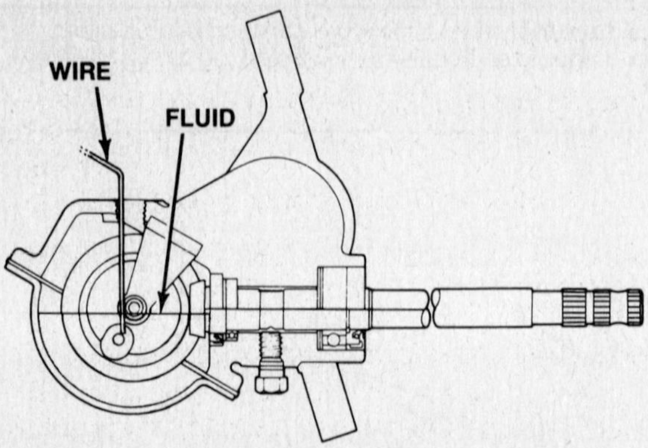

Figure 7-9. Remove the bleeder valve from a Mazda rear steering gear, and insert a wire to check the fluid level. Remove the wire, measure the fluid level, and compare with manufacturer's specifications.

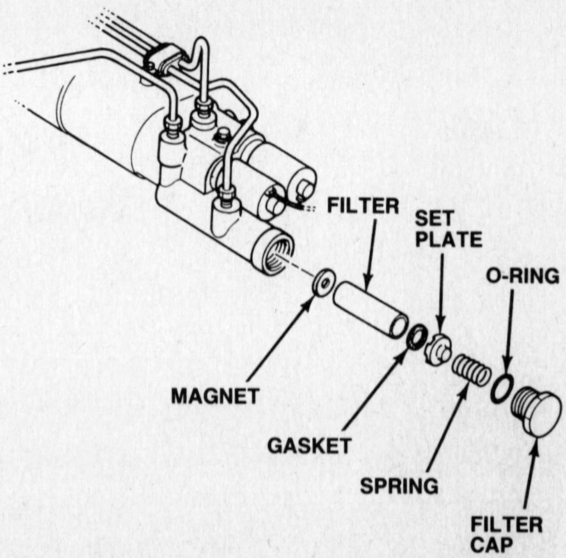

Figure 7-10. The solenoid valve that is mounted on the Mazda rear steering gear has a small oil filter that you should change when you remove the rear steering gear.

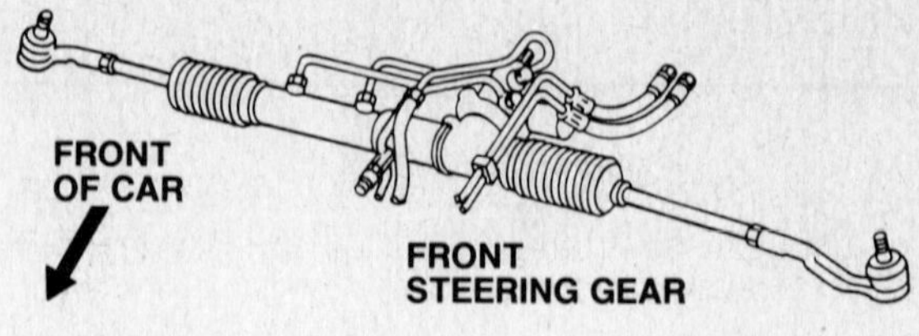

Figure 7-11. The front steering gear on Nissan's four-wheel steering-system is an ordinary hydraulic-assisted rack and pinion steering gear.

connecting the steering angle transfer shaft. The procedure for centering the front steering gear is described in Chapter Four of this *Shop Manual*.

11. Check and adjust wheel alignment, as described in Chapter Fourteen of this *Shop Manual*.

Adjusting rear steering angle

As on the Honda system, you must adjust the gears on the Mazda four-wheel-steering system before you connect the steering angle transfer shaft to the front and rear steering gears, or the four-wheel-steering system will not operate properly. Procedures for centering the front steering gear are in Chapter Four of this *Shop Manual*. Adjust the rear steering gear as follows:

1. Remove the bolt from the rear steering gear, figure 7-8.
2. While looking through the hole, turn the input shaft until you can see the notch in the input shaft through the hole.
3. Install the set bolt in the hole, and seat the end of the bolt in the input shaft notch.
4. Try to turn the input shaft. It should not move.
5. Connect the steering angle transfer shaft to the front and rear steering gears, as described later in this chapter.
6. After installing the steering angle transfer shaft, remove the set bolt to prevent damage to the gear, and reinstall the bolt that normally covers the opening.

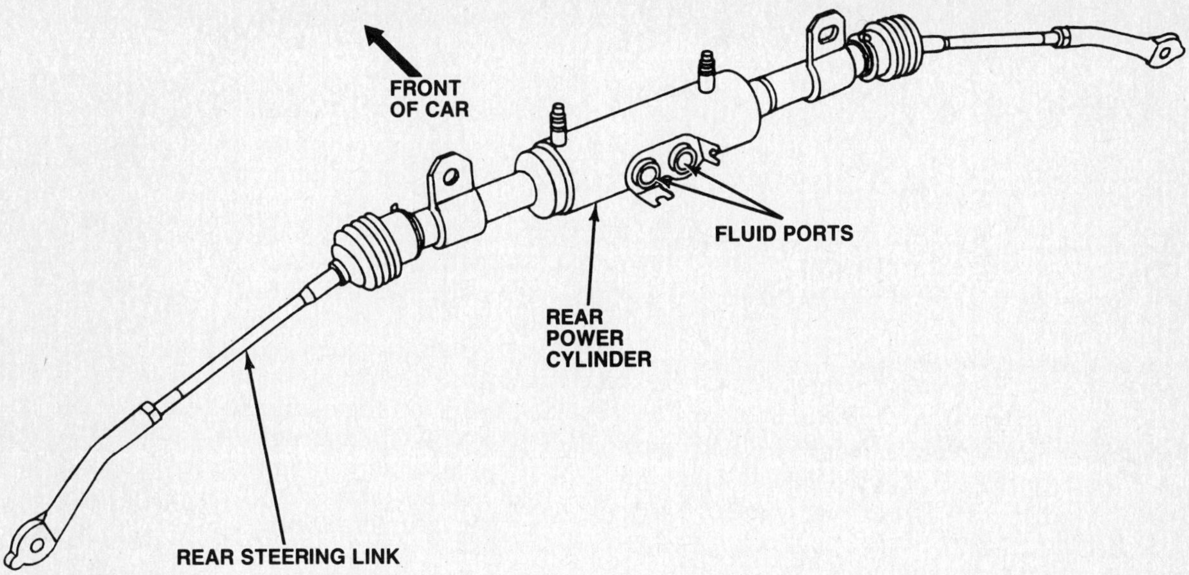

Figure 7-12. Unlike the Honda and Mazda examples, the rear power cylinder on Nissan's four-wheel-steering system is completely hydraulically controlled.

Inspecting the rear steering gear fluid level
Unlike the Honda or Nissan rear steering gear or power cylinder, the Mazda rear steering gear has a fluid reservoir that you should check and fill, if needed.

1. Park the vehicle on level ground. Unscrew the bleeder valve that is on top of the rear steering gear.
2. Insert a wire into the hole, figure 7-9, ensuring that the wire touches the bottom of the steering gear reservoir. Do not allow dirt to enter the gear.
3. Remove the wire, and measure the fluid level. Fluid in the reservoir should be 44 ± 3 mm (1.7 ± 0.1 inch) deep.
4. If the level is too low, check for leaks, and repair any problems. Then refill the gear with the quantity of fluid recommended in the service manual.

Installing an oil pan gasket
If fluid for the rear steering gear is leaking past the oil pan gasket, replace the gasket, figure 7-7. Loosen and remove the bolts that run along the perimeter of the oil pan, and remove the pan and gasket. Install a new gasket, and install the oil pan. Refill the reservoir with power steering fluid, and check the fluid level, as described earlier. Check the oil pan for leaks.

Replacing the solenoid valve oil filter
After you remove the rear steering gear, remove the solenoid valve filter, and replace it. Unscrew the filter cap on the solenoid valve, figure 7-10, and remove the filter, O-rings, spring, magnet, and set plate. Clean the spring, magnet, and set plate. Install the magnet, filter, rubber gasket, set plate, and spring, along with a new O-ring, and filter cap.

Nissan

The front steering gear on a Nissan four-wheel-steering system is like the gear on a two-wheel system. Nissan uses hydraulic and electronic input to the rear steering gear, and *does not* use a shaft from the front to rear steering gear like the Honda and Mazda systems. This means that the front gear on the Nissan system does not have mechanical modifications, figure 7-11. Service the front steering gear using the procedures in Chapters Four and Six of this *Shop Manual*.

The hydraulic actuator at the rear axle is called a rear power cylinder, rather than a rear steering gear. Service procedures for the power cylinder follow.

Inspecting the rear power cylinder
Run the engine until it reaches normal operating temperature. With the engine idling, check the power cylinder for leaks, figure 7-12, especially at the hydraulic lines and fittings. Inspect the rubber boots that cover the inner steering links (tie rod ends) for leaks, cracks, deterioration, or other damage, and replace parts as needed.

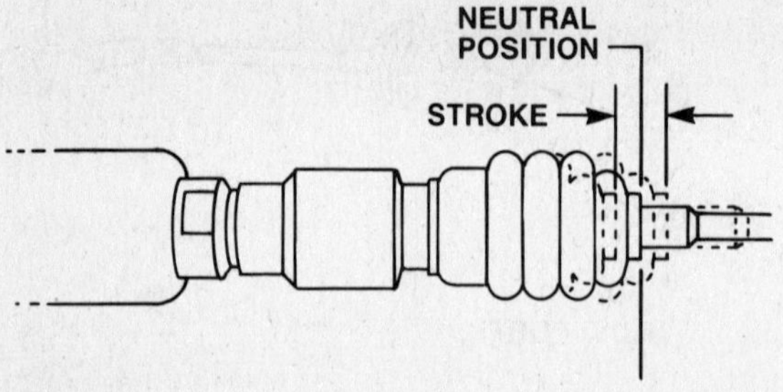

Figure 7-13. Check the rod stroke by measuring the distance the rod extends and retracts while conducting the HICAS self-diagnostic test.

Testing the rear power cylinder
You can check power cylinder operation by performing the following self-diagnostic test and measuring rod stroke.

1. Run the engine until it reaches normal operating temperature, and put the transmission in PARK or NEUTRAL. Raise the vehicle high enough so the wheels are off the ground. Have an assistant sit in the passenger compartment. Turn the engine off.
2. Set the HICAS system in the self-diagnostic mode by starting the engine and turning the steering wheel from left to right at least five times. You must be sure to turn the steering wheel at least 20 degrees from the neutral position. At the same time, press the brake pedal five times within ten seconds after starting the engine.
3. Increase engine speed to 2,000 rpm or more. Do not press the brake pedal for the rest of this test, or the computer will leave the diagnostic test mode.
4. Starting with the steering wheel in the neutral position, turn it 180 degrees, and hold the steering wheel in position.
5. Measure the distance one steering link extends and the other retracts, figure 7-13.
6. Turn the steering wheel 180 degrees in the other direction, and measure the extension and retraction of the steering links.
7. Determine the total rod stroke for each rod by adding the extension and retraction values. Each rod should extend about 1/8 inch (3 mm) and retract 1/8 inch (3 mm), for a total rod stroke of 1/4 inch (6 mm). Compare the results with the manufacturer's specifications.
8. If the rod stroke is not within the manufacturer's specifications, adjust it, as described later in this chapter.

Removing and installing the power cylinder
Removing the power cylinder, figure 7-12, is similar to removing a power rack and pinion steering gear.

1. Position the vehicle on a lift, ensuring that parts will not bend under the weight of the car when it is raised. See Chapter One of this *Shop Manual* for more information on lifting a vehicle.
2. Remove the rear wheels.
3. Remove the steering link (tie rod end) pivots from the rear steering arms by loosening and removing the bolts.
4. Disconnect the hydraulic lines from the power cylinder using a flare nut wrench.
5. Plug the hydraulic lines and the openings on the power cylinder so that dirt will not enter the hydraulic system.
6. Remove the bolts that secure the power cylinder to the vehicle's frame or crossmember. Remove the power cylinder from the vehicle.
7. Before installing a new power cylinder, inspect the mounting brackets to be sure they are not bent.
8. Bolt the power cylinder in place in the vehicle.
9. Wipe the hydraulic line fittings and the openings on the power cylinder to remove any dirt. Connect the hydraulic lines to the power cylinder, and tighten the fittings.
10. Bolt the pivots at the end of each steering link (tie rod end) onto the rear steering arms.
11. Bleed the system as described later in this chapter.
12. Adjust rear wheel toe, as described in Chapter Fourteen of this *Shop Manual*.

Bleeding the power cylinder
Bleed air from the power steering system as described in Chapter Six of this *Shop Manual* before bleeding the four-wheel-steering system.

1. Raise the car and set the HICAS system in the diagnostic mode, as described in the section titled ''Testing the rear power cylinder''.
2. Starting with the steering wheel in the neutral position, turn it 180 degrees to the right, and hold it there.
3. Attach a bleed hose and container to the right cylinder bleeder valve, figure 7-14. Loosen the valve and allow air and fluid to leave the cylinder. Retighten the valve.
4. Return the steering wheel to the neutral position, and then turn it 180 degrees to the left.

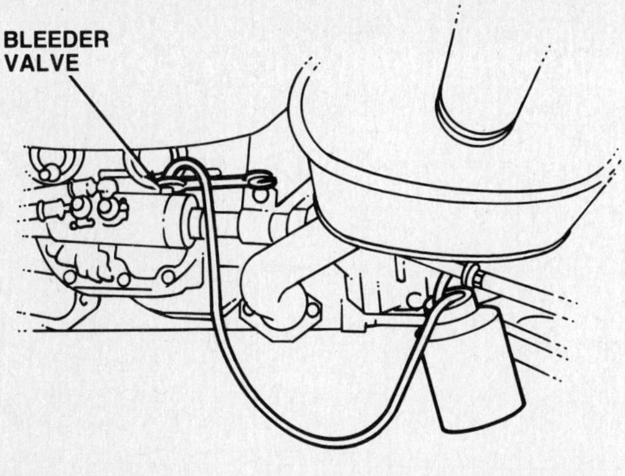

Figure 7-14. Loosen the right and left bleeder valves on the power cylinder, and bleed air from the system.

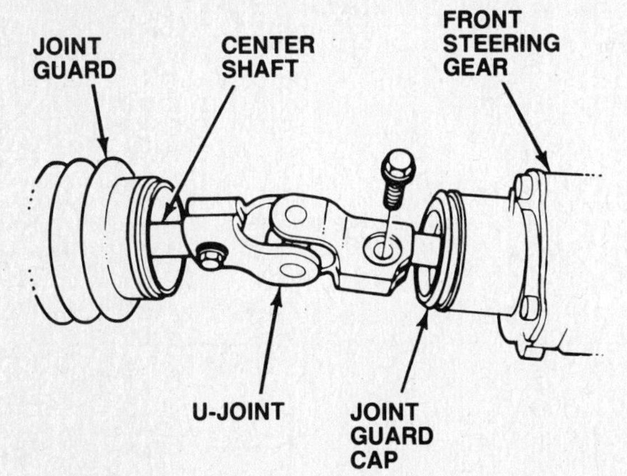

Figure 7-15. The U-joint on the front portion of Honda's center shaft is covered with a joint guard cap and joint guard.

5. Attach a hose and container to the left cylinder bleeder valve. Loosen the valve and allow air and fluid to leave the cylinder. Retighten the valve.
6. Check the fluid reservoir periodically, and top it up as needed. Check for bubbles in the reservoir as well. If there are bubbles, bleed the system again.

STEERING ANGLE TRANSFER SHAFT

Honda and Mazda four-wheel-steering systems use a steering angle transfer shaft to transmit the turning angle of the steering wheel from the front steering gear to the rear steering gear.

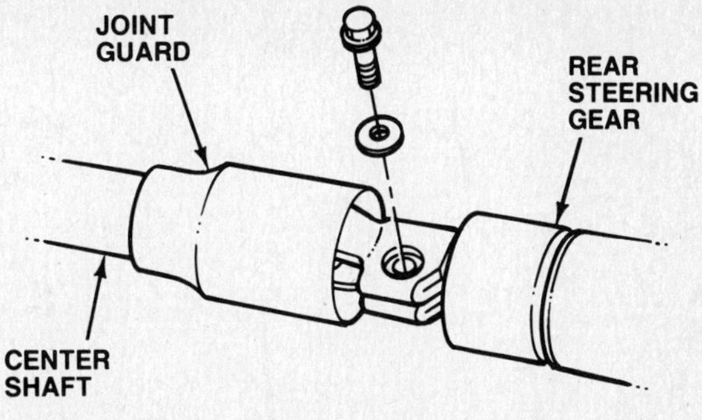

Figure 7-16. Slide the rear U-joint cover forward on the center shaft to expose the joint.

Honda

Honda calls the steering angle transfer shaft a "center shaft". Before disconnecting the center shaft from the front and rear steering gears, install the center lock pin in the rear steering gear to hold the gear in place, as described earlier in this chapter. This ensures that when you reconnect the center shaft, it is oriented correctly with the rear steering gear. Disconnect the shaft as follows:

1. Disconnect the joint guard and cap that covers the U-joint between the center shaft and front steering gear, figure 7-15.
2. Loosen and remove the bolt that secures the U-joint between the front steering gear output pinion and the center shaft.
3. Disconnect the center shaft from the front steering gear pinion.
4. Locate the joint guard that covers the U-joint between the center shaft and rear steering gear, figure 7-16. Slide the joint guard forward.
5. Loosen and remove the bolt that secures the U-joint between the rear steering gear offset shaft and the center shaft.
6. Disconnect the center shaft from the rear steering gear offset shaft.
7. Install the center shaft in the reverse order. Be sure that the front and rear steering gears are centered before you install the U-joint bolts at each end of the shaft.
8. After centering the rear steering gear, remember to remove the center lock pin before driving the vehicle.

Mazda

Mazda uses a steering angle transfer shaft in the four-wheel-steering system. Besides a removal procedure, this section also includes instructions for checking transfer shaft alignment.

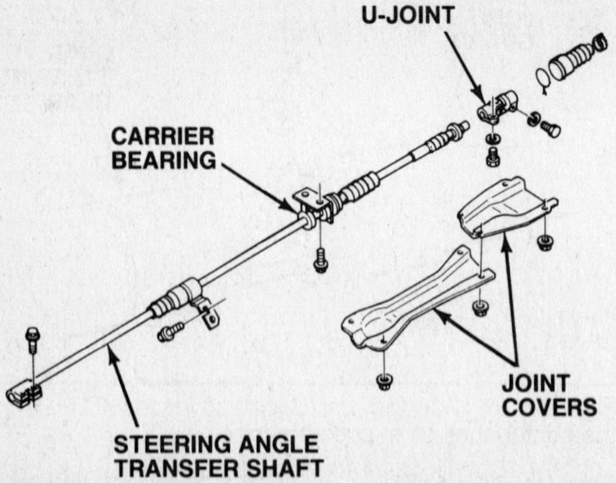

Figure 7-17. When removing the steering angle transfer shaft on the Mazda four-wheel-steering system, remember that it has two joint covers that bolt to the vehicle's frame.

Inspecting, removing, and installing the shaft
Check the boots that cover the transfer shaft U-joints for cracks or other signs of deterioration, figure 7-17. The carrier bearing and U-joints should not bind when you turn them. Check the transfer shaft for bends, and inspect the joint covers for damaged or missing bolts. If needed, remove and replace the transfer shaft.

1. Loosen and remove the screws that hold the rear joint cover in place. Remove the boot clamp and boot that cover the rear transfer shaft joint.
2. Loosen and remove the bolt securing the joint between the rear steering gear input shaft and the transfer shaft.
3. Remove the transfer shaft from the rear steering gear.
4. Remove the boot clamp on the front joint of the transfer shaft, and slide the boot back to expose the joint.
5. Loosen and remove the bolt that secures the joint between the front steering gear input shaft and the transfer shaft.
6. Remove the transfer shaft from the front steering gear input shaft.
7. Loosen and remove the bolts that secure the two transfer shaft mounting brackets to the vehicle's frame.
8. Remove the brackets, and remove the transfer shaft.
9. Replace the shaft in the reverse order. Be sure that the steering gears are centered before installing the shaft.
10. Check shaft alignment as described in the next section. After aligning the shaft, remember to remove the set bolt from the rear steering gear before driving the vehicle.

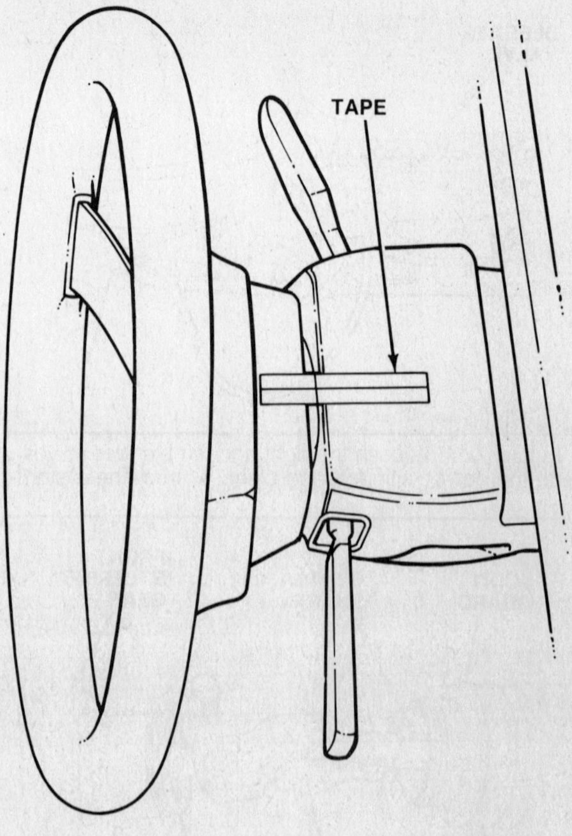

Figure 7-18. When checking steering angle transfer shaft alignment, use tape on the steering wheel and hub to mark the steering wheel's straight-ahead position.

Checking shaft alignment
The steering angle transfer shaft must be properly oriented with respect to the front and rear steering gears, or the four-wheel-steering system will not operate properly. Use this procedure to check the shaft's alignment.

1. Center the steering wheel, as described in Chapter Four of this *Shop Manual*. With the steering wheel in the neutral position, align two pieces of tape on the steering wheel and column, figure 7-18.
2. Drive the vehicle straight ahead, and readjust the tape, if needed, so that when the front wheels are positioned straight ahead, the pieces of tape are aligned.
3. Stop and start the engine. Check that the rear wheels do not turn off center when the engine starts running. If the wheels move, remove the transfer shaft, center the front and rear steering gears, and reinstall the shaft. Always check shaft align-

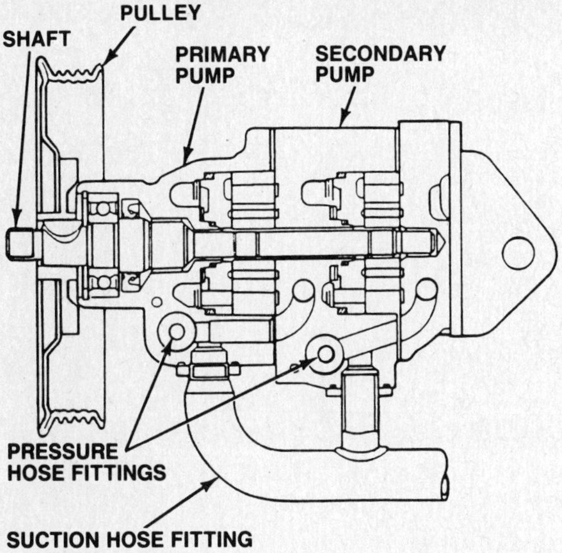

Figure 7-19. Because both the front and rear steering gears use hydraulic controls in the Mazda four-wheel-steering system, the power steering pump size has been increased.

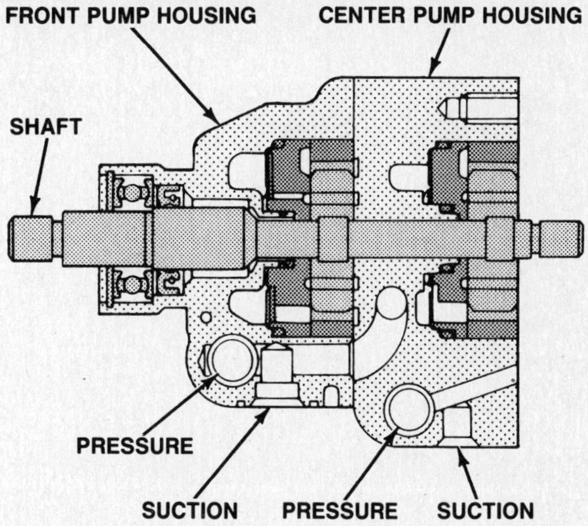

Figure 7-20. The Nissan four-wheel-steering system also uses a larger power steering pump, which is actually two vane-type pumps in a single housing.

ment after disconnecting the steering gear from the transfer shaft or after a wheel alignment.

ELECTRONIC AND HYDRAULIC CONTROLS

In addition to the steering angle transfer shaft, the Mazda four-wheel-steering system uses electronic and hydraulic controls to actuate the rear steering gear. The Nissan system also uses electronic and hydraulic controls, but it does not use any type of mechanical input to actuate the power cylinder. The Chrysler/Mitsubishi four-wheel-steering system is similar to the Nissan system.

Because these systems have electronic and hydraulic controls, they use computerized self-diagnostic systems that will assist you in troubleshooting. Use the manufacturer's service manual to start the diagnostic tests and interpret trouble codes. Chapter Three of this *Shop Manual* contains information on computer diagnostic systems and electrical testing.

POWER STEERING PUMP

Because the rear steering gear in the Honda four-wheel-steering system is not actuated by hydraulics, the power steering pump does not require modifications. However, the Mazda and Nissan systems, which both use hydraulics, have modified power steering pumps to handle the increased volume of fluid.

Mazda

The pump in the Mazda system, figure 7-19, is similar to the one used in their two-wheel-steering system, the difference being that it is actually two power steering pumps in one housing. Inspect and replace the power steering pump as described in Chapter Six of this *Shop Manual*.

When performing the power steering system pressure tests, use the procedures described in Chapter Six of this *Shop Manual*. Pump and front steering gear pressures in the four-wheel system will be slightly higher than pressures in a two-wheel system, so be sure to check the manufacturer's specifications.

Nissan

The Nissan four-wheel-steering system also uses a power steering fluid pump that is actually two pumps combined in a single housing, figure 7-20. Both are vane-type pumps.

Inspect and replace the power steering pump using the procedures in Chapter Six of this *Shop Manual*. Power steering system pressure tests must be performed using HICAS computer self-diagnostics, so you should refer to the manufacturer's service manual.

REAR STEERING LINKAGE

The steering linkage at the rear steering gear or power cylinder is much like the linkage used on rack and pinion steering systems.

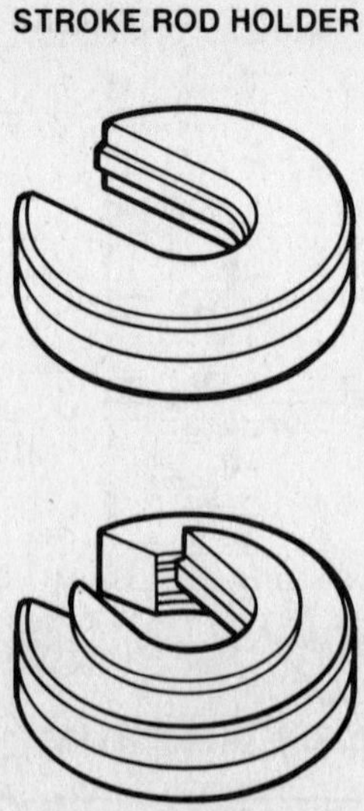

Figure 7-21. The stroke rod holder is a tool that keeps the stroke rod from turning while you are removing the tie rods on a Honda rear steering gear.

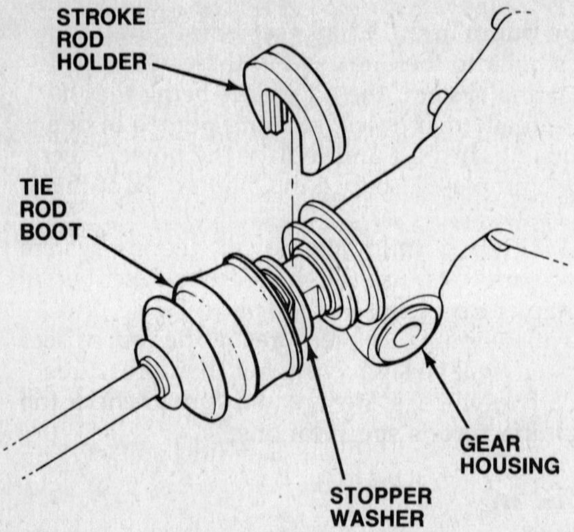

Figure 7-22. Install the stroke rod holder between the tie rod stopper washer and the gear housing.

Honda

To remove and replace tie rods on the rear steering gear, you will need a stroke rod holder, figure 7-21, which is a special tool provided by Honda. This tool holds the stroke rod firmly in place, keeping it from rotating while you are removing the tie rods.

1. Position the lift supports under the vehicle chassis using the manufacturer's recommended lift points, and raise the vehicle.

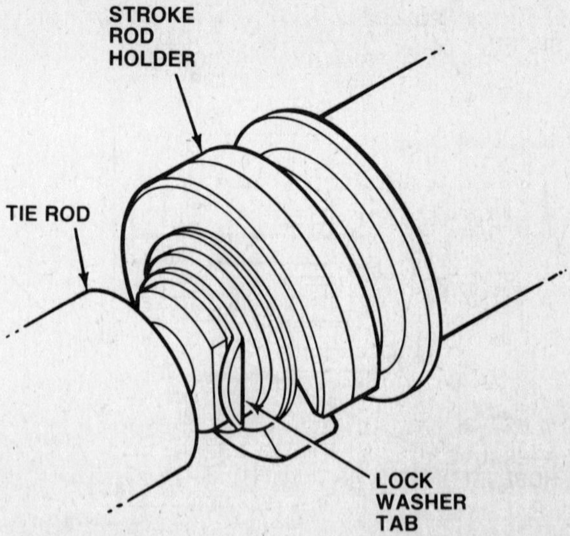

Figure 7-23. As on an ordinary rack and pinion steering gear, you must straighten the lock washer tab before you can remove the tie rod from the rear steering gear of the Prelude.

See Chapter One of this *Shop Manual* for more information on lifting a vehicle.

2. Install the rear steering gear center lock pin as described earlier in this chapter.
3. Loosen the clamps that secure the tie rod boots, and slip the boots back to expose the inner tie rod ends.
4. Install the stroke rod holders, figure 7-22.
   a. Slide the holder between the tie rod stopper washer and gear housing.
   b. Turn the outer and inner pieces of the stroke rod holder until they lock together.
   c. Install the stroke rod holder on the other tie rod as well.
   d. Remove the center lock pin from the rear steering gear.
5. Straighten the tabs on the lock washers, which hold the rear tie rods to the stroke rod, figure 7-23.
6. Remove the stroke rod holder.
7. While holding the stroke rod with a wrench, figure 7-24, remove the tie rod by turning it with another wrench. *Do not rotate the stroke rod*. Remove the other tie rod in the same manner.
8. Before reinstalling tie rods, install a new tie rod stopper washer and lock washer on the stroke rod.
9. Install the tie rod while holding the stroke rod with a wrench.

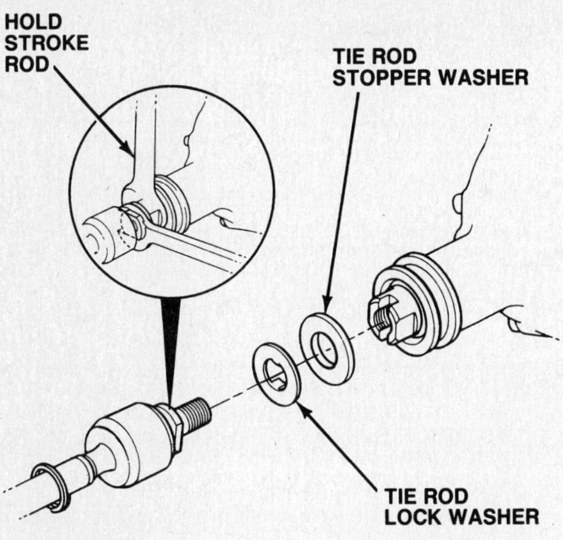

Figure 7-24. Hold the stroke rod with a wrench to keep it from twisting while you turn the tie rod to remove it.

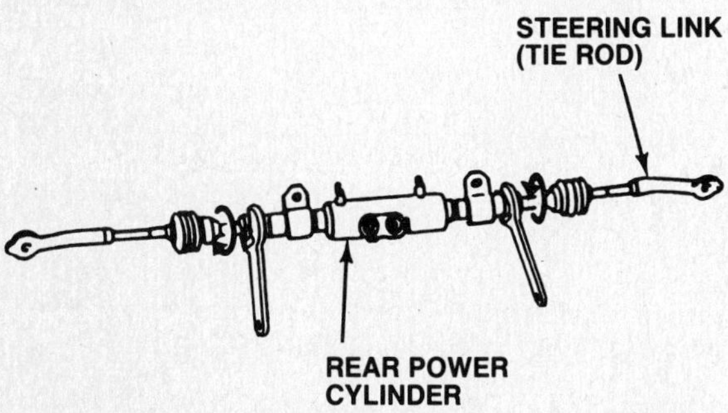

Figure 7-25. When removing steering links (tie rods) from the Nissan rear power cylinder, hold one ball joint housing with a wrench while you turn the other.

10. Install the stroke rod holder again, using the procedures in step 4.
11. Bend the lock washer tab.
12. Remove the stroke rod holder.
13. Replace the tie rod boot and clamp.

Mazda

On the Mazda rear steering gear, you cannot remove and replace the inner tie rod ends. If the inner tie rod ends are bent or worn, replace the entire rear steering gear. If the outer tie rod end ball joints are worn, replace them as you would the tie rod ends on a rack and pinion steering linkage. See Chapter Five of this *Shop Manual* for replacing tie rod ends.

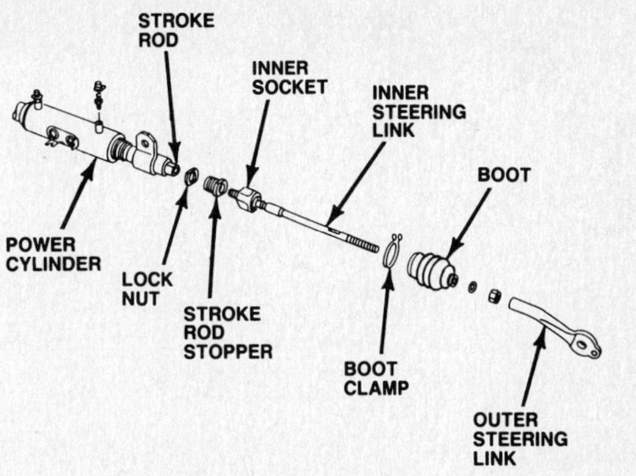

Figure 7-26. The stroke rod stopper holds the stroke rod and steering link together.

Nissan

Nissan calls the rear tie rods "steering links". The outer link is a pivot that connects to the rear steering arms, and the inner link is a ball joint that connects to the power cylinder. If steering links are damaged, replace them using the following procedure.

1. Position the lift supports under the vehicle chassis using the manufacturer's recommended lift points, and raise the vehicle. See Chapter One of this *Shop Manual* for more information on lifting a vehicle.
2. Remove the clamps from the steering link boots, and slide the boots to the outer ends of the links.
3. Remove the bolts that secure the outer steering links to the rear steering arms.
4. Hold one ball joint socket on the power cylinder with a wrench, and use another wrench to turn the ball joint socket on the other side of the power cylinder, figure 7-25. Remove the steering link.
5. Loosen and remove the stroke rod stopper and lock nut, figure 7-26.
6. Remove the other steering link by holding the other side of the power cylinder assembly with a wrench, figure 7-27. Loosen and remove the stroke rod stopper and lock nut.
7. Install new lock nuts and stroke stoppers, if the old ones are damaged.
8. Hold one end of the power cylinder with a wrench while installing the new steering link. Use thread sealant on the ball joint housing threads.
9. Install the other lock nut, stroke stopper, and steering link.

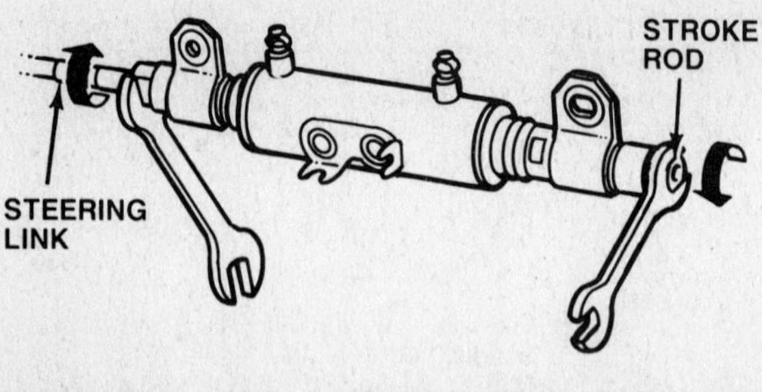

Figure 7-27. Hold one end of the stroke rod with a wrench and turn the steering link on the opposite side of the power cylinder to loosen and remove the steering link.

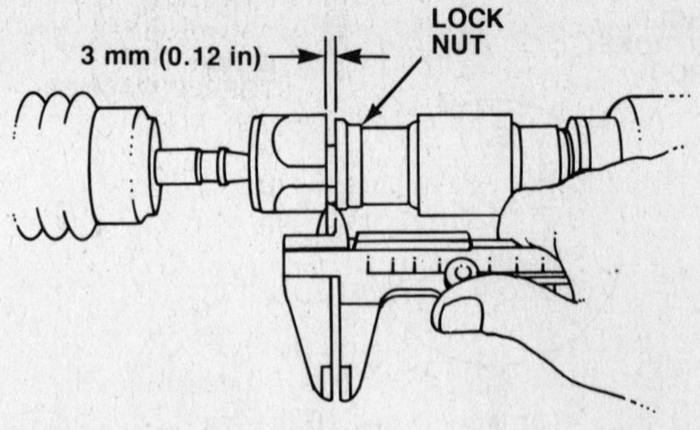

Figure 7-28. You must adjust the stroke rod stoppers after installing the steering links so that the stroke will be the same on each side of the power cylinder.

10. Then adjust the stroke stoppers as follows, figure 7-28:
    a. Loosen the lock nut.
    b. Turn the stroke stopper until the distance between the inner ball joint and stroke stopper is 3 mm (0.12 inches). Use calipers to measure this distance.
    c. Tighten the lock nut.
    d. Adjust the stroke stopper on the other steering link.
11. Torque both ball joint housings to the manufacturer's specifications.
12. Install the bolts that secure the outer steering links to the rear steering arms.
13. Reinstall the steering boots, and secure them with clamps.

PART THREE

Suspension System Service

Chapter

8

Shock Absorber, Strut, and Spring Service

Before repairing suspension system problems, a technician must understand how suspension components work together. Refer to Chapters Eight through Eleven of the *Classroom Manual* and *Shop Manual* for theory and repair procedures for the suspension system. Use Chapter Two of this *Shop Manual* to diagnose suspension system problems.

This section on suspension systems is divided into four chapters. This chapter covers procedures for servicing components that cushion or damp suspension movement, such as shocks, struts, and springs. Procedures for inspecting and replacing other suspension components, such as control arms and trailing arms, are covered in Chapter Nine. Chapter Ten describes basic procedures for servicing axles, the driveline, and driveline joints, and Chapter Eleven contains tips for servicing vehicles with electronically controlled suspensions. (Because these systems have special service procedures for lifting and towing, always check a vehicle to see if it has an electronically controlled suspension system.)

IDENTIFYING SHOCKS, STRUTS, AND SPRINGS

Determine which type of shock, strut, or spring is used on the vehicle you are servicing. Shock absorbers damp suspension movement and may limit suspension travel on some vehicles. The most basic shock absorber uses hydraulic fluid to damp suspension movement, figure 8-1. Vehicle manufacturers generally use basic hydraulic shocks as the original equipment on new cars. Modifications to the basic hydraulic shock absorber include:

- Gas-charged shocks, figure 8-2
- Coil-over shocks, figure 8-3
- Air-charged shocks, figure 8-4
- Variable shocks, figure 8-5.

You will find struts on both the front and rear suspensions of many late-model vehicles. Struts resemble shock absorbers and function in a similar manner. Unlike shocks, however, they support vehicle weight at the upper mount. Most struts have an integral coil spring, but modified struts use a separate spring elsewhere on the suspension. The top of the strut is mounted to the vehicle body, and can be accessed through the engine compartment or trunk. The bottom of the strut is mounted to the wheel knuckle or control arm.

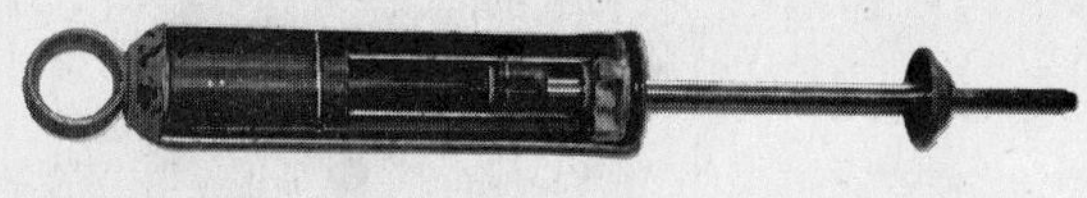

Figure 8-1. This hydraulic shock absorber has a section cut away so that you can see the shock absorber piston and cylinder, which contains hydraulic fluid.

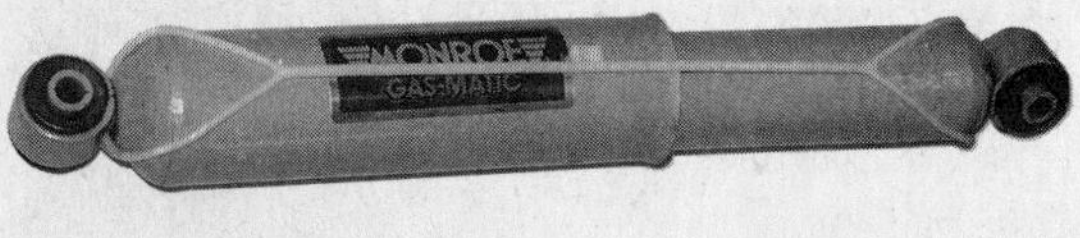

Figure 8-2. Gas-charged shocks use pressurized gas to reduce bubble formation in the hydraulic fluid. A fastener on the shock keeps it compressed until you install the shock on the vehicle.

Figure 8-3. This road racing vehicle uses coil-over shocks to increase suspension system stability. You will find coil-over shocks on some trucks and large vans as well.

There are three basic strut designs:

- MacPherson struts, which may have a solid connection at the wheel hub, figure 8-6, or a ball joint connection, figure 8-7
- Modified struts, figure 8-8
- Pivot-base struts, figure 8-9.

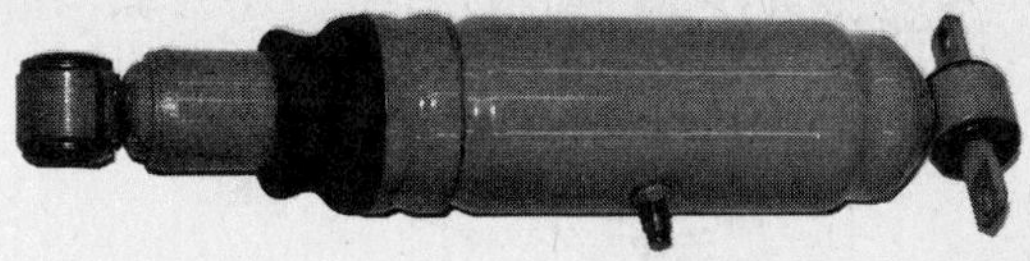

Figure 8-4. Some heavy-duty shocks can be manually charged with air to increase shock stiffness and adjust vehicle height.

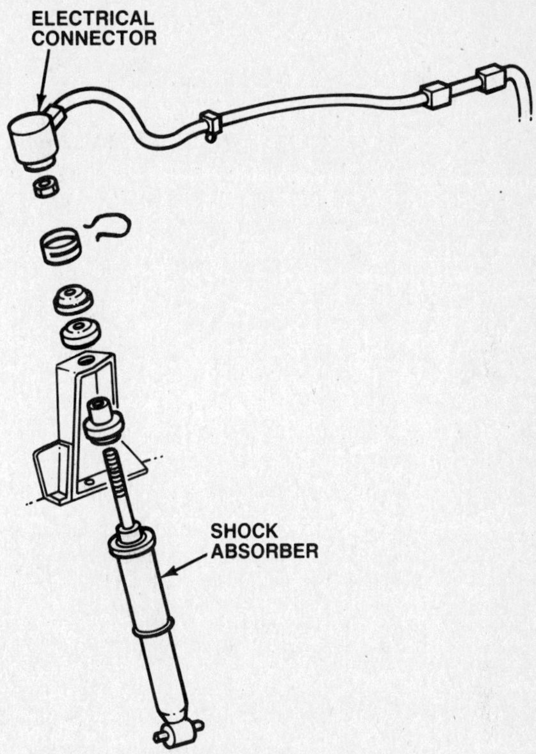

Figure 8-5. Variable shock absorbers are controlled by a suspension computer. Because they are electronically controlled, you will find an electrical connector somewhere on the shock.

As described in the *Classroom Manual*, there are four types of suspension springs used on passenger vehicles:

- Coil springs, figure 8-10
- Leaf springs, figure 8-11
- Torsion bars, figure 8-12
- Air springs, figure 8-13.

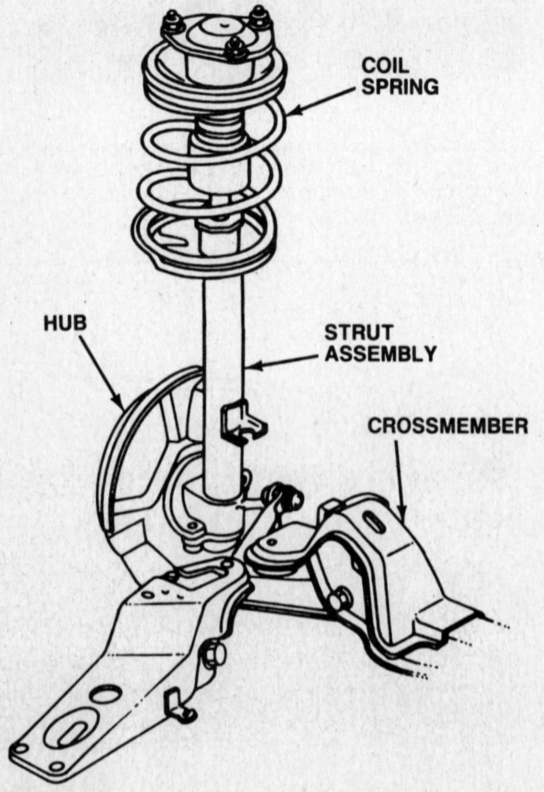

Figure 8-6. The base of this MacPherson strut has a solid, immovable connection at the hub. The coil spring surrounding the strut is the main, load-carrying suspension spring.

Figure 8-7. On this MacPherson strut, the base is connected to the control arm with a ball joint.

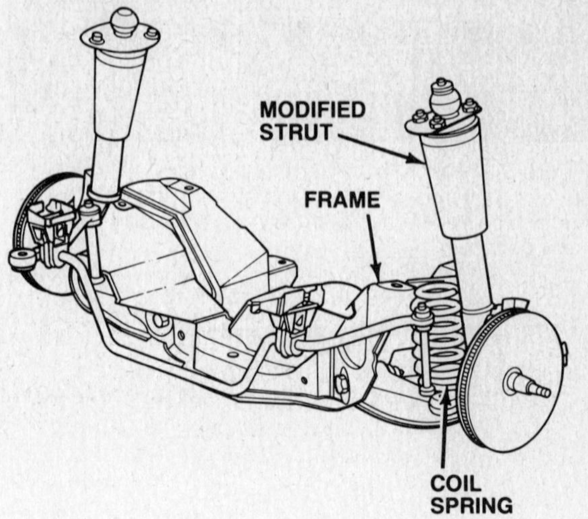

Figure 8-8. Modified struts are similar to MacPherson struts, the only difference being that the suspension spring is not part of the strut assembly. The spring is located elsewhere on the suspension.

SHOCK ABSORBERS

Defective shock absorbers can greatly affect a vehicle's handling, steering, and even braking. Remember that shocks wear gradually, and a customer may not notice when they are worn out. A car with 30,000 to 40,000 miles (48,000 to 64,000 kilometers) is likely due for new shocks, so be sure to check vehicle mileage. If you are working on a vehicle that uses electronically controlled shocks or springs, use Chapter Eleven of this *Shop Manual* as a guide for inspecting and replacing the shocks.

Inspecting Shocks

During the test drive, if the vehicle sways on turns, suspect worn shocks. Wheel hop during braking is also a sign that the shocks should be checked. If suspension components, especially springs, bushings, and ball joints are worn, check the shocks. (Worn shocks allow the suspension springs to oscillate more often, which accelerates spring and suspension component wear.)

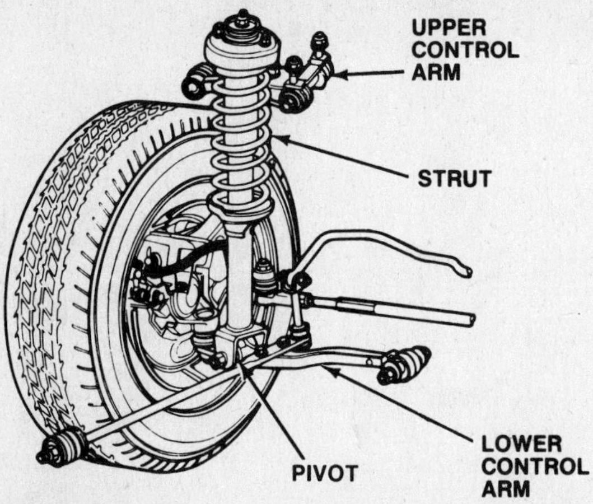

Figure 8-9. Instead of mounting directly to the wheel knuckle like MacPherson struts, pivot-base struts have a pivoting connection at the base. Pivot-base struts are used on strut/SLA suspensions, which have an upper control arm in addition to the strut assembly.

Figure 8-10. Coil springs are used on a number of suspension types from solid axle to independent systems. The coil spring may rest on the axle or axle housing, on the base of a strut, or on a control arm.

Figure 8-11. Leaf springs are usually used on trucks and vans. The leaf springs may be mounted longitudinally, as shown, or transversely.

Figure 8-12. Torsion bars are mounted to the vehicle frame on one end and to a suspension component on the other. Like leaf springs, torsion bars may be oriented longitudinally or transversely.

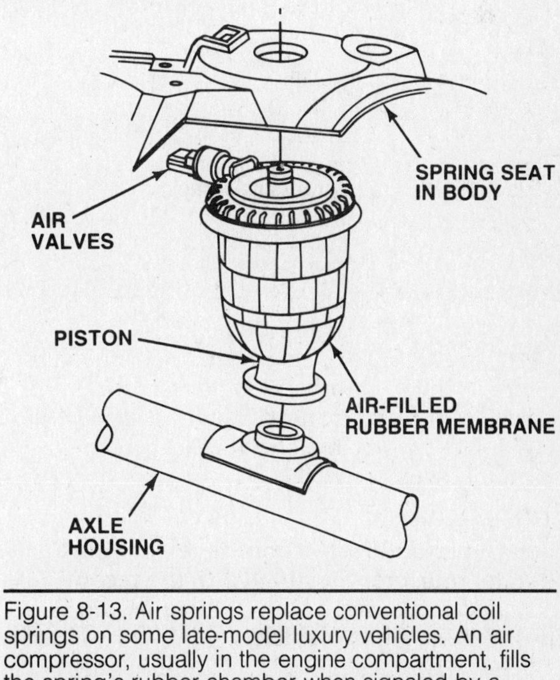

Figure 8-13. Air springs replace conventional coil springs on some late-model luxury vehicles. An air compressor, usually in the engine compartment, fills the spring's rubber chamber when signaled by a computer-controlled solenoid.

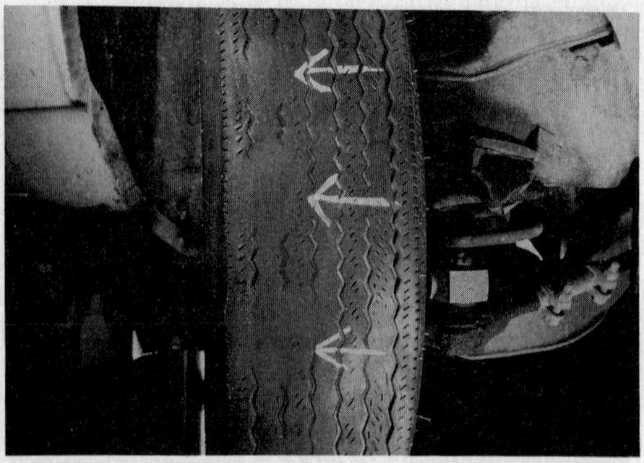

Figure 8-14. Round indentations, or "cupping", is a tire wear pattern that indicates worn shocks. Cupping may occur when bounce is not controlled by the shock absorber.

Figure 8-15. Check for cracks where the shock absorber mounts to the vehicle frame.

Check shock absorbers by performing the bounce test, a visual inspection, and if needed, a bench test. If vehicle ride height is low, suspect sagging springs rather than worn shocks because shock absorbers do not support vehicle weight. Keep in mind, however, that if the springs are worn, the shock absorbers are probably worn as well.

The bounce test
Before performing the bounce test, be sure that all four tires are inflated to the recommended pressures, and that there are no heavy items in the vehicle.

1. Push down hard on one corner of the vehicle two or three times.

Figure 8-16. These bushings on the upper mount hold the shock absorber firmly in place. Check bushings to be sure that they are not cracked and brittle.

2. Watch the jounce and rebound action.
3. The car should stop moving after one or two bounces. If the vehicle bounces excessively, then the shock in that corner is likely worn, and you should replace it. If you replace one shock, you should replace the other shock on the same axle for even handling.
4. Perform the bounce test at all four corners of the vehicle, noting which corners, if any, allow excessive bouncing.

Visual inspection
If you are still unsure about shock absorber condition after performing the bounce test, raise the car on a lift, and perform the following visual inspection:

- Check for abnormal tire wear by running your hand from the inside to the outside of the tire tread. If you find "cupping," figure 8-14, which feels like rounded indentations in the tread, the shocks are worn and should be replaced. Another possible cause of cupping is a tire and wheel that is out of balance.
- Check for cracks on the frame where the shock absorber mounts, figure 8-15. Grab each shock and push it back and forth to check for play, which indicates that the mount bushings are worn, figure 8-16.
- Check for oil on the shock absorber, figure 8-17. A light oily film, called "weepage," is normal. But excessive oil, as shown, indicates that the shock absorber seals are damaged.

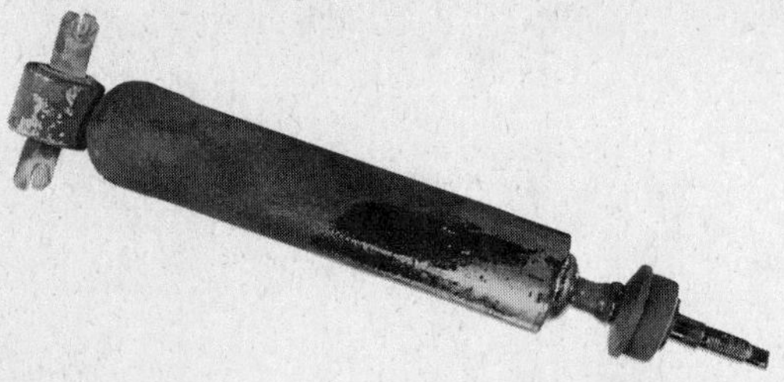

Figure 8-17. If you find oil dripping from the shock, the seals are worn and the shock absorber has lost its damping ability.

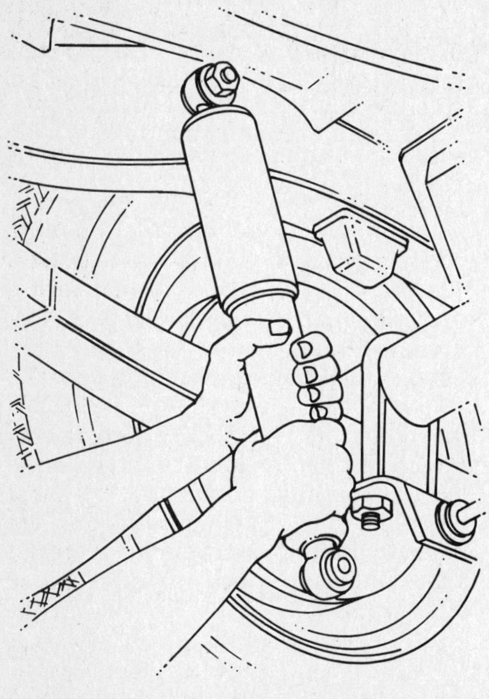

Figure 8-18. Extend and compress the shock absorber. Check for changes in resistance, which indicate that the shock absorber is damaged internally.

- If the suspension uses basic hydraulic shocks, disconnect the shock at the lower end, observing the CAUTION note that follows. Pull to extend, and push to compress the shock, figure 8-18. Check for changes in resistance, which indicate the shock is defective. Do not perform this test if the vehicle uses gas, coil-over, or air shocks.

CAUTION: You must support suspension components and secure suspension springs before disconnecting the shocks on some vehicles. Neglecting to do this can be extremely

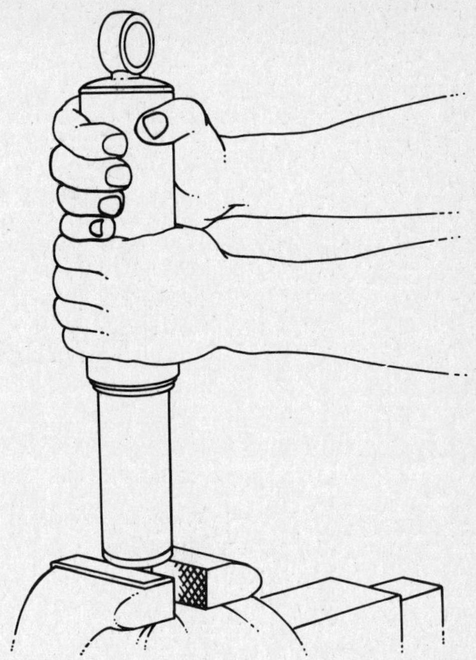

Figure 8-19. Secure the shock absorber in a vise when performing the bench test. Do not clamp the shock absorber housing because it bends easily. Instead, clamp it at the lower mount.

dangerous. Disconnect the shock absorber mounts only after following the safety precautions discussed later in this chapter in the section titled "Removing and Installing Shock Absorbers." Also, follow the safety precautions outlined in Chapter One of this *Shop Manual*.

Bench test

If needed, bench test a basic hydraulic shock absorber. Do not perform this test on gas-charged or coil-over shocks. First, remove the shock absorber from the vehicle, as described later in this chapter in the section titled "Removing and Installing Shock Absorbers."

1. Hold the shock upright as it is installed in the vehicle, which allows air bubbles in the hydraulic fluid to rise to the top. Fully extend and compress the shock three times to remove air bubbles.
2. Clamp the shock in a vise, securing it at the base, figure 8-19. Do not clamp the vise jaws on the shock housing or you may damage the shock.
3. Extend and compress the shock as rapidly as possible.

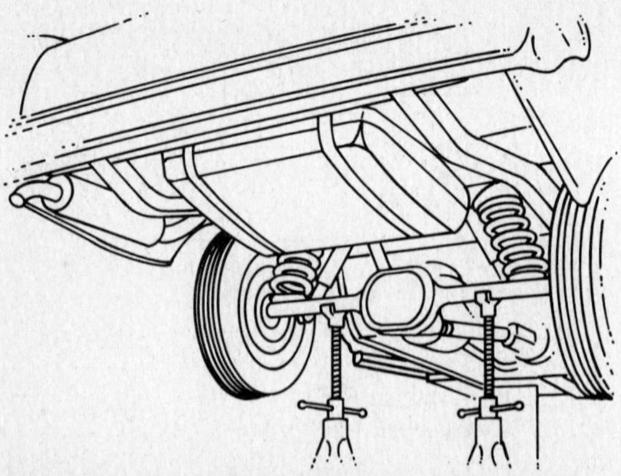

Figure 8-20. If the shock absorbers limit downward travel of suspension components, you must support the components with a jack stand to keep them from dropping when the shocks are removed.

4. It is normal to feel greater resistance when compressing the shock than when extending it. However, you should replace the shock if:
   - The piston skips or jerks
   - There is little or no resistance
   - You hear a scraping noise.

Removing and Installing Shock Absorbers

The average life of a standard hydraulic shock absorber is 30,000 to 40,000 miles (48,000 to 64,000 kilometers). However, if a vehicle is routinely driven on extremely rough roads, you may have to replace the shocks sooner. If you determine that the shocks are worn, use the following procedures to remove and replace them. Be sure to follow the safety precautions discussed in Chapter One of this *Shop Manual* when working on any suspension component.

Removing shocks

Replacing shock absorbers is usually not difficult. If one shock is worn, replace the other shock on the same axle to ensure even handling.

If the vehicle uses manually filled, heavy-duty air shocks, release the air in them. If the vehicle has an automatic leveling system as part of an electronically controlled suspension, be sure to disable it. The owner's and service manuals have instructions for doing this. See Chapter Eleven of this *Shop Manual* for lifting

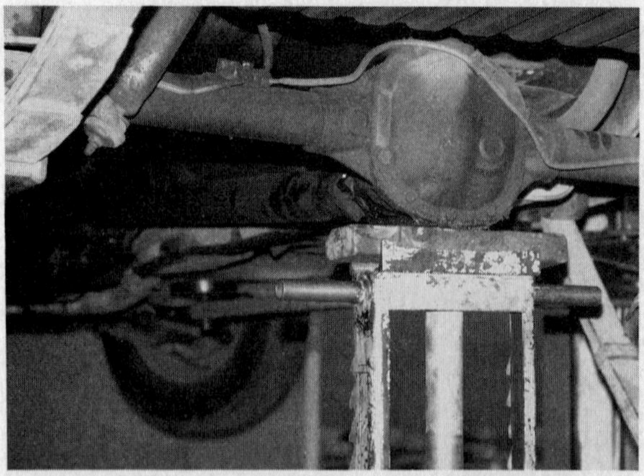

Figure 8-21. If one is available, use a bumper jack to support the axle before removing shock absorbers.

vehicles with electronically controlled suspensions.

1. Position the vehicle on a lift, and ensure that components will not bend under the weight of the car when it is raised. See Chapter One of this *Shop Manual* for lifting a vehicle. Loosen the wheel lug nuts, and lift the vehicle.
2. Remove the tire and wheel for easier access to the shock absorber.
3. On some vehicles, shock absorbers limit downward suspension travel. For these vehicles you must support the suspension using one of the following methods or the manufacturer's recommended procedure before removing shock absorbers.
   - Support the axle, beam, or lower control arm with long jack stands, figure 8-20, or a bumper jack, figure 8-21.
   - If you do not have extra long jack stands, remove the shocks with the vehicle's wheels on the ground. Raise the vehicle slightly to remove weight from the wheels, but do not lift the wheels off the ground. You will have to work under the car to remove and install the shock absorbers, but the ground will support the tires, wheels, and suspension when the shocks are removed.
4. You will find four types of shock absorber mounts — bayonet, bar, ring, and integral-stud, figure 8-22. Disconnect the upper and lower shock absorber mounts. You may have to apply penetrating oil to the fasteners that secure the shock. If one is available, use a "double square" socket, which is specially designed for removing and installing the nuts on shock absorbers.

SHOCK ABSORBER MOUNTS

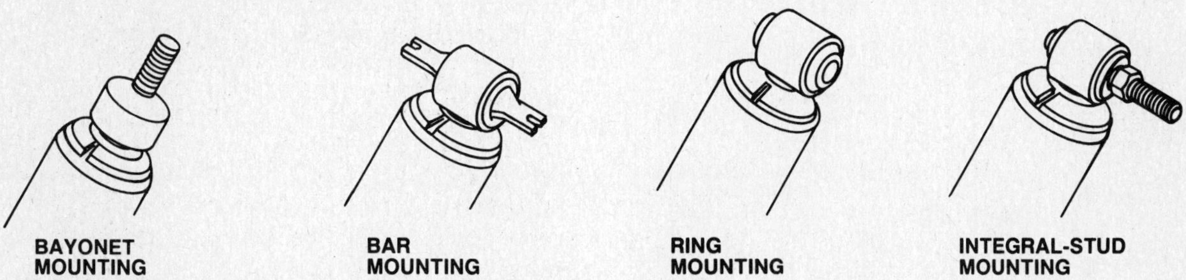

Figure 8-22. There are four types of shock absorber mounts. On many shock absorbers, the top and bottom mounts are different.

Figure 8-23. To remove a shock absorber that has a ring mount, remove the nut, and slip the shock absorber off the stud.

These sockets are available from most major tool manufacturers.

- To remove a bayonet mount, hold the stud that extends from the shock absorber with locking pliers. Loosen the nut with a wrench. Remove the nut, washer, and bushing from the stud, and push the stud out of the mounting hole.
- To remove a bar mount, remove the nuts at both ends of the bar. You may have to hold the bolt head with a wrench to keep it from turning while you are removing the nut. Remove the bolts from the shock and suspension component, and set them aside.
- To remove a ring mount, figure 8-23, loosen and remove the nut on the bolt or stud. Remove the bolt, or slip the shock off the stud.
- To remove an integral-stud mount, loosen and remove the nut on the stud. Then slip the stud out of the suspension component, and remove the shock.

5. If the shock absorber is inside a coil spring, carefully guide the shock out of the spring, and remove it. There is usually a hole in the lower control arm through which you can remove the shock, although on some vehicles you must remove the shock from the top of the spring.

Installing shocks

Compare the new shock absorbers with the old ones to be sure that you are installing the correct shocks. Pay special attention to the type of mounts and the shock absorber extended length, which is usually printed on the box. Aftermarket parts manufacturers have catalogs that specify the correct shocks for most domestic and import vehicles.

Before installing a basic hydraulic shock absorber, prime it to remove air bubbles from the hydraulic fluid. (Do not attempt to prime a gas-filled or coil-over shock absorber.) Hold the shock upright as it is installed in the vehicle, and extend it fully. Turn the shock upside down and compress it. Repeat this procedure several times.

Support suspension components with jack stands, if needed, while you are installing the new shocks. Extend the shock absorber, and attach the upper and lower mounts, turning the nuts finger tight. During installation, do not hold the shock housing with vise grips, as you can bend the metal and ruin the shock. Also, never hold the shock absorber piston rod. The piston rod moves through the seal while the shock is operating, and any nicks on the piston rod will cut the seal. Lower the vehicle, and tighten the nuts or bolts to the manufacturer's recommended torque. Bounce the front and rear of the vehicle to ensure that the shocks are operating properly.

If you are installing gas-filled shocks, there is a fastener that keeps the shock absorber compressed. *Do not* remove this fastener until after

Figure 8-24. Check the nuts that secure the upper strut mount to the vehicle body. In this example, the three outer nuts secure the strut to the body, and the center nut holds the strut assembly together, keeping the spring compressed.

you have secured the shock absorber's top and bottom mounts to the vehicle suspension and frame. First secure one shock absorber mount. Then position suspension components with a jack until you can secure the other mount. After connecting the upper and lower mounts, remove the fastener on the shock absorber, which allows the shock to expand.

Another way to install gas-filled shocks is to secure one shock absorber mount, and then *loosen* the fastener, which allows the shock absorber to expand slightly. When the shock has expanded enough so that it is even with the other mount, tighten the fastener to keep the shock from expanding further, and secure the other shock absorber mount. Remove the fastener, and allow the shock to fully extend. Lower the vehicle and tighten the nuts or bolts to the manufacturer's recommended torque. Bounce the front and rear of the vehicle to ensure that the shocks are operating properly.

STRUTS

A strut is similar to a shock absorber, but is reinforced because it also must support vehicle weight. A fluid-filled cartridge inside the strut housing damps suspension oscillations like a shock absorber. This cartridge is removable in some systems, but for many late-model cars, if the cartridge is worn, you must replace the entire strut assembly.

Inspecting Struts

Check strut condition using the shock absorber bounce test described earlier in this chapter. If the vehicle resists bouncing, the strut shaft may be bent or corroded. If the vehicle bounces excessively, the struts are worn and should be replaced. (Many manufacturers recommend that you replace both struts on a single axle, even if only one is worn.)

During the test drive, in addition to the symptoms described for shock absorbers, check for vibration during acceleration, which indicates that strut mounts may be loose. If the vehicle is hard to steer or if you hear squeaking noises during turns, the strut upper bearing may be worn or corroded, or the strut rubber bushing may be worn.

During the visual inspection, *do not* disconnect the lower strut mount to check strut movement, and *do not* attempt to perform the bench test, as described in the shock absorber inspection procedure. Instead, inspect the strut for leaks, bends, and dents. Check that the nuts at the upper strut mounts are tight, figure 8-24. Note the ride height at all four corners of the vehicle. If the vehicle is not level, the strut springs are likely worn.

Removing the Strut Assembly

Before removing the struts, be sure to secure suspension springs.

WARNING: If you are working on a suspension system with modified struts, the springs are not integral with the struts. Therefore, you must use the procedures and precautions given later in the section titled "SPRINGS" to secure suspension springs. Then you can safely remove the struts.

1. Position the vehicle on a lift ensuring that components will not bend under the weight of the car when it is raised. See Chapter One of this *Shop Manual* for lifting a vehicle. Loosen the wheel lug nuts, but do not lift the vehicle.
2. Locate the bolts that secure the upper strut mount to the vehicle body. For front axle struts, the top mounts are in the engine compartment, figure 8-25. For rear axle struts, you can reach the upper mount bolts through the hatch or trunk, figure 8-26. However, on some vehicles you must remove the back seat to access the rear strut mounts in the passenger compartment.
3. Remove the dust cap that covers the strut upper mount. Mark one of the studs protruding from the upper mount, and mark the corresponding stud hole, figure 8-27, so that you can match them up when you reinstall the strut.
4. Loosen the nuts that secure the upper strut mount to the vehicle body. Remove all but one nut, leaving the remaining nut finger-tight.

Figure 8-25. The upper mounts on front struts can be accessed from the engine compartment. On this vehicle, a rubber cap covers the center nut.

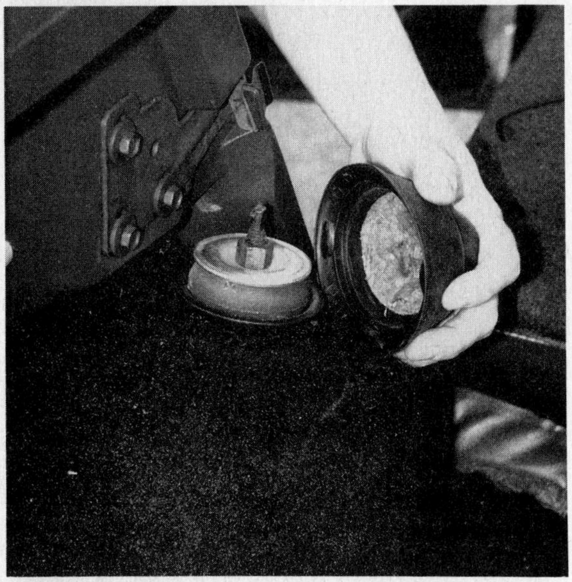

Figure 8-26. The upper mounts on rear struts are accessed from the hatch, trunk, or under the rear passenger seat.

CAUTION: *Do not* remove the center nut on the upper strut mount. The center nut holds the upper spring seat in place and keeps the coil spring compressed.

5. Raise the vehicle on the lift. Remove the tire and wheel for easier access to the strut base.
6. If you are working on a modified strut suspension, support the axle, beam, or lower control arm with jack stands, and secure the suspension springs, as described in the "SPRINGS" section of this chapter. Supporting the axle, beam, or lower control arm will keep the spring compressed while you remove the strut.

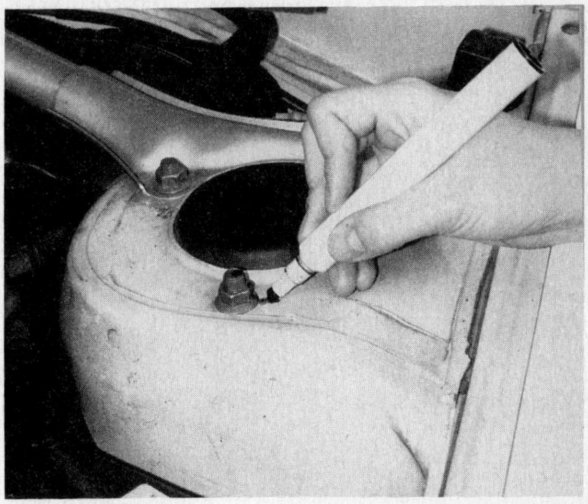

Figure 8-27. Mark a stud and the corresponding hole before removing the strut. On some vehicles, the upper strut mount is tapered and must be repositioned in the same way during reassembly or the wheel alignment camber adjustment will be off.

7. If you are working on a strut/SLA suspension, you may have to disconnect the ball joint between the long knuckle and upper control arm before removing the strut. See Chapter Nine of this *Shop Manual* for information on disconnecting ball joints.
8. If brake hoses are fastened to the strut, disconnect the hose from the strut assembly. (If you have to open the brake hydraulic system in order to remove the strut, you must bleed the brakes and add new fluid after completing this job. Be sure that you include this in your estimates for labor time and cost.) On some vehicles, the brake caliper prevents access to the lower strut mount, and you must remove it. If you remove the caliper, be sure to support it using rope or wire. Do not allow the caliper to hang by the brake line, as this will damage the line and fittings. Chapter Twelve of this *Shop Manual* has basic service information for the brake system if you need it. Also disconnect any electrical connectors for an electronic suspension.
9. Place the hooks of the spring compressor on the strut spring, figure 8-28. Turn the bolts to compress the spring about one or two inches. This will relieve some of the spring tension at the base of the strut, making the lower strut mount easier to disconnect.
10. Remove the bolts that secure the lower strut mount to the wheel knuckle, figure 8-29. Then separate the wheel knuckle and strut assembly by pulling them apart.

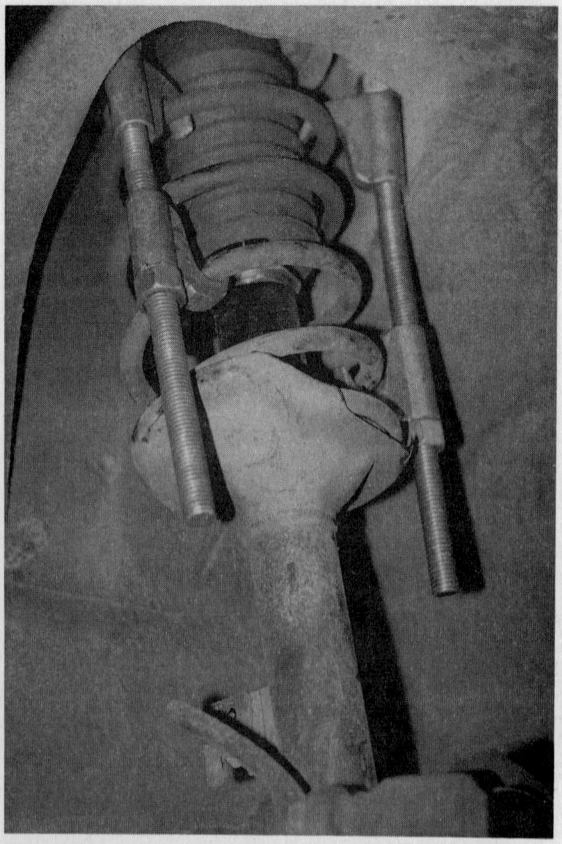

Figure 8-28. Although it is not absolutely necessary, you can compress the integral strut spring before removing the strut from the vehicle. This relieves some of the spring pressure on the lower strut mount.

Figure 8-29. On most MacPherson and modified struts, a bracket that extends from the base of the strut bolts to the wheel knuckle.

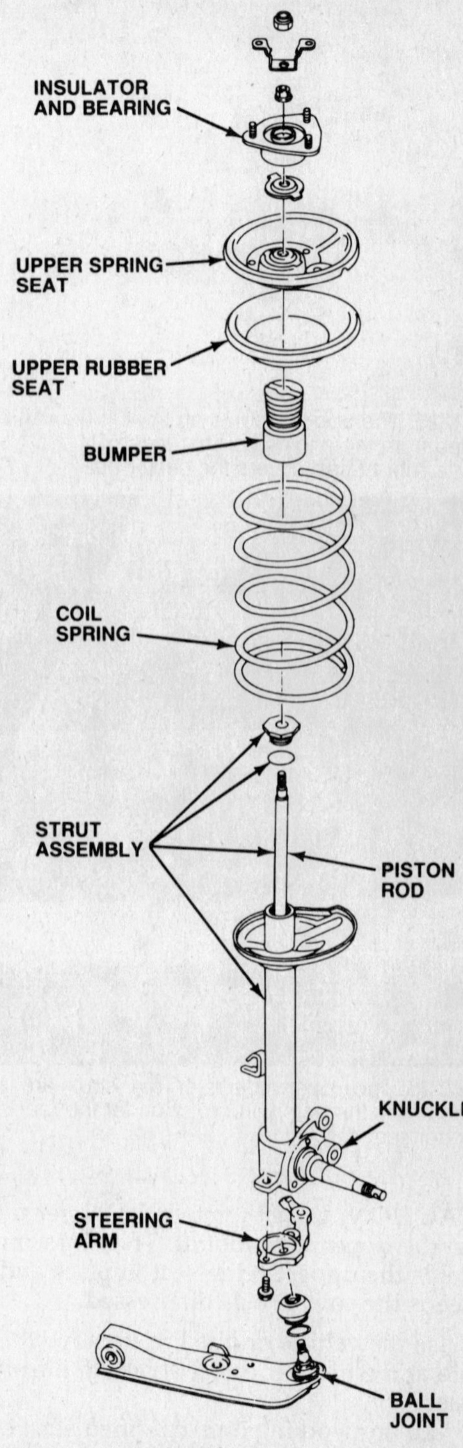

Figure 8-30. On some strut systems, the wheel knuckle and strut cannot be separated. To remove this type of strut you must separate the ball joint between the wheel knuckle/lower strut mount and the lower control arm.

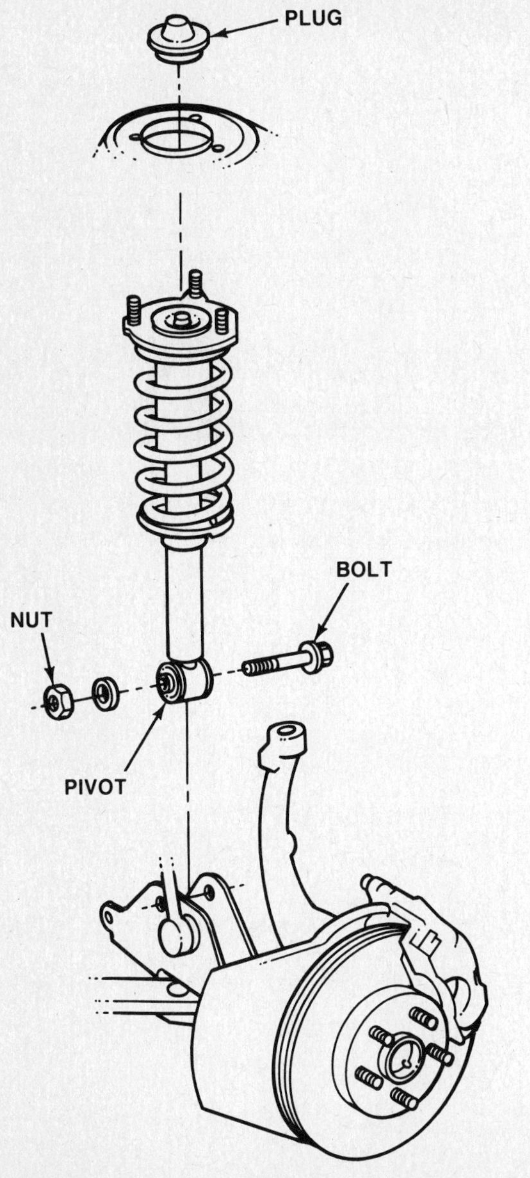

Figure 8-31. If the strut is connected with a pivot, remove the nut that secures the through-bolt, and remove the bolt.

- On some vehicles, the strut and wheel knuckle are an integral assembly, and you must remove the wheel knuckle as part of the lower strut mount, figure 8-30. Separate the ball joint or joints that connect the wheel knuckle to the lower control arm using one of the procedures described in Chapter Nine of this *Shop Manual*. Then separate the lower control arm from the wheel knuckle/lower strut mount.
- If the vehicle uses pivot-base struts, locate the pivot at the base of the strut.

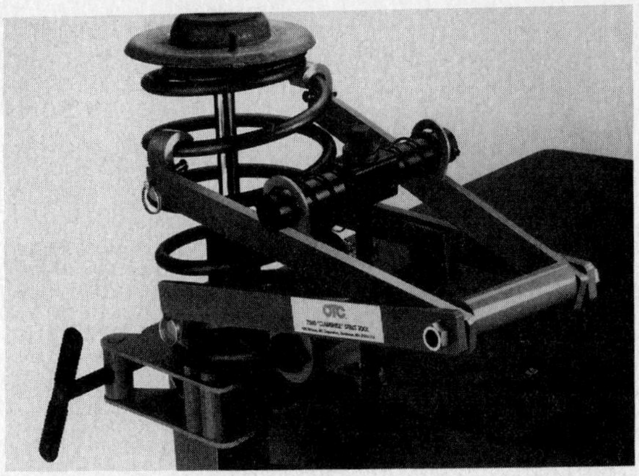

Figure 8-32. This OTC strut spring compressor clamps the strut firmly in place. After placing the forks over the spring coils, use an impact wrench to turn the bolt and compress the spring.

Hold the bolt head with a wrench, and loosen and remove the nut, figure 8-31. Separate the strut and lower control arm.

11. Remove the last nut from the upper strut mount, and remove the strut from the vehicle. If you remove both struts, mark the left and right ones so that you can install them in the same place later. (They are not usually interchangeable.) To avoid confusion, many technicians will remove and replace one strut at a time.

Disassembling the Strut

Service procedures vary depending on the type of strut used on the vehicle. If the vehicle manufacturer does not provide strut rebuild kits, remove the old strut and replace it, using the procedures given later in this chapter. Usually, the springs are replaceable. For modified struts, refer to the ''SPRINGS'' section later in this chapter to remove and replace coil springs. For MacPherson and strut/SLA suspensions, compress and remove the integral coil spring, as follows, before disassembling the strut.

Compressing and removing a strut coil spring
A MacPherson strut or strut/SLA suspension has a highly compressed, load-carrying, integral coil spring. Use a spring compressor that is specially designed for strut springs. Any improvised methods for strut spring removal are extremely dangerous and are not recommended by vehicle manufacturers or professional technicians.

1. Mount the strut in a vise, figure 8-32, and place the spring compressor forks over the

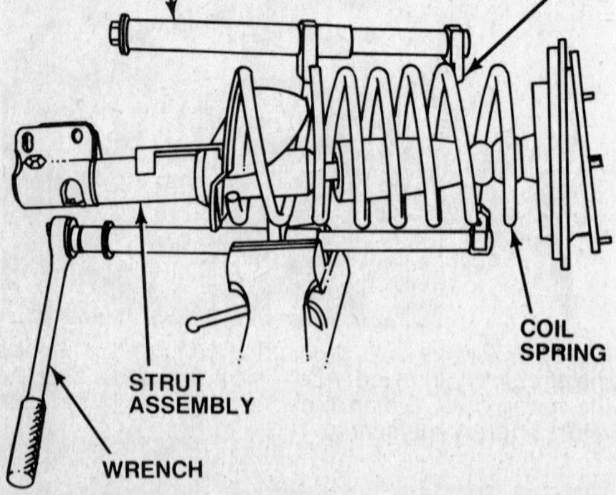

Figure 8-33. If you are using a manual fork-type compressor, place the forks on the upper and lower spring coils. Turn the long bolt with a wrench to compress the spring.

coils at each end of the spring, figure 8-33. If the shop has a pneumatic strut spring compressor, mount the strut in the press, figure 8-34.

2. Compress the spring until there is clearance between the piston rod nut and the upper spring seat, figure 8-35. For a manual-type spring compressor, tighten the clamp with a wrench to compress the coil spring. *Do not* use an impact wrench to tighten the spring compressor, unless specified by the spring compressor manufacturer.
3. With the spring compressed and secure, remove the retaining nut at the center of the upper spring seat, figure 8-36.
4. Remove the upper spring seat and any bushings, bearings, or insulators that are under the seat, figure 8-37.
5. Gradually release spring tension. On the manual-type spring compressor, loosen the clamp that secures the spring. On the pneumatic-type spring compressor, use the manufacturer's instructions to release spring tension. Remove the spring, figure 8-38. If you removed both struts from the vehicle, mark the left and right struts and springs, as they are not interchangeable.
6. Remove rubber bushings or bump stops, as well as the dust cover, if any, from the strut, figure 8-39.

Figure 8-34. This strut spring compressor is air-powered. Secure the strut in the upper and lower brackets on the compressor.

Installing a new strut cartridge

If you are working on a vehicle that uses struts with replaceable cartridges, use this procedure to replace the cartridge. On most late-model vehicles, if the cartridge is worn, you must replace the entire strut assembly.

1. After compressing and removing the strut coil spring, use a special wrench to remove the nut at the top of the strut, figure 8-40. Remove any seals that are under the nut. Discard these seals, and replace them with new ones, as they prevent hydraulic fluid from leaking. New seals usually are included in a strut cartridge rebuild kit.

Figure 8-35. The upper bracket of this strut spring compressor moves downward when air pressure is applied, compressing the coil spring.

2. Pull the old cartridge out of the strut housing, figure 8-41. For some vehicles you will have to use a twisting motion to unscrew the cartridge from a threaded mount at the base of the strut housing.
3. Install the new cartridge, figure 8-42, and twist it in place. Some manufacturers suggest that you add a small amount of lightweight oil to the strut housing to help dissipate heat generated by the shock cartridge.
4. Install the new seals and replace the nut, tightening it to the manufacturer's specifications.

Inspecting strut hardware
Before assembling the strut, inspect the upper spring seat and rubber bushing. If either are worn, replace them with new ones. Check the bump stop and dust cover for wear. Inspect the upper bearing, figure 8-43, if there is one, and replace it if damaged. Finally, inspect the

Figure 8-36. Remove the strut retaining nut after compressing the spring. This is the nut at the center of the upper spring seat.

Figure 8-37. After you remove the retaining nut, you can remove the upper spring seat and bearing from the strut assembly.

Figure 8-38. Gradually release spring tension, and remove the spring.

coil spring, using the spring inspection procedure described later in this chapter. Many of these replacement parts are available in strut rebuild kits.

Assembling the Strut

After discarding damaged hardware, assemble the strut.

1. Mount the strut in a vise, as described earlier in this chapter. Extend the strut to its full length.
2. Place the coil spring on the strut, seating the spring in the lower spring seat.
3. Place the dust cover, bump stop, rubber bushing, and upper spring seat over the spring.
4. Compress the spring until the threaded end of the rod clears the top of the upper spring seat.
5. Secure the upper spring seat with the strut retaining nut, and tighten the nut to the manufacturer's recommended torque.
6. Gradually release tension on the spring.
7. Check that the spring is properly positioned in the upper and lower spring seats.
8. Remove the strut from the vise.

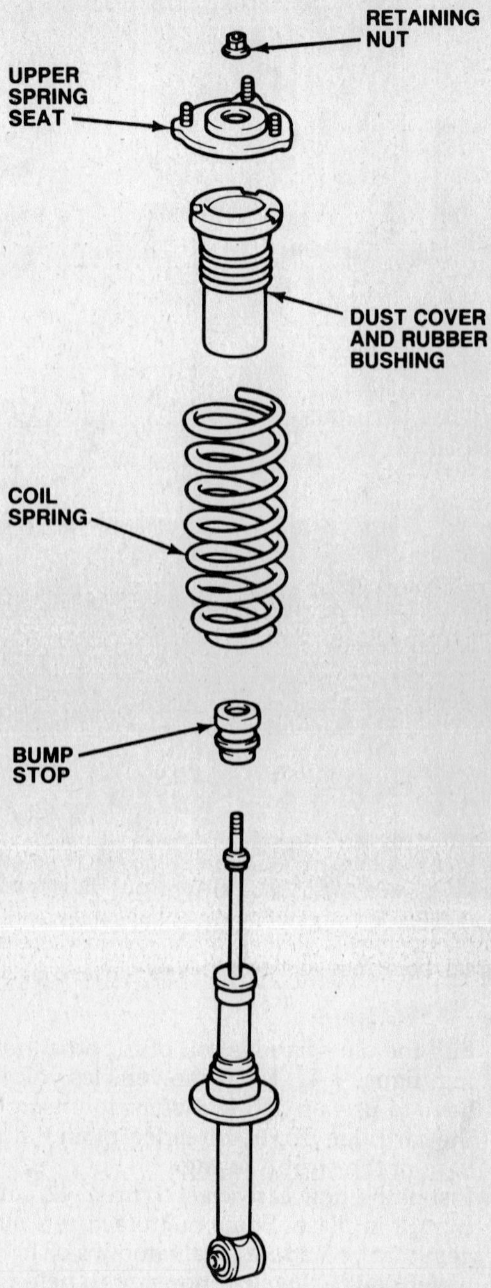

Figure 8-39. After removing the strut's upper spring seat, you can remove the spring, bump stops, bushings, and insulators.

Installing the Strut

If you are working on a vehicle with modified struts, and the suspension springs are worn, replace the springs before installing the strut. Procedures for installing coil springs are given later in this chapter. Install a strut as follows:

1. Place the spring compressor hooks over the coils on the strut spring. Use a wrench

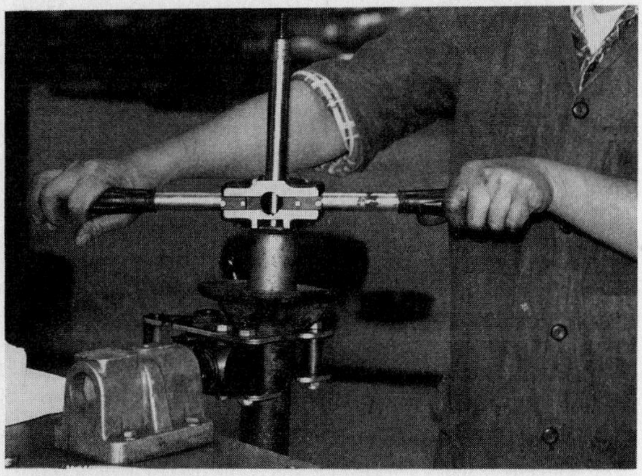

Figure 8-40. This wrench is specially designed for removing the strut cartridge nut.

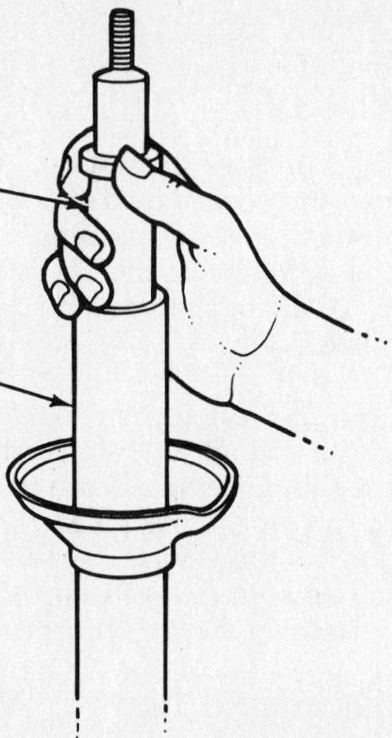

Figure 8-41. Some struts have a removable shock absorber cartridge, but on most late-model vehicles, you must replace the entire strut when it is worn.

to turn the bolt, and compress the spring about one or two inches to relieve pressure on the upper and lower strut mounts.

2. Position the upper strut mount on the vehicle body. If you are installing a rebuilt strut, be sure to align the stud and stud hole that you marked earlier. If you are installing a new strut, check the manufacturer's instructions for aligning the stud with the proper stud hole or camber settings may be off. Tighten the nuts that secure the upper mount to the vehicle body.

Figure 8-42. After adding oil to the strut housing, install a new strut cartridge.

3. Connect the lower strut mount to the wheel knuckle, and tighten the bolts to the manufacturer's recommended torque.
   - If the strut and wheel knuckle are an integral assembly, connect the ball joint or joints that join the strut and lower control arm. Use the procedures described in Chapter Nine of this *Shop Manual* for more information on installing ball joints.
   - If the vehicle has a strut/SLA suspension, position the pivot at the base of the strut on the lower control arm. Tighten the bolt that secures the strut pivot to the lower control arm.

Figure 8-43. This strut uses an upper bearing. If the bearing is worn, replace it before reassembling the strut.

4. If you disconnected other control arms, such as on a strut/SLA suspension, connect the ball joints. Then remove the spring compressor, and torque the castle nuts on the ball joints.
5. If you removed the brake caliper or unclipped the brake line from the strut, replace them. Also attach any connectors for an electronically controlled suspension. If you opened the brake hydraulic system, be sure to bleed the brakes and add fluid. Chapter Twelve of this *Shop Manual* contains information on servicing the brake system.
6. Install the wheels, lower the vehicle, and torque the wheel nuts. Torque the nuts and bolts at the upper and lower strut mounts. Be sure to use the manufacturer's recommended torque specifications. Replace the dust caps that cover the upper strut mounts.
7. Bounce the car. Check and adjust wheel alignment, as described in Chapter Fourteen of this *Shop Manual*.

Figure 8-44. Bump stops keep the suspension components and frame from hitting each other when the spring and shock can no longer limit suspension travel. If bump stops are worn, the springs may be weak or broken.

SPRINGS

Suspension springs support the vehicle's weight and absorb road shock to provide a comfortable ride for passengers. As discussed in Chapter Eight of the *Classroom Manual*, there are a variety of suspension spring types and locations.

After determining whether the vehicle uses coil springs, leaf springs, or torsion bars, inspect the springs and the hardware used to mount them. Replace worn or broken springs using the procedures given in the following sections. If the vehicle uses electronically controlled air springs, see Chapter Eleven of this *Shop Manual* for more information.

Inspecting Springs

A number of things cause suspension spring wear, but the primary one is friction-generated heat, which results from normal spring use. As the springs sag and weaken, they cannot support the weight of the vehicle, so the chassis lowers. This change in chassis height, or "ride height," affects control arm angle, which in turn, causes steering and wheel alignment problems. When worn or broken springs do not absorb road shock, other suspension components do, which eventually damages ball joints, bushings, and other suspension components. Use the following procedures to inspect suspension springs.

Visual inspection

A careful visual inspection will help you determine if suspension springs should be replaced. Before the inspection, check and adjust tire pressures. Park the vehicle on level ground. If

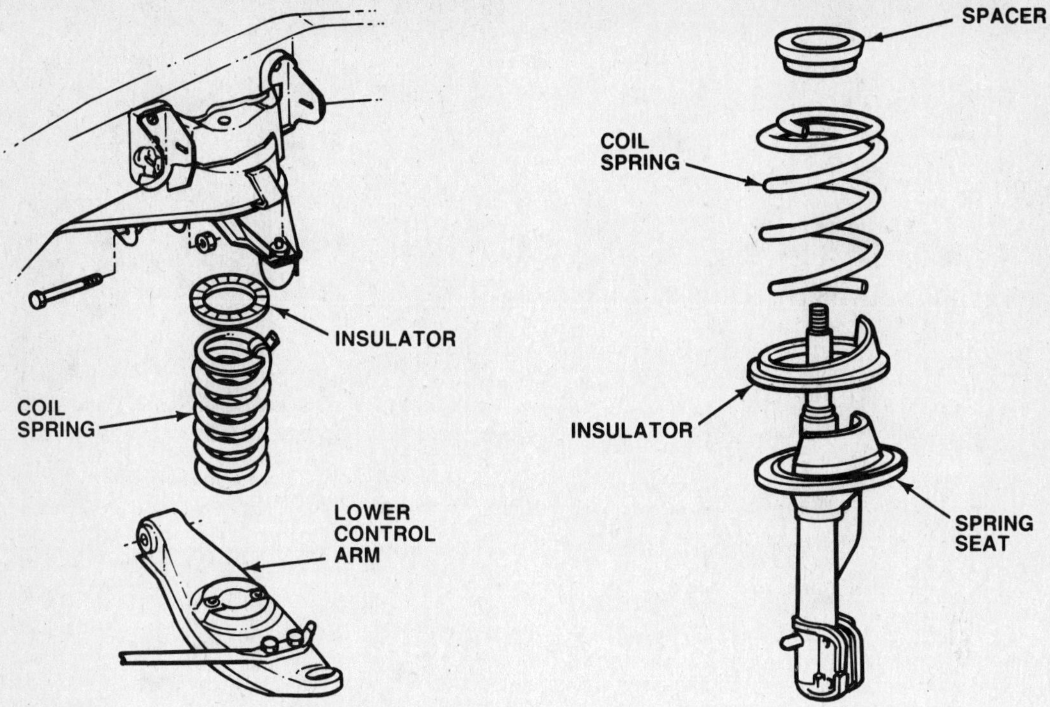

Figure 8-45. A spring insulator is placed between the spring seat and coil spring to reduce noise. Insulators may be used on control arm and strut coil springs.

there is anything in the car that may weigh it down on one side, remove the load before inspecting the springs.

- Bounce the car at each corner, as in the shock absorber bounce test described earlier in this chapter. If the springs are severely worn, the chassis will hit the rubber bump stops or the vehicle frame. You may hear a thumping noise as suspension components hit the frame.
- Walk around the vehicle, noting if height is uneven between the front and rear bumpers or between each side of the vehicle. If ride height is uneven, especially from side to side, the springs are likely worn.
- Use the procedure in Chapter Fourteen of this *Shop Manual* to measure vehicle ride height. If the ride height is not within specifications, the springs are likely worn.
- Measure the distance between each control arm and the ground. They should be the same distance from the ground.
- Inspect bump stops, figure 8-44. If they are worn, it is a good indication that the springs are worn as well.
- Check springs for obvious physical damage, such as cracks in the metal, broken brackets, worn or missing bushings, or other missing hardware.
- Check for scratches, nicks, or severe rust pits in the springs, which weaken them, eventually leading to failure.
- Check for signs that indicate the springs were intentionally heated with a torch. As described in Chapter Eight of the *Classroom Manual*, heat weakens suspension springs. Some people heat suspension springs to lower ride height without realizing that they risk suspension spring failure.

If you are checking coil springs, look for these additional items:

- Look for polished surfaces between the coils, which indicate that the coils are contacting each other, or ''clashing,'' during spring compression.
- If the suspension uses coil spring insulators, be sure that the insulators lay flat and are centered between the spring and spring seat, figure 8-45.
- Look for temporary repairs made by other technicians. One example is a rubber ''doughnut'' that fits between the spring and seat to correct vehicle ride height. Another is a rubber or metal spacer that is inserted between spring coils to prevent the

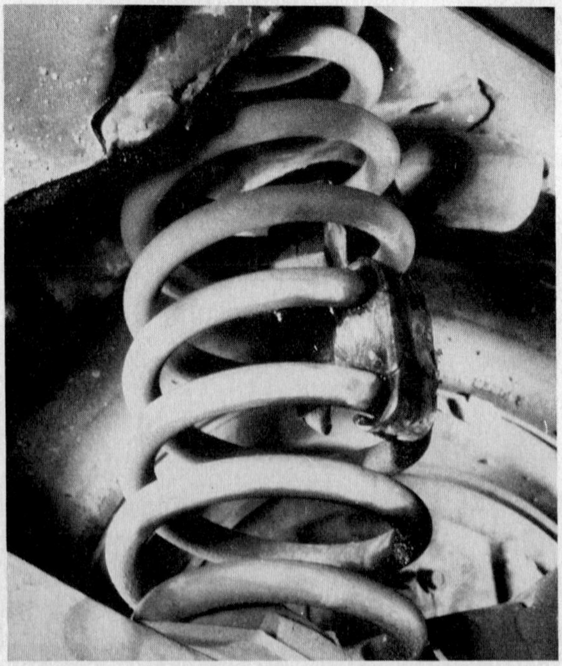

Figure 8-46. Some technicians use these rubber inserts to temporarily repair springs, but in the long run it is best to replace defective springs.

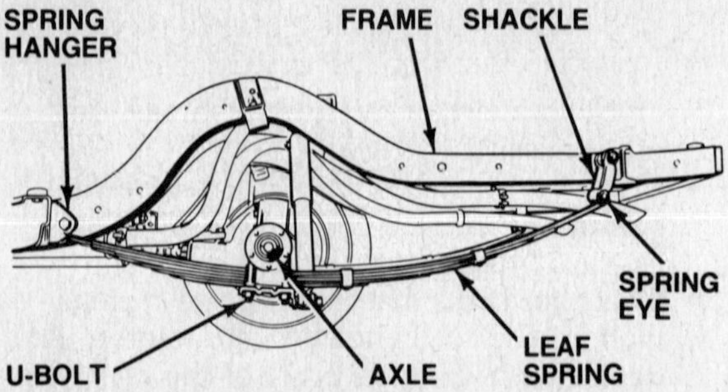

Figure 8-47. Check the leaf springs and the hardware that holds them in place. Be sure that the spring hanger and shackle are securely connected to the vehicle frame.

springs from compressing normally, figure 8-46. It is good practice to replace defective springs rather than use these temporary repairs.

If you are checking leaf springs, figure 8-47, look for these additional items:

- Be sure that the shackles are securely attached to the leaf springs and frame, and check that the nuts securing the U-bolts to the axle are tight, figure 8-48.
- Check rubber bushings, which reduce spring vibration. Shackles and spring hangers have rubber bushings in the spring "eyes," figure 8-49. If the bushings are worn or missing, you will hear a squeaking noise when you bounce the car. Some manufacturers also place a rubber bushing between the leaf spring and U-bolt, which you should inspect for wear.
- Check the condition of the leaves near the U-bolts. Because there is not much distance between the leaves at this point, this area is subject to a great deal of stress, and you may find cracks.
- Some leaf springs have plastic or zinc pads between the leaves, or rubber pads at the ends of the leaves to reduce friction. Check the condition of these spacers or pads.

Figure 8-48. U-bolts secure the leaf springs to the axle. Check the nuts on the ends of the U-bolts to be sure they are tight. Otherwise the leaf springs will move around on the axle.

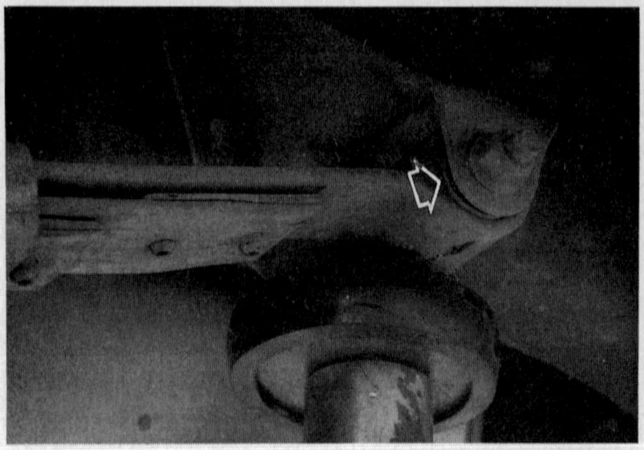

Figure 8-49. Check the rubber bushings in the spring eyes. Also check the shackle and spring hanger at the vehicle frame to be sure that they are mounted securely.

If you are checking torsion bars, figure 8-50, look for these additional items:

- Torsion bars are marked for the left and right sides of the car, and are not interchangeable. Check that they are properly placed.

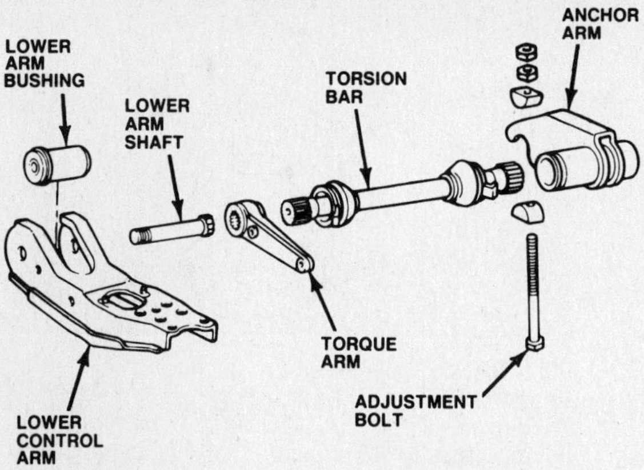

Figure 8-50. Be sure that the torsion bars are mounted securely to the frame at the anchor arms and to the suspension at the torque arms.

- Check the torsion bar adjustment bolt, which is located near the anchor arm. The bolt should hold the torsion bar firmly in place.
- On some torsion bars, there is a seal that covers the area where the torsion bar and anchor arm meet. This seal should not be ripped or damaged.
- If the torsion bar is housed in an axle tube, check that the tube is not bent and the tube seals are in good condition. If you suspect damage, remove the torsion bar, as described later in this chapter. Check that the splines on the ends of the torsion bar are properly greased and are not stripped.

Servicing Coil Springs

As discussed in Chapter Eight of the *Classroom Manual*, there are a number of things that affect suspension spring performance. They include:

- Spring material
- Spring rate and frequency
- Spring end construction
- Spring length
- Base coil diameter.

Although two springs may look alike, they may not have the same properties, so use a reliable source when choosing replacement springs for a customer's vehicle. The manufacturer's part number is often stamped in the metal at the base of the spring, although some manufacturers put the spring part number on a tag that falls off long before the spring has to be replaced. If this is the case, use an aftermarket parts catalog, which will give you the correct spring part number.

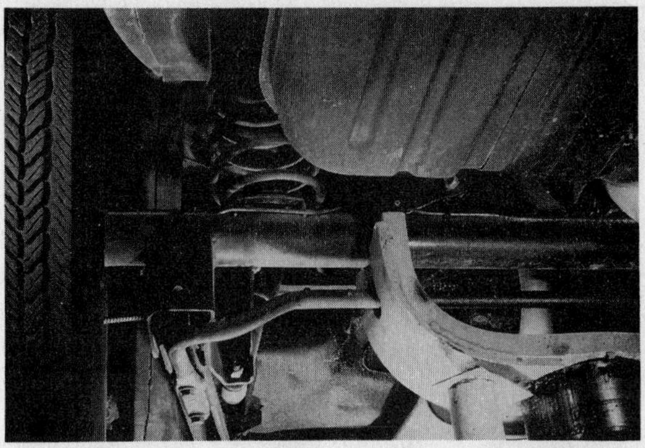

Figure 8-51. Place a jack under the lower spring seat, and raise the jack to compress the spring. In this example, the lower spring seat is on the beam axle.

Before replacing springs, determine which suspension components compress them. If you remove a suspension part that compresses the spring, you risk injury to yourself and damage to the vehicle. Always secure the springs using one of the methods described in the following section.

Securing coil springs
You must secure coil springs so they will not fly out from between suspension components while you are removing parts. There are two common ways to secure coil springs, and the method that you use depends on the type of suspension system you are working on and the tools that are available. *Never* attempt to remove coil springs without securing them first.

Safety clips
Clips are easy to use, and are recommended when removing the coil springs between the upper or lower control arm and frame on an SLA suspension or on a beam axle. Check the rating of the spring clips to be sure that they can hold the spring you are removing.

1. Disconnect any components that may prevent the spring from being removed, such as tie rod ends or brake lines.
2. Place a jack under the suspension component that contains the lower spring seat. This may be a lower control arm or a beam axle.
3. Raise the jack to compress the spring, figure 8-51.
4. Place spring clips around the coils according to the spring clip manufacturer's instructions, figure 8-52. Long clips can usually secure five coils, and short ones can secure four. Adjust the jack height, as needed, to position the clips.

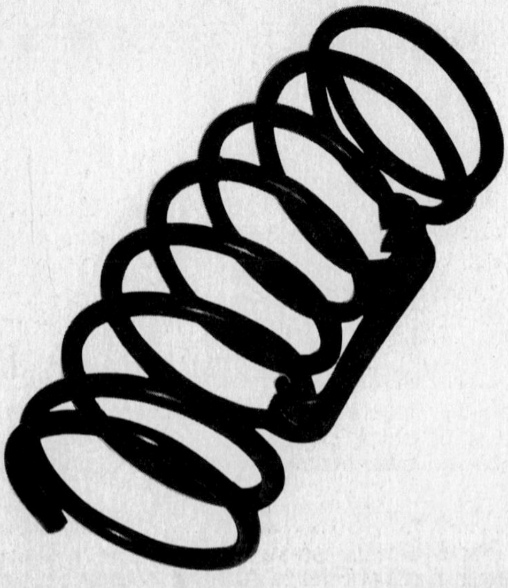

Figure 8-52. After compressing the spring, place clips around the coils using the spring clip manufacturer's recommended procedure. The clips will keep the spring compressed after you remove the jack.

5. Slowly lower the jack to release pressure from the spring. The clips should fit snugly on the coils, keeping the spring compressed. As you lower the jack, check the clips to be sure they are secure.
6. Remove the spring using one of the procedures given later in this chapter. Be sure to handle the compressed spring carefully.
7. To remove the clips, place the spring in a compressor, such as the one described in the next section. Compress the spring, and remove the clips.

Compressor plates

Use compressor plates, figure 8-53, if the lower end of the coil spring is easily accessible from under the vehicle. Compressor plates often are used when removing coil springs from between the upper control arm and frame and on some modified strut suspension systems.

1. Place the upper and lower compressor plates on the upper and lower spring coils. Be sure that the compressor plates are rated to withstand the spring's force.
2. Slip the threaded rod through the thrust washer, ball nut, and lower plate. Turn the rod to thread it through the hole in the upper compression plate, and install the pin into the ball nut to secure the rod.
3. Check the position of the plates to be sure that they are secure on the coils.
4. Use a socket wrench to turn the threaded rod, which slowly brings the plates together. Continue turning the rod until the spring is compressed so that you can remove it. Use one of the procedures given in the following section to remove the coil spring.
5. After removing the spring and tool from the vehicle, slowly unscrew the threaded rod, and remove the compressor plates from the spring.
6. Compress the new spring to the same length that the old one was compressed before installing it.

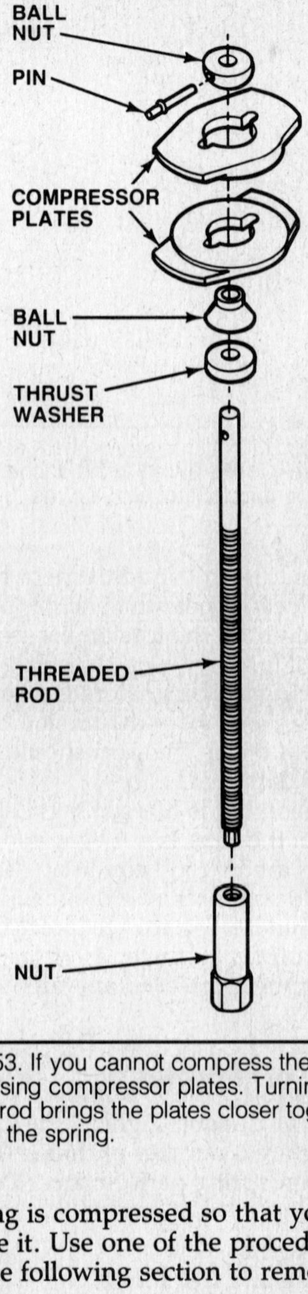

Figure 8-53. If you cannot compress the spring with a jack, try using compressor plates. Turning the threaded rod brings the plates closer together to compress the spring.

Removing and installing coil springs

Coil springs may be located between the upper or lower control arm and frame on an SLA suspension system, or between a trailing arm,

Figure 8-54. Some spring seats have holes in them to allow for water drainage. These holes are used by some manufacturers to properly position the springs.

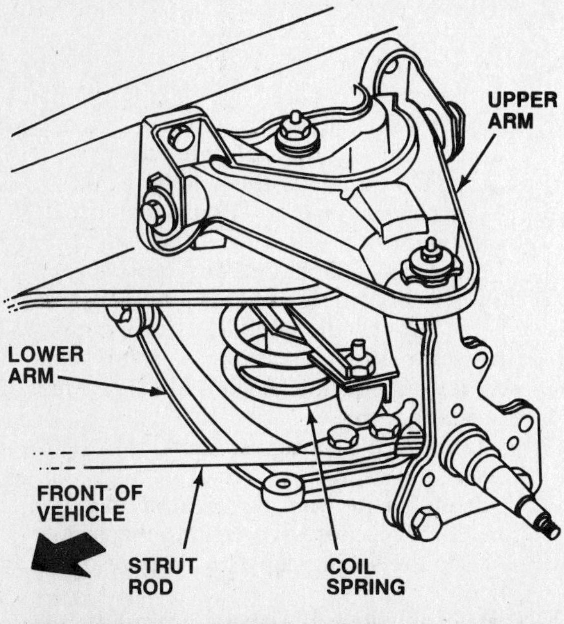

Figure 8-55. On most SLA (short-long-arm) suspension systems, the coil spring is positioned between the lower control arm and frame.

I-beam, or axle and the vehicle frame or crossmember. Be sure that you have tools, such as safety clips or compressor plates, to secure the coil springs before removing and installing them. To remove coil springs:

1. Position the vehicle on a lift, ensuring that components will not bend under the weight of the car when it is raised. See Chapter One of this *Shop Manual* for lifting a vehicle. Loosen the wheel lug nuts, and lift the vehicle.
2. Remove the tire and wheel, as this normally makes it easier to access suspension components.

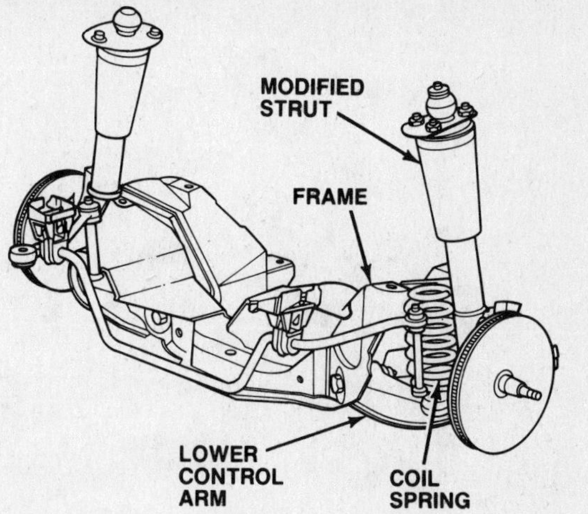

Figure 8-56. On this modified strut suspension where the spring rests between the frame and lower control arm, you cannot use a jack to compress the spring. Instead, use compressor plates.

3. Support the axle, beam, or control arm with extra long jack stands.
4. If the shocks limit downward suspension travel or hinder access to the springs, remove them, as described earlier in this chapter.
5. Use the instructions in the following sections to remove and install coil springs on various suspension types.

Note: When installing coil springs, remember that some are designed to be placed only on the left or right side of the vehicle. Also check for special grooves or holes in the spring seats, figure 8-54. Position the springs according to the manufacturer's instructions, using the grooves or holes as a reference point. Check spring insulators, and replace them if they are worn.

Coil springs with control arms

If the coil spring is positioned on the lower control arm of a short-long-arm (SLA) suspension, figure 8-55, or between the lower control arm and the vehicle frame as on this modified strut suspension, figure 8-56, use this procedure to remove them. Rarely will you find the coil spring between the upper control arm and the frame, although some Saabs have this setup.

1. Disconnect the outer tie rod ends, if necessary, as described in Chapter Five of this *Shop Manual*.
2. Place a jack under the lower control arm, and lift the control arm until the lower ball

Figure 8-57. This coil spring rests between a solid axle and the vehicle frame. Notice the part number tag on the coil spring.

joints are unloaded. The lower ball joints will move easily when weight is removed from them.

3. Secure the coil spring with safety clips or compressor plates, as described earlier in this section.
4. Disconnect the ball joints at the lower control arm. Chapter Nine of this *Shop Manual* has complete instructions for disconnecting suspension ball joints. Also, disconnect the strut rod if one is attached to the lower control arm, as described in Chapter Nine of this *Shop Manual*.
5. Lower the jack, and pull the control arm down. Remove the old spring. Compress a new spring, and secure it with spring clips. Place the new spring in the control arm spring seat.
6. Use the jack to raise the lower control arm. Connect the ball joint or joints, as described in Chapter Nine of this *Shop Manual*.
7. Continue raising the control arm until the spring is compressed enough so that you can remove the clips. Remove the clips from the spring coils.
8. Slowly lower the jack to remove pressure from the spring.
9. Connect the outer tie rod end as described in Chapter Five of this *Shop Manual*, if you disconnected them.
10. Install the wheels, shock absorbers, and antiroll bars.
11. Lower the vehicle, and bounce it to settle the springs. Torque the wheel lug nuts. Then check ride height and wheel alignment if needed, as described in Chapter Fourteen of this *Shop Manual*.

Coil springs with solid or semi-independent axle

Use this procedure to remove the coil springs in semi-independent and nonindependent solid axle suspensions where the spring is positioned between the trailing arm, I-beam, or axle and the frame or crossmember, figure 8-57.

1. After removing the shock absorbers as described earlier, secure the coil springs with clips or compression plates. If you are using safety clips, use a jack to lift the suspension component and compress the spring.
2. If there are retaining bolts or retaining plates at the top or bottom of the springs, remove the bolts or plates after compressing the spring, figure 8-58.
3. Lower the jack slowly to lower the I-beam, axle, or trailing arm, and remove the spring. Do not allow the axle to drop, as you can damage other suspension components.
4. Check spring insulators, and replace them if they are worn. Compress a new spring and secure it with safety clips or a compression plate. Place it in the spring seat, and raise the jack to lift the I-beam, axle, or trailing arm.
5. Fasten the coil spring retaining bolts or retaining plates, if used, to the spring.
6. With the suspension component and spring still compressed with the jack, remove the spring clips or compressor plates.
7. Install the shock absorbers, antiroll bar, and wheels.
8. Lower the vehicle and bounce it to settle the springs. Torque the wheel lug nuts. Check ride height and wheel alignment if needed, as described in Chapter Fourteen of this *Shop Manual*.

Servicing Leaf Springs

Like coil springs, leaf springs sag when they age. As described in Chapter Eight of the *Classroom Manual*, leaf springs usually are made of metal alloy strips, layered one on top of the other. Lighter, monoleaf springs are made of fiberglass or a plastic composite material. Leaf springs may be mounted longitudinally or transversely on the vehicle.

As with coil springs, you must remove the shock absorbers and support the weight-

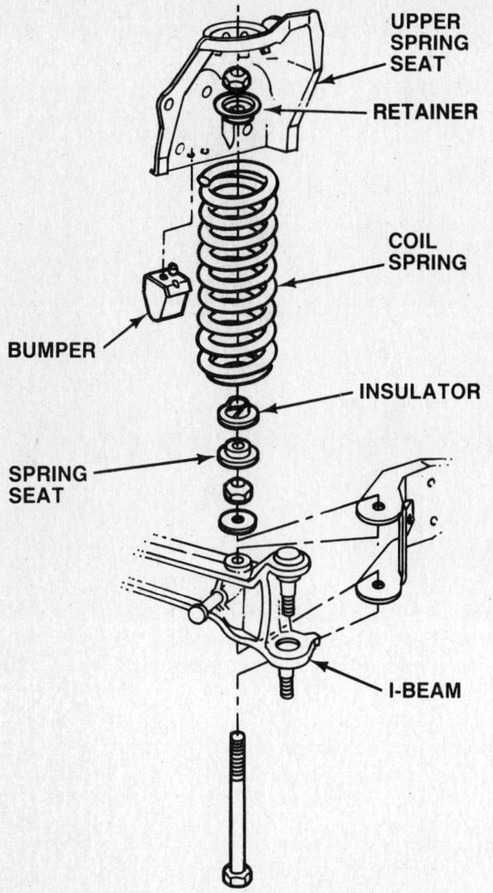

Figure 8-58. The coil springs on an I-beam suspension may have retainers that secure the springs to the axle and frame.

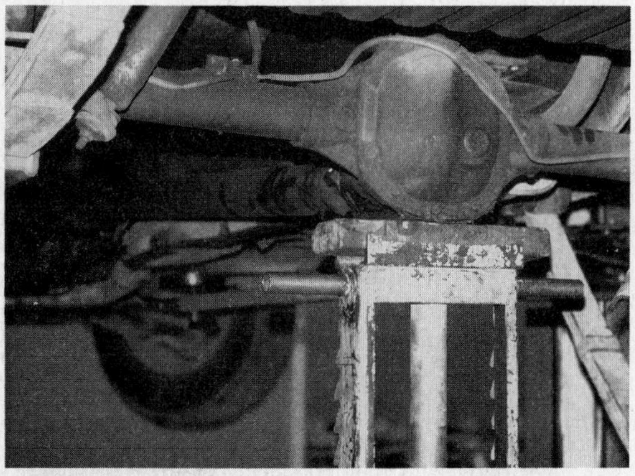

Figure 8-59. Use jack stands or a bumper jack, as pictured, to support the axle while you remove the shocks and leaf springs.

carrying suspension components before removing leaf springs. For longitudinal leaf springs, if you replace the spring on one side of the vehicle, you should replace the spring

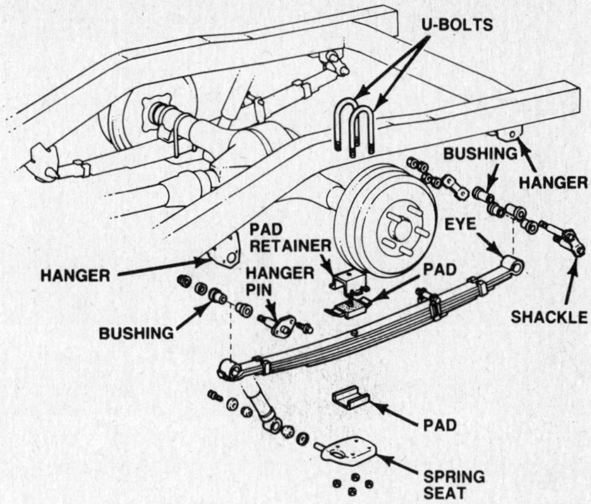

Figure 8-60. A leaf spring is connected to the axle with U-bolts, and to the frame with a spring hanger at one end and a pivoting shackle at the other.

on the other side as well. Most vehicle manufacturers suggest that you remove and install one spring at a time.

Removing leaf springs
Before removing leaf springs, remove the shock absorbers, as described earlier in this chapter. Then use the procedures described in the following sections to remove longitudinal and transverse leaf springs.

1. Position the vehicle on a frame contact lift, ensuring that components will not bend when the vehicle is raised. See Chapter One of this *Shop Manual* for lifting a vehicle. Loosen the wheel lug nuts, and lift the vehicle.
2. Remove the tire and wheel, as this usually makes it easier to access suspension components.
3. Support the control arm, axle, or beam with extra long jack stands or a bumper jack, figure 8-59.
4. Remove the shock absorbers, as described earlier in this chapter. Then remove the longitudinal or transverse leaf springs using one of the following procedures.

Longitudinal leaf springs
Longitudinal leaf springs run parallel to the side of the vehicle. Both ends of the spring attach to the vehicle frame, but one end is mounted to a pivoting shackle so that the spring can lengthen and compress during jounce and rebound, figure 8-60.

1. After raising the car, supporting the axle or other suspension component with a

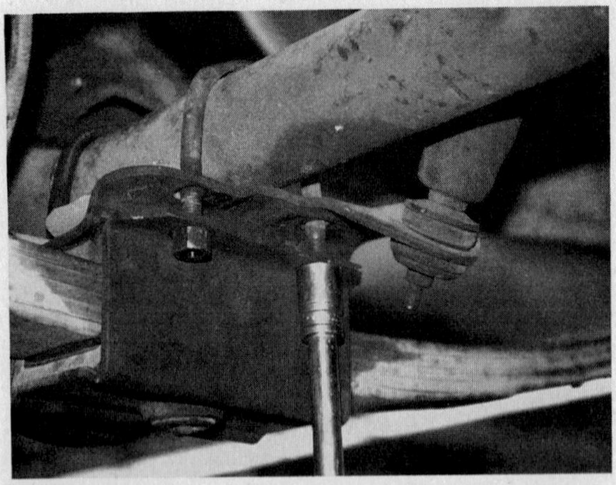

Figure 8-61. The leaf springs are held to the vehicle's axle with U-bolts. Remove the nuts on the U-bolts, and remove the leaf springs.

jack, and removing the shock absorber, place a jack under the leaf spring to support it, or have an assistant help you support the leaf spring.

2. Remove the U-bolts that secure the leaf spring to the axle by loosening and removing the nuts that hold the U-bolts in place, figure 8-61.
3. Remove the spring seat and pad, if used. Some vehicles use two plates, one above and one below the leaf spring. Others use a single plate under the leaf spring and a rubber spacer on top. If there are any rubber insulators between the spring and the retaining plates, remove and discard them if they are worn.
4. Use the jack to slightly lower or raise the axle and move it away from the leaf spring. Continue to support the leaf spring.
5. Remove the nuts that secure the front and rear of the leaf spring to the frame. On one end of the spring, a bolt or stud slips through the spring eye, figure 8-62, and mounts to the vehicle frame on a spring hanger. On the other end, a bolt or stud slips through the eye and into a shackle that can pivot forward and backward, allowing the leaf spring to lengthen and shorten.
6. With the help of an assistant, lower the leaf spring to the floor, using the jack if needed. Continue to support the axle or other suspension component.
7. Inspect the spring hanger and shackle, which are attached to the vehicle frame. Be sure that they are not bent or weakened by rust. Check the retaining plates that secure the leaf springs at the axle. They should not be bent either. If they are, replace them.

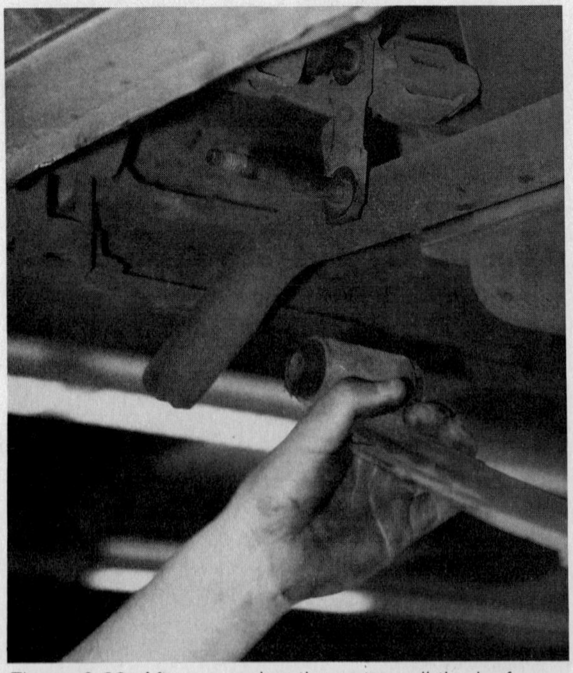

Figure 8-62. After removing the nuts, pull the leaf spring off the spring hanger and shackle.

8. Inspect the rubber insulators. If these are worn, replace them with new ones.
9. If you plan to reuse the old leaf springs, inspect them, as described later in this chapter, before installing them.

Transverse leaf springs

A transverse leaf spring runs parallel to the axle and connects to a suspension component at each wheel. Before removing a transverse leaf spring, remove the tire, wheel, and shock from both sides of the vehicle. Disconnect the antiroll bar from both control arms. To remove the fiberglass, transverse leaf spring on a late-model Corvette front suspension, you need a special leaf spring compressor tool, figure 8-63. Use the following procedure to remove this type of transverse leaf spring:

1. Remove the spring protectors that cover each end of the spring. Four bolts secure the protectors to the vehicle frame.
2. Install the leaf spring compressor tool, figure 8-63. Place the clamps on each end of the spring. Then turn the bar that runs between the clamps to compress the spring.
3. Disconnect the lower control arm from the steering arm at both wheels. Use the procedure described in Chapter Nine of this *Shop Manual* for disconnecting ball joints.
4. Remove the nuts that hold the spring retainers to the frame on each end of the spring, and remove the retainers. The spring retainers look like flattened U-bolts.

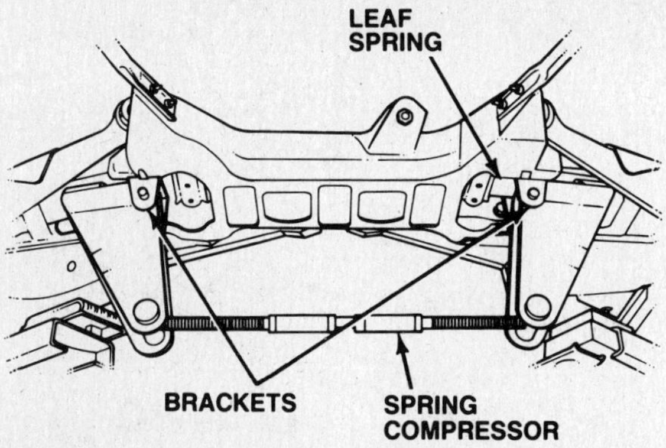

Figure 8-63. On a late-model Corvette front suspension, you will need a special tool to compress the transverse leaf spring before removing it. Front leaf springs were used on Corvettes starting in 1984 and are still used today.

5. Gradually release the tension on the spring by turning the bar on the compressor tool counterclockwise. Remove the compressor tool.
6. Pull the lower control arms down, and remove the spring ends. Pull the spring out from one side of the vehicle to remove it. Support the spring while you are pulling it out to prevent dragging it across the other suspension components. Deep scratches can damage the spring.
7. Inspect the spring and components as described in the next section.

Inspecting leaf springs
If you plan to reuse leaf springs, inspect them as follows:

- Check the leaves near the center bolt, figure 8-64. If they are cracked, do not reuse the spring.
- If the spring has rubber-type leaf separators, check them for wear, and replace them if needed. To replace leaf separators, clamp the leaves at the center bolt with a C-clamp. Remove the center bolt. Install new separators, and replace the center bolt. Use an extra-long center bolt to secure the springs, and then cut off the excess with a hacksaw.
- Leaf springs with multiple leaves have clips that hold them together. If any of these clips are broken, install new ones, if available, or replace the entire spring.
- Finally, check the condition of the rubber bushings inside the spring eyes. If the bushings are worn or missing, remove and replace them. Often these bushings are held in place with metal tabs. To remove the bushings, straighten the tabs with pliers. Press the old bushings out, and press in new ones. Bend the tabs back to hold the bushings in place.

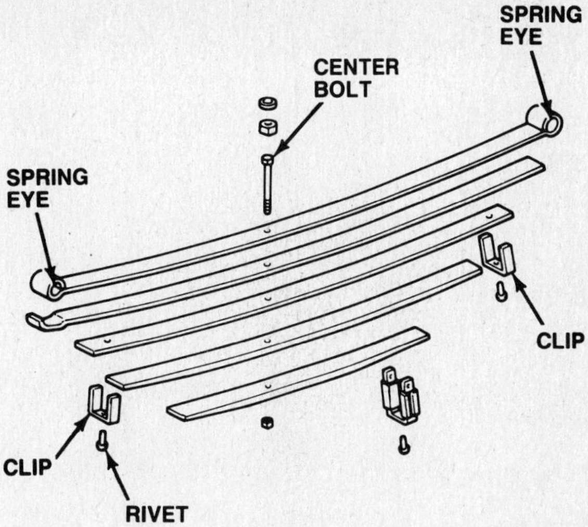

Figure 8-64. The center bolt and clips hold the spring leaves together. If the leaves, clips, or center bolt are damaged, replace or repair the spring.

Installing leaf springs
After inspecting the springs and hardware, use one of the following procedures to install longitudinal or transverse leaf springs.

Longitudinal leaf springs
Because leaf springs are heavy, you may need a jack to support the spring while you are installing it. Also, you may need the help of an assistant.

1. Position the leaf spring so that one of the eyes lines up with the spring hanger on the vehicle.
2. Place the leaf spring eye on the spring hanger stud. Place a washer and nut on the stud, but do not tighten the nut.
3. Line up the other end of the leaf spring with the spring shackle, and put it on the shackle. Place a washer and nut on the bolt or stud, but do not tighten the nut.
4. Position the retainer plates and any insulators on the leaf spring and axle. Align the spring's center bolt with the axle. If necessary, use a jack to move the axle and spring in order to get them into position.
5. Place the U-bolts around the spring and axle, and secure the U-bolts with nuts. Tighten the nuts on the U-bolts, spring

Figure 8-65. Use the manufacturer's recommended torque value when tightening the bolt at the leaf spring shackle.

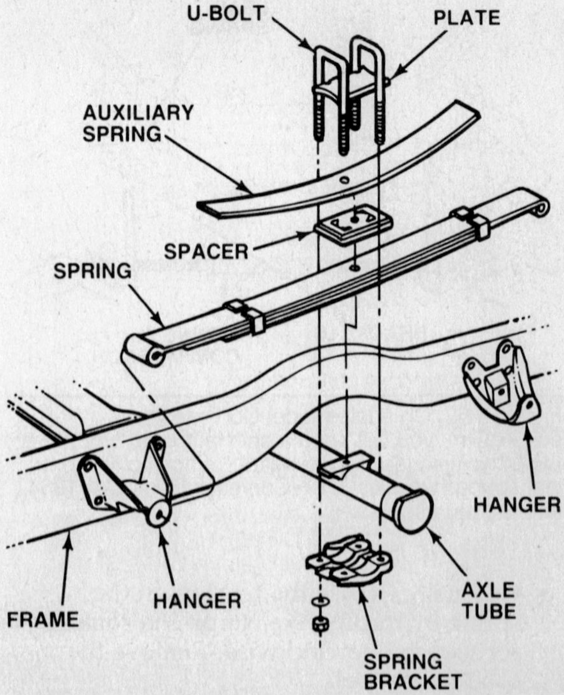

Figure 8-66. Some vehicles are equipped with auxiliary leaf springs from the factory, but you can purchase auxiliary leaf spring kits from an auto supply store for aftermarket installation.

hanger, and shackle to the manufacturer's recommended torque, figure 8-65.

6. Install the shock absorbers, antiroll bar, and wheels.
7. Lower the vehicle and bounce it to settle the springs. Then check and adjust wheel alignment if needed, as described in Chapter Fourteen of this *Shop Manual*.

Transverse leaf springs

To install a transverse leaf spring using a spring compressor, use the following procedure:

1. Slide the spring into position so that each end is near a lower control arm. Support the spring while you are sliding it in place to avoid scratching it.
2. With the help of an assistant, position the lower control arms to seat the spring ends.
3. Clamp the spring compressor on the leaf spring. Turn the bar that runs between the clamps to compress the spring.
4. Install the spring retainers, but do not fully tighten the bolts.
5. Connect the lower control arm ball joints to the steering arm, and torque the nuts in place, as described in Chapter Nine of this *Shop Manual*.
6. Gradually release tension on the spring, and then remove the spring compressor.
7. Bolt the spring protectors at each end of the leaf spring to the vehicle frame. Torque the spring retainer bolts.
8. Install the shock absorbers, antiroll bar, and wheels.
9. Lower the vehicle, and bounce it to settle the springs. Torque wheel lug nuts. Then check and adjust wheel alignment if needed, as described in Chapter Fourteen of this *Shop Manual*.

Installing auxiliary leaf springs

Some manufacturers install auxiliary leaf springs on pickup trucks and 4WD vehicles to increase the suspension's weight carrying capacity, figure 8-66. Aftermarket auxiliary leaf spring kits are available as well. To install an auxiliary leaf spring kit, support the vehicle as described earlier in the section titled "Removing leaf springs." Then use the following guidelines:

1. Remove the U-bolts at the axle, and discard them. Refer to the procedure earlier in this chapter on removing longitudinal leaf springs for instructions on removing the U-bolts.

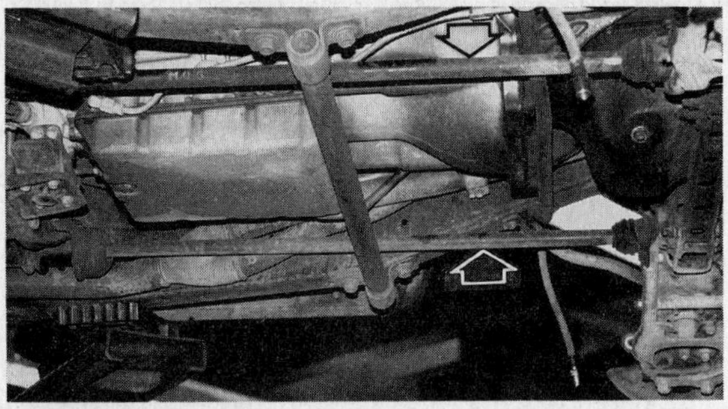

Figure 8-67. Torsion bars twist and untwist to absorb the forces placed on the vehicle's suspension system. They are mounted to the vehicle frame at one end and to a suspension component at the other.

2. Place the spacer block that comes with the kit on top of the leaf springs that are already installed in the vehicle.
3. Place the auxiliary springs on top of the spacer block.
4. Install the retainer plate and U-bolts, but use the U-bolts that come with the auxiliary spring kit. These U-bolts are longer than the original ones. Install and tighten the nuts to the manufacturer's recommended torque.

Servicing Torsion Bars

Like coil and leaf springs, torsion bars will wear over time because they must twist and untwist to absorb the forces from the vehicle's wheels. But unlike coil or leaf springs, some torsion bars can be adjusted or ''tensioned'' to compensate for sag.

Like leaf springs, torsion bars are used on a variety of suspension types and can be oriented either longitudinally, figure 8-67, or transversely. We will cover adjustment, removal, and installation procedures for the following torsion bar applications:

- Longitudinal torsion bars, SLA front suspensions
- Transverse torsion bars, SLA front suspensions
- Transverse torsion bars, trailing arm front suspensions
- Transverse torsion bars, trailing arm rear suspensions.

Adjusting torsion bars

Some torsion bars can be adjusted to increase or decrease tension and set vehicle ride height. For information on measuring ride height, see Chapter Fourteen of the *Classroom Manual* and *Shop Manual*.

Figure 8-68. Most torsion bars have an adjuster nut or bolt that you turn to increase or decrease tension on the bar.

Longitudinal and transverse torsion bars, SLA front suspensions

If ride height is not properly set, inspect the suspension components, and adjust the torsion bars if needed.

1. Drive the vehicle onto a level alignment rack, or park the vehicle on level ground.
2. Turn the torsion bar height-adjusting nut or bolt clockwise, figure 8-68, to raise the vehicle ride height and to tension the bar. Turn it counterclockwise to lower the ride height and to loosen tension.
3. Bounce the vehicle to allow the suspension to settle. Then check the ride height, as described in Chapter Fourteen of the *Shop Manual*.

Transverse torsion bars, trailing arm rear suspensions

To adjust torsion bars on a rear trailing arm suspension, first you must remove the torsion bars, as described later in this chapter. Then position the trailing arms for correct vehicle ride height, and install the torsion bars:

- On some vehicles, such as the Chrysler Monaco, you must insert a special tool through the shock absorber upper and lower bolt mounts, and hold them apart a specified distance for correct vehicle ride height, figure 8-69. Then, insert the torsion bars into the axle tubes.
- For Volkswagens with a torsion bar rear suspension, you must insert the torsion bar into the axle tube, and use a protractor to measure the angle of the spring plate

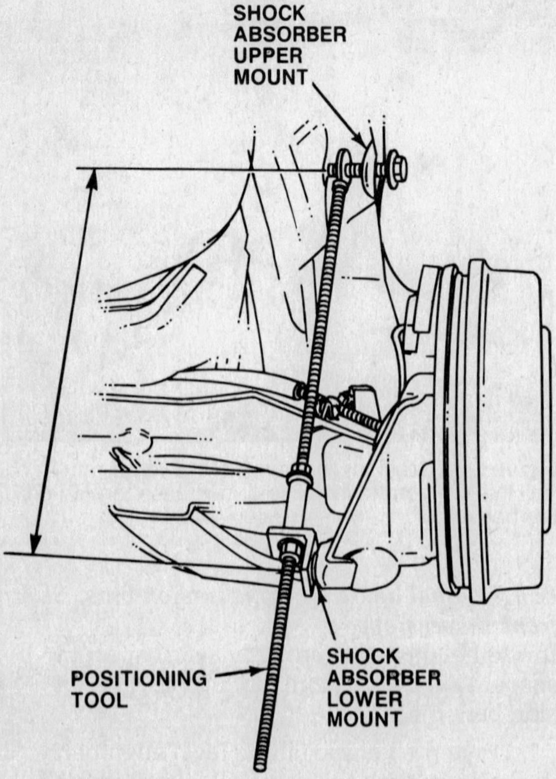

Figure 8-69. On some Chrysler vehicles, you must use this tool to position the suspension components before installing the torsion bars. Doing this will ensure a proper ride height setting.

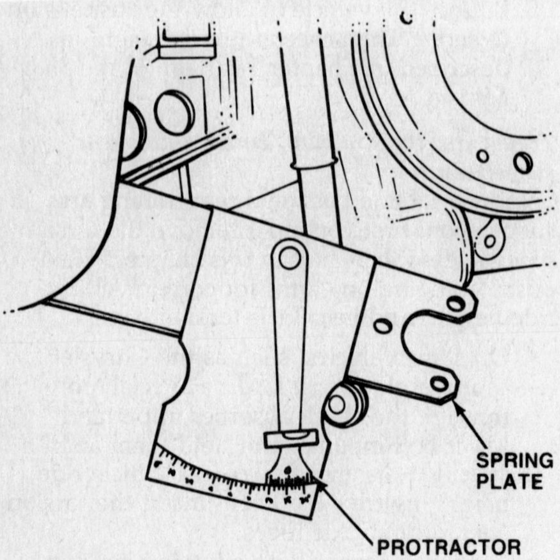

Figure 8-70. On this Volkswagen rear suspension, the spring plate must be at an angle specified by the vehicle manufacturer to be at the correct ride height.

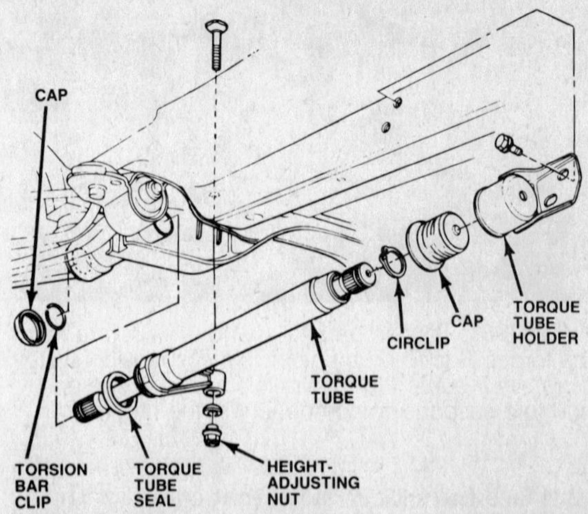

Figure 8-71. On this Honda Civic, the front torsion bars are housed in torque tubes.

with respect to the vehicle body, figure 8-70. If this angle does not meet the manufacturer's specifications, remove the torsion bar, reposition it, and take the measurement again. Repositioning the torsion bar one spline at the inner end changes the angle by nine degrees, and repositioning it one spline at the outer end changes the angle by eight degrees and ten minutes. The torsion bar can be repositioned until the right combination for the correct spring plate angle is achieved.

Removing torsion bars

If torsion bars are damaged or weak, or if you are setting the ride height on a vehicle with a rear trailing arm suspension, remove the torsion bars using one of the procedures in the following section.

Longitudinal torsion bars, SLA front suspensions

On some vehicles, such as Hondas built in the mid- to late-1980s, the torsion bars are protected by torque tubes, figure 8-71. To remove the torsion bars on this type of vehicle:

1. Position the vehicle on a lift, and ensure that components will not bend under the weight of the car when it is raised. See Chapter One of this *Shop Manual* for lifting a vehicle. Loosen the wheel lug nuts, lift the vehicle, and remove the front wheels.
2. Remove the height-adjusting nut by turning it counterclockwise with a wrench.

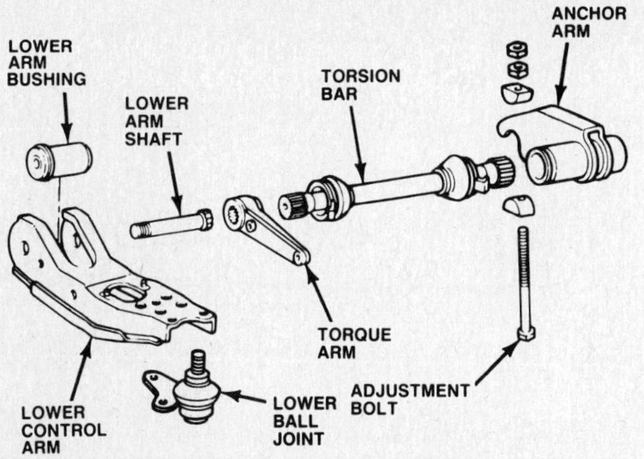

Figure 8-72. After turning the adjuster bolt to relieve tension on the torsion bar, simply unbolt and remove the anchor arm, torsion bar, and torque arm as an assembly.

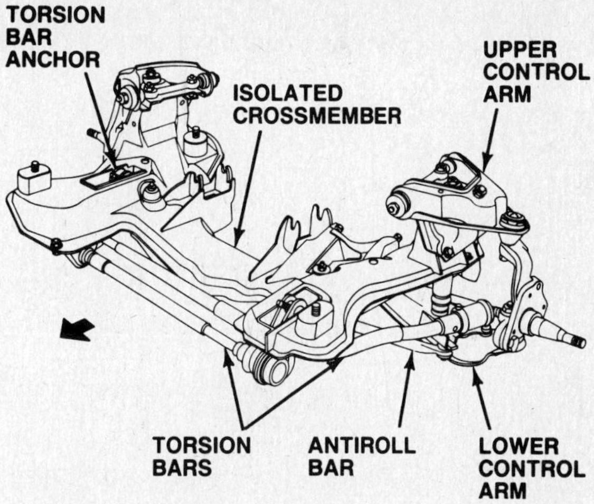

Figure 8-73. Chrysler used this unique transverse torsion bar front suspension on some RWD vehicles from 1976 through 1989.

3. Remove the bolts that secure the torque tube holder bracket, and remove the bracket and cap. Remove the cap at the other end of the torque tube as well.
4. Use circlip pliers to remove the circlips at each end of the torsion bar. Tap the end of the torsion bar with a hammer to push the bar out of the tube.

Most front suspension torsion bars are not housed in a torque tube, figure 8-72. Use the following procedure to remove them:

1. Position the vehicle on a lift ensuring that components will not bend under the weight of the car when it is raised. See Chapter One of this *Shop Manual* for lifting a vehicle. Loosen the wheel lug nuts, lift the vehicle, and remove the front wheels.
2. Remove the adjuster bolt lock nut, and measure the exposed threads on the adjuster bolt for later reference when you are installing the torsion bar.
3. Loosen the adjuster bolt by turning it counterclockwise with a wrench until there is no tension on the torsion bar.
4. Remove the nuts and bolts that secure the torque arm to the lower control arm.
5. Slip the anchor arm off the adjusting bolt. Then remove the anchor arm, torsion bar, and torque arm as an assembly.
6. Pull the torque arm and anchor arm from the torsion bar.

Transverse torsion bars, SLA front suspensions
The Chrysler transverse torsion bar SLA front suspension is a unique suspension spring setup, figure 8-73. Because transverse torsion bars were used on a number of Chrysler vehicles for many years, we will describe the removal and installation procedures briefly. For more information, refer to the Chrysler shop manual for the vehicle that you are servicing.

1. Position the vehicle on a lift, and ensure that components will not bend under the weight of the car when it is raised. See Chapter One of this *Shop Manual* for lifting vehicle. Loosen the wheel lug nuts, lift the vehicle, and remove the front wheels.
2. Loosen both height-adjusting nuts by turning them counterclockwise with a wrench until there is no tension on the torsion bars. Remove the adjuster bolts.
3. Raise the lower control arms with jacks until the distance between each end of the torsion bar is $2^7/_8$ inches (73 mm) from the crossmember.
4. Remove the brackets that connect the antiroll bar to the lower control arm.
5. Remove the two bolts that secure the bracket holding the torsion bar to the lower control arm.
6. Remove the bolts that hold the pivot cushion and bracket in place on the frame. Then remove the torsion bar and anchor arm as an assembly from the vehicle. Remove the friction plates.
7. Remove the anchor arm from the end of the torsion bar.

Transverse torsion bars, trailing arm front suspensions
This type of front suspension was used on Type 1 Volkswagens, figure 8-74, which includes the Beetle and Karmann Ghia.

1. Separate the ball joints between the torsion arms and knuckle, as described in

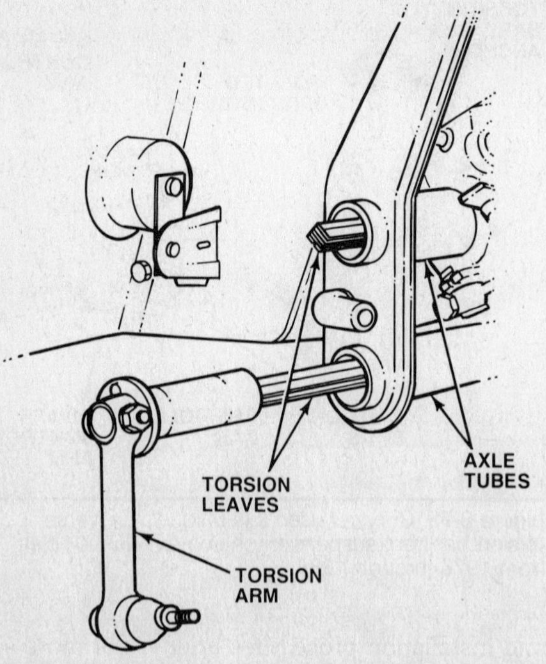

Figure 8-74. This older Volkswagen front suspension uses two sets of transverse torsion leaves, which are housed in a front axle beam.

Chapter Nine of this *Shop Manual*, and remove the steering knuckle and brake drum as an assembly.
2. Remove the antiroll bar.
3. Loosen and remove the lock nuts on the torsion arm set screws, and loosen the set screws.
4. Pull the torsion arms out of the axle tubes on one side of the axle. Remove the grease seals on the torsion arms.
5. Loosen the nut that locks the Allen screw in place at the center of the axle tube. Remove the Allen screw.
6. Go to the other side of the axle, and pull the torsion arms. The torsion leaves, which are connected to the torsion arms, will come out as well.

Transverse torsion bars, trailing arm rear suspensions
On this suspension, torsion bars are housed inside the rear axle tube. The procedure below is for Chrysler vehicles with transverse trailing arm rear suspensions, figure 8-75, but the Type 1 Volkswagen rear suspension is similar.

1. Position the vehicle on a lift, and ensure that components will not bend under the weight of the car when it is raised. See Chapter One of this *Shop Manual* for lifting a vehicle. Loosen the wheel lug nuts, lift the vehicle, and remove the rear wheels.

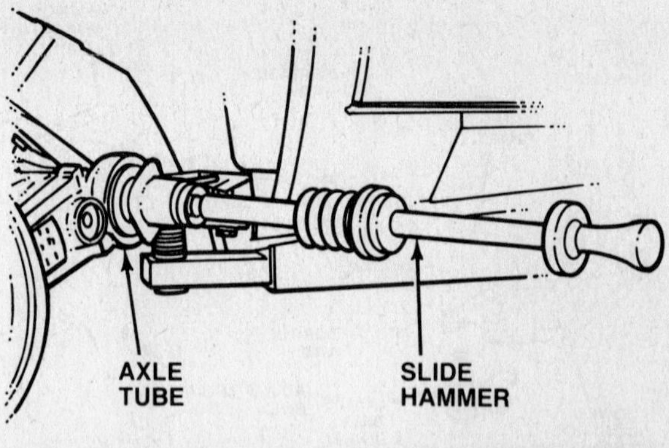

Figure 8-75. You will need a slide hammer to remove the rear torsion bars from the axle tubes on this Chrysler vehicle.

2. Support the trailing arms with jacks, and remove the shock absorbers, as described earlier in this chapter.
3. Pry off the caps from the ends of the torsion bar housings, and remove the retaining clips from the torsion bars with pliers.
4. Use a slide hammer or other tool to pull the torsion bars from the axle tubes. You may have to adjust the position of the trailing arm with the jack to decrease tension on the torsion bar.

Installing torsion bars
Replacement torsion bars are not interchangeable between the left and right sides of the car. Check the ends of the bars for an "R" or "L" for right and left, and install the torsion bars on the correct side of the vehicle.

Longitudinal torsion bars, SLA front suspensions
To install torsion bars that are protected by a torque tube, use the following procedure:

1. Place a new torque tube seal around the end of the torque tube, and apply grease to the seal and the inside of the torque tube.
2. Grease the splines at both ends of the torsion bar, and insert the torsion bar into the torque tube. Align the marks on the torsion bar and torque tube.
3. Push the torsion bar until the splines reach the lower control arm end of the torque tube. Align the marks on the torsion bar and lower control arm, and push the torsion bar in place.

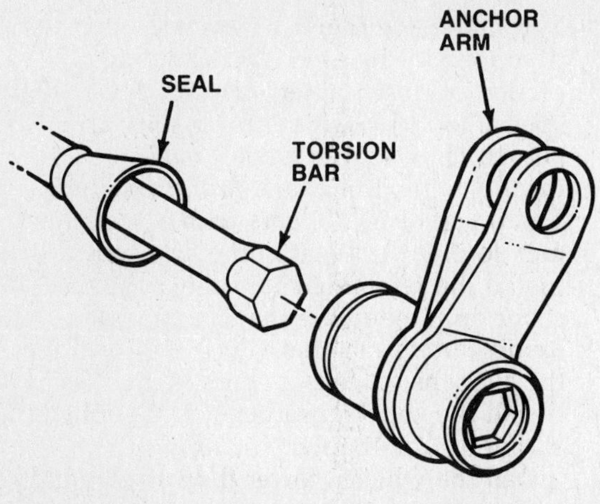

Figure 8-76. Torsion bars usually have splines or a hex-head that fits in the anchor arm to ensure that the torsion bar will not slip when it twists.

4. Install the clip that secures the torsion bar at the lower control arm end. Install the protective cap. Then install the circlip and cap bushing at the other end of the torsion bar.
5. Place the torque tube holder bracket on the end of the torque tube, and bolt it in place on the frame. Install the height-adjusting nut, but do not tighten it.
6. Install the wheels, lower the vehicle, and torque the wheel lug nuts.
7. Adjust the torsion bar as described earlier. Check vehicle ride height as described in Chapter Fourteen of this *Shop Manual*.

To install a torsion bar that is not housed in a torque tube, use the following procedure:

1. Apply grease to the torsion bar splines.
2. Align the marks on the torsion bar and anchor arm, and press the anchor arm onto the torsion bar.
3. Align the marks on the torsion bar and torque arm, and press the torque arm in place.
4. Begin installing the assembly on the chassis by placing the anchor arm over the adjusting bolt.
5. Install the bolts that hold the torque arm to the control arm, and tighten the bolts to the manufacturer's recommended torque.
6. Install the bolt and height-adjusting nut, which secures the anchor arm to the frame. Turn the nut clockwise until the bolt protrudes from the anchor arm. Use the measurement that you took earlier to adjust the bolt.
7. Install the wheels, lower the vehicle, and torque the wheel lug nuts.
8. Adjust the torsion bar as described earlier. Check vehicle ride height as described in Chapter Fourteen of this *Shop Manual*.

Transverse torsion bars, SLA front suspensions
After checking the torsion bars for damage and replacing worn bushings, install the torsion bar as follows:

1. Coat the hex end of the torsion bar with grease, figure 8-76, and install the anchor arm and cover on the end of the torsion bar.
2. Position the torsion bar at the lower control arm, and position the anchor arm at the crossmember. Install and tighten the bolts that secure the anchor arm to the crossmember. Install the adjusting bolt at the anchor arm, but do not tighten it.
3. Bolt the pivot cushion and bracket to the frame, but do not tighten the bolts that hold them in place.
4. Install and tighten the bolts that secure the torsion bar ends to the lower control arms. Tighten the bolts to the manufacturer's recommended torque.
5. Check that the anchor arm is fully seated on the crossmember. Install the friction plates, and tighten the pivot cushion bracket bolts.
6. Install the antiroll bar.
7. Tension the torsion bar slightly by turning the adjuster nut clockwise.
8. Replace the wheels, lower the vehicle, and torque the wheel lug nuts.
9. Adjust the torsion bar as described earlier. Check vehicle ride height as described in Chapter Fourteen of this *Shop Manual*.

Transverse torsion bars, trailing arm front suspensions
Use this procedure to install the torsion leaves in the axle tubes of Type 1 Volkswagen front suspensions.

1. After checking the leaves for cracks or other damage, grease them liberally with multi-purpose grease.
2. Check the number of leaves, and position them so that the recesses in the leaves align with the torsion arm set screws.
3. Insert the leaves in the axle tube, aligning the recess in the center of the leaves with the hole for the Allen screw.
4. Install the Allen screw and lock nut.
5. Replace the grease seals on each end of both axle tubes.

6. Install the torsion arms, their set screws, and the lock nuts.
7. Install the steering knuckle and brake drum assembly by pressing the torsion arm ball joints into the knuckle. Install the antiroll bar.
8. Replace the wheels, lower the vehicle, and torque the wheel lug nuts.
9. Adjust ride height. Check ride height as described in Chapter Fourteen of this *Shop Manual*.

Transverse torsion bars, trailing arm rear suspensions

This procedure is for Chrysler vehicles with transverse torsion bars on a trailing arm rear suspension. The Type 1 Volkswagen rear suspension is similar.

1. Use the manufacturer's recommended tool to position the suspension components. You may have to use a jack to do this. As described in the torsion bar adjustment section of this chapter, you must use a special tool to position the trailing arm before inserting the torsion bar.
2. Insert the torsion bars into the axle tubes. Tap the ends of the bars with a brass hammer so that they seat in the axle tubes.
3. Install the retaining clips and protective cover on the ends of the axle tubes.
4. Remove the tools that you used to position the suspension components.
5. Install the shock absorbers, as described earlier in this chapter.
6. Install the wheels, lower the vehicle, and torque the wheel lug nuts.
7. Check ride height as described in Chapter Fourteen of this *Shop Manual*.

Chapter

9

Suspension Linkage Service

This chapter describes service procedures for the suspension linkage, such as control arms, trailing arms, antiroll bars, and strut rods. Service procedures for ball joints and bushings, which allow movement between these parts, are also covered in this chapter. Use the information in Chapters Nine and Ten of the *Classroom Manual* to determine the type of suspension system you are working on. Then use the information in this chapter to inspect, remove, and install suspension parts, as needed.

SUSPENSION BALL JOINTS

Ball joints are used between suspension components that must pivot. They are found between the control arms and wheel knuckles of front and some rear independent suspensions. Many rear independent suspensions do not need a wide range of movement, so metal or rubber bushings, rather than ball joints, are used. Procedures for servicing suspension bushings are covered later in this chapter.

SLA and strut/SLA suspensions have two ball joints at the wheel knuckle, one at the upper and one at the lower control arm, figure 9-1 (top). MacPherson and modified strut suspensions usually have only one ball joint between the lower control arm and steering knuckle, figure 9-1 (bottom).

There are two types of ball joint: load-carrying and non-load-carrying. Load-carrying ball joints are used on control arms that support suspension springs and vehicle weight. Non-load-carrying ball joints do not support vehicle or suspension component weight. Load-carrying and non-load-carrying ball joints are constructed differently and have different service procedures. Use the following procedures to identify, inspect, lubricate, separate, and install ball joints.

Identifying Ball Joints

A load-carrying ball joint is mounted on the control arm that supports the suspension spring or torsion bar. Use the illustrations that follow to help you identify ball joints:

- In an SLA suspension, either control arm may support the spring, which means that either the upper or lower ball joint may be the load-carrying joint, figure 9-2.
- In most strut/SLA suspensions, the spring and strut place the weight on the lower control arm, and the lower ball joint is the load-carrying joint, figure 9-3.
- In MacPherson strut suspensions, the upper strut bearing carries vehicle weight,

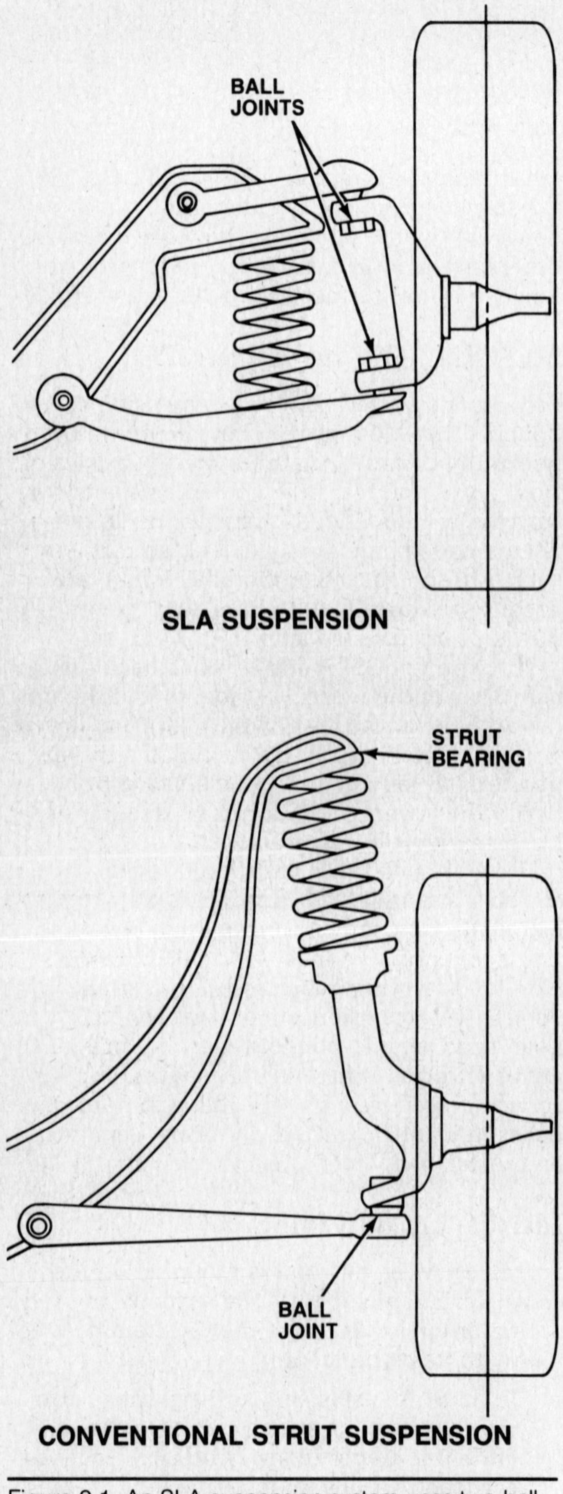

Figure 9-1. An SLA suspension system uses two ball joints, one at the upper arm and one at the lower arm. On a conventional strut system, a single ball joint is used at the lower arm, and a bearing is used at the top of the strut.

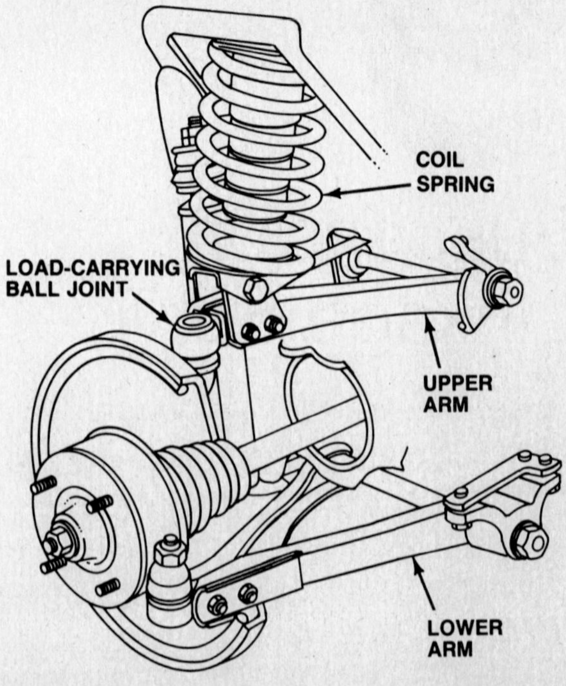

Figure 9-2. When the suspension spring rests on the upper arm, the upper ball joint is the load-carrying joint.

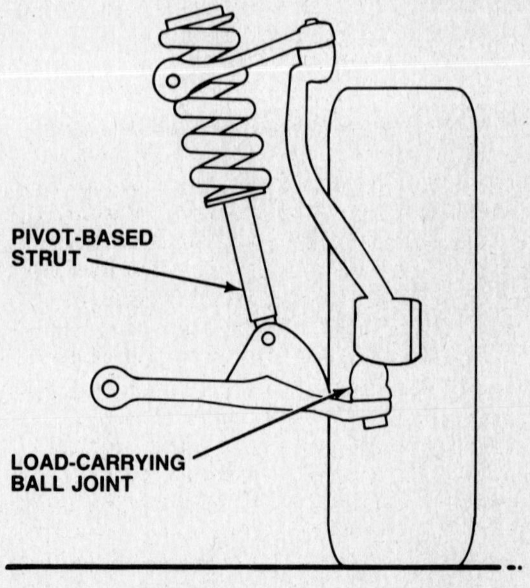

Figure 9-3. On this strut/SLA suspension, the lower ball joint carries vehicle weight.

and the lower control arm ball joint is non-load-carrying, figure 9-4.

- In modified strut suspensions, the spring is usually on the lower control arm, so the lower control arm ball joint is the load-carrying joint, figure 9-5.

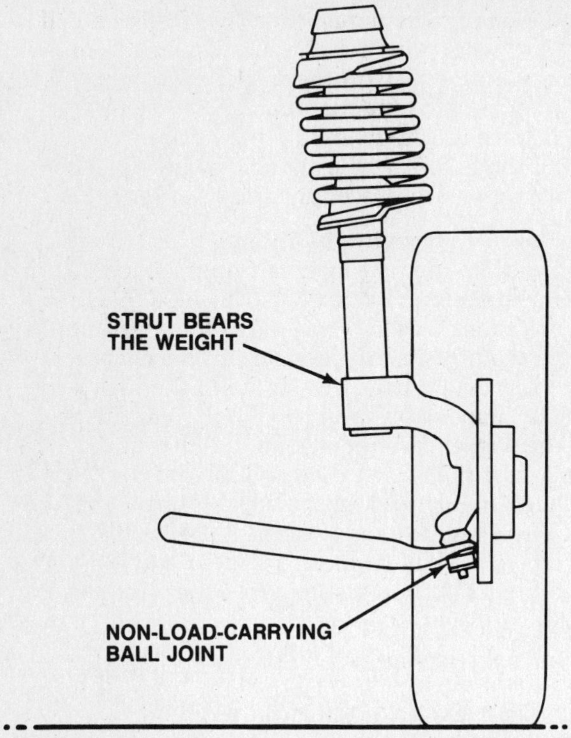

Figure 9-4. The upper portion of MacPherson struts supports vehicle weight, and the lower ball joint is non-load-carrying.

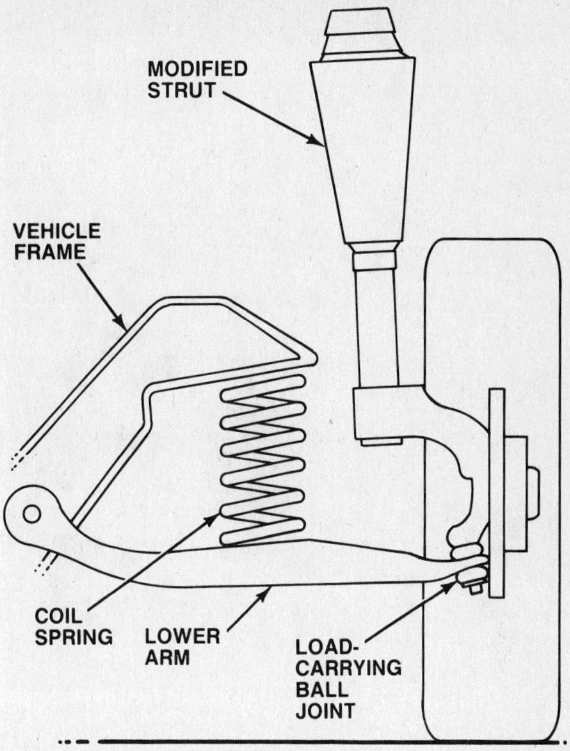

Figure 9-5. On a modified strut suspension, the load-carrying ball joint is determined by the position of the suspension spring. Here, the lower ball joint carries the load.

Inspecting Ball Joints

The best time to inspect ball joints for damage and wear is while you are lubricating the chassis. Replace worn ball joints using the procedures given later in this chapter because a worn ball joint will affect wheel alignment. Rather than holding the wheel firmly in place, a worn ball joint will allow play at the steering knuckle, which can lead to dangerous handling problems.

Check for cracks or breaks in the control arm around the ball joint, especially around a load-carrying ball joint. If the control arm is damaged, replace it as described later in this chapter. Check ball joint dust covers for tears or cracks. Contaminants, such as dirt and water, will enter the ball joints if the covers are damaged, causing accelerated wear. Replace the joint if the cover is damaged.

Checking for wear

Check ball joints for wear indicators. A wear indicator is a fitting at the base of the joint that protrudes when the ball joint is in good condition, figure 9-6. As the joint wears, the fitting recedes into the ball joint housing until it is flush with the base of the joint.

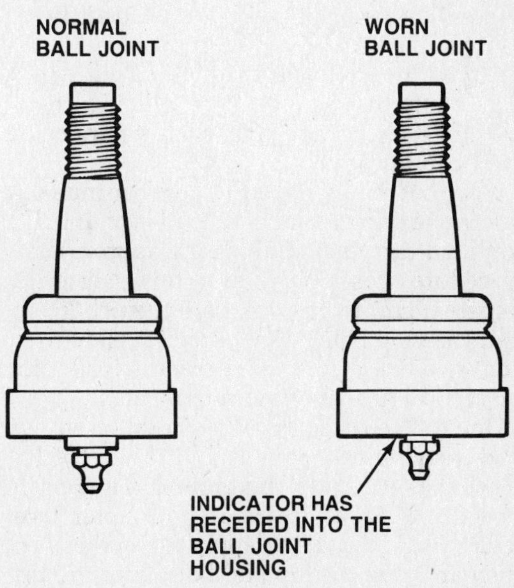

Figure 9-6. The wear indicator protrudes from the base of the ball joint when the joint is in good condition. When the indicator is flush with the base, the ball joint is worn.

Figure 9-7. If there is excessive play when you push and pull the top and bottom of the tire, the non-load-carrying ball joints are worn. Be sure to tighten wheel bearings or apply the brakes before performing this test.

Ball joints with wear indicators must be checked while the vehicle is on the ground at normal ride height. (Do not unload these joints to check for play.) If you unload ball joints with wear indicators, the fitting will recede into the joint, and you will incorrectly identify the joint as worn when it may be in good condition.

If the ball joints do not have wear indicators, determine which joints are load-carrying and non-load-carrying, and use the appropriate procedures described later in this chapter to inspect them. In general, you must check non-load-carrying ball joints for lateral (side-to-side) movement, and load-carrying ball joints for vertical (up-and-down) movement.

One test to determine the condition of non-load-carrying ball joints is to lift the vehicle until the wheels are off the ground. Tighten the wheel bearing, as described in Chapter Twelve of this *Shop Manual*, or apply the brakes. Then push and pull the top and bottom of the tire while checking for movement, figure 9-7. If you detect play, the ball joints may be worn, and you should investigate further. If the wheels are difficult to align or if the vehicle is not steering and handling properly, measure ball joint wear using the procedures that follow.

Many states require that you measure ball joint wear with a dial indicator, and record the amount of play on the work order. If the ball joint exceeds the manufacturer's tolerances, replace it. The following procedures give methods for measuring ball joint wear, although shop procedures may vary.

Non-load-carrying ball joints
Separate the non-load-carrying ball joint from the suspension component using a pickle fork or other method. Procedures for separating ball joints are covered later in this chapter.

Check the ball joint stud and the socket in the suspension component for wear. Inspect the socket carefully for an out-of-round condition. If the socket is worn, cracked, or damaged in any way, replace the control arm or wheel knuckle, as described later in this chapter. If the ball joint stud is in good condition, use this tension test to check the joint for internal wear.

1. After separating the ball joint stud from the suspension component, thread the castellated nut back onto the stud.
2. Thread another nut onto the stud so that it butts against the castellated nut.
3. Use a torque wrench calibrated in inch-pounds to turn the outer nut on the ball stud.
   - If the torque reading is between 25 and 100 inch-pounds (3 and 11 Nm), the joint is in good condition. Use the manufacturer's specifications for exact torque values.
   - A reading above or below this range indicates that the joint is damaged or worn.

Load-carrying ball joints
Because this test measures play with a dial indicator, any other possible sources of play should be eliminated before conducting the test. Check and adjust wheel bearings, as described in Chapter Twelve of this Shop Manual, before conducting this test on load-carrying ball joints.

1. Unload the ball joint as described below. Be sure the jack is as close to the ball joint as possible.
   - For ball joints that carry weight at the lower control arm, place the jack under the lower control arm near the wheel knuckle, and raise the jack, figure 9-8.
   - For ball joints that carry the weight at the upper control arm, place the jack under the frame or crossmember. Before lifting the vehicle, wedge a block

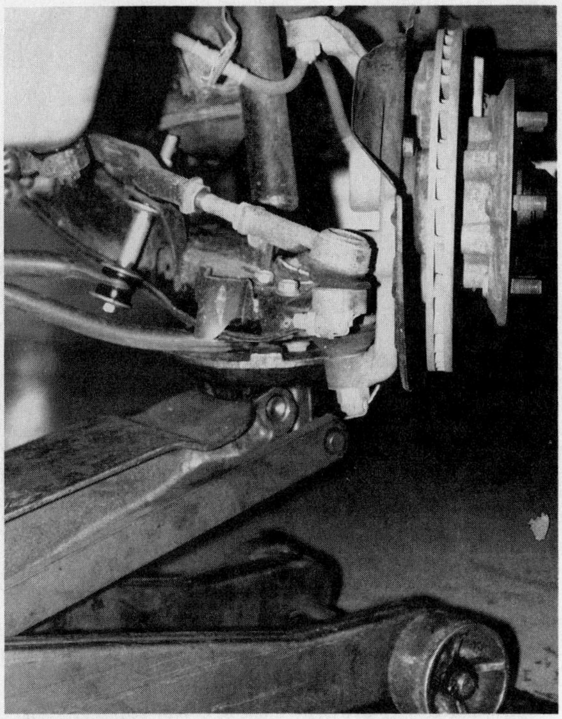

Figure 9-8. Place a jack as close to the ball joint as possible. Use the jack to lift the lower control arm, which unloads the ball joint.

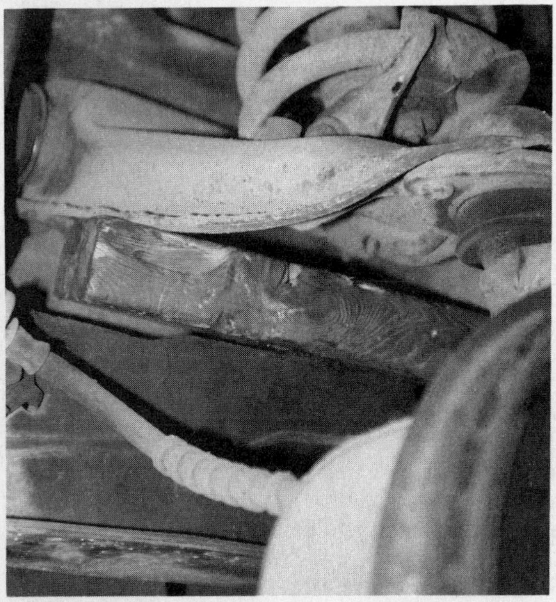

Figure 9-9. Support the upper control arm with a block of wood before lifting the vehicle at the frame. This will unload the upper ball joint.

between the upper control arm and the vehicle frame, figure 9-9. Then lift the vehicle until the wheel is off the ground. The block will position the upper control arm so that the ball joint is unloaded.

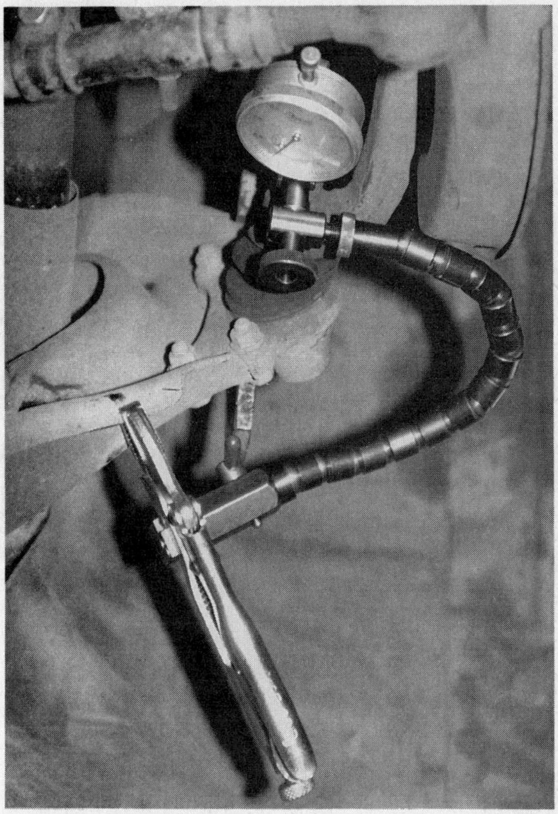

Figure 9-10. Place a dial indicator on the control arm, as shown, to measure ball joint play.

2. Clamp the dial indicator to the control arm. Position it near the load-carrying ball joint.
3. Place the dial indicator plunger near the ball joint, figure 9-10.
4. Reset the dial indicator to zero.
5. Brace a pry bar against the floor and under the wheel to move the wheel fully upward and downward. Measure movement with the dial indicator. Compare this distance with the manufacturer's specifications for ball joint play. The total range of movement is the ball joint's axial (vertical) play. Most specification charts will list the maximum allowable axial play. If play exceeds the specified amount, replace the ball joint.

MacPherson strut suspensions do not have a load-carrying ball joint. If there is vertical play at the wheel of a MacPherson strut system, then the upper bearing, not a ball joint, is likely worn. See Chapter Eight of this *Shop Manual* for information on disassembling and replacing strut components.

On modified strut systems, the ball joint at the lower control arm is the load-carrying joint. Support the lower control arm with a

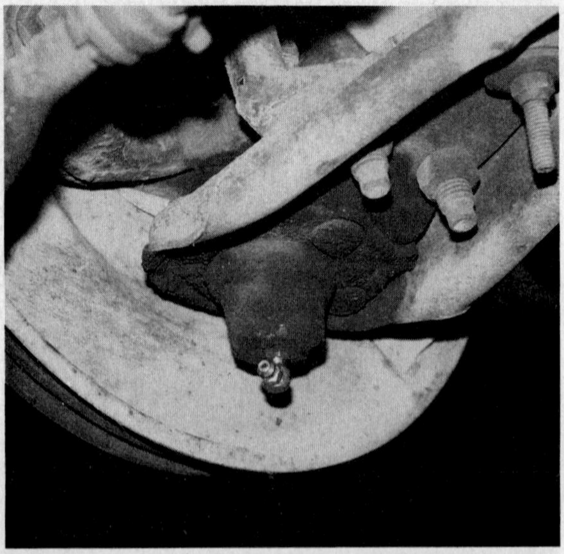

Figure 9-11. Check ball joints for grease plugs or fittings. This grease fitting matches the tip of the grease gun, so that grease can be injected into the joint.

jack when inspecting this ball joint. Support the vehicle frame to unload and check the bearing at the top of the strut.

Lubricating Ball Joints

Some late-model vehicles use permanently sealed ball joints, which cannot be greased. If the ball joints have grease plugs or fittings, they can be lubricated to reduce wear as follows:

1. Raise the vehicle on a lift.
2. Locate ball joints, and determine if they are load-carrying joints, as described earlier in this chapter. You will have to unload load-carrying ball joints to properly grease them.
3. If the ball joints have grease plugs, remove the plugs and install grease fittings. These fittings match the grease gun, figure 9-11.
4. Wipe dirt from the grease fittings with a cloth. If you do not, dirt will enter the joint when you inject grease.
5. If you are using a power grease gun, check that the delivery pressure is low. If you inject grease at high pressure, you will break ball joint seals.
6. Check that the lubricant in the grease gun is the one recommended by the vehicle manufacturer.
7. Place the grease gun attachment over the fitting, and add lubricant slowly until the boot feels firm. Do not over-fill the boot or fill it too quickly.

Figure 9-12. Notice that the ball joint is pressed into the control arm. The ball joint joins the control arm and knuckle at the base of the strut.

8. Use a jack to unload load-carrying ball joints. Move the wheel from side-to-side to distribute the grease inside the joint.
9. After greasing ball joints, remove the grease fittings and install grease plugs. If you wish, as a courtesy to the customer, you may leave the fittings in place for the next chassis service.
10. Wipe any excess grease from the joint.

Separating and Installing Ball Joints

If ball joints are damaged or worn, remove them and install new ones, as described in the following procedures. After installing new ball joints, check wheel alignment, as described in Chapter Fourteen of this *Shop Manual*. Wheel position will likely change once the new joints are in place.

A very important safety-related step in ball joint removal is securing suspension springs. Because the springs exert a great deal of force on suspension components, you risk damaging the vehicle and injuring yourself and others if you attempt to remove ball joints without securing the springs. Chapter Eight of this *Shop Manual* has procedures for securing suspension springs on a number of different suspension types. Be sure to refer to this chapter before separating ball joints.

On most vehicles, the ball joint is mounted in the control arm, and the ball joint stud inserts into a socket in the steering knuckle to join the knuckle and control arm, figure 9-12. A castle nut and cotter pin secure the ball stud.

To remove and install ball joints, raise the vehicle on a lift, and remove the tire and

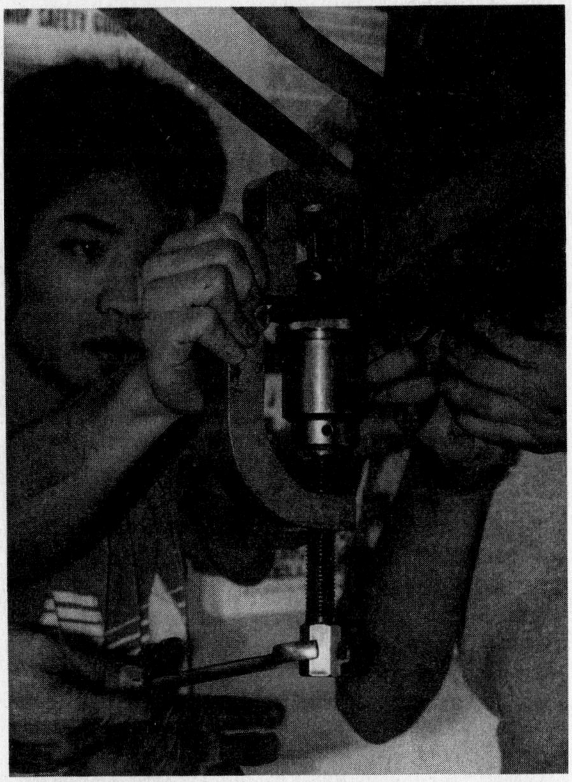

Figure 9-18. Place the top of the ball joint press over the stud and the base of the press at the base of the ball joint. Turn the large bolt to press the ball joint into the control arm.

3. If you removed the coil springs, install them using the procedures in Chapter Eight of this *Shop Manual* as a guide.
4. Install the ball joint stud into the steering knuckle socket. Thread the castle nut onto the stud, and torque the nut to the manufacturer's specifications. If the holes in the nut and stud do not line up when you try to insert the new cotter pin, do not loosen the nut. Instead, tighten it until you can insert the pin. Insert the new cotter pin, and bend the ends back.
5. Lubricate the ball joint with the manufacturer's recommended grease, as described earlier in this section.
6. Install the wheel and tire, and lower the vehicle.
7. Align the wheels, as described in Chapter Fourteen of this *Shop Manual*.

BUSHINGS

Like ball joints, bushings allow flexibility between two suspension components, and may be made of metal or rubber. Metal bushings are extremely rigid, while rubber bushings allow some flexibility and absorb vibration. Manufacturers tune suspensions using bushings of varying sizes and hardness. Always use the manufacturer's specified replacement bushings to ensure that the vehicle's suspension will function as designed.

Figure 9-19. Check control arm bushings at the pivots where the control arm attaches to the vehicle frame.

Locating and Inspecting Bushings

Bushings are in a number of places on the suspension system, but the most common are as follows:

- On front and rear independent suspensions, bushings are used at the control arm pivots, where the control arm connects to the frame, figure 9-19. Trailing arms, which are like control arms, have bushings at the pivots where they mount to the vehicle frame.
- On some rear independent suspensions, the joint between the control arm or trailing arm and knuckle does not have to be as flexible as the joint on a front suspension. Pivots with bushings usually replace ball joints in this location, figure 9-20.
- On some vehicles, a strut rod is connected between the lower control arm and frame. Strut rods stabilize the control arm by preventing fore-and-aft movement during braking. Bushings used on the ends of strut rods damp vibration, figure 9-21.
- If the vehicle has an antiroll bar, you will find bushings between the bar and the vehicle frame, figure 9-22, as well as on the ends of the bar. Some manufacturers mount the antiroll bar with special spacer bars, which also have bushings.
- Check around the differential or transaxle on a rear-wheel-drive or four-wheel-drive

Figure 9-20. Most rear independent suspensions use a pivot and bushing between the control or trailing arm and knuckle, rather than a ball joint.

Figure 9-21. A strut rod bushing damps vibration from the fore-and-aft movements of the lower control arm during braking.

vehicle. Rubber mounts are sometimes placed here to absorb vibration.

If control arm bushings are worn, caster and camber alignment settings will change, and the tires will have lateral scrub wear. A worn strut rod bushing is difficult to detect, as these bushings wear from the inside out. One sign of strut rod bushing wear is a popping noise and forward or rearward movement of the strut rod or control arm during braking. Grasp the strut rod with your hand, and push it back and forth, checking for movement.

While you are lubricating chassis components and inspecting ball joints, visually check bushings for wear. Push and pull suspension parts to check for play. Use a pry bar to lever some of the larger components, figure 9-23. They should be mounted solidly in place.

Figure 9-22. The antiroll bar is mounted to the control arms and vehicle frame with bushings and brackets.

Figure 9-23. Use a pry bar to move suspension components and check bushings, as on this radius arm. If there is play, the bushing is worn.

Metal bushings will squeak when you move the suspension component.

Prod rubber bushings with a screwdriver to check for oil contamination. Oil will soften bushings and break down the rubber. Also check bushings for off-center wear and cracks. Small cracks are a sign of normal wear, but bushings with large cracks are damaged, and should be replaced.

Removing and Replacing Bushings

Whenever you remove a suspension component for some other service, check bushings for wear, and replace them while you have the component removed from the vehicle. If worn bushings are severely affecting driveability, use the procedures in this *Shop Manual* to remove the parts necessary to access the worn suspension bushing.

Mount the component in a vise, if needed, before removing the bushing. You will probably have to press the bushing from the component with a press and adapter that fits over the

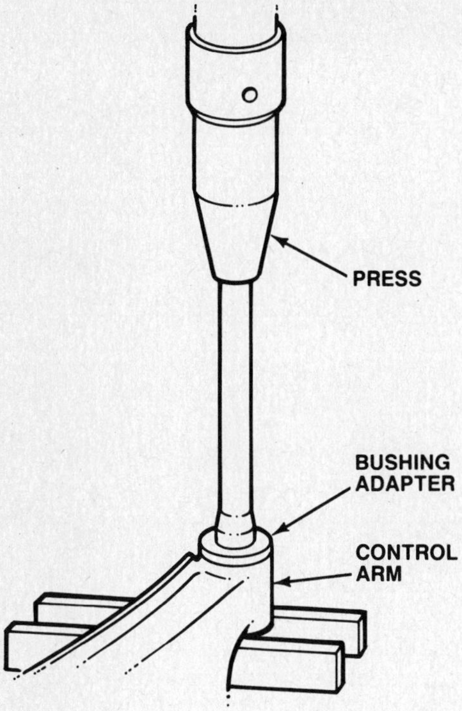

Figure 9-24. Many manufacturers have special tools or adapters that fit over suspension component bushings. Use the adapter or special tool to press the bushing out.

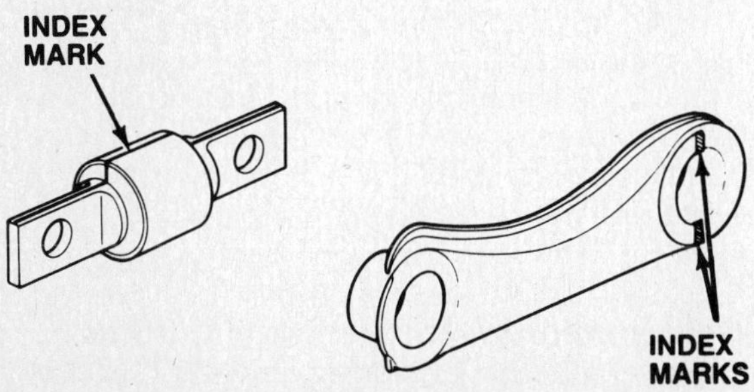

Figure 9-25. Check the bushing and suspension arm for index marks. If you find them, use the manufacturer's instructions to install new bushings. Be sure that the index marks line up.

bushing, figure 9-24. Usually you can use the same special tool to press a new bushing in place, but be sure to install the bushing straight into the hole. Some manufacturers place index marks on the bushing and suspension component. These marks must line up to ensure proper bushing installation, figure 9-25.

After pressing the bushing into the suspension component, install the part. Do not torque the nuts or bolts while the vehicle is on

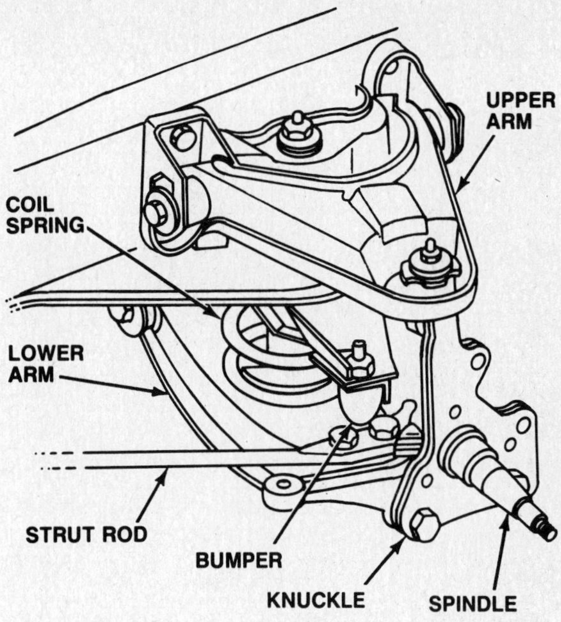

Figure 9-26. The knuckle connects the wheel to the rest of the suspension system. On non-driven axles, the wheel mounts on a spindle that extends from the knuckle.

the lift. Tighten them once the vehicle is on the ground. (Torquing these nuts while the vehicle is on a lift may cause the bushings to twist.)

KNUCKLES

The "knuckle" is a cast component that connects the wheel to the rest of the suspension system, figure 9-26. Depending on the vehicle you are working on, steering and brake system parts may also be connected to the knuckle.

A "steering arm" extends from the knuckles at the front wheels (and from the knuckles at the rear wheels, if the vehicle has four-wheel steering). Tie rods or other steering linkage parts connect to the steering arms, figure 9-27. The steering arm may be bolted to the knuckle, or it may be a cast part of it. A knuckle with a cast steering arm is called a "steering knuckle".

Non-driven wheels mount on wheel spindles, which extend from the knuckles, figure 9-27. Wheel bearings support the wheel on the spindle. The spindle is usually a cast part of the knuckle, but it can be a bolt-on component on some vehicles.

Driven wheels are supported by the axle shafts, hubs, wheel bearings, and axle flanges, figure 9-28. You can read more about adjusting

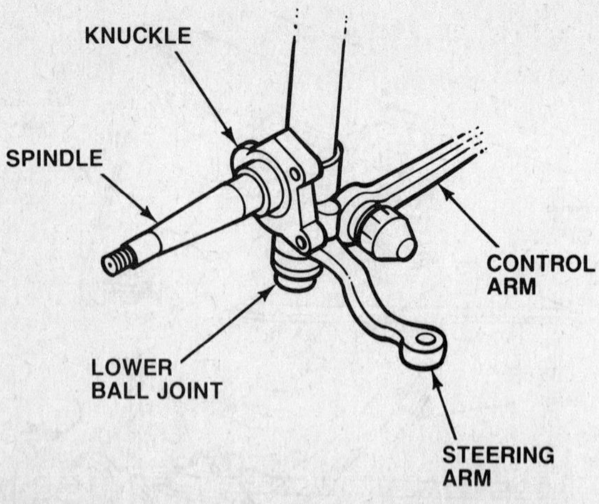

Figure 9-27. The steering arm is an extension of the knuckle. The steering arm has a socket for connecting the tie rod end ball stud.

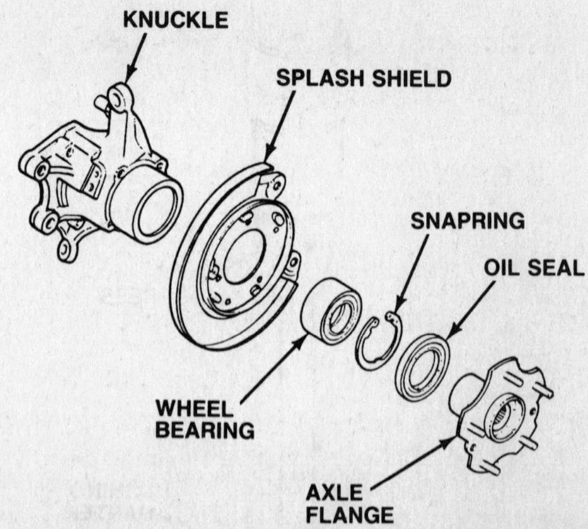

Figure 9-28. On a driven axle, the axle shafts, knuckle, and wheel bearings support the wheel.

and replacing wheel bearings in Chapter Twelve of this *Shop Manual.*

Steering axis inclination (SAI) is a wheel alignment measurement that is determined by the design and position of the knuckle. SAI is not usually adjustable. If you determine that SAI does not meet the manufacturer's specifications when you are performing a wheel alignment, the knuckle or control arm may be bent, and you must replace parts as needed. Chapter Fourteen of this *Shop Manual* has more information on measuring SAI and other wheel alignment angles.

To check the knuckle, raise the vehicle on a lift. Remove the wheel and brake components, as described in Chapter Twelve of this *Shop Manual*, if needed. Inspect the spindle and steering arm, as described in the following sections. Then check the ball joints and pivots that connect the knuckle to the rest of the suspension, and replace any defective parts. Finally, inspect and replace the knuckle, if necessary, using the following procedures.

Inspecting the Knuckle

Loose connections between the knuckle and suspension components affect steering control and wheel alignment. If the knuckle is bent, SAI will be affected. Check for loose connections and bends, as well as cracks, rust, and other damage, and replace the knuckle as needed.

On non-driven wheels, inspect the spindle for bends and wear, especially where the wheel bearings are seated. Check for cracks, rust, and other damage. If the threads on the end of the spindle are damaged or rusted, replace the spindle. Some trucks and many older vehicles use a spindle that bolts onto the knuckle, figure 9-29. On these vehicles, simply unbolt the spindle, and replace it with a new one. On most late-model vehicles, however, the spindle is an integral part of the knuckle, and you must replace the entire knuckle if the spindle is damaged.

Check the steering arms at the front knuckles (and the rear knuckles if the vehicle has four-wheel steering). Inspect for bends, cracks, or other damage. If the arms are bent, the vehicle's steering geometry will be incorrect. You must replace damaged steering arms.

Bolt-on steering arms are easy to replace, figure 9-29. Simply use a wrench to remove the bolts, and remove the steering arm. However, most vehicles have steering arms that are a cast part of the knuckle, and you must remove and replace the entire knuckle to replace the steering arm.

Check the ball stud between the steering arm and tie rod or other steering linkage component. If there is play, replace the ball stud. Chapter Five of this *Shop Manual* has procedures for servicing steering linkage components.

Removing the Knuckle

As discussed in Chapter Eight of the *Classroom Manual*, there are four types of knuckle-to-suspension connections, depending on the suspension system you are working on:

- Ball joints (the most common) are used on SLA, strut/SLA, and strut suspensions.

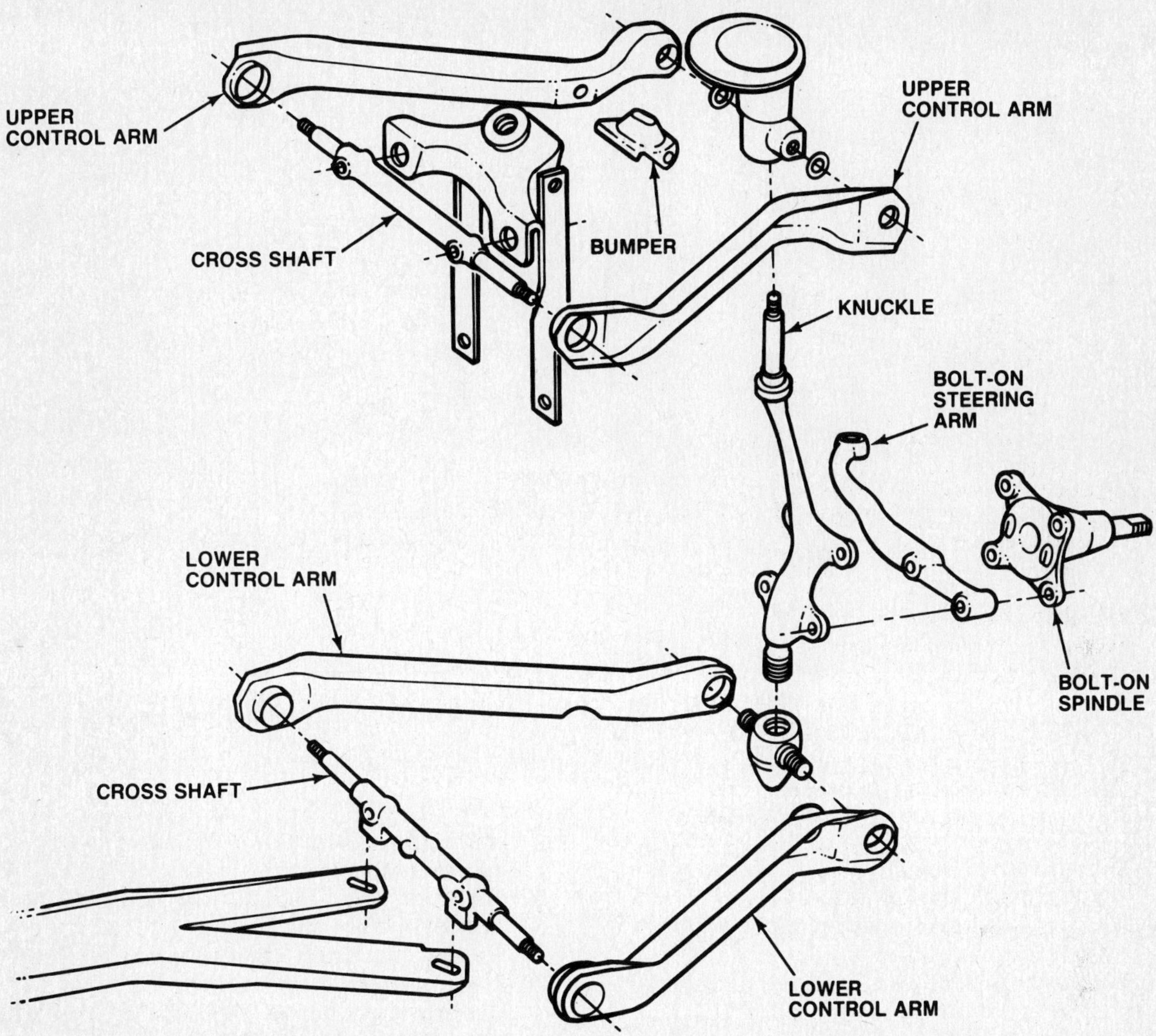

Figure 9-29. Unlike most late-model vehicles, this 1963 Rambler American uses a spindle and steering arm that bolt to the knuckle. On most late-model vehicles, the spindle and steering arm are a cast part of the knuckle.

- Isolated strut base is used on MacPherson and modified strut systems.
- Pivoting bushings are used on some rear suspension links where minimal movement is needed at the wheel, and on some strut/SLA suspensions to connect the base of the strut to the knuckle.
- Kingpins are used on I-beam and twin I-beam suspensions of some late-model trucks and vans.

If you determine that the knuckle is damaged, use one of the following procedures to disconnect the knuckle from the suspension.

Ball joint-knuckle connection
If the control arm and wheel knuckle are connected with ball joints, separate the ball joints as described earlier in this chapter, and replace the knuckle:

1. With the vehicle on a lift, remove the tire and wheel, brake caliper, and other brake components, if needed as described in Chapter Twelve of this *Shop Manual*.
2. Compress and secure the suspension spring using safety clips or compressor plates, as described in Chapter Eight of this *Shop Manual*. Once the spring is secure, remove the shock absorber, if needed.
3. Use a pickle fork or other method to disconnect the tie rod end ball stud from the steering arm, as described in Chapter Five of this *Shop Manual*.

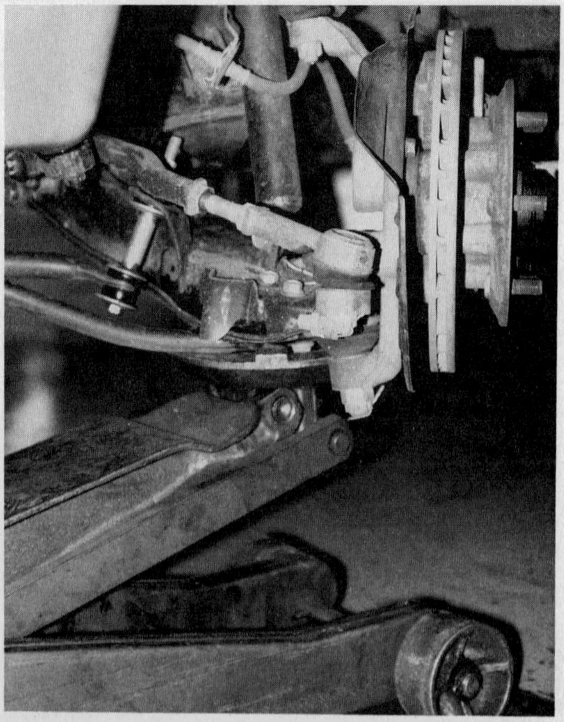

Figure 9-30. Most late-model vehicles use ball joints between the knuckle and control arms. After securing the suspension spring, support the suspension component that contains the load-carrying ball joint, and separate the joint.

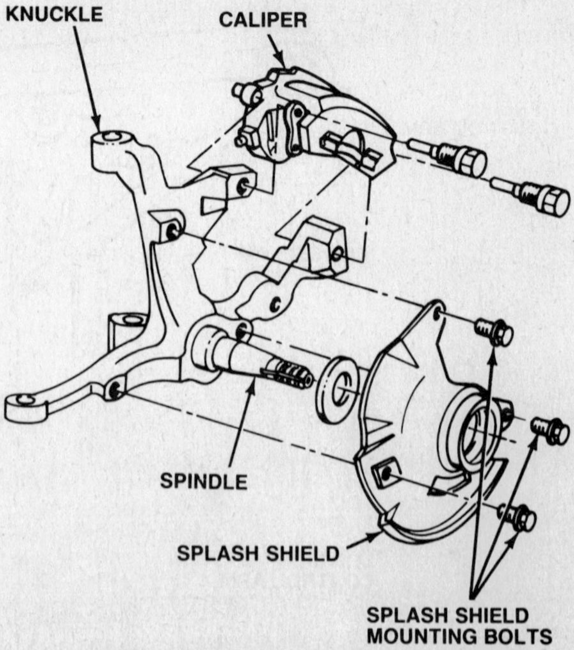

Figure 9-31. The splash shield keeps water and dirt from entering the brake assembly. If you are replacing the old knuckle with a new one, remove the splash shield so that you can install it on the new knuckle.

4. If a strut rod is connected to the control arm, unbolt and remove it.
5. Before separating the ball joints, support the control arm that has the load-carrying joint, figure 9-30, as discussed earlier in this chapter in the section titled "SUSPENSION BALL JOINTS".
6. Separate the ball joints using the procedures described earlier in this chapter.
7. Inspect the knuckle for damage. If the sockets for ball joint studs are out-of-round or if they look polished, replace the knuckle. Excessive play between the ball stud and socket will eventually lead to ball stud breakage.
8. If the knuckle is damaged, remove the splash shield. Loosen and remove the nuts and bolts securing the splash shield to the knuckle, figure 9-31. Install these components on the replacement knuckle.
9. Before installing the new knuckle, check that the suspension spring is properly positioned on the control arm. Use a jack to compress the suspension spring.
10. Position the knuckle at the control arms. Insert the control arm ball studs through the sockets on the replacement knuckle. Install the castle nuts on the ball studs, and tighten them to the manufacturer's recommended torque. Install cotter pins.
11. If you removed the strut rod, reconnect it at the control arm.
12. Remove the safety clips or compressor plate from the suspension spring, as described in Chapter Eight of this *Shop Manual*. Install the shock absorber if you removed it earlier.
13. Install any brake parts that you removed, and install the tire and wheel.
14. Lubricate the ball joints as described in Chapter Eight of this *Shop Manual*, and check and adjust wheel alignment as described in Chapter Fourteen of this *Shop Manual*.

Strut base-knuckle connection

Raise the vehicle on a hoist, supporting the weight of the car at the vehicle frame so the suspension components hang. If you are removing the knuckle on a modified strut system, be sure to secure the suspension spring, as described in Chapter Eight of this *Shop Manual*.

Loosen and remove the bolts that secure the isolated strut base to the knuckle, as described in Chapter Eight of this *Shop Manual*, figure 9-32. Then separate the lower control arm ball joint, and remove the knuckle.

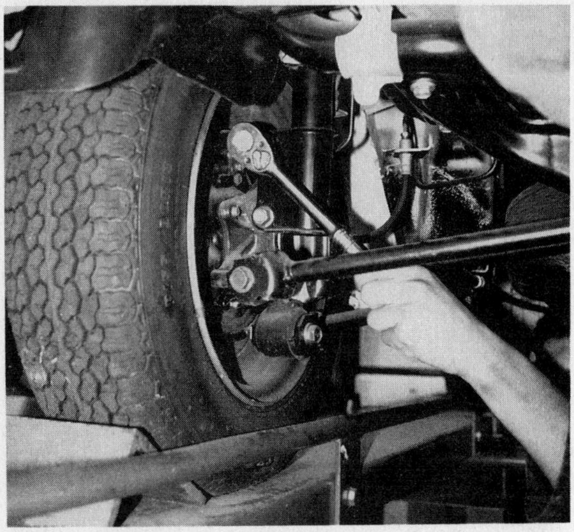

Figure 9-32. Most struts bolt directly to the knuckle, and you can simply remove the bolts to disconnect the lower strut mount.

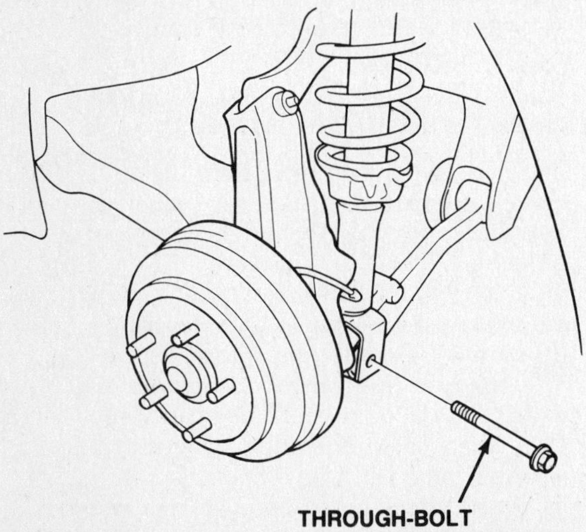

Figure 9-33. A through-bolt connects a pivot-base strut or other link to the wheel knuckle.

Pivoting bushing-knuckle connection
This connection is used on some rear suspensions at the strut base, such as strut/SLA suspensions where the knuckle and control or trailing arm are a single unit. Pivots are also used on suspension links between the knuckle and vehicle frame.

Compress and secure the strut suspension spring. Disconnect pivots by loosening and removing the through-bolts, figure 9-33. After removing the through-bolt, separate the link or strut from the knuckle. Check pivot bushings, and replace them if needed, as described earlier in this chapter.

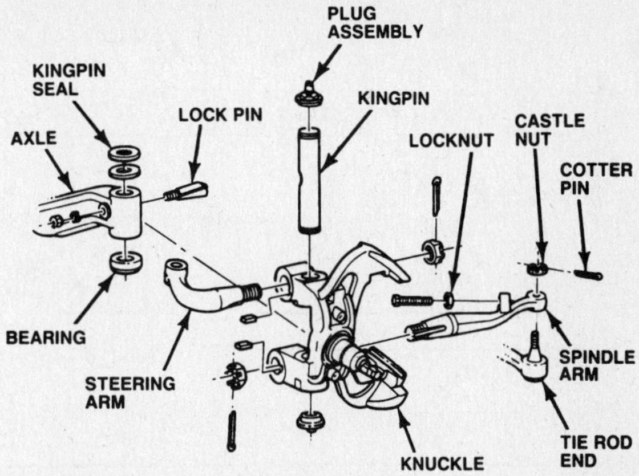

Figure 9-34. Many large, late-model trucks and vans use kingpins at the wheel knuckle. Kingpins provide a stronger connection than ball joints, but they can pivot only on one axis.

Kingpin-knuckle connection
Raise the vehicle on a lift, and inspect the connection between the kingpin and knuckle for looseness. Hold the tire at the top and bottom, and push and pull it. If you detect play, the kingpin bushings are worn, and you should replace them.

As with ball joints, you can use a dial indicator to check the amount of play at the kingpin. Clamp the dial indicator to the vehicle frame, and place the indicator plunger on the inside of the tire at the bottom. Push and pull the tire, noting the dial indicator reading. If play exceeds 1/4 inch (6 mm) for a 16-inch (405-mm) wheel, 3/8 inch (9 mm) for a 17- to 18-inch (430- to 460-mm) wheel, or 1/2 inch (13 mm) for a wheel 18 inches (460 mm) or larger, remove the kingpin and replace the bushings. Remove and replace the knuckle, if needed, as follows, figure 9-34:

1. With the vehicle on a lift, remove the tire and wheel, brake caliper, and other brake components, as described in Chapter Twelve of this *Shop Manual*.
2. Secure the suspension spring using one of the procedures described in Chapter Eight of this *Shop Manual*.
3. Loosen and remove the nuts and bolts securing the splash shield. Remove the splash shield.
4. Remove any other brake system parts that may be blocking access to the knuckle.
5. Separate the ball stud that connects the tie rod end and steering arm, as described in Chapter Five of this *Shop Manual*.

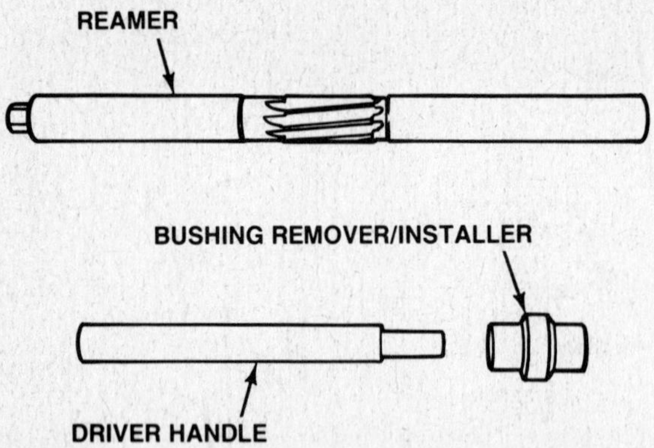

Figure 9-35. These tools are provided by Ford to remove and install the kingpin bushings on F-series trucks.

6. A lockpin secures the kingpin in the housing. Locate the lockpin and nut, and remove the nut. Some kingpins use two lockpins. Use a drift and hammer to push the lockpins out.
7. Remove the caps or grease plugs at the top and bottom of the kingpin.
8. Use a brass drift and hammer to push the kingpin from the steering knuckle.
9. After removing the kingpin, inspect it for wear, and replace it if needed.
10. Separate the knuckle from the axle. If you are replacing the entire knuckle, continue to step 12. If you plan to reuse the knuckle, remove the bushings. These bushings usually are made of bronze, and you will need special tools provided by the manufacturer to drive the bushings from the knuckle, figure 9-35. Install new bushings. Many machine shops will remove and install kingpin bushings if you do not have the proper tools.
11. The vehicle manufacturer will specify the proper bushing diameter, and you must ream the new bushings to this size so that the kingpin fits properly. Many machine shops will provide this service if you do not have the proper tools.
12. Position the new knuckle at the axle, and insert the kingpin using a brass drift, if needed. Install the caps or grease plugs at the top and bottom of the kingpin.
13. Insert the lockpin or pins in the kingpin housing, using a brass drift, if needed. Install the lockpin nuts.
14. Connect the tie rod end to the steering arm, and install any brake system components that you removed earlier.
15. Compress the suspension spring, and remove the clips or plates securing it.
16. Install the tire and wheel.
17. Check and adjust wheel alignment, as described in Chapter Fourteen of this *Shop Manual*.

CONTROL ARMS

Control arms connect the wheel knuckle to the vehicle's frame or crossmember. Pivoting bushings, ball joints, or kingpins connect the control arm to the knuckle, as described in the last section. The other end of the control arm connects to the frame or crossmember with pivots.

Control arms are used on all front independent suspensions. Front non-independent suspension systems do not use control arms. Independent rear suspensions use trailing and semi-trailing arms that extend from the frame toward the rear of the vehicle. Service procedures for control arms, trailing arms, and semi-trailing arms differ slightly. We will discuss trailing and semi-trailing arm service later in this chapter.

Inspecting Control Arms

There are two basic control arm designs used with SLA, strut, and strut/SLA suspensions. The double-pivot (wishbone) type connects to the vehicle frame in two places. The single-pivot type connects to the frame in one place, and often uses a strut rod, as described later in this chapter, for added stability, figure 9-36.

Bushings at the control arm pivots may be made of metal or rubber. You can visually inspect control arm bushings for wear while the vehicle is on a lift. Use a flashlight and small pry bar to check for cracked or distorted bushings. One indication of worn control arm bushings is when you cannot accurately set wheel alignment angles because of excessive play.

Some double-pivot control arms mount to a shaft, called a "cross shaft", that threads through the control arm pivots and bolts to the frame. To check bushing wear at the cross shaft pivots, use a screwdriver to prod them, checking for brittle or oil-soaked bushings, figure 9-37. If the bushings are damaged, remove the control arm, and replace the cross shaft bushings.

The outer end of the control arm, which connects to the knuckle, has a ball joint or bushing. Inspect these parts as described earlier in this chapter.

Check for rubber bump stops, which limit suspension arm travel, figure 9-38. If bump stops are worn or cracked, replace them.

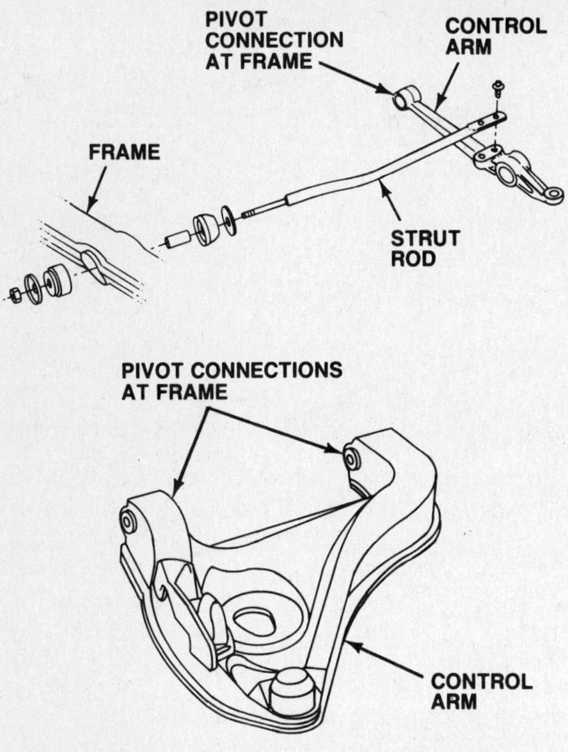

Figure 9-36. The control arm may connect to the vehicle frame at one or two points. If it connects to the frame at only one point, then a strut rod is used to provide added stability.

Figure 9-37. Use a screwdriver to prod control arm cross shaft bushings. If bushings are brittle or contaminated and swollen with oil, replace them.

Bump stops may be attached to a bracket that bolts to the frame, or they may be press-fit onto a metal extension on the suspension component or frame. Worn bump stops indicate that

Figure 9-38. A bump stop keeps suspension components that travel too far from hitting the vehicle frame. If the bump stop is worn or cracked, check springs, struts, and shocks for wear.

suspension arms are traveling farther than they are supposed to, and you should check springs, struts, or shocks, for wear. If the bump stop is missing, install a new one.

Removing and Installing Control Arms

In most cases, you will not have to replace control arms unless they are bent in a collision. Bushings, cross shafts, and bump stops are prone to wear, however. Replace them by removing the control arm first, as follows:

1. Place the vehicle on a lift, and remove the tire and wheel.
2. Locate the suspension springs:
   - For suspensions that use a torsion bar at the lower control arm, disconnect the torsion bar from the control arm, as described in Chapter Eight of this *Shop Manual*.
   - For suspensions that use coil springs, compress and secure them, and disconnect the shock absorbers, as described in Chapter Eight of this *Shop Manual*.
3. Unload the load-carrying ball joint on the control arm using the procedures described earlier in this chapter. This ensures that the control arm is supported. If you plan to remove a lower control arm, remove the brake caliper, rotor, antiroll bar, and strut rod, if equipped, as described later.
4. Remove the cotter pin and castle nut from the ball joint stud, and separate the ball joint, as described earlier in this chapter. If the control arm is attached to the knuckle with a pivot bushing, simply loosen the

Figure 9-41. This VW front suspension is unusual in that it uses two sets of trailing arms. (Volkswagen calls them torsion arms.) Most trailing arms are used on rear suspensions.

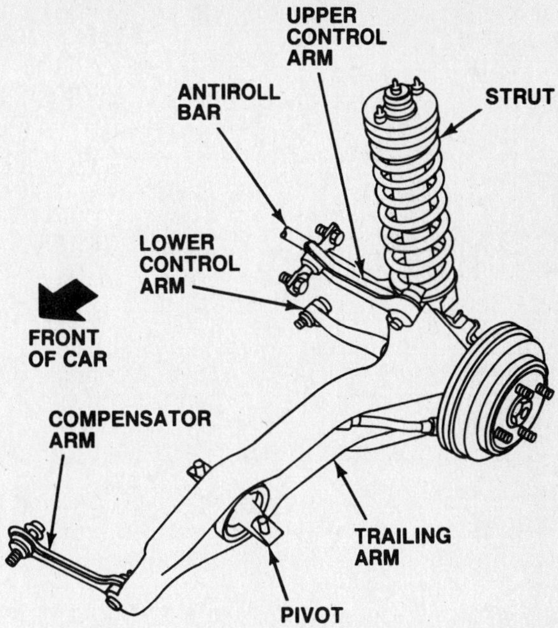

Figure 9-42. The trailing arm is not susceptible to much wear or damage. Check the bushings at trailing arm pivots.

bushing, alternating from side to side, to keep the cross shaft centered. Install the cross shaft on the vehicle frame, but do not tighten the bolts to the manufacturer's recommended torque until the vehicle is on the ground at normal ride height.

TRAILING AND SEMI-TRAILING ARMS

Trailing and semi-trailing arms are like control arms, except that they are used at the rear suspension and extend toward the rear of the vehicle. Volkswagen uses trailing arms, or torsion arms, on the *front* suspensions of early-model Type I (Beetle and Karmann Ghia) and Type II (Bus) vehicles, and we will discuss these trailing arms briefly in this section. But in general, trailing and semi-trailing arms are used on a variety of rear suspension types from independent strut suspensions to semi-independent strut or torsion bar suspensions, and independent rear-engine suspensions.

Volkswagen Front Trailing (Torsion) Arms

Volkswagen Type I vehicles and Type II vehicles use torsion arms on the front suspension, figure 9-41. These arms extend from the front axle tubes and connect to the knuckle with kingpins (1965 and earlier model Type I and 1961-67 Type II) or ball joints (1966 to 1974 Type I or 1968 and later Type II).

To remove torsion arms that are connected to the knuckle with ball joints, separate the ball joints using the procedures described earlier in this chapter. Remove the antiroll bar from the lower torsion arm as described later in this chapter. Loosen and remove the torsion arm set-screw lock nuts, remove the set screws, and pull the torsion arms from the axle tubes. Inspect ball joints, and check the torsion arms for bends. Reinstall the arms in the reverse order, and check wheel alignment.

If you are servicing a Volkswagen that uses kingpins at the knuckle, use the procedure described earlier in this chapter to remove the kingpins, and separate the knuckle and torsion arms. Then separate the torsion arms from the axle as described in the preceding paragraph. It is a good idea to check and adjust wheel alignment after reinstalling torsion arms.

Inspecting Rear Trailing and Semi-Trailing Arms

Rear trailing and semi-trailing arms are sturdy and generally not susceptible to damage unless the vehicle has been in a collision, figure 9-42. Check trailing arms for stress cracks or severe rust. The bushings, located at pivot points, are susceptible to wear, and you should visually inspect them when the vehicle is on a lift. Replace bushings, as described later in this chapter, if they are cracked or distorted. If you remove the trailing arms for any reason, take

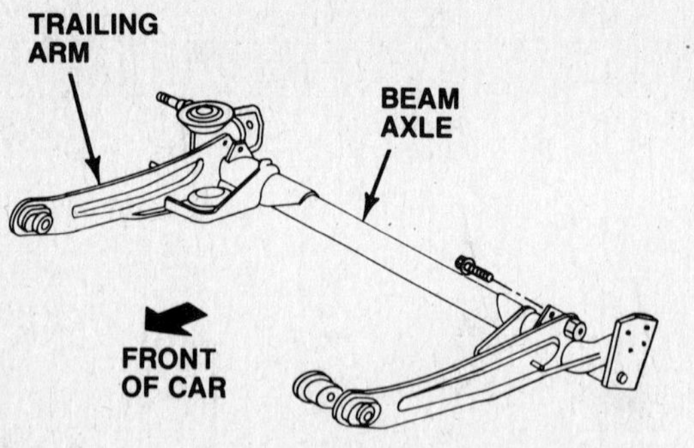

Figure 9-43. If the trailing arms are welded to a beam axle, check the welds to be sure they are in good condition.

the opportunity to check and replace the bushings, as they are easier to inspect and remove when the trailing arms are out of the vehicle.

On some solid axle and semi-independent suspensions, the trailing arms are welded to the axle, figure 9-43. Check the condition of the welds to be sure the trailing arms are secure.

Remove trailing arms and replace the bushings using the following procedures. On some vehicles, you can replace the bushings with the trailing arms in the vehicle, although it is usually easier to remove the trailing arms first.

Removing and Installing Rear Trailing and Semi-Trailing Arms

Many front-wheel drive cars use a solid rear axle trailing arm suspension, and others use independent trailing arms. Many larger, rear-wheel-drive vehicles use solid axle trailing arm suspensions, although they too may use independent trailing arms. Rear-engine vehicles and some rear-wheel-drive vehicles use semi-independent trailing arm suspensions.

Most manufacturers suggest that you replace trailing arms as a pair, but remove and install only one trailing arm at a time, especially on RWD solid axle rear suspensions. Because suspension systems vary slightly from manufacturer to manufacturer, use the procedures in this section as a guide to remove and install rear trailing arms, but refer to the manufacturer's shop manual for the vehicle you are servicing.

Independent trailing arm suspension, FWD and RWD

If you are working on a rear suspension that has a strut, modified strut, strut/SLA, or coil spring suspension, use the following procedure to remove and install trailing arms.

1. With the vehicle on a lift, remove the tire and wheel for easy access to the suspension system.
2. If the vehicle uses modified struts with a remote coil spring or has a coil spring suspension, use the procedures in Chapter Eight of this *Shop Manual* to compress and secure the spring before you disconnect the suspension arm or strut mount. Keep a jack or jack stand under the trailing arm while you are removing it.
3. Remove the antiroll bar, as described later in this chapter. Disconnect any other suspension links that attach to the trailing arm, especially on a strut/SLA suspension.
4. When disconnecting the trailing arms on a RWD vehicle, use a jack or jack stands to support the differential. Some manufacturers suggest disconnecting the rear driveshaft from the differential before disconnecting the trailing arms, as the differential may tip slightly when the arms are removed. See Chapter Ten of this *Shop Manual* for more information on servicing the driveline.
5. Disconnect the lower strut mount from the trailing arm, using the procedures in Chapter Eight of this *Shop Manual* as a guide, or disconnect the trailing arm from the wheel knuckle.
6. Remove the through-bolt or bolts that secure the trailing arm pivot to the frame. Remove the trailing arm, and replace bushings, if needed.
7. Install the trailing arms by reversing the removal procedure. After assembling the rear suspension, lower the vehicle to the ground and torque the through-bolts. Do not tighten these bolts while the vehicle is on the lift.

Solid axle trailing arm suspension, FWD

Some FWD cars have a solid rear axle with trailing arms welded onto each end. If a trailing arm is damaged, you must remove and replace the entire rear axle. However, you can remove and install bushings in these trailing arms without removing the axle from the vehicle.

1. Raise the vehicle on a lift, and remove the tire and wheel for easier access to the suspension components.
2. Disconnect brake lines between the axle and frame, so they will not be damaged when the axle is lowered.
3. Support the axle with a jack, and disconnect the shock absorbers. Compress the

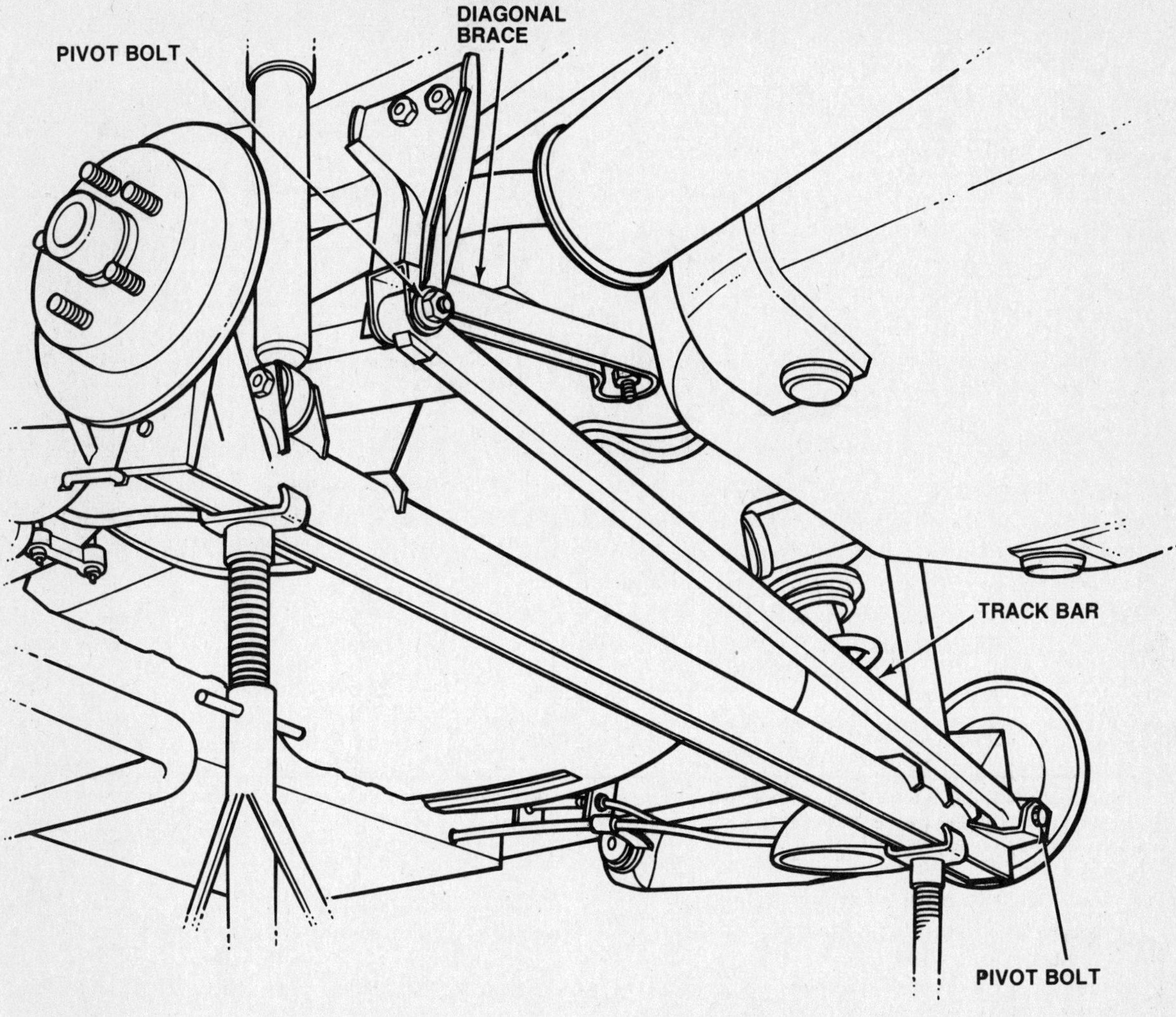

Figure 9-44. On a solid axle rear suspension with trailing arms, a Panhard rod or track bar helps stabilize the axle.

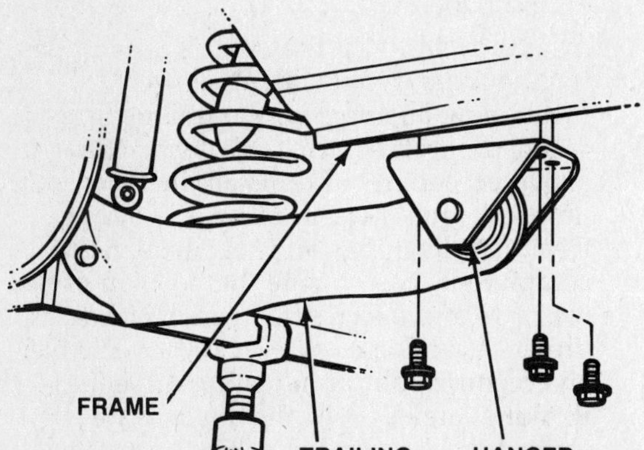

Figure 9-45. Remove the bolts that secure the trailing arm hanger bracket to the vehicle frame, and lower one side of the axle to access the trailing arm bushing.

suspension springs, and secure them as described in Chapter Eight of this *Shop Manual*.

4. If the vehicle uses a track bar, disconnect the bolt that secures the bar to the axle, figure 9-44.
5. Lower the jack slightly, to lower the axle.
6. Remove the bolts that secure the trailing arm hanger bracket to the vehicle frame, figure 9-45. Remove the hanger bracket to expose the bushing on the trailing arm, and replace the bushing, if needed.
7. If you are removing the axle and trailing arms as an assembly, remove the suspension springs. Support the other side of the axle with a jack, and remove the hanger bracket on the other side as well. Lower the axle, and make any needed repairs.

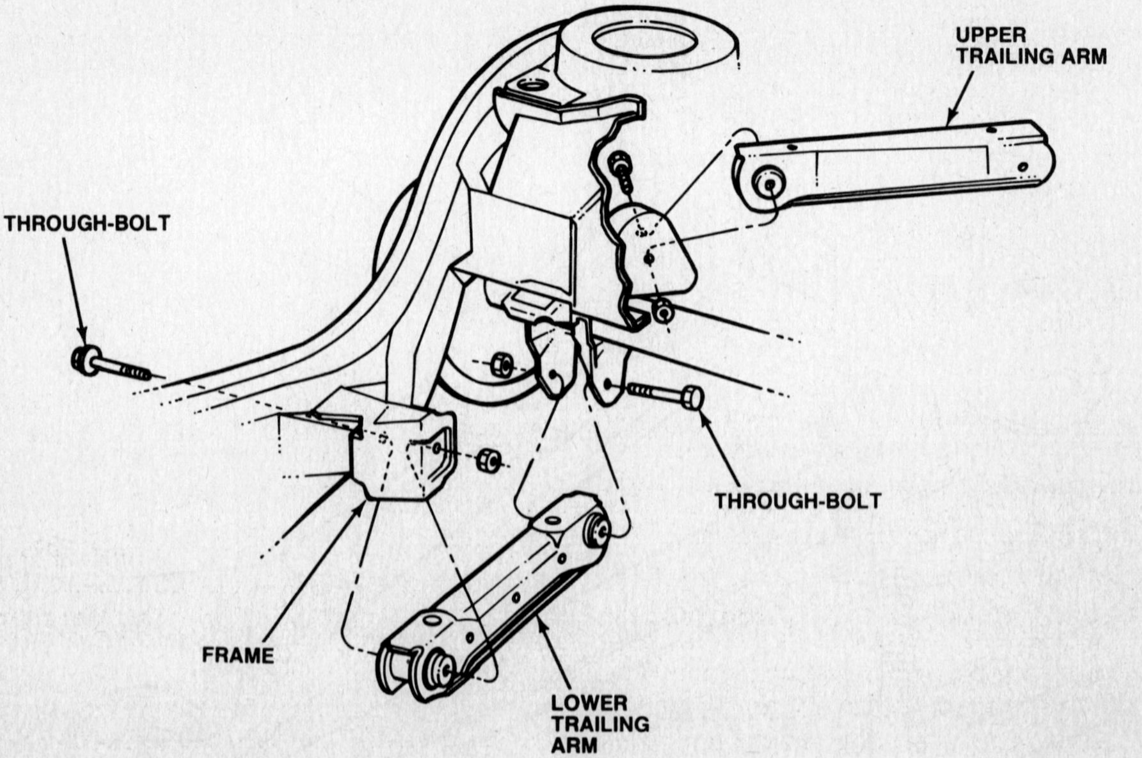

Figure 9-46. Once the suspension springs are secure, removing trailing arms is simply a matter of disconnecting the through-bolts at the frame and wheel.

8. Use a press to remove and install bushings, as discussed later in this chapter.
9. Reinstall the trailing arms and axle, reversing the removal procedure. Torque through-bolts *after* lowering the vehicle.

Solid axle trailing arm suspension, RWD
Remove and install one trailing arm at a time. Removing both trailing arms is dangerous because the rear axle may tip or rotate, causing driveshaft damage and alignment problems. In fact, you can injure yourself if the axle is not properly supported.

Rear-wheel-drive trailing arm suspensions often have additional components, such as torque arms or upper control arms, to help reduce axle windup, as discussed in Chapter Ten of the *Classroom Manual*.

1. Raise the vehicle on a lift, and remove the tires and wheels for easier access to suspension components.
2. Raise the rear axle slightly with a jack, secure the suspension springs, and disconnect the shock absorbers, as described in Chapter Eight of this *Shop Manual*.
3. Lower the axle enough to remove the springs, but keep the jack in place to support the axle.
4. If the suspension uses an antiroll bar, disconnect it, as described later in this chapter.
5. Remove through-bolts that connect any suspension links from the frame to the trailing arm, and remove the links.
6. Remove the bolts that connect the trailing arm to the frame and wheel knuckle or axle, figure 9-46, and remove the trailing arm from the vehicle. If the vehicle uses upper trailing arms, remove them as well.
7. Install new bushings in the trailing arms.
8. Install the trailing arms, reversing the above procedure, and install the bolts, but do not torque them in place.
9. Replacing bushings and rear suspension components can change the position of the rear axle and differential, which affects driveline working angle. If the vehicle has an eccentric bolt for adjusting driveline working angle, check the angle as described in Chapter Ten of this *Shop Manual*, and turn the eccentric bolt, as needed, to adjust the angle. If the driveline working angle does not meet manufacturer's specifications, severe driveline vibration may result.

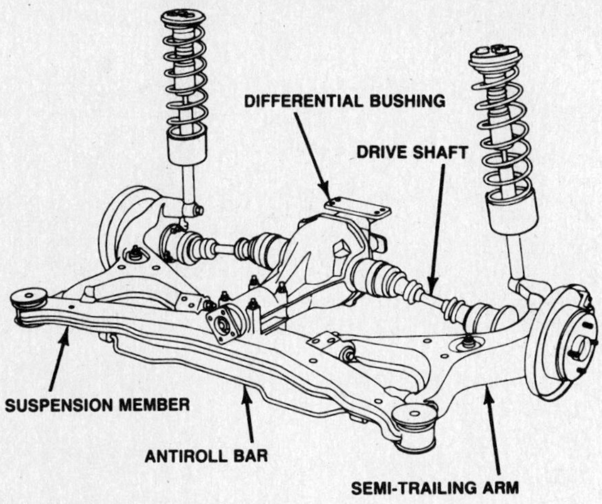

Figure 9-47. This rear-wheel drive Infiniti M30 uses a semi-trailing arm suspension at the rear axle.

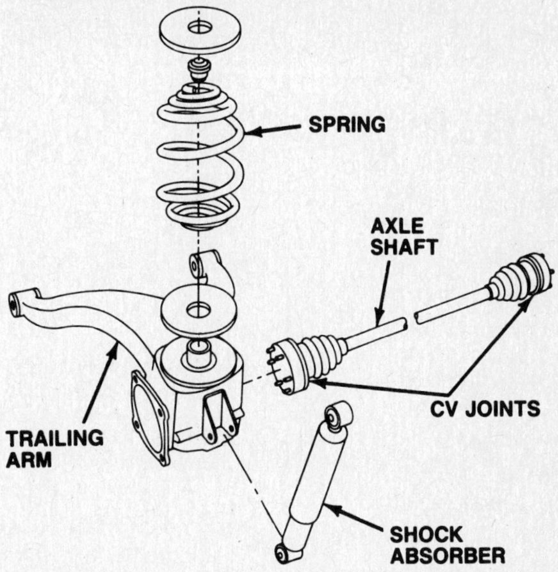

Figure 9-48. The VW Vanagon is a rear-engine vehicle that uses a semi-trailing arm suspension at the rear axle.

10. Install the tires and wheels, and lower the vehicle to the ground. Torque the trailing arm through-bolts to the manufacturer's specifications.

Semi-trailing arm suspension, RWD and rear-engine vehicles
The Infiniti M30 has a semi-trailing arm rear suspension that uses half-shafts extending from the differential, through the trailing arms, and to the driven wheels, figure 9-47. The inner ends of the semi-trailing arms connect to a crossmember or to the vehicle frame. The rear-engine VW Vanagon has a similar, although less sophisticated setup, figure 9-48. On the VW, the half-shafts extend from the transaxle. Use the following procedure to service this type of semi-trailing arm suspension:

1. With the vehicle on a lift, remove the tire and wheel for easier access to the suspension system.
2. If the vehicle uses MacPherson struts, disconnect the strut from the wheel knuckle after compressing the strut spring.
3. If the vehicle uses springs and shocks, compress the suspension spring and secure it, as described in Chapter Eight of this *Shop Manual*. Then disconnect the lower shock absorber mount.
4. Using the procedures in Chapter Ten of this *Shop Manual*, remove the axle shaft. This includes removing the wheel hub assembly so that you can remove the axle from the trailing arm.
5. Remove the antiroll bar, if there is one, as described later in this chapter.
6. Remove the through-bolts that secure the inner semi-trailing arm pivots to the suspension member or frame, and remove the semi-trailing arm from the vehicle.
7. Install new bushings in the semi-trailing arms, if needed.
8. Install the semi-trailing arms, reversing the removal procedure, but install the through-bolts finger-tight.
9. Lower the vehicle to the ground, and torque the through-bolts to the manufacturer's specifications.

Removing and Installing Trailing Arm Bushings

As with control arm bushings, to remove and install trailing arm bushings, you must remove the trailing arms from the vehicle and use a special press and adapter. For larger bushings, some manufacturers recommend that you use a mild (1 percent) soap solution during installation to help ease the bushing into the trailing arm pivot. Never hit the end of a bushing with a hammer, as you may distort the rubber.

Some manufacturers place index marks on the bushings and trailing arm pivots, which are used to properly position the bushings during installation. Other bushings have flanges that must be a specified distance from the trailing arm pivot, figure 9-49. Check the replacement bushing and trailing arm for alignment marks. If there are any, refer to the

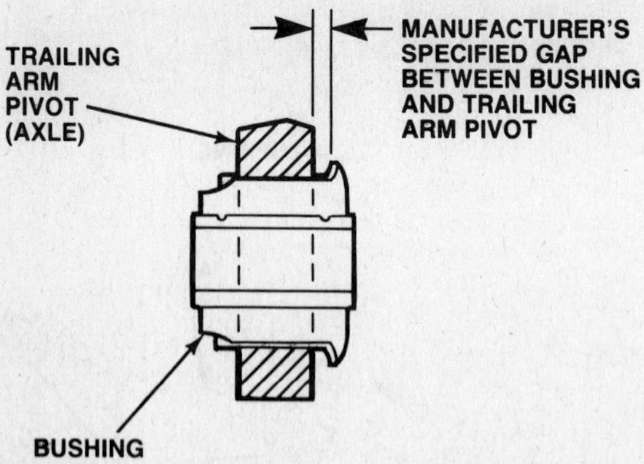

Figure 9-49. If the bushing has a rubber or metal flange, be sure the flange is the correct distance from the trailing arm pivot.

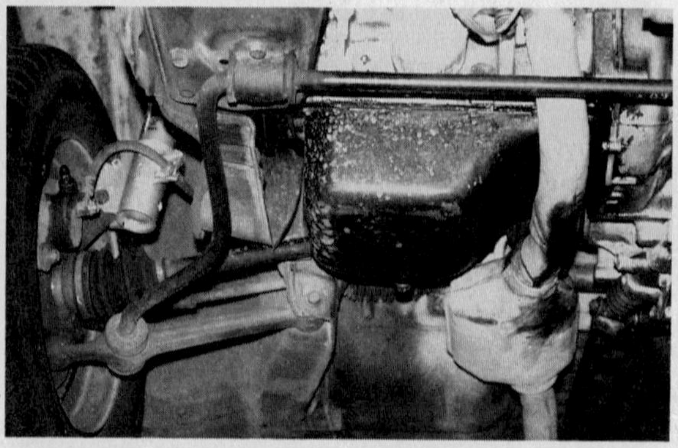

Figure 9-50. The antiroll bar may be found on front and rear suspensions. Brackets, bushings, and bolts hold the antiroll bar in place.

manufacturer's service manual for procedures on how to properly install the bushing.

ANTIROLL BARS AND OTHER SUSPENSION LINKS

Vehicle manufacturers use a number of secondary suspension components to fine-tune, stabilize, or strengthen the suspension system. The names of these components vary from manufacturer to manufacturer, and not all of these parts are used on all suspension systems.

Antiroll Bar

When a vehicle turns, centrifugal force transfers most of the vehicle weight to the outer wheels, which causes body roll. Antiroll bars help distribute some of the vehicle weight from the outer to the inner wheels during turns, reducing body roll. Antiroll bars are used on front and rear suspensions.

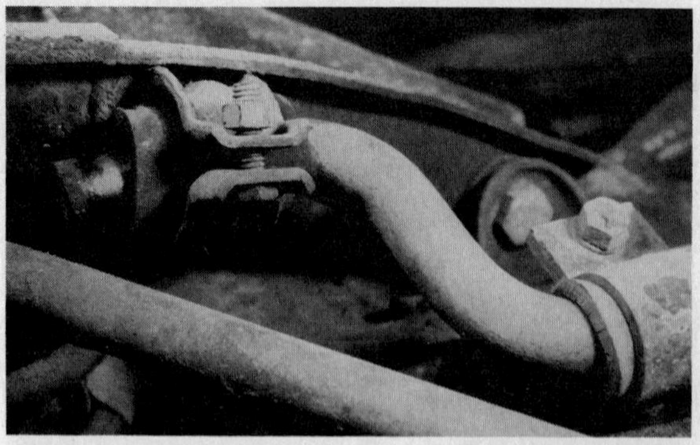

Figure 9-51. The bushings on this antiroll bar are worn and cracked, and should be replaced. Notice that the bracket holding the antiroll bar is not fully tightened.

Antiroll bars are also called "sway bars", "anti-sway bars", "stabilizer bars", or "roll bars" in automotive literature. Each end of the antiroll bar connects to the lower control arms or knuckle extensions with bushings, bolts, and brackets, figure 9-50. The middle section of the antiroll bar attaches to the vehicle frame with brackets and bushings. Sometimes small links, called "spacer bars", are used between the antiroll bar and vehicle frame, lower control arm, or knuckle extension.

An antiroll bar is not prone to wear or damage, and unless it is extremely old, it is likely in good condition. The bushings on the ends of the antiroll bar and between the antiroll bar and frame may wear, figure 9-51, and you should check them using the bushing inspection procedure described earlier in this chapter.

Use the following procedure to replace the bushings, or remove the old antiroll bar and install a new one:

1. Position the vehicle on a lift, and ensure that components will not bend under the weight of the car when it is raised. See Chapter One of this *Shop Manual* for lifting a vehicle.
2. Remove the tire and wheel, if necessary, to gain access to the ends of the antiroll bar.
3. Remove the nut and bolt that braces the antiroll bar and bushings to the lower control arm or knuckle extension. Remove the bracket at the other wheel as well.
   - If you are replacing the bushings, slide them off the ends of the antiroll bar, and press new ones in place. Reconnect the antiroll bar at the lower control arm or knuckle extension. Torque the bolts, and lower the vehicle. If you replace a bushing on one side of the antiroll bar, you should replace the bushing on the other side as well.

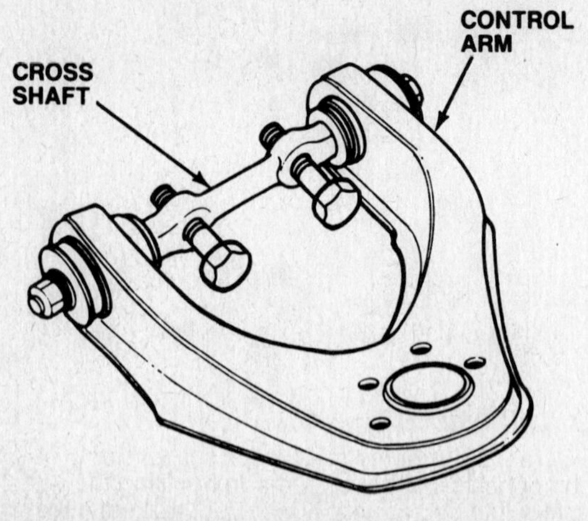

Figure 9-39. After you remove the bolts securing the cross shaft to the frame, remove the control arm and replace the pivot bushings.

nut on the end of the through-bolt, and remove the through-bolt. Remove the suspension spring.

5. Next, separate the control arm from the vehicle frame, which may connect at one or two pivot points. Remove the nuts that secure the cross shaft to the frame, figure 9-39, and remove the control arm. On a single-pivot control arm, remove the through-bolt.

Inspect the control arm for stress cracks, severe rust, bends, or other damage. Replace bushings and bump stops if needed, as described later in this section, or replace the entire control arm. To install the control arm:

1. Position the inner end of the control arm at the vehicle frame. For a double-pivot control arm, bolt the cross shaft in place. For a single pivot control arm, install the through-bolt. Do not fully tighten the bolts.
2. If you removed the suspension spring, keep it secured, but place it in the spring seat. Connect the torsion bar if you disconnected it earlier.
3. Put the ball joint stud into the socket on the knuckle. Install the castle nut, and tighten it to the manufacturer's specifications to seat the ball joint. Install a new cotter pin.
4. Raise the control arm with a jack to compress the spring, so you can remove the clips or other device securing the spring.
5. Install the shock absorber, and connect the antiroll bar and strut, if you disconnected them earlier.

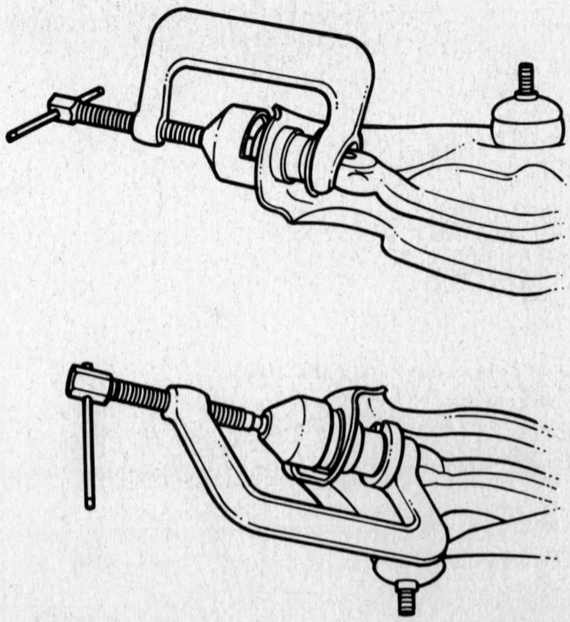

Figure 9-40. Unless the bushing is severely worn, you will need an adapter and press to remove the bushing from the control arm pivot. Use a similar adapter to install a new bushing.

6. Install any brake system components that you may have removed, and install the tire and wheel.
7. Lower the vehicle to the ground, and tighten the bolts securing the inner control arm pivots to the manufacturer's specifications.

Replacing Control Arm Bump Stops and Bushings

After removing the control arm from the vehicle, replacing bump stops and bushings is not a difficult task. Bump stops are usually held to the frame or control arm with a bracket and one or two bolts, figure 9-38. Remove the bolts, and replace the bump stop with a new one. Some bump stops fit over an extension on the frame, and you simply pull them off and push new ones in place.

To replace control arm bushings, you must first remove the control arm from the vehicle. For rubber bushings, use a press to press the bushings out of the control arm pivot, figure 9-40. You may need a special adapter for the press to remove and install the bushings. After installing the control arm in the vehicle, do not tighten the bushing pivot bolts until the vehicle is on the ground at normal ride height.

When installing new metal bushings, use an adapter and press to apply pressure to each

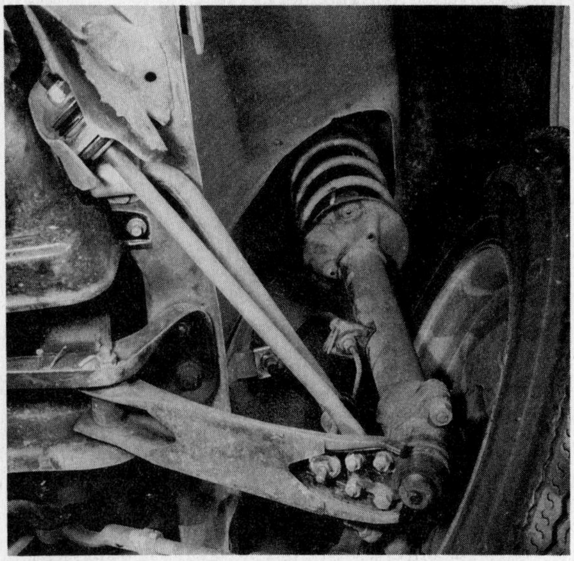

Figure 9-52. Check the bushings between the strut rod and vehicle frame. Replacing the bushing is simply a matter of disconnecting the strut rod, and installing a new bushing.

4. Remove the bolts that hold the antiroll bar bushing and bracket to the vehicle frame.
   - If you are replacing the bushings, place new ones under the brackets, and loosely install the bolts. Lower the vehicle so the weight is on the wheels, and torque the bolts in place.
   - If you are removing the antiroll bar, continue with this procedure.
5. Remove the antiroll bar from the vehicle. If spacer bars are used between the antiroll bar and lower control arms or knuckles, you should unbolt them and replace these bushings as well.
6. Install a new antiroll bar by securing the bar to the vehicle frame with brackets and bushings. Do not torque the bolts.
7. Install the bushings, spacer bars, and antiroll bar onto the lower control arms or knuckles, but do not tighten the bolts. Be sure to install new bushings in pairs. Make any needed adjustments to position the antiroll bar while the bolts are loose.
8. Lower the vehicle, and torque the bolts that secure the antiroll bar to the vehicle frame and lower control arms or knuckles.

Strut Rod

If the lower control arm connects to the vehicle frame at only one point, a strut rod may be used to support the structure for added stability. Strut rods brace the lower control arm against the vehicle frame and keep the control arm from swinging during braking. The strut rod may be located in front of or behind the control arm.

Bushings on the end of the strut rod damp control arm movement, figure 9-52. These bushings are prone to wear, and you should inspect them as described earlier in this chapter. Strut rod bushing wear can affect caster, which is a wheel alignment angle. In fact on some vehicles, caster is adjusted by turning a threaded strut rod. Use the following procedure to disconnect the strut rod from the frame and control arm, and replace the bushings:

1. Raise the vehicle on a lift, and remove the tire and wheel to gain access to the strut rod, if needed.
2. Locate the nut that secures the strut rod to the vehicle frame. Loosen and remove the nut.
3. Remove the nuts and bolts that secure the other end of the strut rod to the control arm.
4. Slide the strut rod out.
5. Remove the bushings and install new ones. Check the manufacturer's procedures for installing strut rod bushings. If you install the bushings incorrectly, you may not be able to adjust caster.
6. Install the nut that secures the strut rod to the frame, checking the bushing to be sure that you do not pinch it.
7. Install the nuts and bolts that secure the other end of the strut rod to the lower control arm.
8. Lower the vehicle, and torque the nuts to the manufacturer's specifications. Check and adjust wheel alignment, as described in Chapter Fourteen of this *Shop Manual*.

Radius Arm

The radius arm is a large link used on the front twin I-beam suspensions of some Ford trucks. Its function is similar to that of the strut rod, described earlier, except that it braces the outer end of the I-beam, rather than the control arm, against the vehicle frame, figure 9-53. Use the following procedure to remove the radius arm and replace the bushings, if needed:

1. Raise the vehicle on a frame hoist, and place extra-long jack stands under the axle.
2. Remove the front tires for easier access to the radius arms.
3. If the vehicle has an antiroll bar, remove it, using the procedure described earlier in this chapter.

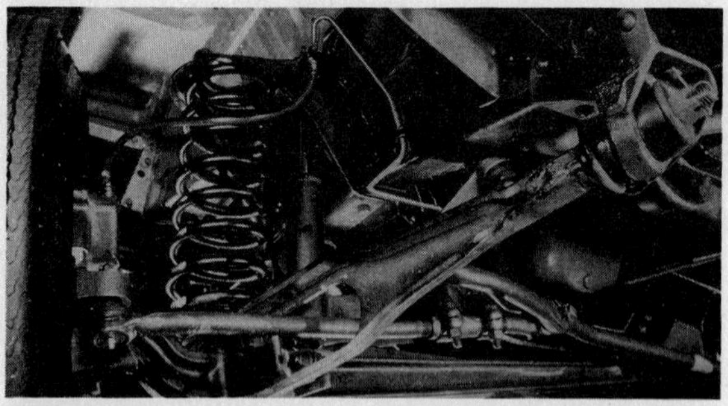

Figure 9-53. The radius arm braces the twin I-beam against the vehicle frame for added stability on some Ford trucks.

4. Secure and remove the suspension spring, as described in Chapter Eight of this *Shop Manual*. Remove the shock absorber, if needed.
5. Remove the lower spring seat, and remove the bolt that secures the radius arm to the axle.
6. Loosen and remove the nut, washer, and insulator at the other end of the radius arm where it bolts to the frame. Remove the radius arm and bushings, if worn. Press new bushings on the radius arm.
7. Position the radius arm at the axle, and install the through-bolt. Tighten it finger-tight.
8. Install the other end of the radius arm onto the bracket at the frame. Tighten the nut to the manufacturer's torque specifications.
9. Install the spring seat, spring, and shock absorber, as described in Chapter Eight of this *Shop Manual*.
10. Install the antiroll bar, as described earlier in this chapter.
11. Install the tires and wheels, lower the vehicle, and tighten the bolt that secures the radius arm to the axle. Be sure to use the manufacturer's torque specifications.
12. Check and adjust wheel alignment, as described in Chapter Fourteen of this *Shop Manual*.

Trailing and Lateral Links

Independent rear suspensions use a number of links to stabilize rear suspension components. These links are located between the knuckle or trailing arm and the frame or crossmember. Bushings on the ends of these links damp suspension movement. Depending on whether the links are positioned transversely or laterally, they may be called "trailing" or "lateral" links, figure 9-54, although there are a number of other names used by vehicle manufacturers.

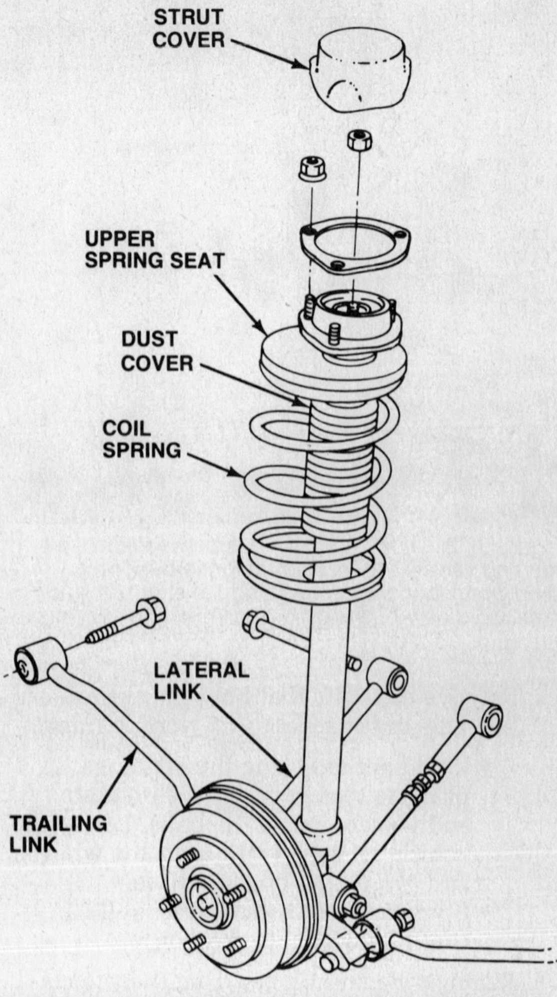

Figure 9-54. This Nissan Axxess uses a number of suspension links between the wheel knuckle and vehicle frame. Note that one of the links has an adjuster to set rear-wheel toe.

Removing and replacing links is simply a matter of loosening the nuts and removing the bolts that secure the link to the knuckle or frame. After removing a link, inspect it for damage and bends. Check the bushings at each end for wear. If worn, press the old bushing out, and press a new one in place. Then reinstall the link.

Check rear wheel toe, as it may have changed with the installation of new bushings. Some rear wheel suspension links have adjusters for setting rear wheel toe. For more information on rear wheel alignment, see Chapter Fourteen of this *Shop Manual*.

Figure 9-55. The Panhard or track rod is used on solid-axle rear suspensions to reduce side-to-side movement.

Panhard Rod (Track Rod)

The Panhard rod, or track rod, stabilizes a solid rear axle. It connects at some point on the axle and at the vehicle frame. Pivots on each end of the rod allow for normal suspension movement, figure 9-55.

The Panhard or track rod is connected to the frame or crossmember and the axle with through-bolts. To remove the rod, simply remove the through-bolts. Check the bushings at each end of the rod for wear and replace them, if necessary, before installing the Panhard or track rod.

Torque Arm

The torque arm is located between the differential and transmission on some solid axle, rear suspensions, figure 9-56.

To remove the torque arm, lift the vehicle and support the rear axle with a jack. Remove the bolt that secures the Panhard rod to the frame, if one is used. Support the rear axle with a jack. Disconnect the lower shock mount, compress and secure the coil springs, lower the rear axle, and remove the springs, as described in Chapter Eight of this *Shop Manual*. Remove the bolts that secure the torque arm to the differential and transmission, and remove the torque arm.

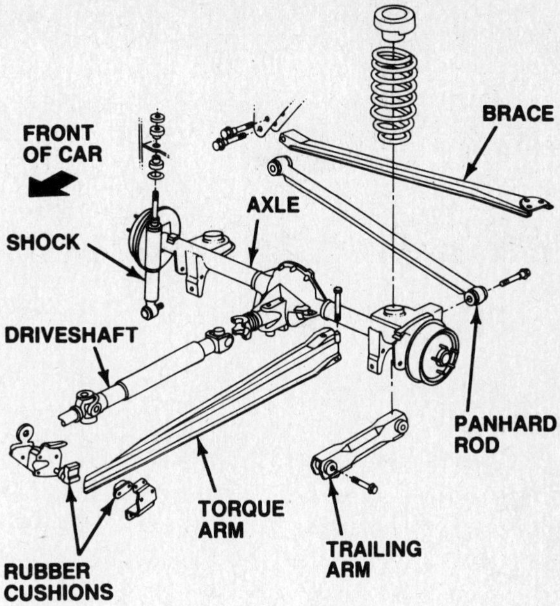

Figure 9-56. The torque arm, used on some solid axle rear suspensions, reduces the effects of axle windup.

Check the bushing at the front portion of the torque arm where it mounts to the transmission. If the bushing is worn, replace it before installing the torque arm.

Chapter

10

Basic Axle and Driveline Service

Although the driveline is not part of the suspension system, it can affect vehicle handling if parts are damaged. For example, a bent axle or worn driveline joint may cause vibration and handling problems. Many shops that service suspensions also perform minor driveline repairs.

Use this chapter, along with the information in Chapter Eleven of the *Classroom Manual*, to service axles and axle shafts, driveshafts, and driveline joints.

AXLES

Chapter Eight of the *Classroom Manual* describes three types of axles: stub axles, dead axles, and drive axles. Stub axles, used at the non-driven wheels of independent and semi-independent suspensions, are actually extensions of the wheel knuckle, and are sometimes called "spindles", figure 10-1. Service procedures for stub axles are in Chapter Nine of this *Shop Manual* in the section titled "KNUCKLES". Non-driven axles are often called "dead axles", and drive axles are called "live axles" when used to describe solid drive axles on RWD vehicles.

Non-Driven Axles

Many late-model, FWD vehicles use a dead rear axle, figure 10-2, which is a solid, non-driven axle. When you inspect solid rear axles on FWD vehicles, check the welds between the axle and rear trailing arms or suspension members. If you find cracks, repair or replace the axle, as described later in this chapter.

Heavy- and medium-duty RWD trucks use I-beam axles, and some Ford light-duty trucks use twin I-beam axles at the front suspension. Inspect beams for stress cracks. On I-beam and twin I-beam axles, the wheel knuckle and axle are connected with kingpins or ball joints. Check kingpins and ball joints as described in Chapter Nine of this *Shop Manual*.

The inner end of a twin I-beam connects to the vehicle frame and is free to pivot. One I-beam connects to the frame with a through-bolt and bushing, figure 10-3. The other beam connects with a bracket and bushing, figure 10-4. Check the bracket and bushings for wear or damage and replace them, if needed.

Twin I-beam axles have radius arms that run between the I-beam and vehicle frame. Radius arms help stabilize the I-beam during braking. Check the bushing between the radius arm and vehicle frame for wear, figure 10-5.

Figure 10-1. Stub axles are used at non-driven wheels. The part on which the wheel hub rides is sometimes referred to as the "spindle".

Figure 10-2. A dead axle is a non-driven axle that runs from one wheel to the other. This dead axle is used at the rear of a front-wheel-drive vehicle.

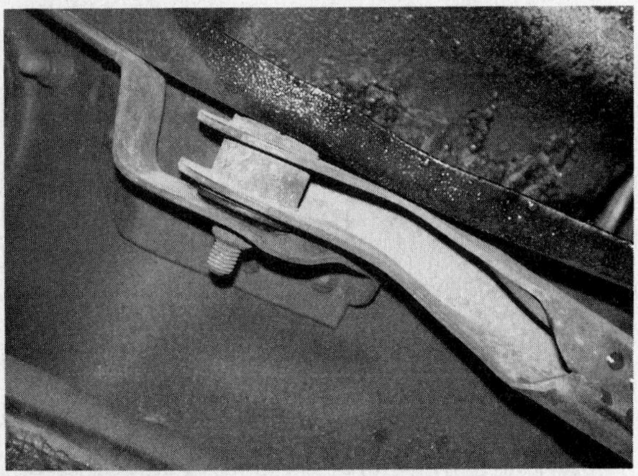

Figure 10-3. The inner end of a twin I-beam pivots at the vehicle frame. I-beams provide strength to the front axle, and the pivots allow for independent suspension movement.

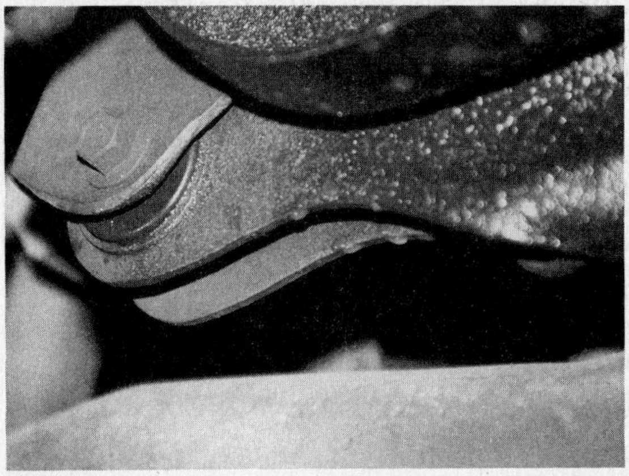

Figure 10-4. This twin I-beam inner pivot connects to the vehicle frame with a bracket, bushing, and through-bolt.

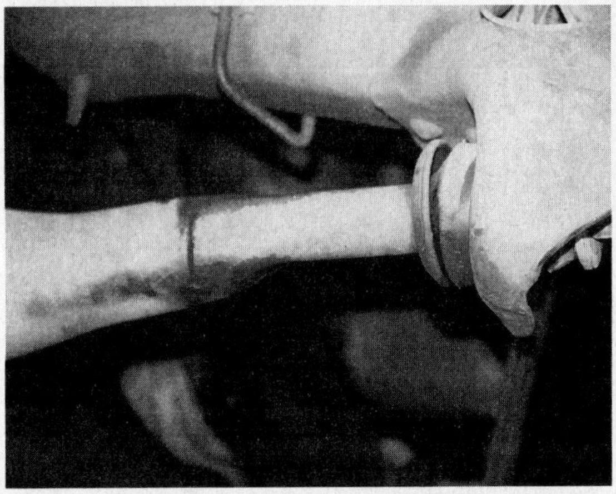

Figure 10-5. A bushing between the radius arm and vehicle frame damps vibration when the vehicle is slowing down or accelerating. Inspect this bushing for cracks.

Removing and installing a rear axle from a FWD vehicle

Many FWD vehicles, especially light-weight FWD vans, use a solid rear axle, figure 10-2. Use the following procedure as a guide to remove a solid rear axle on a FWD vehicle. Remember to turn off the electronic suspension system if there is one.

1. Loosen the rear wheel nuts. Position the lift supports under the vehicle chassis using the manufacturer's recommended lift points, and raise the vehicle. See Chapter One of this *Shop Manual* for more information on lifting a vehicle. Remove the rear tires and wheels.
2. Support the rear axle with a jack and jack stands.
3. Remove brake hardware running from the vehicle frame to the wheel hubs or axle, including parking brake cables, brake

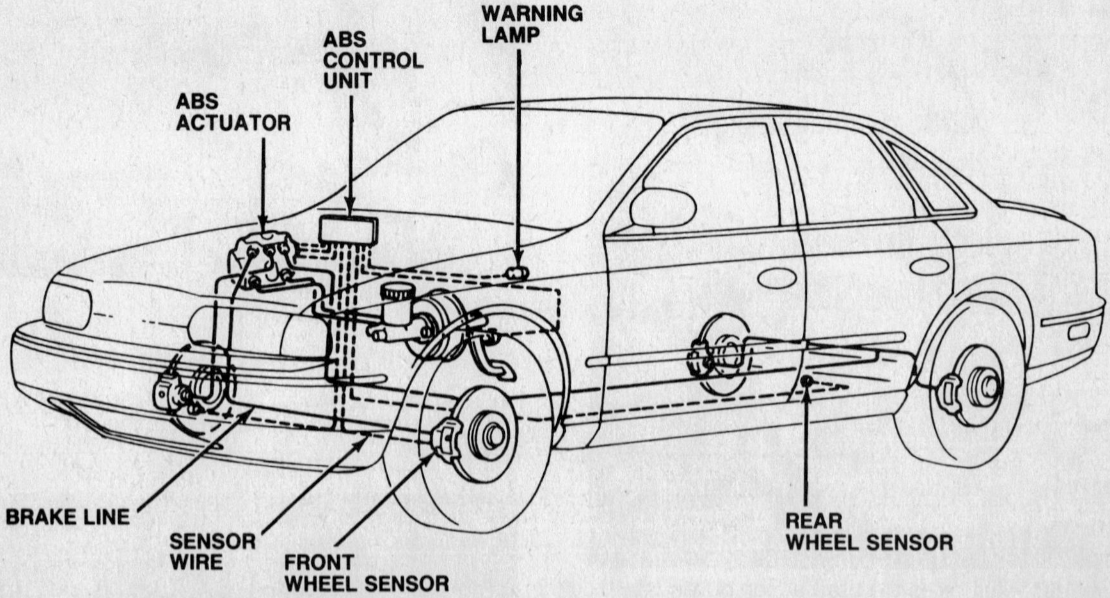

Figure 10-6. The dashed lines on this illustration represent wiring harnesses, and the solid lines represent hydraulic lines on this braking system. Be sure to disconnect the wires and hydraulic lines at the rear axle before lowering it.

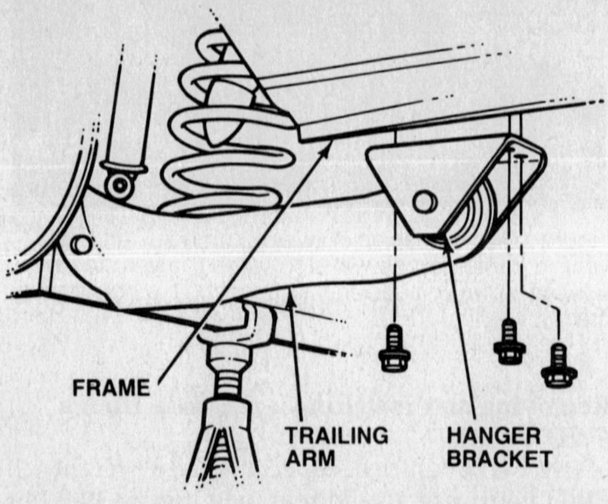

Figure 10-7. Secure the suspension springs and disconnect the shocks. Then disconnect the brackets or bolts that secure the rear axle trailing arms to the vehicle frame.

lines, and proportioning valve hardware. If you neglect to do this, you will damage this hardware when the axle is lowered.

4. If the vehicle has an electronic suspension system or anti-lock brakes, figure 10-6, be sure to disconnect sensors and non-flexible lines at each wheel hub, or you may damage the electronic or hydraulic lines when you lower the axle.
5. Secure the suspension springs with safety clips or spring compressor plates, as described in Chapter Eight of this *Shop Manual*.
6. Remove the through-bolts that secure the track bar or Panhard rod to the axle, as described in Chapter Nine of this *Shop Manual*. Secure the track bar or Panhard rod with wire so it does not hang down while you are removing the axle.
7. Disconnect the lower shock absorber mounts, using the procedures in Chapter Eight of this *Shop Manual* as a guide, if needed.
8. With the trailing or control arms still attached to the frame, lower the axle just enough to remove the suspension springs.
9. Loosen and remove the bolts that secure the trailing or control arms to the vehicle frame, figure 10-7. Be sure to disconnect the suspension arms on both sides of the axle.
10. Lower the axle to the ground with the jack.

To install the axle, use a floor jack to raise it into position. Be sure that the suspension springs are properly positioned. Hold the axle in place with jack stands, and tighten the bolts that secure the trailing or control arm brackets to the vehicle frame. Connect the shock absorbers and the track bar or Panhard rod. Reconnect any electrical connectors and hydraulic

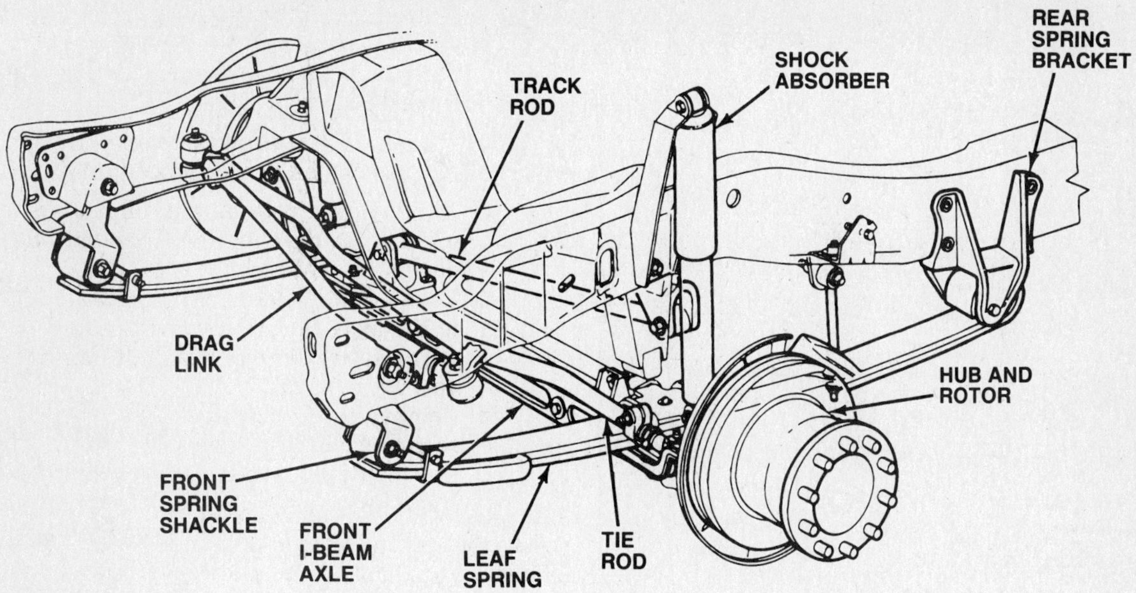

Figure 10-8. This solid I-beam front axle is used on a number of heavy-duty trucks. The track rod provides lateral stability.

lines, if you disconnected them earlier. Install the tires and wheels, and lower the vehicle.

Tighten the suspension pivot-bushing bolts to the manufacturer's recommended torque. Torque the wheel nuts as well. Some vehicle manufacturers suggest that you check and adjust wheel alignment, as described in Chapter Fourteen of this *Shop Manual*, after replacing the axle.

Removing and installing a front I-beam or twin I-beam

The solid I-beam front axle is used on a number of medium- to heavy-duty trucks because it can support heavy loads. A twin I-beam front axle, used on some light-duty Ford vehicles, combines the sturdiness of a beam axle, while allowing for independent front suspension movement. To remove a front I-beam, figure 10-8, or twin I-beam, figure 10-9, use the following guidelines:

1. Loosen the front wheel nuts. Position the lift supports under the vehicle frame using the manufacturer's recommended lift points, and raise the vehicle. See Chapter One of this *Shop Manual* for more information on lifting a vehicle. Remove the front tires and wheels.
2. Disconnect the tie rods from the front steering knuckles by breaking the taper at the tie rod ends. Use the procedures in Chapter Five of this *Shop Manual* as a guide.
3. Disconnect the wheel knuckle from the I-beam. The knuckle-to-beam connection uses either ball joints or kingpins. Procedures for disconnecting ball joints and kingpins are in Chapter Nine of this *Shop Manual*.
4. Disconnect any brake hardware from the axle.
5. Place a jack under the I-beam, or use two jacks if you are working on a vehicle with twin I-beams.
6. Disconnect the lower shock mounts, as described in Chapter Eight of this *Shop Manual*.
7. For a twin I-beam suspension with coil springs, use the steps that follow. For a solid front I-beam suspension, skip this step and proceed to step 8.
   a. Use the jack to lift the I-beams, and compress the suspension springs. Secure the springs with spring compressor plates or safety clips, as described in Chapter Eight of this *Shop Manual*.
   b. Remove the screws that hold the spring retainers in place.
   c. If the vehicle uses radius arms and an antiroll bar, remove them, as described in Chapter Nine of this *Shop Manual*.
   d. Lower the I-beams just enough to remove the suspension springs.
   e. On twin I-beams, one I-beam connects to the vehicle frame with a through-bolt. While still supporting the beam

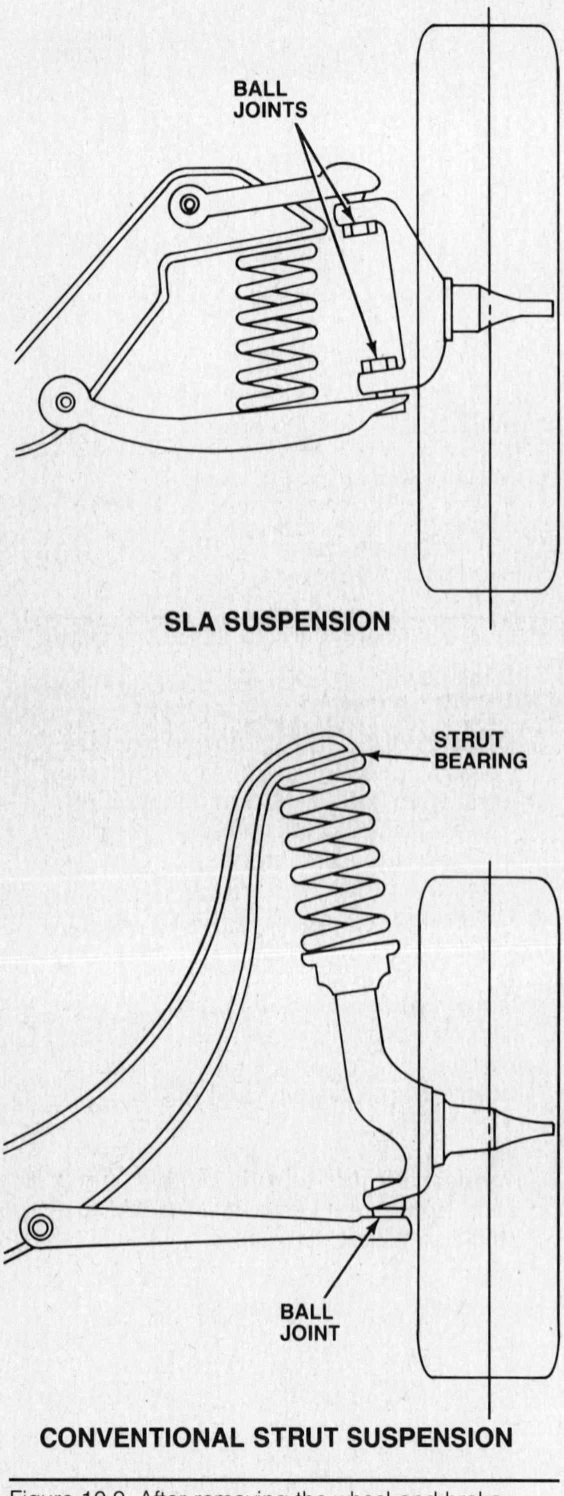

Figure 10-9. After removing the wheel and brake hardware, disconnect the twin I-beams from their pivot points at the vehicle frame.

with jack stands, remove the through-bolt, and lower the axle. The other I-beam connects to the frame with a bracket. Support the beam with jack stands, and remove the bolts that secure the bracket to the frame. Lower the axle.

8. If the vehicle uses a solid I-beam with leaf springs use the following procedure:
   a. If the vehicle uses an antiroll bar, remove it, as described in Chapter Nine of this *Shop Manual*.
   b. Remove the bolt securing the track rod to the axle. Secure the track rod to the frame so it does not hang down while you are working.
   c. Remove the U-bolt nuts. The U-bolts secure the leaf springs to the axle.
   d. Lower the axle, by lowering the jack stands.

To install the I-beam or twin I-beams, use a jack to lift the I-beam in place. For a twin I-beam, be sure that the suspension springs are properly positioned, and install the spring retainers. Connect the twin I-beam inner pivots, and install the radius arms and antiroll bar. For a solid I-beam, secure the leaf springs to the axle with the U-bolts, and tighten the U-bolt nuts. Install the track rod and antiroll bar, if used. For both types of I-beam, connect the lower shock mounts and replace any brake hardware that you removed earlier. Attach the wheel knuckle to the I-beam by securing the kingpin or ball joint connections. Compress the springs with the jack, and remove the compressor plates or spring clips. Connect tie rods, install the tires and wheels, and lower the vehicle to the ground.

While the vehicle is on the lift, be sure to install suspension through-bolts finger tight. After lowering the vehicle, tighten the through-bolts to the manufacturer's recommended torque.

The Ford 4WD Explorer uses a twin I-beam front suspension. On this setup, one of the I-beams contains the differential, figure 10-10. Before removing the twin I-beams on this vehicle, you must disconnect the driveshaft joints, as discussed later in this chapter.

Drive Axles

Drive axles may be located inside a solid, rear-axle housing, or they may be part of an independent front or rear suspension. On an independent suspension system, axle shafts, or "half-shafts", connect the transaxle and wheel hub. CV joints or U-joints on each end of the axle shaft allow for changes in suspension angle. Use the following procedures to

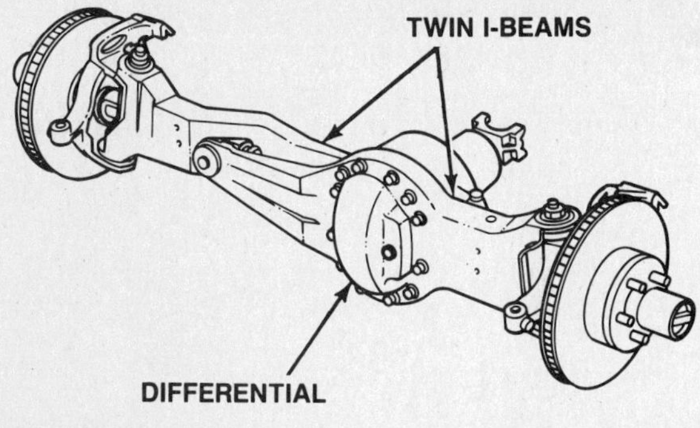

Figure 10-10. This Ford twin I-beam front axle is used on the 4WD Explorer. One of the I-beams contains a differential. To remove this axle, you must also disconnect the driveshaft.

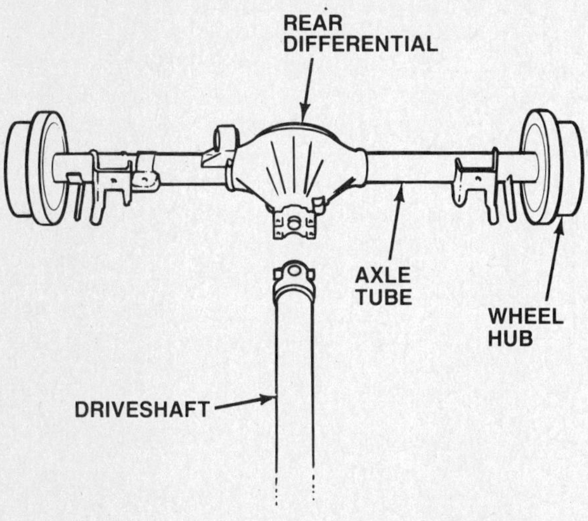

Figure 10-11. Axle shafts are inside the axle tubes. The shafts have a splined or other fitting with the differential gears and wheel hubs. The shafts transmit torque from the differential to the rear wheels.

inspect, remove, and install drive axles. Procedures for servicing driveline joints are given later in this chapter.

Solid axles
Solid drive axles are classified by how the axle shafts are supported by bearings in the axle housing. There are three types of solid drive axles on RWD vehicles: full-floating, three-quarter floating, and semi-floating, which are described in Chapter Ten of the *Classroom Manual*.

Axle tubes, which are part of the axle housing, extend from the differential to the rear wheel hubs, figure 10-11. Axle shafts inside these housings mesh with the differential gears and connect to the rear wheel hubs to turn the vehicle's wheels.

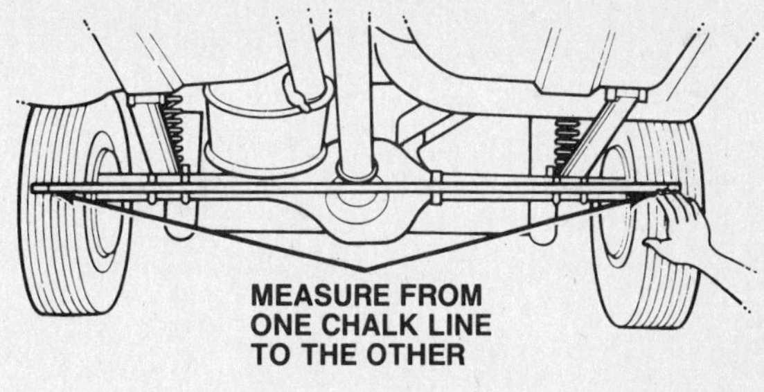

Figure 10-12. Measuring the distance between two chalk marks on the rear wheels can help you determine rear wheel toe.

Because solid rear axles include the rear differential, you must disconnect the driveshaft before removing the rear axle, as described later in this chapter. After installing a solid rear axle, check and adjust the driveshaft working angle, which is also described later in this chapter. Neglecting to do this will cause driveshaft vibration in some vehicles.

Checking a solid rear axle
Rear axles are sturdy and not prone to much damage or wear. They can become misaligned if the vehicle is in a collision. Check the axle shaft housing for dents, which may keep the axle shafts inside from turning freely. If the rear wheels are not bent, and the tires have unusual wear patterns, use the following procedure to check the rear axle. If you have wheel alignment equipment, see Chapter Fourteen of this *Shop Manual*. (Checking rear wheel alignment angles with an alignment machine will also give you information about the condition of the rear axle.) To check the rear axle without alignment equipment, use the following procedure:

1. Raise the vehicle on a lift, supporting it at the frame so that the rear axle hangs freely.
2. Use chalk to mark the center of the tread on each rear tire.
3. Rotate the rear tires so the chalk marks are toward the front of the vehicle. Using a tape measure, measure the distance between the two marks on the tires, figure 10-12. Record this distance.
4. Rotate the wheels so the chalk marks are facing the ground. Measure and record the distance between the marks.
5. Rotate the wheels until the chalk marks are toward the rear of the vehicle. Measure and record the distance between the marks.
6. Calculate toe-in or toe-out by subtracting the rear reading from the front reading.

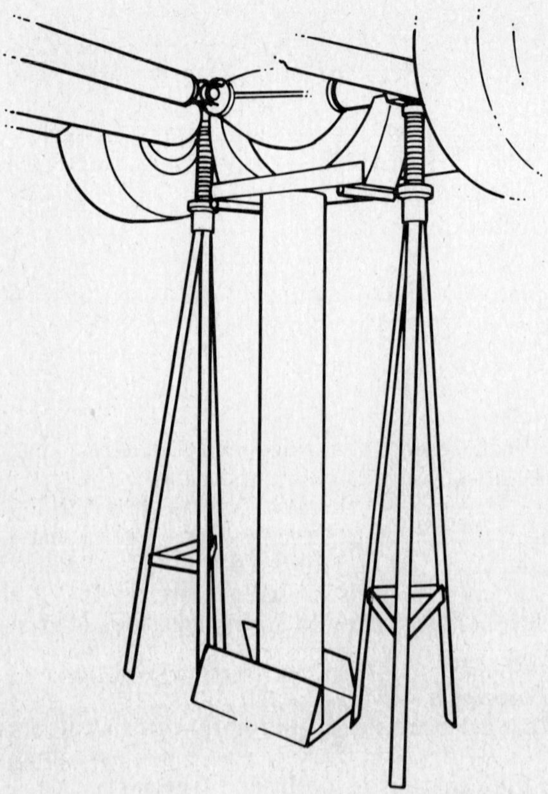

Figure 10-13. When you are removing a solid rear axle, be sure to support it with jack stands. Do not allow the axle to tilt while the driveshaft is connected to it.

Compare your result with the manufacturer's specifications for rear toe.

7. Calculate camber by averaging the front and rear readings and then subtracting the center reading from the average. Again, compare results with the manufacturer's specifications for rear camber.
8. If the rear axle alignment settings do not meet specifications, the axle may be bent and you should check it further. If aligning the rear wheels does not help, and the problem is severe, replace the rear axle.

If fluid is leaking from the differential housing, refer to the manufacturer's service manual before making repairs. Disassembling and repairing the rear axle is beyond the scope of this book.

Removing and replacing a solid rear axle
Coil or leaf springs are used with solid rear drive axles. Use the following procedure as a guide to remove the axle:

1. Loosen the rear wheel nuts, and raise the vehicle on a lift, supporting it at the frame. Remove the rear tires and wheels.
2. Support the rear axle with jack stands. Place one jack stand on each side of the differential, figure 10-13.
3. Many manufacturers recommend draining the differential fluid before removing the rear axle. Remove the differential drain plug, and drain the fluid into a container.
4. If the vehicle has anti-lock brakes, be sure to disconnect sensors and hydraulic lines at each wheel hub. Disconnect any brake hardware from the axle. Some manufacturers suggest removing the brake drums.
5. Mark the driveshaft and differential with chalk before disconnecting them, so that you can realign the parts during reassembly. Disconnect the driveshaft from the differential, using the procedure given later in this chapter in the section titled "DRIVESHAFTS".
6. Disconnect any other hardware between the axle housing and vehicle frame, such as torque arms or U-bolts.
7. Disconnect the lower shock mounts.
8. Raise the rear axle slightly to compress the suspension springs, and secure them with spring compressor plates or safety clips, as described in Chapter Eight of this *Shop Manual*.
9. Depending on the type of suspension system you are working on, you may have to disconnect the suspension arms from the axle or frame. On most vehicles, disconnect these components by removing the through-bolts or brackets that secure them. Refer to Chapter Nine of this *Shop Manual*, if necessary.
10. Lower the axle and differential just enough to remove the suspension springs and retainers.
11. Lower the axle to the ground using the jack.

To install the rear axle, position it near the vehicle frame with jack stands. With the suspension springs in place, install the spring retainers, if used. If brackets hold the suspension arms to the frame, tighten the bolts that secure them. Connect shock absorber mounts, the torque arm, and any U-bolts, if used. Connect the driveshaft to the differential, using the alignment marks that you made earlier. Connect any brake hardware that you removed. Refill the differential with the manufacturer's recommended fluid if you drained it earlier. Install the tires and wheels, and lower the vehicle to the ground.

Be sure to install suspension through-bolts finger tight while the vehicle is on the lift. After lowering the vehicle, tighten the through-

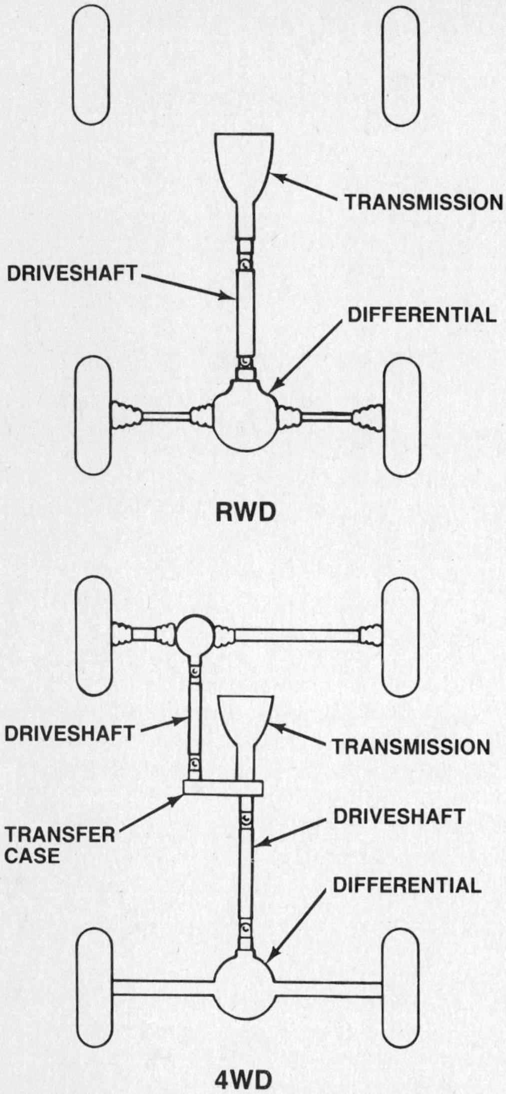

Figure 10-14. Driveshafts transfer torque from the transmission or transfer case to the front and rear differentials. Four-wheel-drive vehicles use two driveshafts.

bolts to the manufacturer's recommended torque. Be sure to check and adjust the driveshaft working angle, as described later in this chapter.

Axle shafts (half-shafts)
Axle shafts transfer torque from the transaxle or differential to the driven wheels. Use the procedures in the manufacturer's service manual to remove axle shafts from a solid rear axle, as these procedures are beyond the scope of this book. The following paragraphs describe how to inspect and service axle shafts used with independent suspensions.

Independent suspensions normally use CV or U-joints at the axle shafts, allowing free suspension movement. Bent shafts or worn shaft joints will result in suspension system problems. To remove axle shafts, use the procedures in the "DRIVELINE JOINTS" section of this chapter. Inspect the axle shafts using the following procedure.

Checking axle shafts
With the vehicle on a lift, check the axle shafts for bends, cracks, or other structural damage. If the driver notices vibration when driving the vehicle at high speeds, it may be due to bent axle shafts. Some tow truck drivers use a J-hook on the axle shafts during towing without realizing that this will bend the shafts. If a vibration problem occurs shortly after a vehicle has been towed, suspect bent axle shafts.

Do not attempt to straighten a bent axle shaft, as this will cause stress cracks and weaken the shaft. Instead, replace the shaft with the manufacturer's recommended part.

Check axle shafts and the chassis components near them for signs of rubbing. If the axle shafts are rubbing, check suspension springs for wear, as discussed in Chapter Eight of this *Shop Manual*.

If you remove an axle shaft, check the splines on both ends. Splines should not be worn, broken, or chipped. If they are, replace the shaft.

Some Ford vehicles are equipped with a two-piece axle shaft. If the coupler on a two-piece shaft wears out, replace both pieces and the coupler with a single solid shaft, which is available from Ford. The two-piece shaft is no longer available or recommended by the manufacturer.

DRIVESHAFTS

On front-engine, RWD cars, the driveshaft transfers torque from the engine and transmission to the rear differential, figure 10-14. On 4WD cars, driveshafts transfer torque from the transfer case to the front and rear differentials. Various types of joints are used on driveshafts, although cross-and-yoke U-joints are the most common. A few driveshafts use constant velocity (CV) joints, and some four-wheel-drive vehicles use ball and trunnion joints.

When the vehicle is carrying heavy loads or when it is traveling over uneven surfaces, the driveshaft angle may change. The U-joints and slip yokes in the driveshaft allow for these angle changes while ensuring even torque transfer from the engine or transfer case to the differential, as described in Chapter Eleven of the *Classroom Manual*.

Figure 10-15. This rubber coupling reduces vibration between the differential and driveshaft.

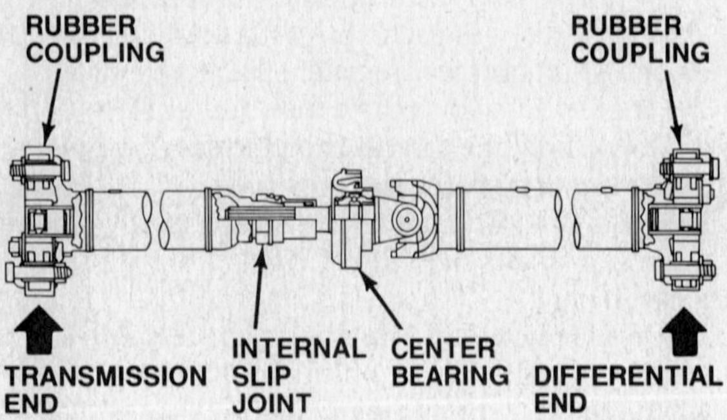

Figure 10-16. Some European and Japanese cars use rubber couplings on the driveshaft to damp vibration and balance the shaft. A slip joint allows for driveshaft length changes.

Inspecting a Driveshaft

With the vehicle on a lift, check the driveshaft for bends, and replace it if needed.

A driveshaft is installed in a balanced position and at a specified working angle so that while spinning at high rpm, it will not vibrate. Before disconnecting driveline joints to remove the driveshaft, make alignment marks on the shaft and differential or transfer case so that you can reinstall the shaft in the same position. Some manufacturers scribe index marks in the driveshaft, and others machine special grooves in it so that there is only one way the driveshaft can be installed.

Many driveshaft joints have grease fittings, and you should lubricate them according to the manufacturer's recommended service interval, which is usually about every six months. Service driveshaft joints more frequently if the vehicle is operated in dusty, muddy, sandy, or wet conditions. Check driveshaft joints for wear, as described later in this chapter in the section titled ''DRIVELINE JOINTS''. Some driveshaft joints are sealed and lubricated for life. The only service they require is a periodic inspection for wear.

Rather than U-joints, some vehicles have rubber couplings between the driveshaft and differential and transmission, figure 10-15. These rubber bushings damp driveshaft vibration. Because the driveshaft mounts solidly to both the differential and transmission on these vehicles, a slip joint near the driveshaft's center bearing allows for driveshaft length changes, figure 10-16.

Figure 10-17. The driveshaft U-joint connects to a flange on the differential. Remove the nuts and bolts, and separate the shaft from the differential.

Removing and Replacing a Driveshaft

If you determine that the joints are worn or the driveshaft is bent, remove and install the driveshaft using the following steps as a guide:

1. Raise the vehicle on a lift, and place a drain pan under the transmission where it connects to the driveshaft.
2. Use chalk to make alignment marks on the driveshaft and differential yokes. Use these marks to correctly position the driveshaft during reassembly.
3. Remove the nuts and bolts securing the driveshaft joint to the pinion yoke on the differential, figure 10-17. Then pull the driveshaft and differential yokes apart, figure 10-18. To separate a rubber coupling, simply remove the bolts that thread through the coupling.

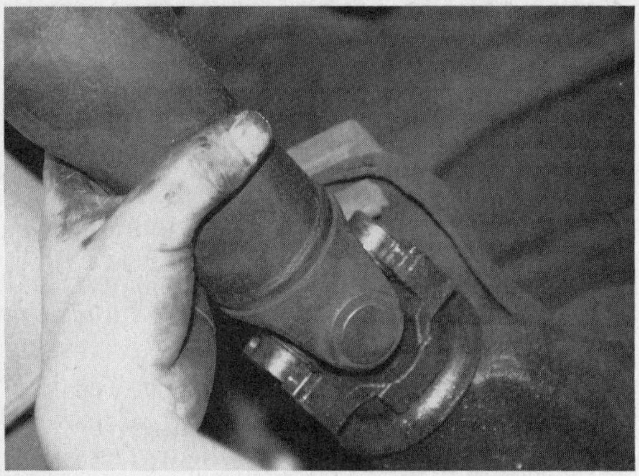

Figure 10-18. Pull the driveshaft away from the differential. If necessary, use a screwdriver to pry them apart.

Figure 10-19. To reduce vibration and provide extra support at the center of the driveshaft, some manufacturers use a center bearing and center support bracket in the driveshaft.

4. Support the free end of the driveshaft with a jack stand. Do not allow the driveshaft to hang.
5. Some vehicle manufacturers use a center bearing and support bracket in the driveshaft, figure 10-19, to reduce vibration and provide extra support at the center of the shaft. Loosen and remove the bolts that hold the support bracket to the vehicle frame. Use another jack stand to support the driveshaft at the center bearing, if needed.
6. Slide the driveshaft back toward the rear of the vehicle, and pull the splined end out of the transmission. Be sure to plug the opening in the transmission so that it does not lose too much fluid.
7. Inspect driveshaft joints for wear, as described later in this chapter.

Before installing the driveshaft, check the differential yoke for damage, figure 10-20.

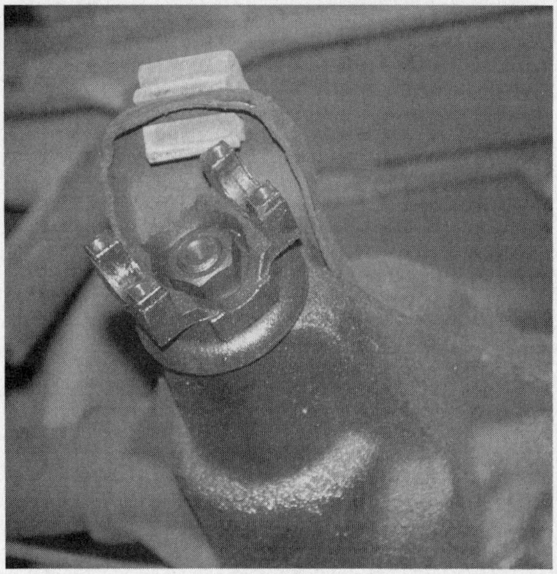

Figure 10-20. Check that the bolt holes in the differential yoke are not stripped or damaged. Also, clean the grooves in the yoke where the driveshaft mounts.

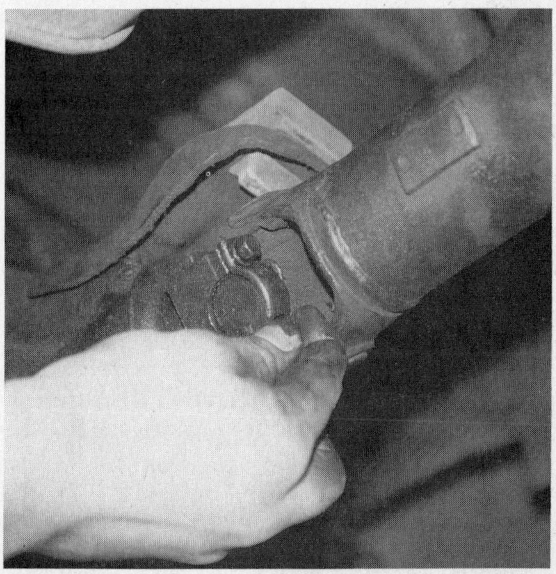

Figure 10-21. Install the bolts that hold the driveshaft to the differential yoke finger tight. Then torque the bolts according to the manufacturer's specifications.

Then lift the driveshaft, and push the splined end of the driveshaft into the transmission. Align the mark on the other end of the driveshaft with the mark on the differential. Install the bolts that hold the driveshaft to the differential, figure 10-21. Torque the bolts to the manufacturer's specifications. Install the center bearing bracket if you removed it earlier. Finally, check the driveshaft working angle, and adjust it if needed, as described in the next section.

DRIVESHAFT ANGLE

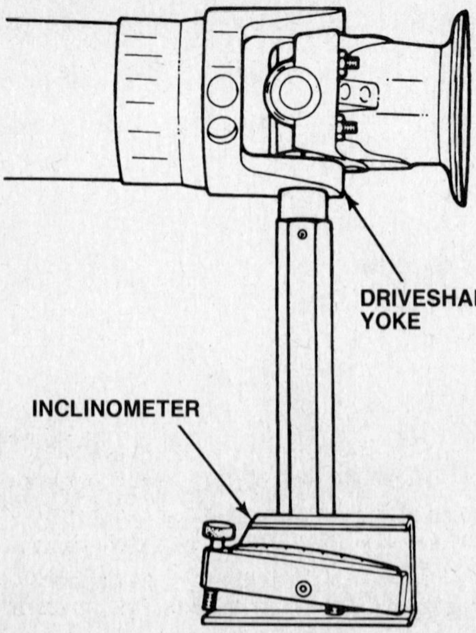

Figure 10-22. An inclinometer is a tool that combines a level and a graduated scale for determining driveshaft angles.

Checking and Adjusting Driveshaft Working Angle

Almost all late-model RWD vehicles use a one-piece driveshaft with a Cardan U-joint at each end to compensate for driveshaft angle changes. Some vehicles with longer driveshafts may use a two-piece driveshaft with three Cardan U-joints. Many luxury vehicles use double Cardan U-joints rather than the single Cardan type to minimize driveshaft vibration.

Most vehicle manufacturers that use two or more U-joints in the driveshaft do not require a driveshaft working angle adjustment, but check the vehicle manufacturer's recommendations to be sure. A general rule of thumb is that the working angles of the rear and center driveshaft joints should be from one to three degrees, and within one degree of each other.

Vehicles that use single Cardan U-joints in the driveshaft cannot compensate for extreme angle changes, and the correct driveshaft working angle is critical. These driveshafts are designed to operate at a predetermined working angle when the vehicle is at normal ride height.

If recommended by the vehicle manufacturer, you should check and adjust the driveshaft working angle after installing a new driveshaft, servicing rear suspension components that change the position of the rear differential, or installing a new engine or transmission:

1. Drive the vehicle onto a level wheel alignment rack, and raise the rack, or be sure that the vehicle is parked on a level surface. To check the driveshaft working angle, the weight of the vehicle must be on the wheels. (You cannot use a regular shop lift, which raises the vehicle at the frame.)
2. Place an inclinometer on the driveshaft, figure 10-22, or where specified by the vehicle manufacturer. You may need a manufacturer's tool to position the inclinometer.
3. Adjust the level so that the bubble is centered in the level indicator.
4. Read the angle on the inclinometer scale, and record the driveshaft angle.
5. Place the inclinometer on the differential yoke, and take another reading, as described above.
6. Place the inclinometer on the transmission yoke, and take another reading, as described above.
7. The vehicle manufacturer will list specifications for driveshaft angles at the transmission and differential if these angles are critical in preventing driveline vibration. Compare your results with the manufacturer's specifications, and adjust the angles if needed:
    - To change the angle at the rear driveshaft joint on a vehicle equipped with leaf springs, place tapered shims between the springs and axle housing. If the vehicle uses coil springs, change the working angle by adjusting the length of the upper control arms or trailing arms. Adjustment methods include turning an eccentric bolt, using shims between the control arm or trailing arm mounting bracket and frame, and repositioning the control arm or trailing arm at a slotted adjustment.
    - To change the angle at the front driveshaft joint, some manufacturers suggest placing shims under the transmission mount until the angle is correct.
8. After making adjustments, check the working angles again. Then test drive the vehicle to be sure that there is no driveline vibration due to an incorrect driveshaft working angle.

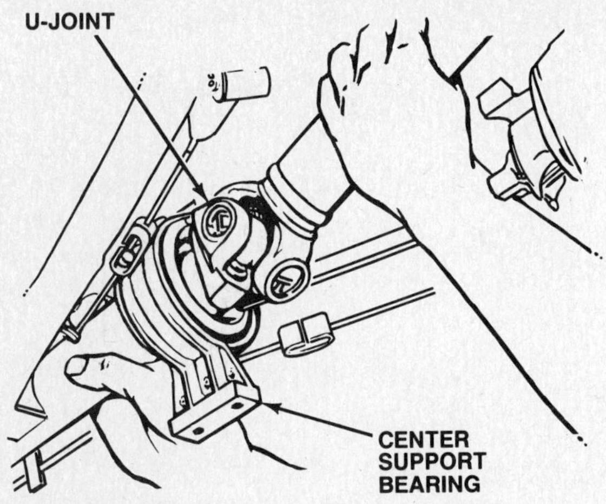

Figure 10-23. Grasp the driveshaft on both sides of the U-joint and twist. If there is play, replace the joint. A clunking noise when you put the transmission in gear is another sign of driveline joint wear.

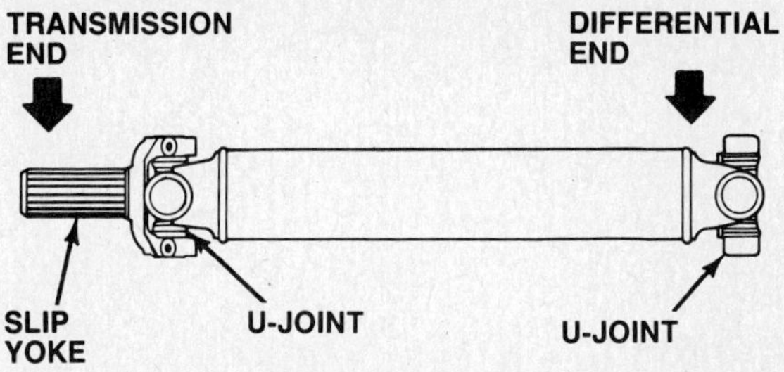

Figure 10-24. A slip joint, located between the transmission and driveshaft, usually is lubricated with transmission fluid. This type of joint allows for driveshaft length changes.

DRIVELINE JOINTS

This section has procedures for inspecting, lubricating, and removing driveline joints. Driveline joints include the U-joints on the driveshafts of RWD and 4WD vehicles, and the CV joints or U-joints on the axle shafts of front- or rear-wheel-drive vehicles with independent suspension.

Universal Joints (U-Joints)

U-joints are usually found on each end of the driveshaft. If there is play in these joints due to wear, the driveshaft will not transfer torque smoothly. If you hear a clunking noise when you put the transmission in gear and begin rolling forward, it is likely that one or both of the driveshaft joints are worn.

Inspecting and lubricating universal-joints
Universal joints will inevitably wear over time, but neglecting recommended periodic lubrication will accelerate joint wear. Most late-model vehicles use permanently sealed and lubricated U-joints. Because these joints are sealed, you cannot lubricate or rebuild them. When they wear, you must discard them. Use the following procedure to inspect and lubricate U-joints at the driveshaft and axle shaft, if they have grease fittings:

1. Position the lift supports under the vehicle chassis using the manufacturer's recommended lift points, and raise the vehicle. See Chapter One of this *Shop Manual* for more information on lifting a vehicle.
2. Check U-joint bolts at the driveshaft. If they are loose, tighten them. Check and tighten the bolts that secure the center carrier bearing bracket, if one is used.
3. Check for rust around the U-joint yoke. If you find a light film of rust, the needle bearings may be wearing, and you should inspect the joint closely for wear.
4. Grasp the driveshaft on either side of the U-joint, and twist the sections in opposite directions, figure 10-23. If you detect movement, U-joint bearings are worn. Use a screwdriver as a lever to pry between the U-joint yokes. Again, if there is movement, the joint is worn and should be replaced as described in the next procedure. If the joint is in good condition, grease it as described in the rest of this procedure.
5. Rotate the driveshaft with your hands until the grease fitting on the U-joint is visible. Wipe the fitting so dirt will not enter the joint when you apply grease.
6. Remove the grease plug, and install a grease fitting if one is not already in the joint.
7. Slowly apply the manufacturer's recommended grease to the U-joint. Most manufacturers recommend a lithium-based grease because it repels water. Forcing too much grease too quickly into the joint may rupture the seals or pop out the end cap. U-joints do not need very much grease.
8. Check the U-joint and the rest of the driveshaft for other grease fittings.
9. If the vehicle you are working on uses a slip joint at the transmission end of the driveshaft, figure 10-24, check the joint for a grease fitting. Usually this slip joint is lubricated by transmission fluid, but on some vehicles there may be a grease fitting. Also check the area around the center bearing for a slip joint and grease fitting. As with the U-joint, apply grease slowly.

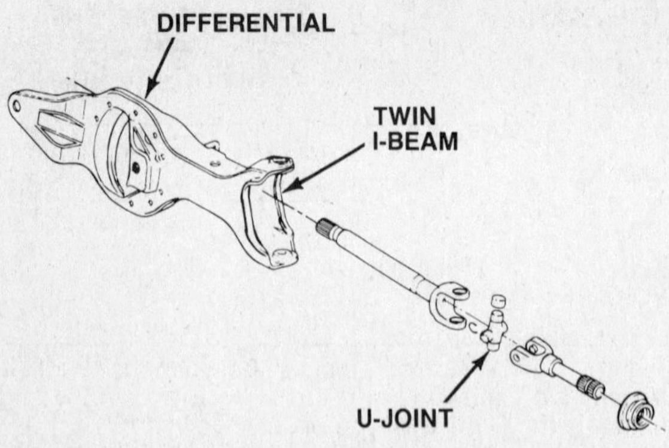

Figure 10-25. Some larger trucks and vans use U-joints on the axle shafts, rather than CV joints. Grease these joints as you would driveshaft U-joints.

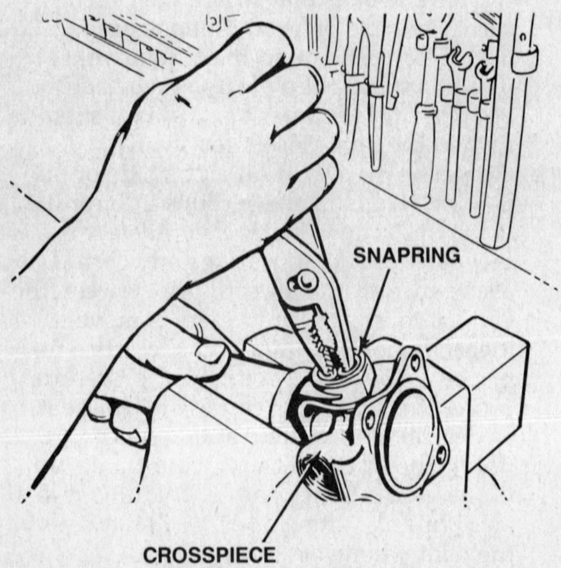

Figure 10-26. Use pliers to remove the snaprings that hold the U-joint crosspiece in place. The snaprings are usually located on the outside of the yoke.

10. Check for U-joints on the axle shafts, and grease them as described earlier, figure 10-25.

Removing and installing universal-joints
To remove the driveshaft from a vehicle, use the procedure given earlier in this chapter in the section titled "DRIVESHAFTS". Then remove and install the U-joint as follows:

1. Place the driveshaft in a soft-jaw vise, being careful that you do not knock off any balance weights that may be on the driveshaft. Remove the fasteners that hold the U-joint crosspiece in place at the driveshaft yoke. The most common fasteners used are snaprings, which are usually located on the outside of the yoke, figure 10-26. Some crosspiece snaprings are located on the inside of the yoke and are a little more difficult to remove. Plastic fasteners are used to secure the crosspiece on some late-model vehicles.

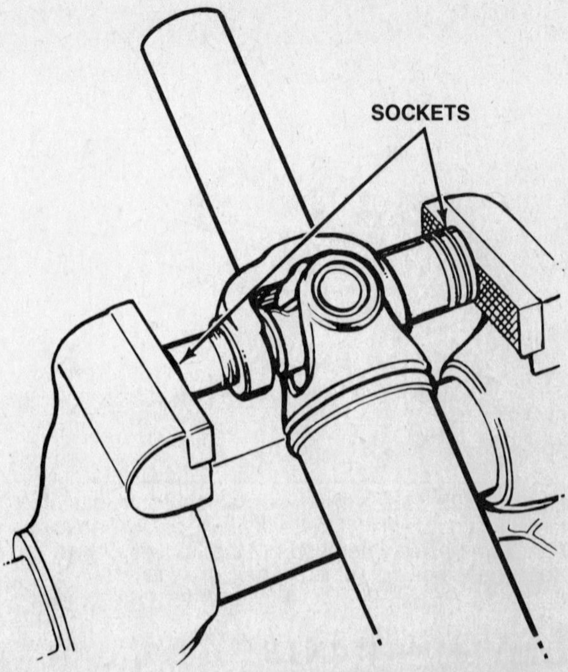

Figure 10-27. Use a C-clamp or vise and sockets that fit over the bearing cups to press the cups from the crosspiece.

2. Use a vise and sockets, figure 10-27, to press the crosspiece bearings and bearing cups from the crosspiece. Some aftermarket tool manufacturers have designed a special tool for removing U-joint bearings. One type is a C-clamp with interchangeable adapters to fit a variety of bearing circumferences. After pressing out all four bearings, remove the crosspiece from the driveshaft yoke.
3. Inspect the driveshaft yoke for wear. If yoke extensions are out-of-round, replace that section of the driveshaft. Discard the old crosspiece and bearings, and use new ones for reassembly.
4. Check the new crosspiece for alignment marks, and check the manufacturer's literature for special installation instructions. Lubricate the bearing cups with extreme pressure (EP) grease, which is sometimes included with the new crosspiece.
5. Insert the crosspiece into the driveshaft yoke, and place the bearings on the crosspiece ends. Press the bearings and bearing

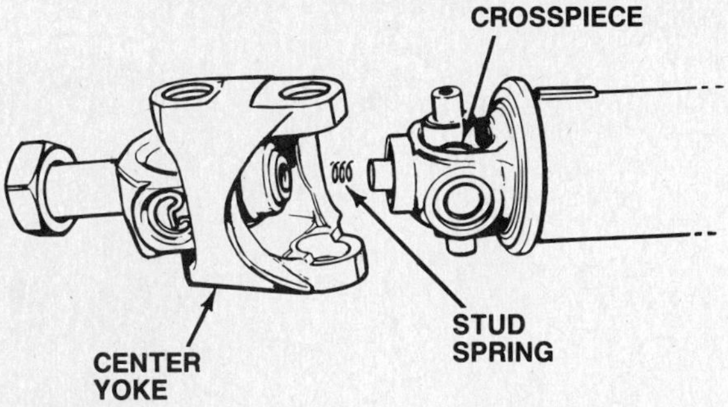

Figure 10-28. Do not lose the stud spring between the driveshaft and center yoke when you separate a double-Cardan U-joint.

cups in place using the vise and socket or C-clamp. If there is resistance while you are pressing, check the position of the bearings, and press again.

6. Install the snaprings.
7. Insert the remaining two crosspiece ends into the yoke on the other half of the driveshaft. Press the bearing cups in place, and install the snaprings.
8. Check the U-joint for flexibility, and install it in the vehicle.

Rather than using one crosspiece to connect two sections of the driveshaft, a double-Cardan U-joint uses two crosspieces and a center yoke to connect two sections of the driveshaft, figure 10-28. This gives the driveshaft additional flexibility, which helps reduce driveshaft vibration. The procedure for disassembling a double-Cardan U-joint is similar to the one for disassembling a single-Cardan U-joint, but you must use a few additional precautions:

- After removing the snaprings, press out the bearing cups. Do not press the bearings too far or you may damage the centering-ball assembly. Remove the other two crosspiece bearings the same way.
- Before removing the crosspiece, press the driveshaft toward the center yoke to compress the stud spring. Then carefully work the crosspiece out of the yoke. Do this slowly so that you do not lose the spring and bearing rollers.
- Remove the other bearings and crosspiece using the same procedures and cautions.

Constant Velocity (CV) Joints

Constant velocity joints, especially on the axle shafts of FWD cars, affect steering, suspension, and driveline performance. It is difficult to detect CV joint wear when the vehicle is on a lift because these joints have a certain amount of play designed into them. Diagnosing CV joint problems usually involves listening for unusual noises while driving the vehicle. See Chapter Two of this *Shop Manual* for information on diagnosing CV joints.

If a worn CV joint is not replaced, then the transaxle, axle shafts, and suspension components are subject to additional stress, which accelerates wear. Use the procedures in the following sections as a guide to inspect, lubricate, and remove CV joints.

Inspecting and lubricating constant velocity joints

Constant velocity joints on front- and rear-wheel-drive axle shafts are expected to last about 100,000 miles, but without proper lubrication, they will wear much faster. Always check CV joint boots for damage to ensure that lubricant remains in the joint.

Outer CV joints (the joints at the wheel end of the axle shaft) usually wear more quickly than the inner joints because they are subject to a variety of axle shaft operating angles. The shaft at the outer CV joint may move as much as 40 degrees to compensate for suspension and steering angles. Normal operating angles at the inner CV joint are only about 10 to 20 degrees.

When checking CV joint boots, raise the vehicle on a lift, and use a trouble light. Wipe grease and dirt from the boot so that you can see any small tears or punctures. Check that the boot clamps are secure. Also, check to be sure that oil or coolant is not leaking on the boots, as this contaminates the rubber, making the boots weak and prone to tears.

Usually, if the boot is damaged, then the CV joint is worn. This is because as a CV joint turns, centrifugal force pushes grease outward, forcing it through even the tiniest of openings in the boot. This lack of lubrication coupled with the constant movement of the joint quickly leads to CV joint failure. In addition, a damaged boot allows dirt, water, and other contaminants into the joint.

If the boot is damaged, remove it. (You may have to remove the axle shaft first, which is described in the following section.) Place a small amount of grease from inside the boot between your finger tips, and rub your fingers together. If the grease feels gritty, replace the CV joint, as described later in this chapter. If the grease feels smooth, you may have caught the problem before any damage was done to the joint. If this is the case, grease the joint and install a new boot, as described later in this chapter.

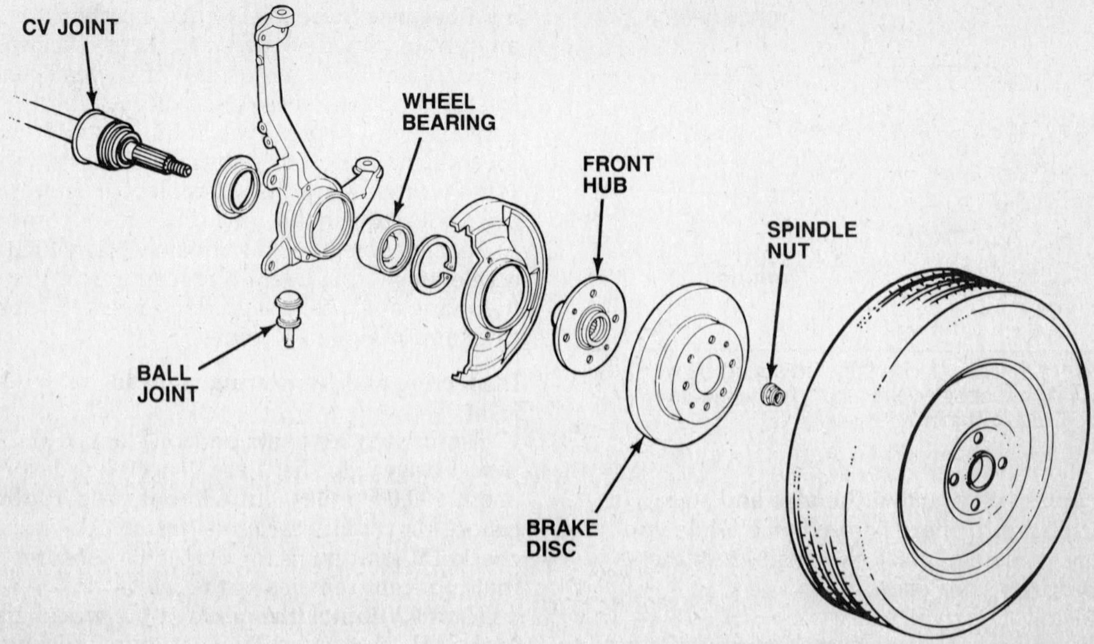

Figure 10-29. The spindle nut secures the wheel assembly to the axle shaft and CV joint assembly. Never use an impact wrench to remove the spindle nut because the vibrations can damage wheel bearings and CV joints.

Removing and replacing the axle shaft assembly

Use this procedure as a guide to remove axle shafts and CV joints, figure 10-29. Note that after completing step 7, the procedure for removing the axle shaft will vary from manufacturer to manufacturer. On some vehicles, you must disconnect the inner CV joint from the transaxle or differential first, and then remove the outer CV joint from the knuckle. However, most vehicle manufacturers suggest that you disconnect the outer CV joint first, which is the method we will cover here.

1. Remove the wheel cover to expose the wheel nuts and spindle or hub nut.
2. With the vehicle on the ground, have an assistant apply the brakes while you loosen the spindle nut with a breaker bar. Do not use an impact wrench, as the vibrations from an impact wrench can damage the CV joint and wheel bearings. Some vehicles have a locking tab that you must pry away from the nut before you can turn it.
3. Loosen the wheel nuts, and raise the vehicle on a lift.
4. Remove the wheels.
5. Drain the fluid from the transaxle or differential, using the manufacturer's recommended procedure.
6. Place a drain pan under the differential or transaxle to catch lubricant that leaks out while you are removing the inboard CV joints.
7. Secure suspension springs, as described in Chapter Eight of this *Shop Manual*. Then disconnect suspension and steering components from the wheel knuckle, using the procedures given in Chapter Nine of this *Shop Manual* as a guide. The procedures from this point will vary from manufacturer to manufacturer, as discussed earlier.
8. Remove the loosened spindle nut. Then pull the wheel hub and knuckle to separate the axle shaft and CV joint from the knuckle, figure 10-30. For some vehicles, you may need a special tool to press the axle shaft from the knuckle. Do not try to pull the axle shaft from the knuckle, as this may cause the joint to break apart. Instead, pull the CV joint housing, if needed. If the vehicle has anti-lock brakes, there may be a toothed gear on the outer CV joint. This gear sends signals to the ABS controller and must not be damaged, so handle it carefully.
9. Disconnect the inner CV joint from the differential or transaxle. Check the inner CV joint for roll pins, Torx head bolts, or 12-point star bolts that secure the CV joint to

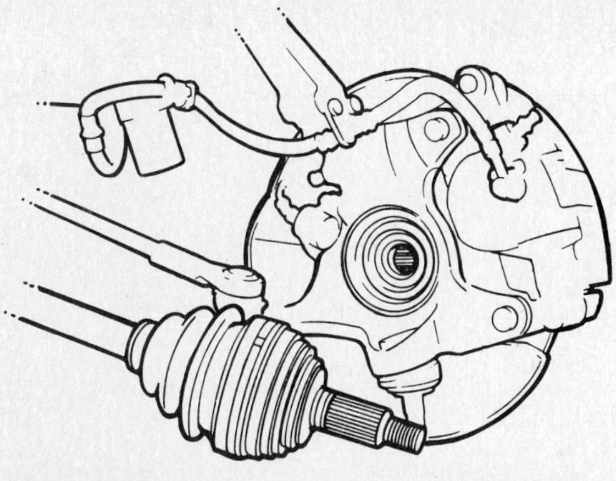

Figure 10-30. The axle shaft has a splined fitting with the wheel assembly. After removing the spindle nut, you can separate the axle shaft and wheel assembly.

Figure 10-31. This inner CV joint is secured to the differential with Torx-head bolts.

the transaxle, figure 10-31. If you find these fasteners, remove them. On some automatic transmissions, you must remove the differential cover, and remove a C-clip on the end of the axle shaft.

10. Grasp the inner CV joint housing and pull it to separate the axle shaft from the transaxle. Pull the housing gradually so the axle shaft oil seal is not damaged. You may want to use a screwdriver to pry the CV joint housing from the transaxle, figure 10-32. If you are working on a car with inner tripod joints, do not pull on the axle shaft, as this may cause the inner CV joint to break apart. Instead, pull the inner CV joint housing while supporting the outer end of the shaft until the shaft releases from the transaxle. *Do not* use a hammer to

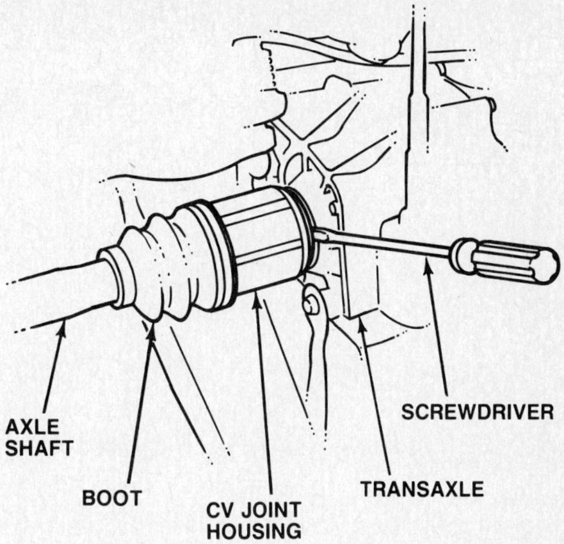

Figure 10-32. Use a screwdriver to pry the inner CV joint housing away from the transaxle. If you pull on the axle shaft, the CV joint may come apart.

separate the CV joint from the transaxle, or you risk damaging the joint.

- On some Fords with automatic transmissions, you must remove only one axle shaft at a time. If you remove both joints, components in the transaxle may rotate slightly and become misaligned.
- Many manufacturers provide special tools that keep transaxle parts aligned while the CV joints are removed. Check the vehicle manufacturer's service manual for specific instructions.

11. Replace worn CV joints. On some vehicles, a CV joint and axle shaft must be replaced as a unit.

Removing and replacing a CV joint

If the CV joints are worn and can be separated from the axle shaft, use the following procedure as a guide to remove and replace them. As mentioned earlier, on some vehicles you must replace the CV joints and axle shaft as a unit.

1. Clamp the axle shaft/CV joint assembly in a soft-jaw vise.
2. Mark the position of the boot on the axle shaft so that you can install the boot in the same place later. Remove the boot clamps, and cut away the boots, figure 10-33. Never reuse old boots.

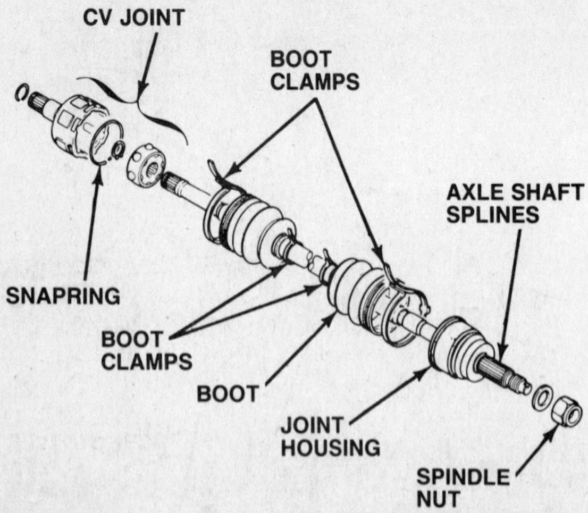

Figure 10-33. Clamps secure the rubber boot on the CV joint housing and axle shaft. The boots keep grease in, and dirt and water out.

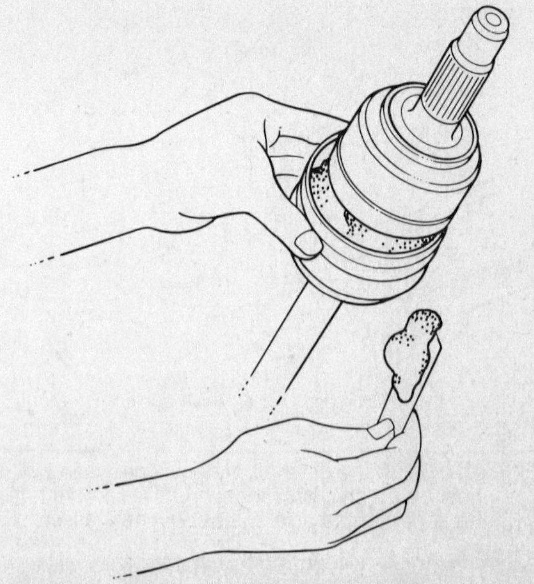

Figure 10-34. Pack the CV joint housing with the manufacturer's recommended grease. Some CV joint boot kits come with grease for this purpose.

3. Locate the snapring that holds the CV joint in place, and remove the snapring with snapring pliers.
4. Remove the CV joint and housing. You may have to tap the CV joint with a soft-face hammer to remove it from the shaft.
5. Check the splines on the ends of the axle shafts. If they are chipped or worn, replace the axle shaft.
6. Before installing a new CV joint, check the part number to be sure that you have the right one.
7. Put tape around the axle shaft splines before installing the new boot. This will keep the splines from ripping the boot.
9. Place a new CV joint boot on the shaft and clamp it in place, using the alignment marks you made earlier. Remove the tape from the axle shaft splines. Install the new CV joint and a new snapring. Install the housing. Tap the joint with a soft-face hammer until you hear a click, which means the joint is in place. Pull on the joint to check it.
10. Pack the CV joint housing with the manufacturer's recommended CV joint grease, figure 10-34. The incorrect grease may contaminate the boot. Often the CV joint boot kit comes with grease. Use the quantity of grease supplied with the boot, and never mix greases.
11. Position the boot end over the joint housing, and install the boot clamp.
    - Two-piece or ''split'' boots are available, but are not recommended by most vehicle manufacturers. These boots are bolted or glued together, and the glue must cure for two to four hours. Bolt-together boots, if not properly installed, will allow CV joint contamination. Single-piece CV joint boots are the best choice, although you must remove the axle shaft to install this type of boot.
    - Some General Motor's FWD vehicles use Hytrel® boots. You will need special replacement clamps and torquing pliers when installing these boots to be sure that they seal properly.
13. Install the axle shaft/CV joint assembly in the vehicle.
14. Check wheel alignment after replacing CV joints.

Chapter

11 Basic Service for Electronically Controlled Suspension Systems

As discussed in Chapter Twelve of the *Classroom Manual*, some manufacturers use electronic controls on their suspension systems. The design of these suspensions vary from manufacturer to manufacturer, but there are some overall similarities that may help you service them.

This chapter covers general service procedures that you can use as a guide when servicing electronically controlled suspensions. For complete diagnostic and repair information, refer to the vehicle manufacturer's service manual.

IDENTIFYING ELECTRONICALLY CONTROLLED SUSPENSIONS

Before you raise the vehicle on a lift or begin suspension system repairs, determine if the vehicle has an electronically controlled suspension. Because these systems use sensors and actuators to adjust the way the vehicle handles, you can seriously damage suspension components if you lift or tow a vehicle without first disabling the electronic controls.

If you have the opportunity to talk to the vehicle owner, ask if the suspension is electronically controlled. Also check the vehicle instrument panel for switches that control suspension handling or check for indicator lights that give the driver information about the suspension system, figure 11-1. Also, check the glove box for the owner's manual, which usually contains precautions and information on disabling the suspension system in the event that the vehicle must be raised or towed.

If this information is not available to you, turn the ignition switch off, and wait a few minutes for the suspension system to stabilize. (The vehicle ride height on some suspension systems will change slightly after you turn off the ignition.) Without raising the vehicle on a lift, check the shock absorbers and suspension springs. Look for air and hydraulic lines as well as electrical connectors and wiring harnesses leading to the springs and shocks, figures 11-2 and 11-3. Be sure to check both the tops and bottoms of springs and shocks, as the location of air, hydraulic, and electrical connectors may vary, depending on the vehicle manufacturer. Air lines are often made of nylon, and suspension system hydraulic lines may look like brake lines.

On electronically controlled suspensions, the shocks and struts usually look like their conventional counterparts. There may be a metal canister that is about the size of an engine oil

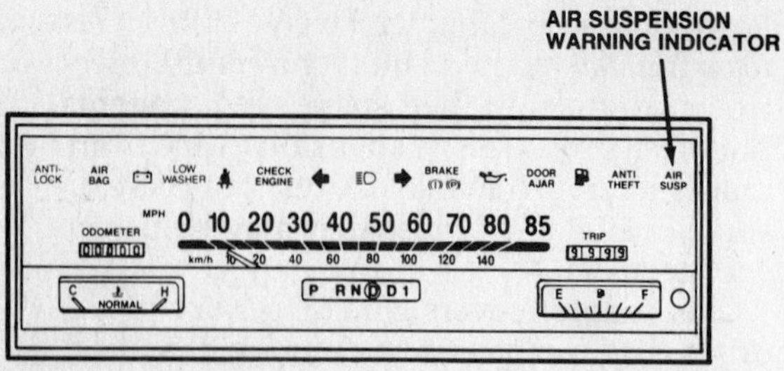

Figure 11-1. Most vehicles have an indicator light on the instrument panel that illuminates when there is a problem with the electronically controlled suspension system.

Figure 11-2. Check the base of the struts or shocks for electrical connectors, which may be covered with a protective rubber cap. This indicates electronic suspension control.

Figure 11-3. Check the tops of the struts for hydraulic lines and electrical connectors, which indicate that the vehicle uses some type of electronically controlled suspension.

Figure 11-4. This metal canister on the strut of the Infiniti Q45 with active suspension is a sub-accumulator. This accumulator stores pressurized fluid and helps damp suspension vibration.

filter attached to each shock or strut, figure 11-4, which serves as an accumulator or pressure damper.

Air springs do not look like coil or leaf springs, but are in the same location as conventional springs, figure 11-5. The Mitsubishi A-ECS system uses struts with regular coil springs, but an additional air spring is located inside each strut coil spring. These cylindrical air bladders are made of black nylon-reinforced rubber.

The locations of the suspension computer, sensors, and other equipment will vary depending on the vehicle manufacturer's design and the type of system used in the vehicle.

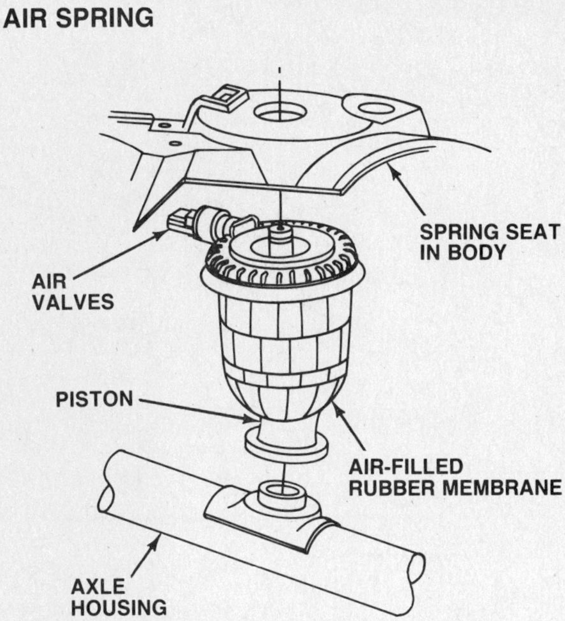

Figure 11-5. This air spring used on the 1990 Lincoln Town Car replaces the conventional coil spring. Look at the control arms and axles where you would find coil springs, and check for a rubber air bladder.

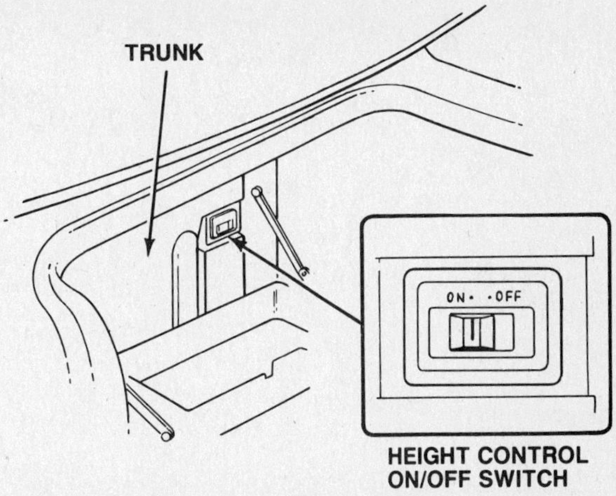

Figure 11-6. Most electronically controlled suspension systems have a switch that will deactivate the suspension's controls, so the technician can safely work on it.

SPECIAL PRECAUTIONS FOR SERVICING VEHICLES WITH ELECTRONICALLY CONTROLLED SUSPENSIONS

If you determine that the vehicle you are working on has an electronically controlled suspension system, check the owner's manual or service manual for special instructions before towing the car, lifting it, or servicing the suspension system.

Lifting and Towing

On some vehicles with electronically controlled suspensions, turning off the ignition switch will not deactivate the suspension. For example, the microcomputer in the 1990 Lincoln Town Car will continue to process suspension height information for 30 minutes after the ignition switch is turned off. If you raise or tow the vehicle without first deactivating the electronic suspension system, sensors will detect that the vehicle ride height is too high. In turn, the microcomputer will open solenoid-controlled air valves to release air from the springs. If too much air is released, the rubber material from which the spring is made will fold over on itself. This causes irreparable damage to the springs, and you will have to replace them.

Another hazard of lifting a vehicle with an electronically controlled suspension is that if the vehicle makes suspension adjustments while it is on the lift, it may shift or fall off the lift, endangering the safety of everyone working in the shop.

Use the manufacturer's service literature or the information in the vehicle owner's manual to deactivate the suspension system before lifting or towing. The most common method is to turn off the ignition switch, and then turn off another switch controlling the suspension system. On the Lincoln Town Car and Continental, the switch to deactivate the electronic suspension system is behind an access panel in the trunk. The Lexus LS 400 with air suspension has a height control ON/OFF switch inside the tool storage compartment in the trunk, figure 11-6.

In an emergency, you can disconnect the battery cables or remove fuses to deactivate the system, but remember that you may also erase trouble codes from the vehicle's computers, as well as radio, mirror, and seat position settings. Remember, disconnect the battery only in an emergency.

For the Infiniti Q45 with active suspension, you must wait two to three minutes after turning the engine off before disconnecting the battery or removing fuses, or the vehicle ride height will change. Infiniti also reminds technicians that once the vehicle is on a lift, you must not start the engine, or the suspension system will re-activate.

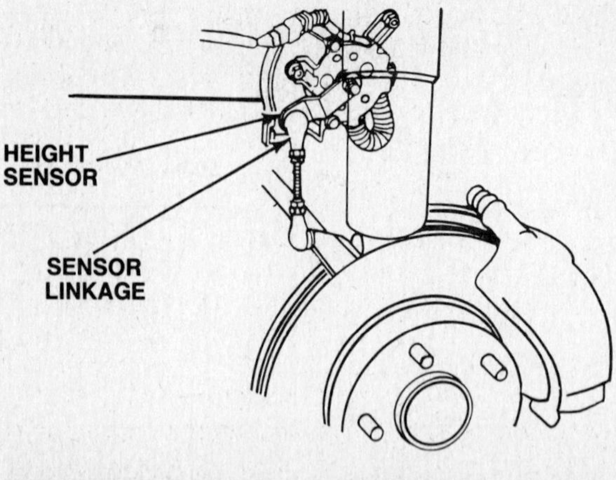

Figure 11-7. This ride height sensor has an adjustable linkage. To set the ride height, turn the nut to lengthen or shorten the linkage until the ride height is correct.

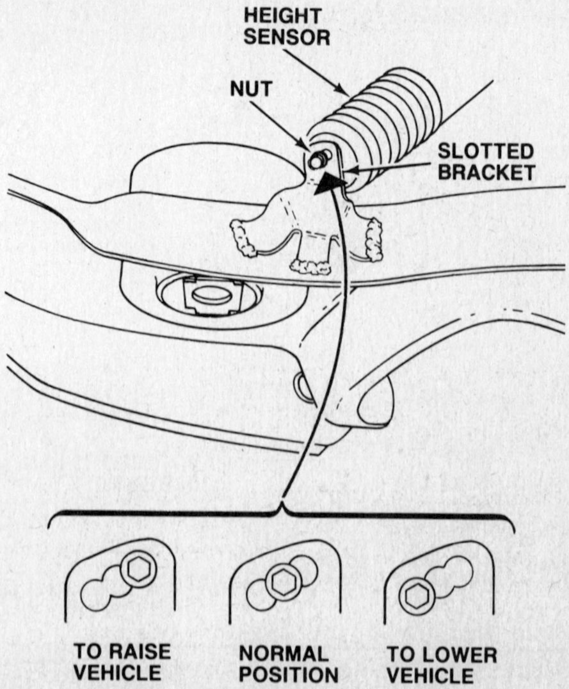

Figure 11-8. Loosen the nut on the end of the ride height sensor, and position the sensor in the appropriate slot on the bracket to adjust ride height.

Wheel Alignment

Wheel alignment procedures for vehicles with electronically controlled suspensions do not differ from procedures used for conventional suspensions. You can find information about wheel alignment in Chapter Fourteen of this *Shop Manual*.

A preliminary step to any wheel alignment is checking and adjusting ride height. Because ride height is adjusted automatically on most electronically controlled suspensions, you must use the manufacturer's recommended procedures to check and adjust ride height before performing a wheel alignment. Many vehicle manufacturers refer to the normal ride height setting on an electronically controlled suspension as "true ride height." The actual vehicle ride height can vary in a vehicle with electronically controlled suspensions depending on which suspension setting the driver is using.

Before measuring "true ride height" on the Lexus LS 400 with air suspension, set the *ride control switch* to NORM (versus SPORT), and bounce the vehicle to settle the suspension. Then start the engine and set the *height control switch* to HIGH. After one minute, set the *height control switch* to NORM, and wait another minute. Set the *height control switch* to HIGH and NORM again, as described above, waiting one minute after changing settings. Cycling through the height control settings twice ensures that the suspension is settled. Be sure to finish the cycle with the *height control switch* in the NORM position. Then measure ride height.

If you must adjust ride height, be sure the *height control switch* is in the NORM position. Loosen the lock nuts on the sensor linkage, figure 11-7, and turn the linkage to lengthen or shorten it, as you would a tie rod end adjuster. When vehicle ride height is correct, tighten the lock nuts.

For 1987 and earlier Lincoln Continentals with air suspension, drive the vehicle onto an alignment rack. Turn the ignition switch on, but do not start the car. After one minute, turn the ignition switch off, and wait for about 20 seconds to allow the suspension to settle to the true ride height. Then turn off the electronic suspension switch, which is located in the trunk, and measure ride height. On these vehicles, to adjust vehicle height you must change the position of the ride height sensor. Loosen the nut, figure 11-8, and move the sensor in the slotted bracket until ride height is correct.

To check and adjust the ride height on the Infiniti Q45 with hydraulically actuated, active suspension, you need a tester called CONSULT. With the vehicle parked on a level surface, the engine running, the transmission in P, and the parking brake OFF, choose "VEHICLE HEIGHT ADJ" and "START" from the menu on the CONSULT tester. Set the *height control switch* to N (Normal), then H (High), and then back to N (Normal). Push the vehicle

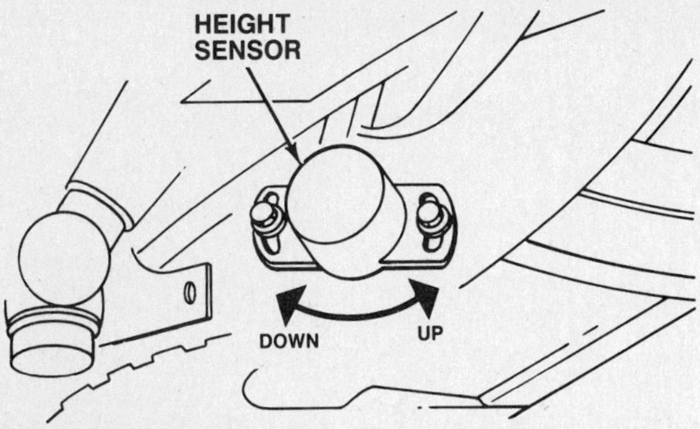

Figure 11-9. Rotate the height sensor on the Infiniti Q45 to raise or lower the vehicle ride height.

back and forth seven to ten feet to settle the suspension. Then measure ride height, and compare results with the manufacturer's specifications. To change ride height on this vehicle, you must loosen the bolt on the vehicle height sensor, figure 11-9, and rotate the sensor until the ride height is correct.

Height-Sensing Brake Proportioning Valves

If the vehicle you are working on uses height-sensing proportioning valves in the braking system, you must not install electronically controlled or any other type of load-leveling shocks, springs, or struts as an aftermarket item. In fact, it is illegal to do so, because devices used to raise vehicle height cause the height sensor in the braking system to misread the position of the vehicle, and the braking system may fail.

Before installing aftermarket load-leveling systems, check the owner's or service manual for information about the brakes.

INSPECTING AN ELECTRONICALLY CONTROLLED SUSPENSION

On vehicles with electronically controlled suspension systems, you need the manufacturer's tester to diagnose complex suspension system problems. However, a brief visual inspection may help you correct simple problems within a few minutes.

Most electronically controlled suspension systems have an indicator light on the dashboard or instrument panel that lets the driver know when the suspension system is not operating properly. For example, on the Lincoln Town Car, a light on the right side of the instrument panel illuminates when the air suspension is defective or when the system has been turned off. The Infiniti Q45 has an indicator light that comes on when the hydraulically actuated, active suspension system is not operating properly. Often, when this light is illuminated or flashing, the computer diagnostic system has trouble codes stored in its memory.

Before checking suspension components, look at the instrument panel indicator lights. If the suspension light is not illuminated, and the bulb and fuse are intact, driveability problems are likely not due to defective electronic controls.

The simplest approach to visually inspecting an electronically controlled suspension is to look at each sub-system individually. In an electronically controlled suspension, you may have a hydraulic or pneumatic sub-system, as well as an electrical sub-system, which includes the computer, sensors, and actuators.

INSPECTING THE HYDRAULIC SYSTEM

Inspecting the hydraulic system on an electronically controlled suspension is not difficult. Check the fluid level. If the level is low after recently filling the reservoir, suspect a leak in the hydraulic system. Check for leaks at hydraulic actuators, such as shocks or struts. Also check the pump and reservoir, hydraulic lines, fittings, and accumulators, if equipped, figure 11-10.

Hydraulic Shocks and Struts

Hydraulic shocks and struts on electronically controlled suspensions look like their conventional suspension system counterparts, except that there is an electrical connector on the top or side, figure 11-11. This connector is used to transmit signals from the suspension computer to the actuator in the shock or strut. When the actuator receives a signal from the computer, the electric motor, valve, or solenoid changes the shock's or strut's damping ability.

Check the body of the shock or strut for leaks. As on conventional shocks and struts, a light oily film, called "weepage" is normal. Large drops of oil are not, and indicate that seals in the shock absorber are not holding the pressurized hydraulic fluid. If you find dripping oil, replace the shocks.

Check near the electrical connector and actuator for hydraulic fluid. As on conventional shocks, you should also inspect the mounts on electronically controlled shock absorbers. If the

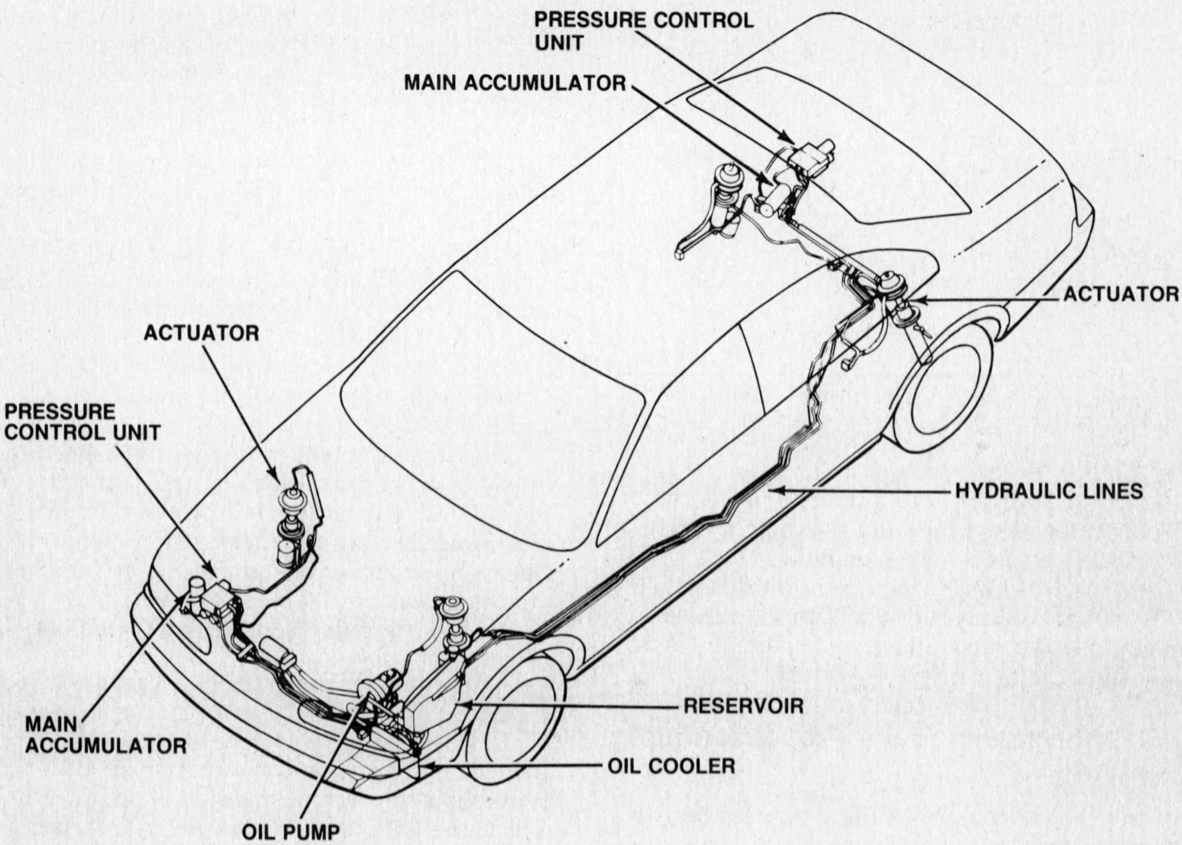

Figure 11-10. The Infiniti Q45 active suspension system has a number of components in the hydraulic system. Check all parts for leaks, paying close attention to fittings.

shock or strut is in good condition but the actuator is not operating properly, you may be able to replace the actuator. However, on some vehicles, you must replace the entire shock or strut because the actuator is inside and cannot be removed.

Most electronically controlled hydraulic shocks and struts are sealed, and hydraulic fluid remains in the shock or strut housing permanently. If you must replace this type of shock or strut, use the procedure for conventional shocks and struts, being sure to disconnect and reconnect the wiring harness connector.

Other Hydraulic System Components

On the Infiniti Q45 with active suspension, electrical connectors and hydraulic lines lead to the struts. The struts have electronically actuated valves that allow pressurized fluid into and out of the strut. Pressurized fluid from a pump is supplied to the struts through hydraulic lines. On this type of system, you must inspect other hydraulic system components in addition to the struts.

Suspension system hydraulic lines look like brake lines, and have flare fittings that are used between the lines and system components. Check hydraulic lines and fittings as you would a brake hydraulic system. Follow the hydraulic lines from each strut to other components in the system. In the Infiniti Q45 suspension, lines lead to main accumulators at the front and rear of the vehicle, subaccumulators at each strut, and a fluid reservoir, cooler, and pump at the front of the vehicle. Besides checking the lines and fittings, be sure to check the various components for fluid leaks. Do not forget to look at air bleeder valves, as they may also leak fluid.

If you find leaks, release pressure in the hydraulic system, using the manufacturer's recommended procedure. For some vehicles, you need the manufacturer's diagnostic tester to do this. Replace defective components as you would in a conventional system. After repressurizing the system, check the components again to ensure that the leaking has stopped. As with power steering and brake hydraulic systems, sometimes reseating a flare fitting

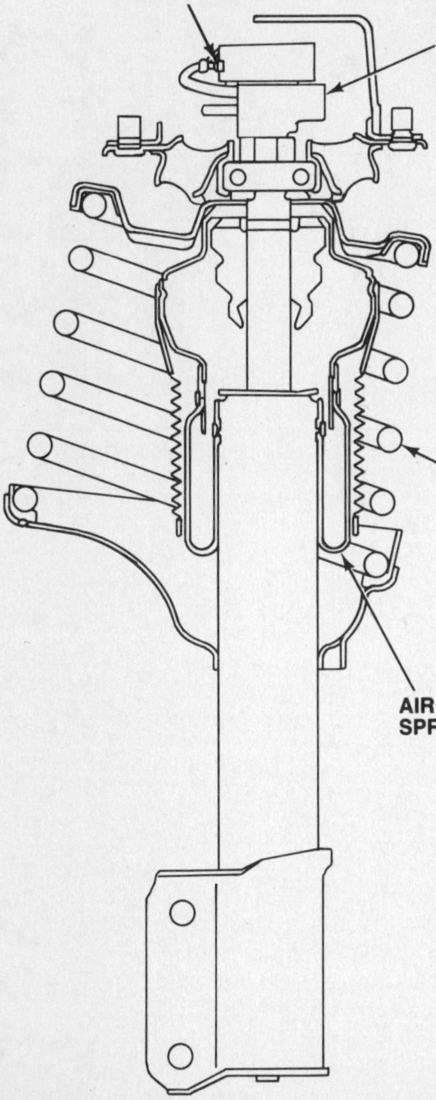

Figure 11-11. This strut has an electric motor that controls a rotary valve inside the strut to adjust damping. An electrical connector at the top of the strut indicates that the suspension has electronic controls.

will stop a leak. Simply release system pressure, loosen and rotate the fitting, and retighten it. Be sure to check the fitting once the system is repressurized.

The Infiniti Q45 active suspension uses a special meter, called CONSULT, figure 11-12, to perform many suspension system repairs. The CONSULT diagnostic tester is used to flush and bleed the hydraulic system, to check for leaks in the high pressure and return line circuits, and to perform a system operation check. Chapter Three of this *Shop Manual* has more information on using diagnostic testers.

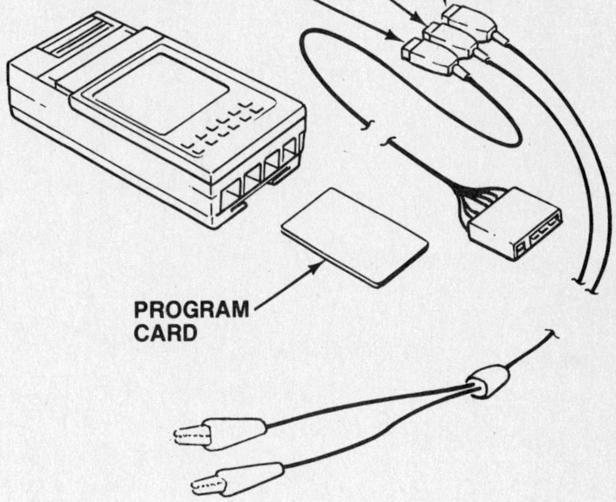

Figure 11-12. This test meter, called CONSULT, is a hand-held computer that helps the technician diagnose suspension problems.

INSPECTING THE PNEUMATIC SYSTEM

Air springs and air shocks are not as popular as hydraulic systems because they use heavy components, such as an air compressor and drier. Using heavy parts reduces vehicle fuel efficiency.

Pneumatic suspension systems (air springs and air shocks) vary in complexity from manufacturer to manufacturer. Some systems have a switch in the car that allows the driver to choose vehicle ride height, and the system simply fills air springs or shocks to raise the chassis. Other, more sophisticated systems use a suspension computer to process information from sensors and switches in the vehicle. As with an electronically controlled hydraulic system, the computer in a pneumatically operated suspension sends signals to actuators that open or close valves, allowing compressed air into or out of the spring or shock. This changes spring rate or shock absorber damping ability for optimum handling.

Diagnostic testers usually are used to determine the cause of driveability problems. However, you can check the instrument panel for indicator lights that illuminate when the suspension system's electronic controls are defective. Also, you should check system components for air leaks.

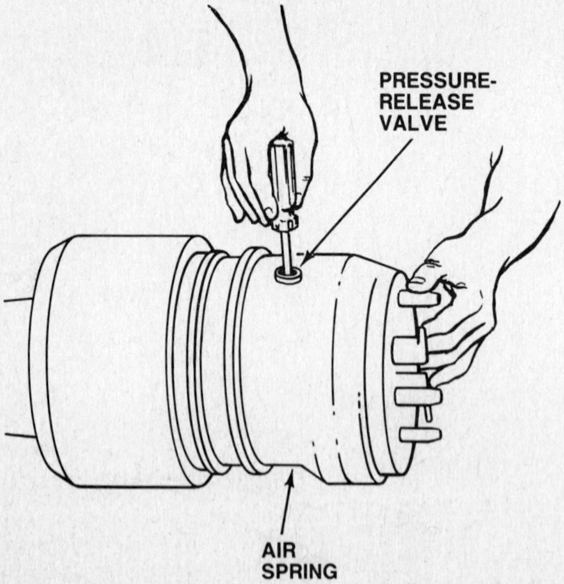

Figure 11-13. Pushing this valve releases air pressure from the Lexus LS 400 air spring.

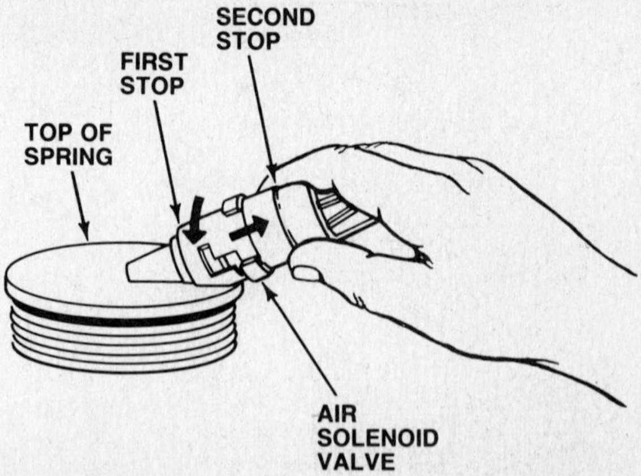

Figure 11-14. When you turn the air solenoid valve on top of the air spring to the first stop, residual air pressure is released from the spring on this 1987 Lincoln Continental.

Air Shocks and Springs

Electronically controlled air shocks and struts look like conventional shocks and struts, except that they have electrical connectors and air lines. The connectors allow signals from the suspension computer to control the actuators. Air lines supply dried, compressed air from the compressor.

Check each shock or strut, and ensure that electrical connectors are in place. Check air line fittings for leaks, as described later in the section titled "Checking the Pneumatic System for Leaks." In general, you can service pneumatic shocks and struts as you would conventional ones after turning off the suspension system, but check the manufacturer's service literature for any special precautions.

Air springs look nothing like their conventional counterparts. They are basically reinforced plastic or rubber air bags that cushion the chassis against road shock. The suspension computer signals the air compressor and air valves to add or remove air pressure from the springs, depending on driving conditions. These springs may be part of a strut assembly, as on the Mitsubishi Galant, figure 11-11, or they may act as coil springs, as on the Lincoln Town Car and Continental, figure 11-5.

Check the rubber or plastic "spring" for pinholes and cracks. If air pressure in the spring drops too low, as may happen if the suspension system is not serviced properly, the spring may be damaged. Low pressure allows the spring to fold over on itself. If this happens, the spring must be replaced.

As with a conventional suspension system, you must secure an air spring before removing any components that hold spring tension. Securing an air spring usually is simply a matter of turning off the electronically controlled suspension, disconnecting the height sensors, and in some cases, releasing residual pressure from the spring according to the vehicle manufacturer's instructions. To release pressure from the air spring on a Lexus LS 400, turn off the air suspension system, and then use a screwdriver to press the small valve on the side of the spring, figure 11-13. On the 1987 Lincoln Continental and 1990 Lincoln Town Car, after turning off the air suspension system, release pressure by turning a solenoid valve on the air spring, figure 11-14.

Checking the Pneumatic System for Leaks

To check for air leaks, apply soapy water to air line joints and fittings while the pneumatic system is pressurized. For example, to check the Lexus LS 400 for air leaks, set the ride height to HIGH, which increases pressure in the pneumatic system. Then apply soapy water to the fittings at the air springs, relief valves, and at the compressor and air dryer, figure 11-15. Soap bubbles will form if air is leaking from these components.

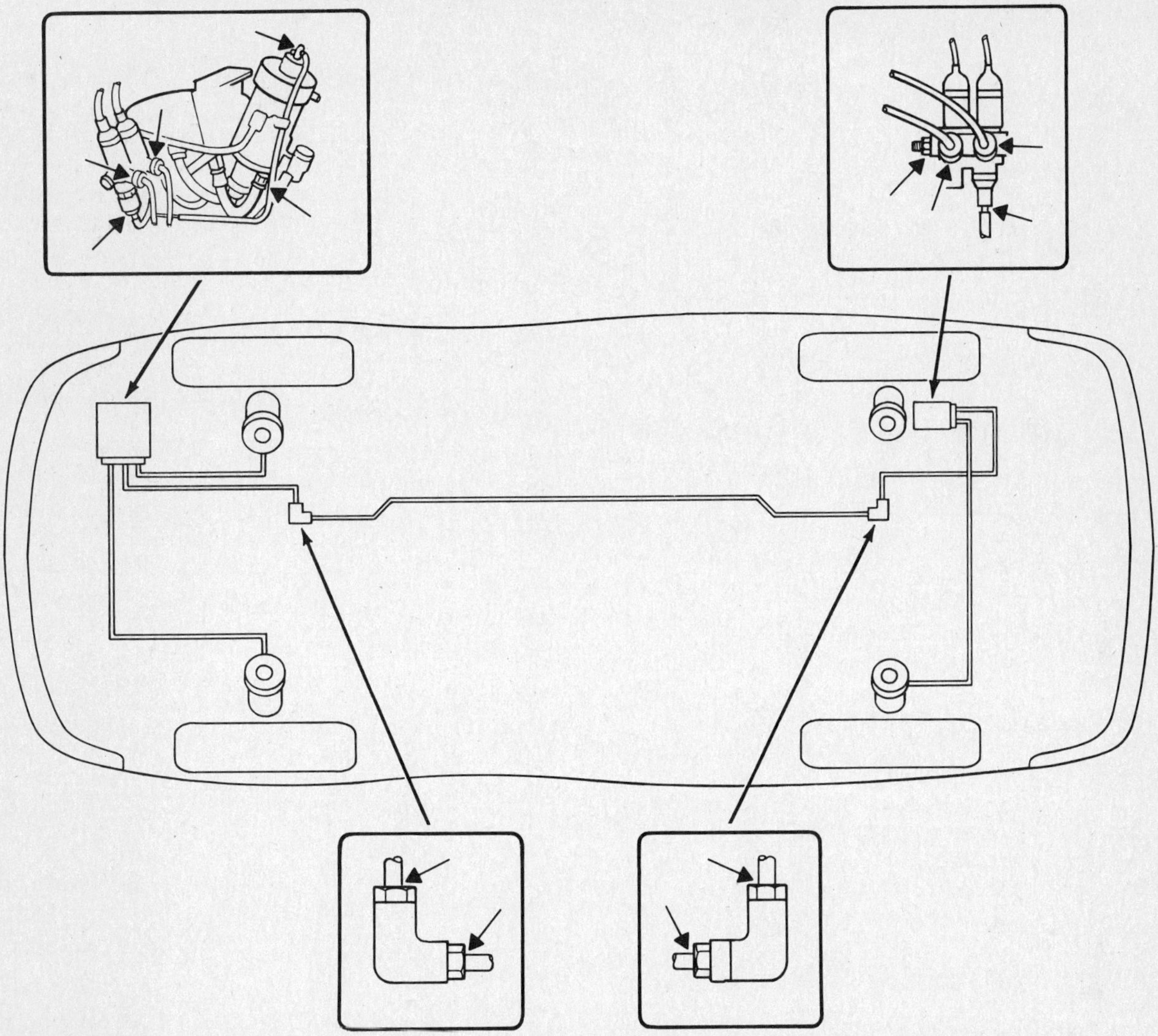

Figure 11-15. Apply soapy water to the air line fittings, indicated with black arrows. If bubbles form, air is leaking from the fitting.

INSPECTING SUSPENSION ELECTRONICS

If the hydraulic or pneumatic systems are not leaking or damaged, then suspect that the vehicles's electrical system or the suspension's electronic controls are defective. Testing electronic controls includes checking the computer system's ability to send, receive, and process signals. It also includes checking the operation of sensors and switches, which send input signals to the computer, and checking actuators, which receive signals from the computer and perform a mechanical action to change suspension feel.

Checking the Vehicle's Electrical System

An electronically controlled suspension system will not operate properly or at all if the vehicle's electrical system is defective. Perform a basic electrical system test, which includes checking battery connections, battery state of charge, alternator output, and system fuses, as discussed in Chapter Three of this *Shop Manual*.

Many electrical problems have been solved by simply replacing a fuse. However, when a fuse blows, it means that current higher than the fuse's amperage value has passed through the circuit. If you find a blown fuse, investigate the situation to determine why excessively high current passed through the circuit.

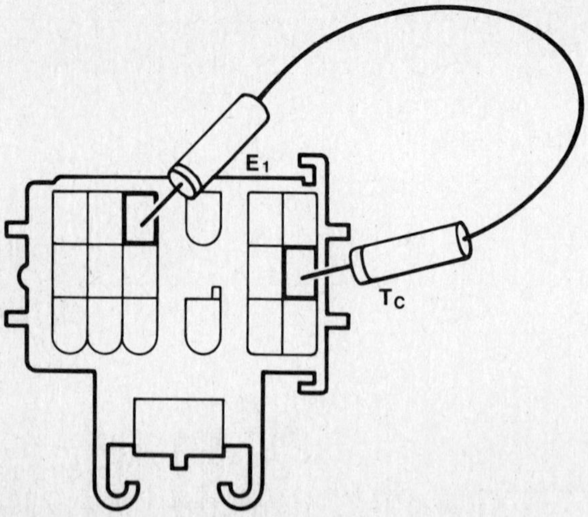

Figure 11-16. Jump the terminals shown on the check connector to retrieve trouble codes from the Lexus LS 400 suspension computer.

Some common causes of excess amperage include a short or ground in the circuit, an internal short or ground in a component, or an accidental ground by a technician when working on a circuit.

If the electrical system is in good shape, but the electronic suspension system is still not operating correctly, then suspect a problem with the suspension computer or electronic controls.

Suspension Computer and Trouble Codes

The heart of an electronically controlled suspension system is the computer, which sends, receives, and processes signals to control the suspension. When the computer or electronic components malfunction, the computer stores trouble codes, which help technicians locate and diagnose problems in an electronically controlled system.

On most vehicles, the driver or technician can tell when the computer has stored a trouble code because the instrument panel indicator light will illuminate or flash. Service manuals tell the owner to have the vehicle serviced when the indicator light flashes, and technicians know that they should check the instrument panel before diagnosing a problem.

The computer stores trouble codes when it detects problems. For example, if a height-control sensor malfunctions, the computer will store a number, let's say "11," in its memory. When a technician reads the trouble codes from the computer's memory, the number "11" will appear. After looking up code "11" in the manufacturer's service manual, the technician will determine that the height-control sensor is faulty and will repair the sensor.

Each vehicle manufacturer uses a different numbering system for trouble codes, so a code "11" from one manufacturer may not represent the same problem as a code "11" from another. Also, manufacturers have various ways of obtaining the trouble codes — at this time, there is no standard method.

Do not attempt to retrieve trouble codes or test electronic equipment, unless you have the manufacturer's procedures. Because electronic components usually use low-amperage electrical signals, they can be damaged if tested improperly. The computer is extremely sensitive to excessive current and is costly to replace. Be sure to use a digital volt-ohmmeter (DVOM) when testing any computerized system. DVOMs have high internal resistance that will protect sensitive computer components from excessive current. Analog meters do not have this high internal resistance.

Describing trouble code retrieval for every electronically controlled suspension system is beyond the scope of this book, but we will look at the Lexus LS 400 with air suspension as an example. You do not need a diagnostic tester to retrieve trouble codes from this system, but you have to know how to locate the diagnostic connector and jump the terminals to access the codes:

1. Turn the ignition switch to the ON position.
2. Jump terminals Tc and E_1 of the diagnostic connector with a jumper wire, figure 11-16.
3. Watch the height control NORM light on the instrument panel. This is where the trouble codes will flash.
4. Count the number of times the light blinks and pauses. The first blinks represent the first digit of a trouble code. After a pause, count the second set of blinks, which represent the second digit. If there are more trouble codes, there will be a long pause. Then the first and second digits of the next trouble code will be displayed. If more than one code is stored in the computer, they will be displayed from lowest to highest numerical value.
5. Be sure to write down the trouble codes on a piece of paper.
6. Use the information in the manufacturer's service manual to understand the codes, figure 11-17. A chart in the manual lists diagnostic codes, what they mean, and

DIAGNOSTIC CODES			
CODE	**SYSTEM**	**BLINKING PATTERN**	**DIAGNOSIS**
—	—		Normal
11	Height control sensor, front RH circuit	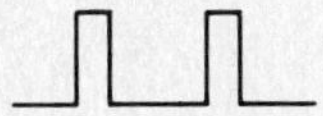	Open or short in height control sensor circuit
12	Height control sensor, front LH circuit		
13	Height control sensor, rear RH circuit		
14	Height control sensor, rear LH circuit		
21	Front suspension control actuator circuit	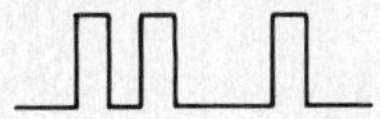	Open or short in suspension control actuator circuit
22	Rear suspension control actuator circuit	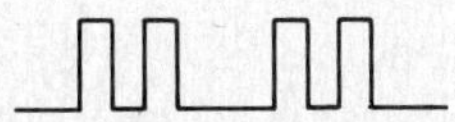	
31	Number 1 height control valve circuit		Open or short in height control valve circuit
33	Number 2 height control valve circuit (Right Side)	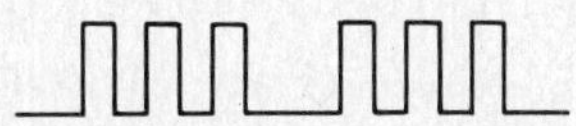	
34	Number 2 height control valve circuit (Left Side)	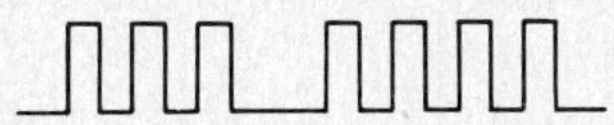	
35	Exhaust valve circuit		Open or short in exhaust valve circuit
41	Number 1 height control relay circuit		Open or short in number 1 height control relay circuit
42	Compressor motor circuit	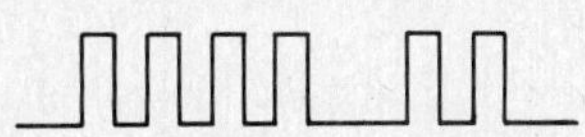	Short circuit in compressor motor or compressor motor locked
51	Continuous current to number 1 height control relay		Current is supplied to number 1 height control relay for approximately 8.5 minutes or longer
52	Continuous current to exhaust valve		Current is supplied to exhaust valve for approximately 6 minutes or longer

Figure 11-17. Count the blinks on the indicator light to retrieve the trouble codes. Then use the manufacturer's service manual to determine what these codes mean.

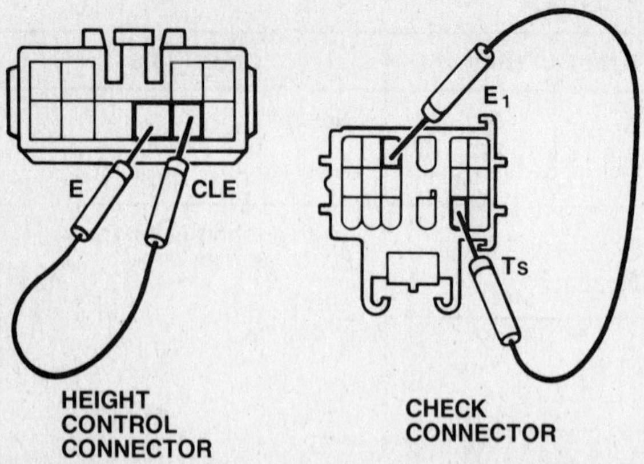

Figure 11-18. After obtaining trouble codes, erase them from the computer's memory by jumping the terminals shown at the height control connector or the check connector.

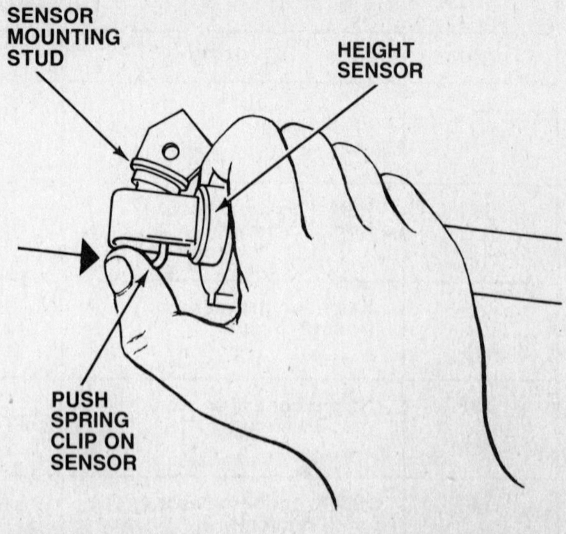

Figure 11-19. To remove the height sensor on the 1990 Lincoln Town Car, push the spring clip back and pull the sensor from the mounting stud.

which circuits are defective. Sometimes the chart will list the pages in the service manual that contain repair procedures. With electronic systems, a common repair is to replace the defective part.

7. After repairing the system, erase the diagnostic codes. You can do this by removing the ECU-B fuse in junction box 1 for at least ten seconds. Or you can jump terminals E and CLE at the height control connector, and terminals E_1 and Ts at the check connector, figure 11-18, while the ignition switch is in the OFF position. Keep these jumpers in place for at least ten seconds. Then turn the ignition switch ON, and disconnect the terminals.
8. After driving the vehicle, check the indicator light again. If it is not flashing, the problem is corrected. If the light flashes after a test drive, check the trouble codes again.

Replacing and Adjusting Electronic Components

Chapter Twelve of the *Classroom Manual* describes how various types of sensors, switches, and actuators operate. Because there are so many electronic suspension system designs, we cannot describe service procedures for all existing systems. Examples for servicing major components in an electronically controlled suspension system are described in the following sections.

Height sensor

Height sensors are used on most, if not all, electronically controlled suspension systems, and there are a few different types of height sensors. The Lexus LS 400 suspension system uses an LED and phototransistor to signal vehicle ride height to the suspension computer, as described in Chapter Twelve of the *Classroom Manual*. Height sensors on the Infiniti Q45 Active Suspension system use a potentiometer to send variable signals to the computer, depending on suspension height. The height sensors on the 1990 Lincoln Town Car with air suspension use magnets and Hall-effect sensors to signal the suspension computer.

Check the electrical connectors at the height sensors to be sure they are secure. Check that the sensor is mounted firmly to the vehicle body, and tighten any loose connections. Next, use the manufacturer's procedures to check and adjust vehicle ride height, as described earlier in this chapter. If you cannot adjust the ride height, or if the sensor is still sending trouble code signals to the computer, replace the sensor using the manufacturer's recommended procedures.

For example, to replace the height sensor on the 1990 Lincoln Town Car, turn off the air suspension system and raise the vehicle on a lift, using precautions noted in the manufacturer's service manual. Some common precautions are listed earlier in this chapter.

Separate the electrical connector between the height sensor and wiring harness. Press the spring clip at the top end of the sensor to release it, figure 11-19, and separate the top of

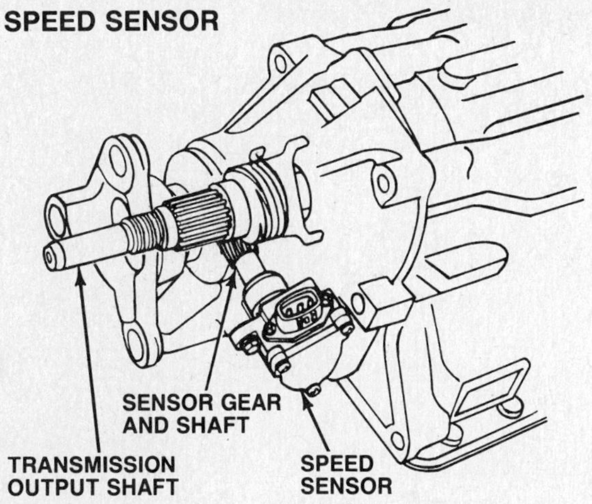

Figure 11-20. The speed sensor on the Lexus LS 400 has a gear that meshes with the transmission output shaft. The gear teeth and the electrical connector should be in good condition.

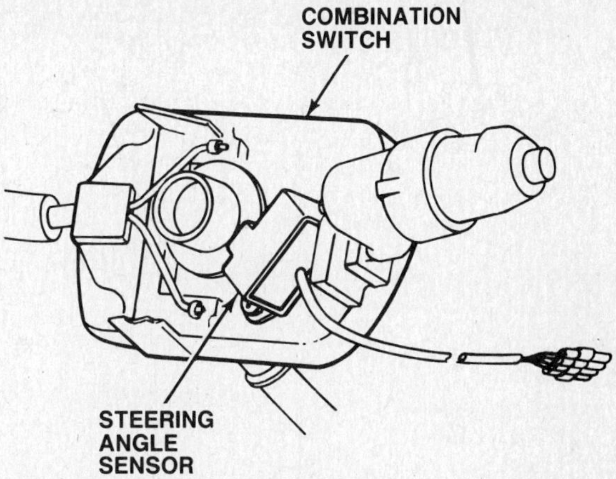

Figure 11-21. The steering angle sensor is connected to the combination switch on the steering column, which you can access by removing the steering wheel and some accessories.

the sensor from the mounting stud. Then do the same for the lower connection of the height sensor. Install a new sensor by simply reversing the removal instructions. Then readjust the ride height.

Speed sensor

The speed sensor tells the control unit how fast the car is going. On the Lexus LS 400 air suspension system, the speed sensor is located at the transmission output shaft, and is the same sensor that sends signals to the engine control computer and the speedometer, figure 11-20. To replace this sensor, remove the transmission control rod, and disconnect the speed sensor electrical connector. Remove the bolt that holds the sensor in place, and remove the sensor, which includes a shaft and small gear that meshes with the transmission output shaft.

Before installing the sensor, assemble the sensor and gear, and lubricate the O-ring on the sensor body with ATF. Torque the bolt that holds the sensor in place, connect the electrical connector, and reattach the control rod. Be sure to inspect and adjust the shift lever position after reconnecting the control rod.

Steering angle sensor

Some electronically controlled suspensions use a steering angle sensor to determine how far and at what velocity the driver has turned the steering wheel so the computer can compensate for sharp turns.

In the Mitsubishi Galant with Active-Electronic Control Suspension (A-ECS), the steering wheel angular velocity sensor is located under the steering wheel, and is part of the combination, or multi-function, switch, figure 11-21. The steering column is a common location for the steering angle sensor.

To remove and replace a steering sensor in the steering column, you must remove the steering wheel and any accessories, as described in Chapter Four of this *Shop Manual*. If the vehicle has an air bag in the steering wheel, be sure to use the precautions described in Chapter Four of this *Shop Manual* before removing the steering wheel.

Disconnect the steering angle sensor wiring harness connector, and remove the screws that hold the sensor in place. Install a new sensor by reversing the removal procedure. Do not get grease or dust on the replacement sensor, because it is delicate and must remain clean and be handled carefully.

G-force sensor

G-force sensors, also called "accelerometers," measure the forces of acceleration in a certain direction. The Infiniti Q45 full-active suspension system uses vertical, lateral, and fore-and-aft G-force sensors. Other vehicles with electronically controlled suspensions can use any combination of G-sensor types.

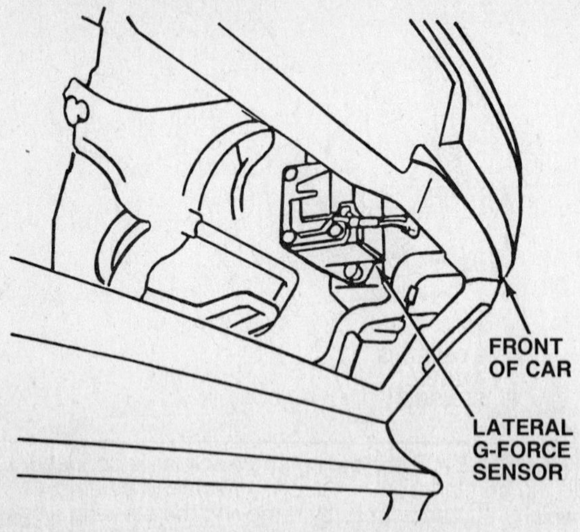

Figure 11-22. When G-sensors are used on an electronically controlled suspension, they can be located in a variety of places. This one is mounted to the vehicle frame behind the front bumper.

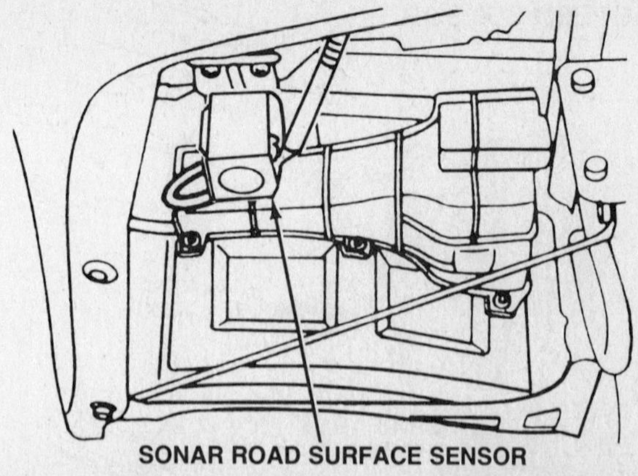

Figure 11-23. The sonar sensor is used by Nissan in the Maxima and the Infiniti M30. If the sensor's reflective surface is damaged, it will not be able to determine the condition of the road.

G-force sensor operation is discussed in Chapter Twelve of the *Classroom Manual*. If diagnostic tests reveal that a G-sensor is defective, replace the sensor. On the Mitsubishi A-ECS system, you must remove the bumper to access the G-sensor, figure 11-22. Then disconnect the wiring harness, remove the screws that hold the sensor to the vehicle body, and remove the sensor. Install a new sensor by reversing the removal procedure.

Sonar sensor
The sonar sensor, which has been used on the Nissan Maxima since 1988 and the Infiniti M30 since 1990, transmits and receives sound waves to detect irregularities in the road surface. The suspension corrects for these irregularities before the vehicle drives over them.

Check the electrical connector between the sonar sensor and wiring harness, and be sure that the sensor is securely connected to the vehicle frame. Also, check the sensor's reflective surface, which should be clean, flat, and even, figure 11-23. If the reflective surface is damaged, replace the sensor by disconnecting the wiring harness, and removing the sensor from the vehicle.

Flow control valve actuators
An air or hydraulic flow control valve actuator receives signals from the suspension computer and regulates the amount of pressurized air or fluid that is allowed to enter and leave shocks or springs. Flow control valves are often solenoids, as discussed in Chapter Twelve of the *Classroom Manual*.

The Mitsubishi Galant with Active-Electronic Control Suspension (A-ECS) uses flow control solenoid valve assemblies to adjust pressure in the air springs. (The hydraulic shock absorbers use a motor and movable piston rod at the top of the strut to control damping effects, as discussed later in this chapter.) The flow control solenoid assembly regulates air flow to the front and rear solenoid valves, figure 11-24. The front and rear solenoid valves control air spring pressure. All three solenoid assemblies can be removed without replacing the air springs because they are mounted remotely from the springs. The flow control solenoid assembly is on the vehicle right side member, the front solenoid valve is at the cowl top panel, and the rear solenoid valve is on the right side of the trunk.

After turning off the electronic suspension and relieving system pressure, disconnect the wiring harness from the solenoid assembly. Then disconnect the air lines leading to the solenoids. The flow control solenoid assembly has nine air lines, and the front and rear solenoid valves have eight. You will need a special tool provided by the manufacturer to disconnect the air lines, figure 11-25. Mitsubishi cautions that you must cover the openings in the solenoid valves and air lines to prevent dirt

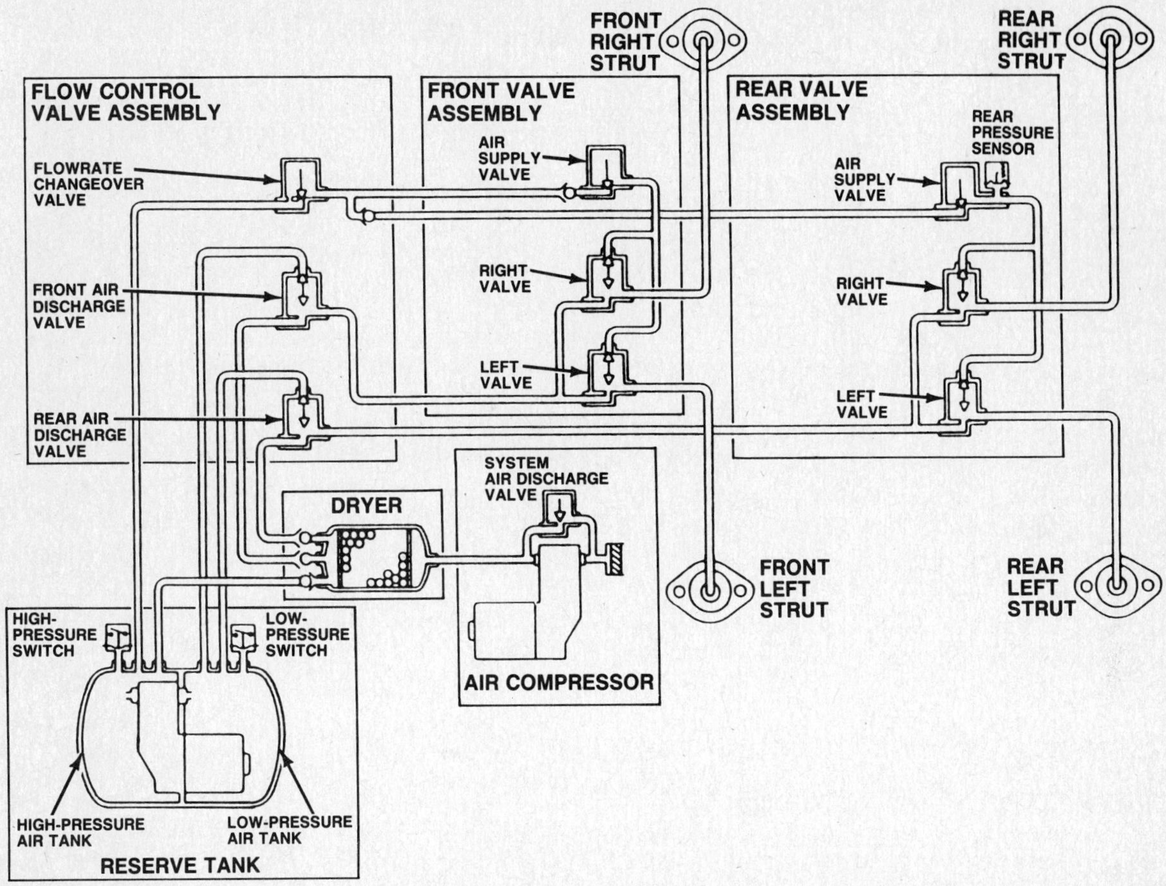

Figure 11-24. Three solenoid valve assemblies control air flow to the front and rear air springs on the Mitsubishi A-ECS suspension.

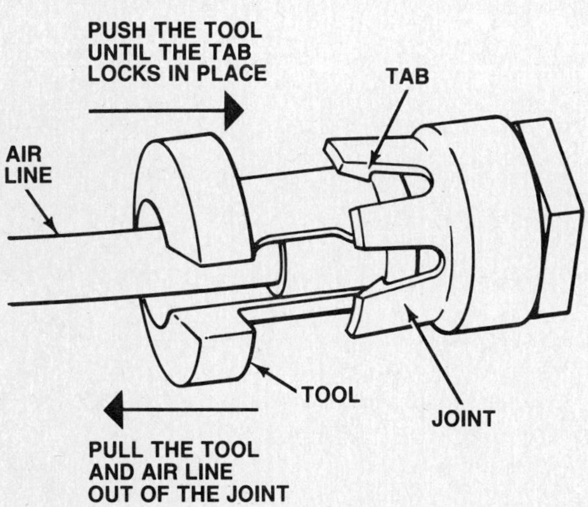

Figure 11-25. You will need a special tool to separate the air lines on Mitsubishi's A-ECS suspension system.

and other materials from entering the pneumatic system. Remove the bolts that secure the solenoid assembly to the vehicle, and remove the solenoid.

When reinstalling the solenoid valves, be sure to match the colors on the ends of the air lines with the color-coded openings on the solenoid valves. Press the air lines in place, and check the pneumatic system for leaks, as described earlier in this chapter.

Motorized actuator

Another type of actuator used in electronically controlled suspensions is the electric motor. After receiving signals from the suspension computer, current is supplied to the motor frame and armature, which creates a magnetic field and causes the motor to turn, as described in Chapter Twelve of the *Classroom Manual*.

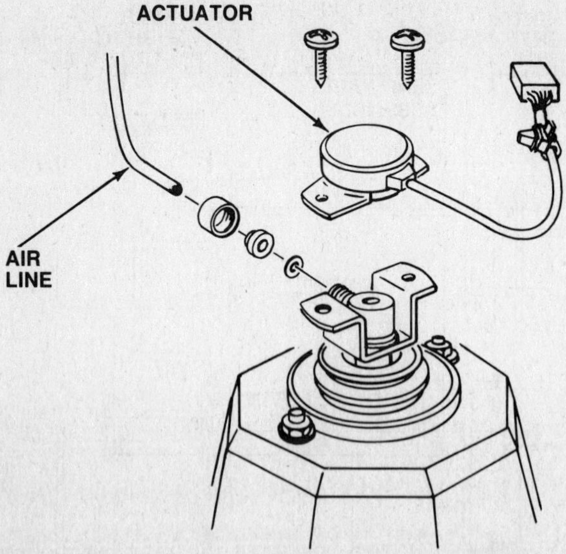

Figure 11-26. The actuator for the electric motor on the strut of a Mitsubishi A-ECS suspension can be replaced without removing the entire strut. If the valve is defective, however, you must replace the strut.

The Mitsubishi Galant with Active-Electronic Control Suspension (A-ECS) uses motorized actuators on the shock absorbers. When the suspension computer signals the motor, the motor moves a control rod that changes the position of a rotary valve inside the shock. The rotary valve's position determines the damping effects of the shock absorber.

If the motorized actuator fails, you can unbolt and replace the actuator at the top of the strut, figure 11-26. If the valve inside the strut is damaged, you must replace the entire strut assembly. After disconnecting the electrical connectors and air lines, use the procedures to remove a conventional strut, as described in Chapter Eight of this *Shop Manual*.

PART FOUR

Wheel and Tire Service and Wheel Alignment

Chapter

12 Wheel and Tire Service

The chapters in Part Four of this *Shop Manual* are about wheel and tire service and alignment. Chapter Twelve contains information you will need to inspect and repair tires and wheels. Chapter Thirteen covers the dynamic aspects of the tire, primarily balance and rotation. Finally, Chapter Fourteen has instructions for aligning the wheels to the manufacturer's specifications.

Some driveability problems caused by faulty wheels or tires, such as pulling or excessive tire noise, are often misdiagnosed as alignment problems. Always check tires and wheels for damage before performing a wheel alignment.

The procedures in this chapter will explain:

- Basic tire and wheel inspection, repair, and replacement
- Wheel bearing inspection, lubrication, adjustment, and replacement
- Basic brake system service, as it relates to repairing tires, wheels, and wheel bearings.

REMOVING AND INSTALLING TIRES AND WHEELS

Removing and installing tires and wheels is one of the most basic services you will perform, and it is relatively easy to do. Included in this procedure are a few tips that will make the job faster and safer for you:

1. Position the lift supports under the vehicle chassis using the manufacturer's recommended lift points, but do not raise the vehicle.
2. If the vehicle has wheel covers, remove them by prying with a screwdriver. Some wheel covers are held in place with a screw located at the center of the cover. The tool to remove it usually comes with the car or is available from the vehicle manufacturer. Be careful that you do not scratch or dent the wheel covers, which are often made of thin plastic or metal. Store the covers in a place where they will not be damaged while you are working on the vehicle.
3. Using a wrench, loosen the wheel lug nuts or bolts one-half turn, but do not remove them. The lug nuts or bolts can have either left- or right-hand threads. For some lug nuts or bolts, you will need a special key or socket, figure 12-1, because an ordinary wrench will not fit the bolt head.
4. Raise the vehicle on the lift. See Chapter One of this *Shop Manual* for more information on lifting a vehicle.

Figure 12-1. Some wheel manufacturers use lug nuts or bolts that cannot be removed with standard or metric sockets. Instead, you need a special key or socket that fits the lug nut or bolt head pattern.

Figure 12-2. This Volkswagen uses lug bolts to secure the wheel to the hub. Most vehicles you will work on use lug nuts instead.

5. Remove the lug nuts or bolts. Most vehicles use lug nuts to secure the wheels, and the wheels fit over lug studs that are pressed into the hub. However, some European cars use lug bolts to secure the wheels, figure 12-2. When you remove lug bolts, you must support the wheel. When the last bolt is removed, the wheel will fall to the ground. If you are removing all four tires and wheels at once, mark them so that you can install them in the same place. Check lug nuts or bolts for rust and cross-threading. Replace any damaged lug nuts, bolts, or studs.

Note: Knock-off wire wheels were used on some sports cars, such as MG, Triumph, and Jaguar, until about 1970. An exposed, single

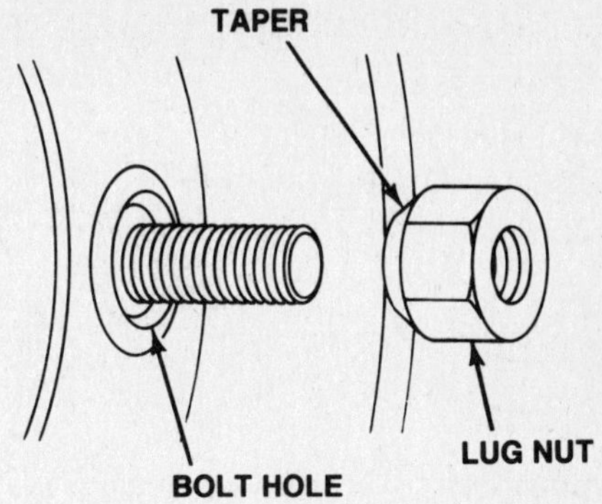

Figure 12-3. The tapered side of the lug nut should fit in the bolt hole. Do not install lug nuts backwards.

nut holds the wheel to the hub. After raising the vehicle, use a brass hammer or raw-hide mallet to turn the nut by hitting it. The wheels on the left side of the car have nuts with left-hand threads so that normal wheel rotation will not loosen them.

6. When you install a wheel, place the wheel on the studs, and loosely install the lug nuts, keeping in mind that some have reverse threads. You must install lug nuts with the tapered side of the nut toward the wheel, figure 12-3. As mentioned earlier, some vehicles use lug bolts. To install lug bolts, hold the wheel against the hub, and align the holes. Then loosely install the lug bolts.
7. Tighten the lug nuts or bolts gradually in a star pattern with a wrench to ensure that the wheel is tightened evenly on the hub, figure 12-4. Do not fully tighten the lug nuts or bolts at this time. Also, do not use an impact wrench because you can easily over-tighten the lug nuts or bolts, which will warp the brake rotor or drum.
8. Lower the vehicle, and use a torque wrench to tighten the lug nuts or bolts to the manufacturer's specifications, again using the star pattern.
9. If the vehicle has wheel covers, press them in place. You may have to hit the rim of the wheel cover with a rubber mallet to seat it. Finally, check and adjust tire pressure.

MOUNTING AND DISMOUNTING TIRES

To dismount a tire from the wheel, you must separate the tire bead from the wheel's bead seat. Shops may use manual and electric tire

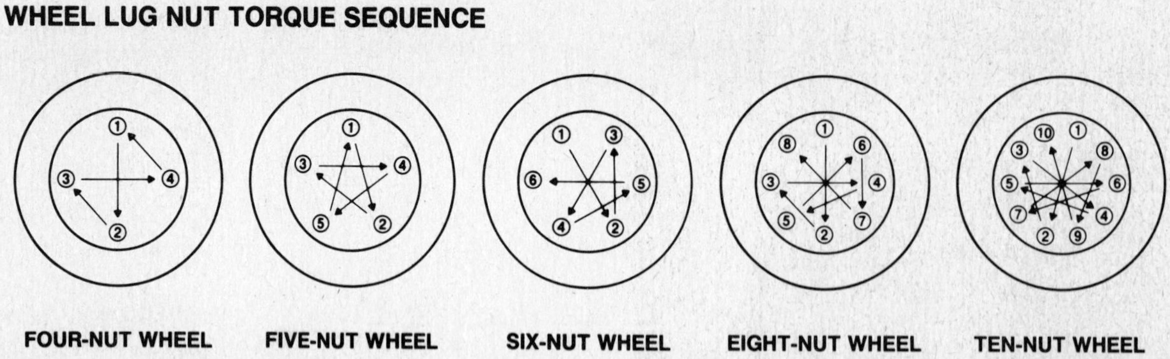

Figure 12-4. These are common patterns for tightening lug nuts or bolts. It is important to use these tightening sequences so that the wheel lies flat on the hub.

Figure 12-5. This pneumatically powered tire changer is used to separate the tire bead from the wheel.

Figure 12-6. Tighten the tire and wheel assembly on the tire changing machine.

changing machines to do this, although pneumatic (air-powered) tire changers are the most popular. A number of aftermarket equipment manufacturers make tire changers, figure 12-5, and each manufacturer has specific instructions for operating their equipment. Use the instructions that come with the machine.

As discussed in Chapter Thirteen of the *Classroom Manual*, wheels can be made of a variety of materials including steel, aluminum, and magnesium alloy, as well as plastic. Because some wheel types have special handling precautions, check the instructions that come with the tire changer for special procedures. Also, wear safety glasses when dismounting and mounting tires.

WARNING: Before removing the tire from the wheel, ask the vehicle owner if he or she used canned tire sealant to temporarily repair the tire. The pressurized gas used as a propellant in some tire sealants is volatile and easily ignited by sparks. Technicians have been seriously injured when a spark from the pry bar ignited the vapors in the tire. If the tire contains sealant, release the air from the tire. Then fill and release air a few times to purge the tire of flammable vapors.

A sample tire changing procedure follows:

1. Remove the tire and wheel assembly from the vehicle. Pry off the balancing weights that clip over the wheel rim, if there are any. (The weights may tear the bead if they get in the way of the tire changer.)
2. Place the tire and wheel assembly on the tire changer with the valve stem pointing toward you. Lock the tire and wheel in place by tightening down the cone-shaped holder, figure 12-6. On some machines, the holder has reverse threads, so you must turn it counterclockwise to tighten it and clockwise to loosen it.

Figure 12-7. The bead breaker on the tire changer forces the bead off the wheel by separating the tire bead and wheel bead seat.

Figure 12-8. After breaking the bead, use the lever to lift the tire bead off the wheel. You must do this to both the upper and lower beads.

3. Use pliers to pull the valve stem from the wheel, which will deflate the tire. Alternatively, you can remove the valve core, which is inside the valve stem, with a special tool as described later in this chapter. If the wheel uses clamp-on valve stems, you cannot pull them out, and you must remove the valve stems later. (This type of valve stem was used on some older vehicles.) In this case, release air from the tire by pressing on the valve core pin.
4. Use the bead breaker on the tire changer to force the upper tire bead toward the center of the wheel, figure 12-7. This breaks the bead seal.

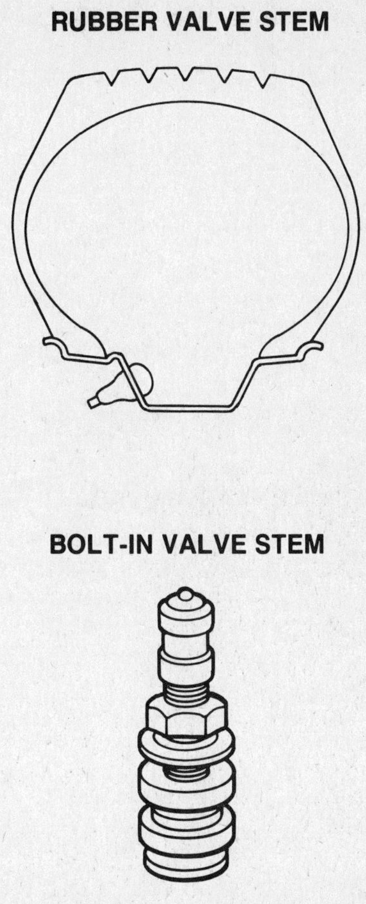

Figure 12-9. There are two types of valve stem. The most common is made of rubber, and it simply fits in a hole in the wheel. As the rubber ages, the seal on this type of stem weakens. The other type of valve stem bolts in place with replaceable rubber washers.

5. A lever-like tool is used to lift the bead. Place the end of the tool under the bead, and place the tool on the tire changer's center head. Press the foot pedal on the tire changer to rotate the tool around the inner circumference of the tire and lift the upper bead over the wheel flange, figure 12-8. The lower bead will rise with the tire and move to the center of the wheel.
6. Press the pedal to rotate the lever around the wheel again, which will lift the lower bead over the wheel flange, separating the tire from the wheel.
7. Before mounting a new or repaired tire on the wheel, remove rust, dirt, and old rubber and lubricants from the area where the wheel bead seats. Use a wire brush to clean steel wheels, and a non-abrasive cleaner on mag or aluminum wheels.
8. Lubricate a new valve stem, and place it in the wheel from the inside, figure 12-9. Pull

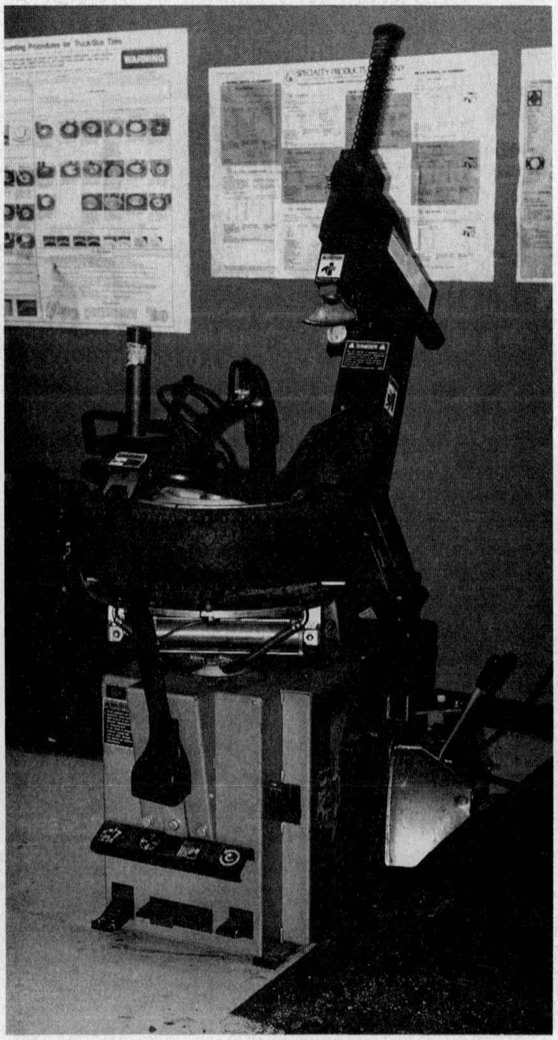

Figure 12-10. Before filling the tire with air, secure it with a safety harness or strap.

on the valve stem from the outside until it seats. You may have to use an installer tool to do this. If the wheel uses a clamp-on valve stem and it is damaged, remove the nut on the inside of the stem, and pull the valve stem out. Install a new valve stem along with its rubber washers. Then tighten the nut. Clamp-on valves are expected to last the life of the vehicle, but you should check them for damage whenever you remove a tire.

9. Before mounting the tire on the wheel, coat the upper wheel bead seat and the tire beads with lubricant. Using lubricant will prevent damage to the bead.

Note: Use lubricant that is compatible with tire rubber and is specially formulated for mounting tires. If you use a water-based lubricant, be sure it contains a rust inhibitor. Do not use silicone-, petroleum-, or solvent-based lubricants because they may allow the tire to slip on the wheel. Solvents are especially dangerous because they can create volatile gases inside the tire, which can explode if ignited.

10. Place the tire over the wheel rim at an angle. Using the lever-like tool, place the lower bead and then the upper bead over the upper wheel flange, reversing the removal steps. Both beads should be in the wheel drop center.
11. Check that the cone-shaped holder is tight. Place a strap around the tire tread, and tighten the strap to hold the tire beads against the wheel. Doing this keeps air from leaking past the beads when you inflate the tire. Some tire changers have a safety harness and strap that you must secure before filling the tire, figure 12-10.
12. Stand away from the tire, and inflate it to about 10 to 15 psi.
13. Continue inflating the tire with short bursts of air until you hear a loud pop, which means that the beads have seated. On car tires, the beads will normally seat at about 30 psi. If it takes more than the maximum recommended pressure stamped on the side of the tire to seat the bead, deflate the tire, relubricate the beads and wheel rims, and try again. However, do not exceed the maximum psi stamped on the tire.

CAUTION: Never have your fingers near the wheel flange when you are inflating the tire, or the bead may pinch your fingers when it seats. Also, *never* use a torch to help seat the bead. Tire rubber contains flammable materials that may ignite when exposed to flame.

14. Remove the safety strap or harness. Install a new valve core, as described later in this chapter.
15. Check and adjust tire pressure, as described later in this chapter. If you repaired a leak, check the tire after it is inflated, using the procedures given later in this chapter.
16. Balance the tire and wheel assembly before mounting it on the vehicle, as described in Chapter Thirteen of this *Shop Manual*.

Dispose of the old tire according to local regulations. Many states charge shops a fee to discard old tires, and often this fee is passed along to the customer.

Figure 12-11. Manufacturers place a sticker in the driver's or passenger's door jamb listing correct tire pressures for the front and rear tires.

INSPECTING AND REPAIRING TIRES AND WHEELS

As part of a routine vehicle inspection, check and adjust tire air pressure, and check tires for damage or wear. By looking for and identifying specific tire wear patterns, usually you can determine the cause of excessive tire wear.

Some types of tire damage are repairable, and procedures for locating a leak, patching a hole, and replacing valve stems and valve cores are included in this section. When you remove a tire for repairs, inspect the wheels for damage as well.

Checking and Adjusting Tire Air Pressure

Check and adjust tire pressure at least monthly. The correct tire pressure ensures that the tread contacts the road as designed, ensuring optimum traction, reduced rolling resistance, and reduced wear.

Tire pressure specifications are given for cold tires, unless otherwise specified. "Cold" means that the tires were recently driven less than three miles, or that the vehicle has been parked for at least one hour. Refer to the vehicle owner's manual for the manufacturer's recommendations.

Tire pressure will change as the vehicle is driven and the tires warm up. As tires warm, the air inside also heats and expands, which increases the pressure inside the tire. If you adjust pressure to the manufacturer's specifications when the tires are hot, the pressure will drop when the tires cool.

Recommended tire pressure specifications are listed in the vehicle owner's manual. Also, most vehicle manufacturers place a sticker listing recommended tire pressures in the driver's or passenger's door jamb, figure 12-11. Check the sticker carefully, as it may list pressures for more than one type of tire. (Many manufacturers equip the same vehicle model with different tires, depending on the model trim level.)

Figure 12-12. This pencil-type gauge is used for checking tire pressures.

Maximum tire pressure is listed on the sidewall, and is not the pressure you should use when adjusting tire pressure.

Use a tire pressure gauge. There are two basic types of gauge — the pencil-type and the dial-type. If you work in a shop, it is a good idea to have your own pressure gauge because the gauges available at service stations are over-used and may be inaccurate.

To check the pressure, remove the valve stem cap. Place the pressure gauge over the valve stem, and press down firmly. The air pressure in the tire will change the reading on the gauge, figure 12-12. Take two or three readings to be sure that the gauge is working properly and giving consistent readings.

If tire pressure is low, add air a little at a time, and recheck the pressure. If tire pressure is too high, release air from the tire, and recheck pressure. Pressures should be even. Uneven tire pressures at the front wheels will cause a noticeable steering pull on some vehicles.

Radial tires, which are used on nearly all cars and most trucks, are supposed to bulge slightly at the sidewall when properly filled. Do not exceed the recommended tire pressure in an attempt to fill the bulge, or you risk a blowout. On the other hand, do not underfill these tires, because this allows the sidewall to flex and weaken. Use the manufacturer's specified pressures, and fill the tires when cold.

Inspecting Tires for Wear and Damage

A quick tire inspection may save a customer from a blowout. First, determine the type of tire used on the vehicle. Radial tires are used on nearly all passenger vehicles, and as described earlier in this chapter, they bulge slightly at the sidewalls. Bias-ply and bias-

Figure 12-13. The arrow on this tire sidewall indicates the direction that the tire is supposed to roll. Do not install these tires in the reverse direction.

WEAR INDICATORS

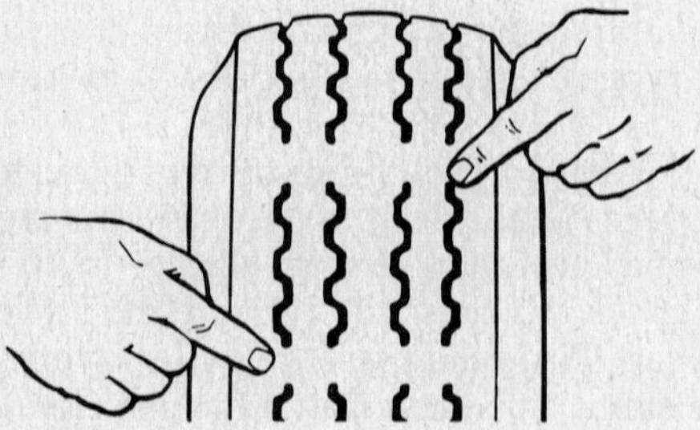

Figure 12-14. Check the tires for solid stripes across the tread. These stripes are wear bars that will show through when it is time to replace the tire.

belted tires do not bulge unless they are underinflated. Use the information in Chapter Thirteen of the *Classroom Manual* to determine the type of tire you are inspecting based on the information on the sidewall. The two front and two rear tires should be the same size and type, or vehicle handling may be unpredictable and dangerous.

Check for directional tires, and be sure that they are installed in the right direction. Often you will find a small arrow on the sidewall that indicates the direction of rotation, figure 12-13. If they are installed in the reverse direction, the vehicle may hydroplane in rainy weather. This is because the sipes in the tire tread direct water to the tread's outer edges when the tires are installed correctly. If they are reversed, water will be directed toward the inside of the tread, creating a layer of water between the tire and the road.

Inspect the tread for wear. If wear bars are showing, figure 12-14, replace the tires. Wear bars are indicators that are about $^1/_2$ of an inch (13 mm) wide. They are made of rubber and

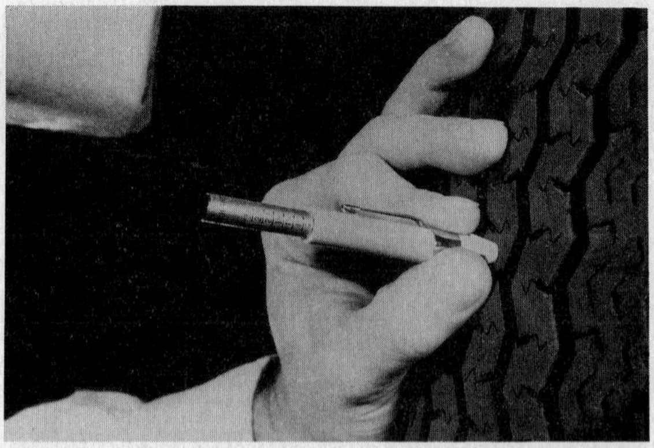

Figure 12-15. A tire tread gauge measures tread depth in increments of $^1/_{32}$ of an inch. If the tread is less than $^1/_{16}$ inch deep, replace the tire.

are molded into the tire during manufacturing. When grooves in the tread are less than $^1/_{16}$ of an inch (1.6 mm) deep, the wear bars will show through, making a series of solid rubber stripes across the tread.

You can also use a tire tread gauge, figure 12-15, which measures tread depth in increments of $^1/_{32}$ of an inch (0.8 mm). If the tread depth is less than $^2/_{32}$ or $^1/_{16}$ inch (1.6 mm) deep, replace the tires. Note that the average life of a car tire is about 40,000 miles (64,000 km).

Check tire sidewalls for bulges, figure 12-16, cracks, and cuts. Because the sidewalls support vehicle weight and are subject to a great deal of flexing, you should replace the tire if the sidewall is damaged, rather than repair it.

There are a number of typical tire wear patterns that indicate defective suspension or steering components, mis-aligned wheels, or tires that are not properly inflated. These wear patterns are most pronounced when the problem has existed for a long time. Sometimes you can determine how a customer drives a vehicle based on tire wear patterns.

Before installing new tires, check the old ones for the following tread wear patterns. If you replace tires without correcting the problems that caused abnormal wear, the new tires will wear just as quickly and with the same wear patterns.

"Heel-toe" wear, figure 12-17, which is common on vehicles that use all-season radials, is a tire wear problem that is fairly easy to correct. This type of wear is inherent in all-season radials because the grooves in the tread are spaced far apart. If heel-toe wear is not severe, remove the tires and wheels, and place them on opposite sides of the vehicle, so that the tires rotate in the reverse direction. This will wear down the high spots on the tires.

Figure 12-16. If you look carefully down the side of this tire, you can see a slight bulge. This indicates ply separation, which weakens the sidewall.

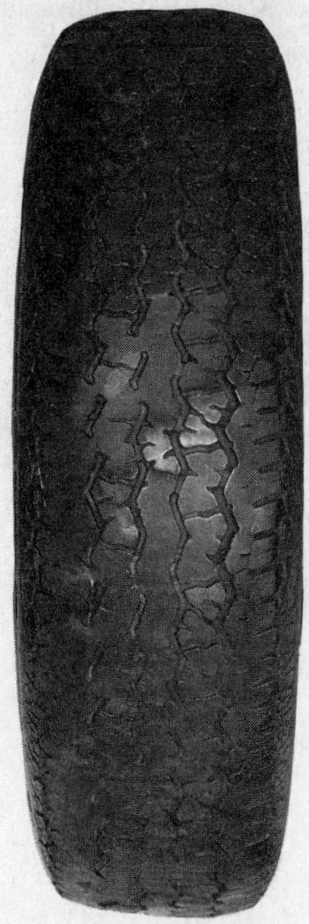

Figure 12-17. Heel-toe wear is fairly common on all-season radials where there are wide spaces between the tread. Rotating the tires according to the manufacturer's recommendations will prevent this type of wear.

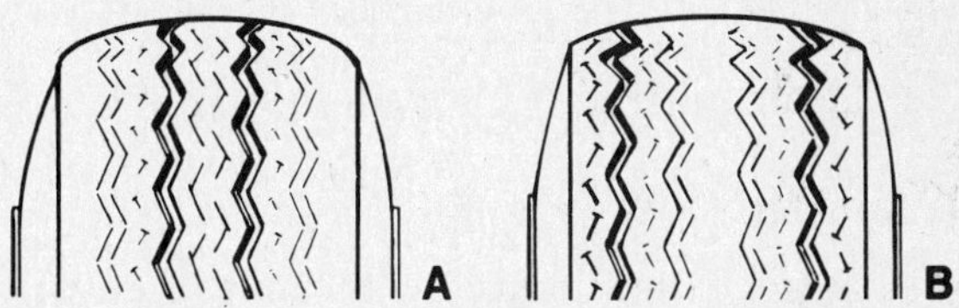

Figure 12-18. Excessive wear on both outer edges or the center of the tread usually is caused by inflation problems. Wear at the outer edges (A) is caused by underinflation, and wear at the center (B) is caused by overinflation.

Wear at both the inner and outer edges of the tread indicates that the tires are underinflated, figure 12-18A. An underinflated tire rides on the outer edges. The center of the tread "collapses" and does not receive much road contact. The obvious solution is to check and correct tire pressure. If you find this type of wear, but tire pressure is correct, wear may be due to high-speed cornering or camber roll, which is caused by excessive caster. Check wheel alignment angles, as described in Chapter Fourteen of this *Shop Manual*, and adjust caster, if needed. If caster is within specifications, you can suggest that the customer purchase lower-profile, high-performance tires, which will better withstand high-speed cornering forces.

If the center of the tread is worn, figure 12-18B, and both outer edges have plenty of tread, the tire is overinflated. Overinflation causes the tire to fill like a balloon, so that the tire rides on the center of the tread, and the outer edges do not contact the road as they should.

Wear on the inner edge of the tread is caused by excessive negative camber or toe, while wear on the outer edge indicates excessive positive camber, toe, or toe-out on turns, figure 12-19. These tire wear problems are corrected by checking and adjusting wheel alignment angles.

Cracks in the tread can be caused by a number of problems from underinflation to manufacturing defects. Also, cracks will appear in the tires if the vehicle is driven at sustained high speeds. If you find this type of tire damage, recommend that the customer install tires with a higher speed rating.

Figure 12-19. These extreme examples of tire wear patterns show wear on the outer tire tread due to excessive positive camber or toe settings.

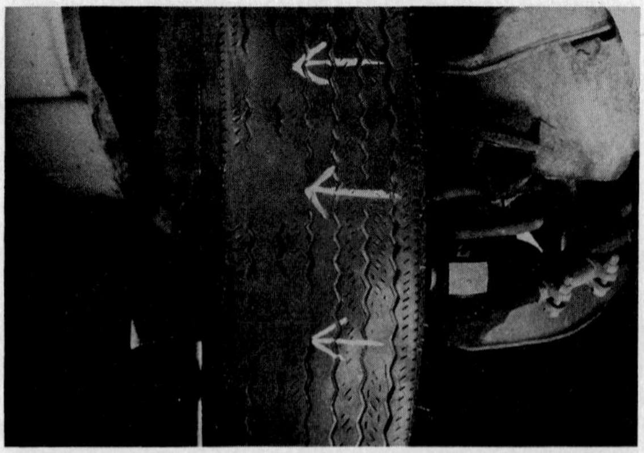

Figure 12-20. Cupping may be caused by a number of suspension or tire and wheel defects, from worn shocks to out-of-balance or out-of-round tires or wheels.

Another type of wear pattern is cupping, figure 12-20, which can be described as ''spotty'' wear. Cupping may appear if the tire and wheel assembly is not balanced, if a wheel is bent, if shocks are worn, or if steering and suspension parts are damaged. Determining the cause of cupping may be difficult, and often it requires more extensive repairs to correct.

Finally, on some front-wheel drive vehicles, the rear tires will display a vertical wipe pattern, which is a result of dog tracking. If the rear axle is not parallel to the front axle, the rear wheels will be ''dragged'', and they will display an abnormal wear pattern.

Inspecting the Wheels

A damaged wheel may allow air to leak from the tire and can cause driveability problems. If the wheel is bent, it will not mount squarely on the hub. After dismounting the tire from the wheel, check the welds on the wheel's drop center. The welds should be intact, and there should be little or no rust.

Check bolt holes for an out-of-round condition, which will allow the wheel to shift on the hub while rotating. Replace the wheel if the bolt holes are damaged.

Check the threads on the lug nuts or bolts. If the threads are damaged or worn, or if a bolt breaks, replace all of the bolts because the others are probably damaged as well.

To remove studs that are pressed in the hub, you must remove the hub from the vehicle. Then remove the metal ''shoulder'' that is on the edge of each stud. (This shoulder helps hold the stud in place.) Swage this metal shoulder off with a special cutting tool before pressing out the stud. If you neglect to do this, you may enlarge the stud hole, and the new stud will not fit tightly. Then use a press to remove the stud. Never use a hammer to remove or install studs, because you may damage the hub. After installing the studs, check that they are fully seated before installing the wheel.

Locating Leaks

To find a leak in a flat tire, remove the tire and wheel assembly from the vehicle, and inflate the tire to the manufacturer's recommended pressure. Check for objects in the tread that may have punctured the tire. Check the sidewalls for cuts and punctures as well.

If the leak is not immediately obvious, prepare a soap-and-water solution, and brush this solution over the tire tread and sidewalls. Soap bubbles will form at the source of the leak. Apply soap and water to the valve stem and wheel rim as well, and check these areas carefully. Mark leaks with a felt-tipped pen, crayon, or chalk before dismounting the tire. Repair leaks as described in the next section.

If you are working on an older vehicle that uses spoke wheels, there is either an inner tube in the tire, or the spoke holes are sealed with mastic to keep air from leaking out. If you do not find leaks in the tire, air may be leaking through the spoke holes. Consult a specialist who can apply mastic to repair leaking spoke wheels.

Repairing Tires

After locating the leak, inspect the tire for damage and wear. If damage or wear is extensive, replace the tire, as repairs to large holes are usually weak and will eventually lead to a blow-out.

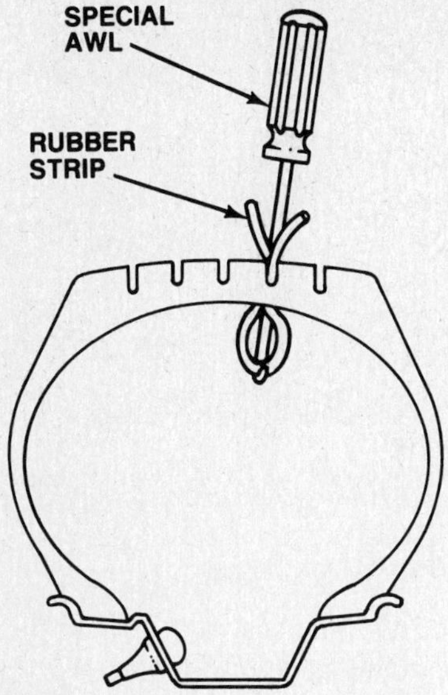

Figure 12-21. A small hole in the tire tread can be repaired from the outside with rubber strips, tire cement, and a special awl that is used for inserting the rubber strip into the tire.

Some standard rules of thumb that you can use to determine if a tire is repairable are as follows. Replace the tire if:

- It has more than two leaks
- The leaks are less than 12 inches (305 mm) apart
- The holes are larger than 1/4 inch (6 mm); many shops will replace a tire if the hole is larger than 1/16 inch (1.6 mm)
- The tire has a hole in the sidewall
- The tire is worn and near the end of its useful life.

If you find a leak at the bead, remove the tire, and check the wheel flange for rust or corrosion. Rust at the flange prevents the bead from seating properly and may be the source of a leak. Clean the wheel flange with a wire brush, remount the tire, and check for leaks again. If the tire still leaks, the bead is weak, and you should replace the tire. Dismounting and mounting a tire weakens the bead.

If a tire is leaking at the sidewall, replace the tire. In addition, because the sidewall flexes so much during normal use, it does not provide a good surface for a patch, and the repair is not likely to last long.

Using sealants that are available in pressurized cans to repair damage to a tire is not a professional tire repair method. These sealants are injected through the valve stem, and are supposed to seal holes in the tire. Aside from the fact that it usually does not provide an effective repair, the sealant may settle unevenly in the tire, causing imbalance. Also, the chemicals used in some of these sealants may rust the wheel, and the pressurized gases in some are flammable. This method of repairing tires is not recommended for the do-it-yourselfer or professional.

Repairs to the tread can be long-lasting if done properly. For small punctures, you may be able to repair the tire from the outside, but the best repairs are made by dismounting the tire from the wheel, and making the repair at the inside of the tire. Both methods are described in the following procedure. If the valve stem or core is leaking, refer to the procedure later in this chapter to replace them.

Repairing a tire from the outside

Radial tire manufacturers recommend that you repair radial tires from the inside, so use this procedure for bias ply tires only. Remove objects imbedded in the tire tread by deflating the tire until pressure is below 15 psi (103 kPa). Use pliers to pull the object from the tread, and inspect the hole. If the puncture is small, you may be able to repair it from the outside.

Kits are available for making these repairs, and include an awl, rubber strips to fill the hole, and tire cement or glue.

1. Clean the puncture with an awl or rat-tail file. This will remove dirt and roughen the edges around the puncture, making a good bonding surface for the rubber strip and cement. Be careful when cleaning the puncture. The hole may not go straight through the tire, and using an awl to clean it may actually damage the tire even more.
2. Coat a piece of rubber stripping with tire cement, and use the awl to insert it into the hole, figure 12-21. The rubber strip should fit tightly, acting as a plug. Remove the awl.
3. Cut off excess rubber stripping so that the plug is even with the tire tread. Allow a few minutes for the cement to set.
4. Inflate the tire to the manufacturer's recommended pressure. Check for leaks before putting the tire and wheel assembly back on the vehicle.

Repairing a tire from the inside

Remove objects imbedded in the tire tread by deflating the tire until pressure is below 15 psi (103 kPa), using pliers to pull the object from

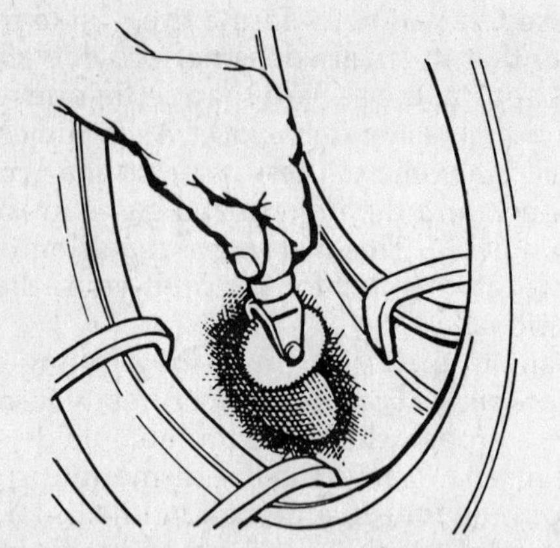

Figure 12-22. After applying the patch, "stitch" it in place with a stitching tool.

the tread. Inspect the hole, and repair it from the inside as follows:

1. Dismount the tire, as described earlier in this chapter.
2. Inspect the inside of the tire for damage. If the only damage you find is the puncture, then repair the tire.
3. Clean the area around the hole with solvent, which removes dirt and oily residue that would prevent the patch from adhering to the inside of the tire.
4. Use a wire brush to roughen the surface of the rubber liner so the patch will adhere. Roughen an area that is slightly larger than the patch. Brush or blow away any rubber particles or dirt.
5. Apply the cement that comes with the tire patch kit to the tire's inner rubber liner. Allow the cement to dry.
6. Pick a patch that will cover the hole and extend at least 1/4 inch past it all the way around. When repairing radial tires, be sure to use a patch designed for radial tires. Be sure to align the arrow on the patch with the tire bead.
7. Remove the covering from the sticky side of the patch, and put the patch over the hole. Use the roller or other patch applying tool that comes with the patch kit to "stitch" the patch in place, figure 12-22.
8. Mount the tire on the wheel, and inflate it to the manufacturer's recommended pressure. Check for leaks before putting the tire and wheel assembly back on the vehicle.

Replacing the valve stem and valve core

The valve stem is expected to last the life of the tire, and when you replace the tire, you

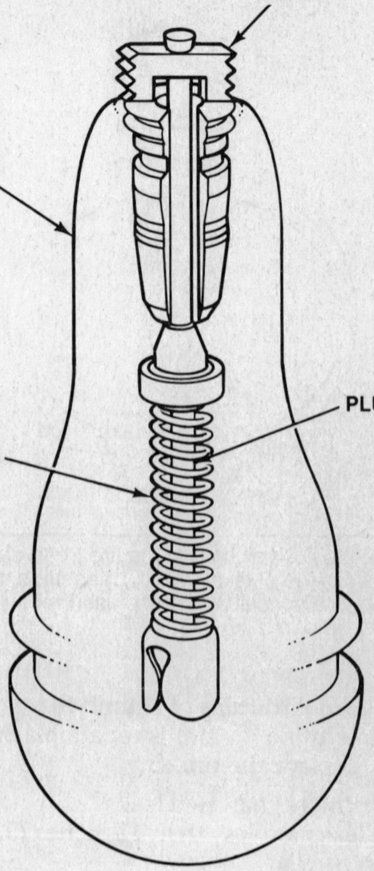

Figure 12-23. The valve core is the metal insert in the stem that actually contains the valve to let air into and out of the tire.

should also replace the valve stem. If air is leaking from the stem, replace it. The procedure for replacing the valve stem is included with the instructions for mounting and dismounting tires, as described earlier in this chapter. You cannot replace the valve stem without removing the tire.

The valve core also may be the source of a leak, and you can replace the core without dismounting the tire. The core is a metal insert in the valve stem, figure 12-23. A rubber or metal cap over the valve stem and core keeps the core clean and helps hold in air should the core leak. If the core is old and allowing air to leak out of the tire, use a special tool to remove the core and install a new one. The tool fits over the outside of the core, figure 12-24. Unscrew the core from the stem, and allow the tire to deflate. Screw a new valve core in place, and inflate the tire.

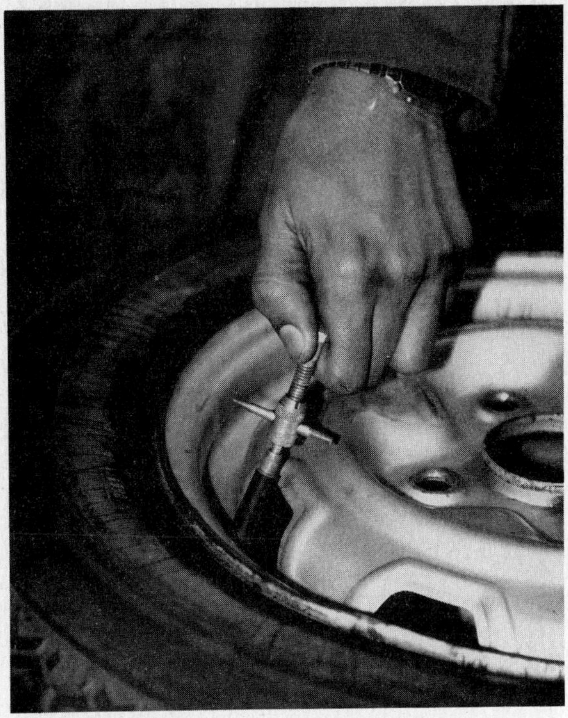

Figure 12-24. Remove and install valve cores with this special tool. After inserting the tool over the core, simply unscrew the core from the stem.

SELECTING CUSTOM WHEELS AND TIRES

The size and type of tires and wheels are part of the vehicle manufacturer's suspension design. When customers decide to replace stock tires and wheels, give them advice on how using custom tires and wheels will affect driveability and suspension wear.When choosing wheels, you should consider the bolt pattern, offset, width, and diameter. When choosing tires, you should consider tire size, tire type, tire load and speed ratings, rubber composition, and the type of vehicle on which the tires will be used.

New wheels must have the same bolt pattern as the stock ones. Chapter Thirteen of the *Classroom Manual* describes how wheel manufacturers designate bolt patterns. If customers choose wheels that are not available in the stock wheel bolt pattern, they can replace the existing wheel hubs with custom-made hubs that are drilled and fitted with lugs to accommodate a different wheel bolt pattern. These custom-made hubs are expensive and are not available for all vehicle models, but do offer an alternative when replacing stock wheels on some vehicles.

The new wheels should have the same offset as the stock wheels or accelerated wheel bearing and drivetrain wear may result. Wheel offset helps determine where vehicle weight is supported on the wheel and tire, so changing offset can also affect handling and braking. Offset is described in Chapter Thirteen of the *Classroom Manual*. For FWD cars, the customer should choose wheels with the same or more positive offset than the stock wheels for CV joint, brake, and suspension component clearance.

The width of the new wheel should be close to the width of the stock wheel. Like offset, wheel width affects weight distribution across the tire tread. As discussed in Chapter Thirteen of the *Classroom Manual*, the inner wheel bearing is larger than the outer one because it is designed to carry most of the vehicle weight. Using wider wheels with the same offset gives the effect of negative camber and transfers more weight to the inner wheel bearing, contributing to accelerated inner wheel bearing wear. Finally, using a wider wheel may simply cause clearance problems. Even if the tire and wheel fit in the wheel well when the vehicle is standing still, there may be suspension or tire clearance problems during vehicle operation.

The diameter of the new wheel should be the same as the diameter of the stock wheel for a number of reasons. If the customer wants to increase wheel diameter, the tire and wheel may not fit in the wheel well. If the customer wants to decrease wheel diameter, ground clearance problems may result. Finally, a change in wheel diameter will affect speedometer calibration. A sensor at the transmission output shaft counts the number of times the output shaft turns. This information, along with gear ratio and wheel size is used to calculate vehicle speed, and is displayed on the speedometer. If you change wheel diameter, the speed displayed on the speedometer will not be the actual vehicle speed. Some manufacturers make replacement speedometer gears that you can use to compensate for different wheel diameters. If you do change the diameter of the wheel, then change the profile of the tire so that the overall tire and wheel diameter remains nearly the same, figure 12-25.

Changing the diameter of the tire and wheel also may affect drivability on some vehicles. This is because the engine control computer receives information from the speedometer about vehicle operating conditions. If the tire and wheel diameter is changed, the computer will receive inaccurate information. Driveability, fuel economy, and emission problems may result. Also, if the vehicle has an anti-lock braking system (ABS), changing the tire and wheel diameter may affect the way the braking

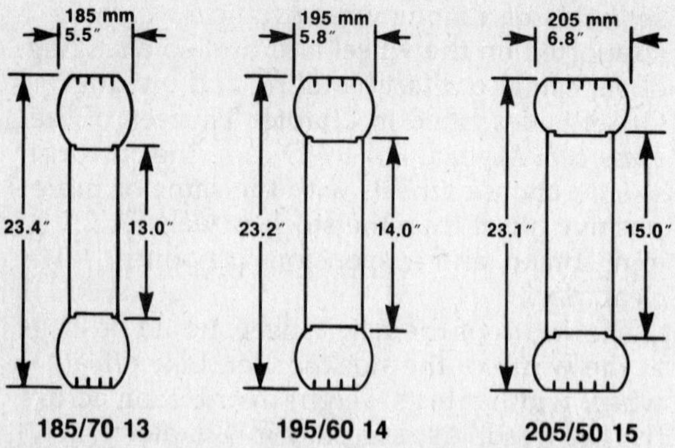

Figure 12-25. If you want to change the aspect ratio and width of the tire for improved performance, you will have to change the wheel diameter as well to maintain the same overall tire outer diameter.

system computer handles information. This can have serious repercussions if a customer is involved in an accident due to ABS failure.

Use the information on the tire sidewall when choosing new tires. Chapter Thirteen of the *Classroom Manual* describes how to read tire sidewall information. Whenever possible, use the same tire specifications recommended by the vehicle manufacturer when replacing tires.

Do not mix different sizes or types of tires on the same axle. Radial and bias-belted tires have different slip angles, and mixing them on the same axle — or the same vehicle for that matter — will give unpredictable handling results.

If you install snow tires for customers, remind them that these tires are designed for use at low speeds in the snow and ice. Using snow tires for high-speed driving in warm weather will heat the tread, which damages the tire, and may create high-speed vibrations. Studded snow tires can damage the pavement, and are prohibited in some areas.

Be aware of tire maximum load and speed ratings, which are indicated on the sidewall. Exceeding either maximum is dangerous. When inspecting the old tires, note if they have cracks across the tread, which is one indicator of high-speed driving. You should recommend that the customer use tires with a higher speed rating.

Tire width and diameter should match the wheel dimensions. For example, if the wheel has a 13-inch (330-mm) diameter, the tire inner diameter should also be 13 inches (330 mm). Tire manufacturers have allowable ranges within which the tire dimensions can vary from the wheel dimensions. If these ranges are exceeded, the tire bead may not seat properly, and may separate from the wheel during driving.

The type of rubber used to make tires affects tire adhesion. Performance tires are made with softer rubber compounds, which help tires adhere to the road. A customer who wants improved road holding may want to try tires made of softer rubber. However, this rubber heats up and wears out faster than general all-purpose street tires, which are made of a rubber compound that compromises some adhesion for durability.

Be sure that directional tires are installed so that they are rolling in the right direction. Most directional tires have arrows on the sidewall that indicate the direction of tire rotation. You can rotate directional tires on the same side of the vehicle, as long as the front and rear axles use the same size tires.

As a general rule, for improved handling, use wider tires with shorter sidewalls (lower aspect ratio). Maintain the original tire outer diameter by using larger diameter wheels, figure 12-25.

BASIC WHEEL BEARING SERVICE

Because wheel bearing condition may affect vehicle handling, basic service instructions are covered in this section. Remove wheel bearings if:

- You determine they are worn
- They require repacking as part of routine maintenance
- You must remove them to replace a leaking grease seal
- You must remove them to make other repairs.

The procedures used to remove and reinstall wheel bearings vary with the type of bearing that is used. Detailed instructions for servicing adjustable dual wheel bearings, figure 12-26, sealed wheel bearings, figure 12-27, and solid axle wheel bearings, figure 12-28, are given later in this chapter.

Except for sealed bearings, all wheel bearings share a number of common service procedures that should be performed whenever the bearing is removed for any reason. The following sections provide some general bearing service tips, and describe how to clean wheel bearings, inspect them for wear and damage, and pack them with grease.

General Bearing Replacement Tips

When you replace an unsealed wheel bearing, leave the new part in its package until you are

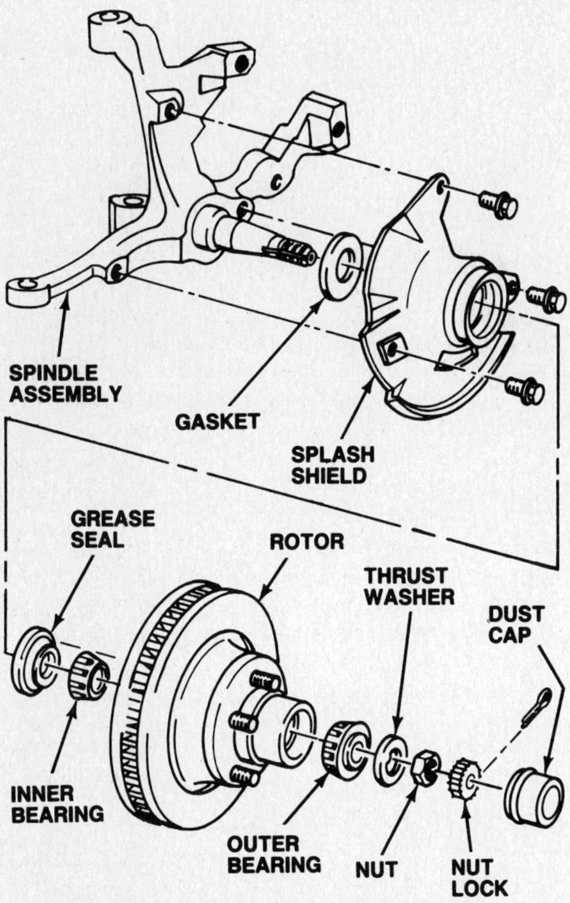

Figure 12-26. This typical adjustable dual wheel bearing assembly uses tapered roller bearings.

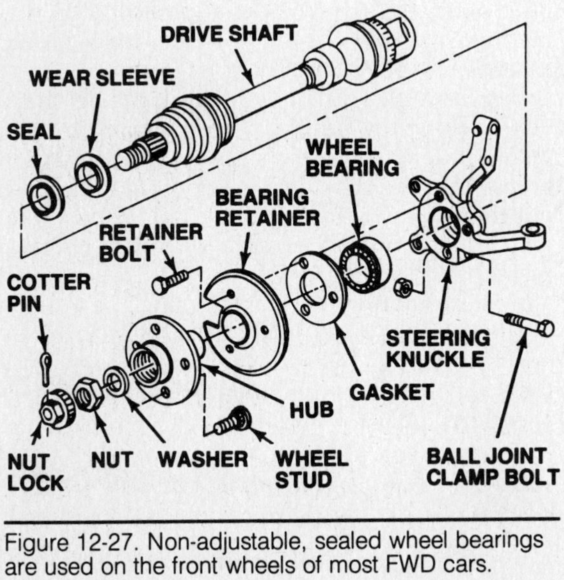

Figure 12-27. Non-adjustable, sealed wheel bearings are used on the front wheels of most FWD cars.

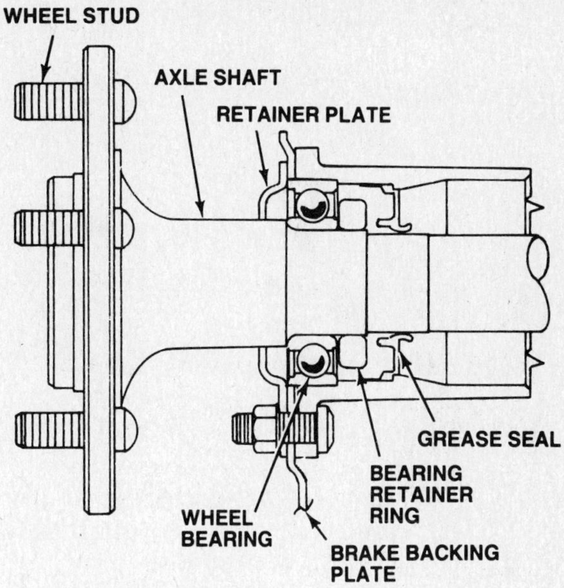

Figure 12-28. Solid rear axles usually have straight roller bearings or ball bearings.

ready to install it because the wrapper is designed to keep the bearing clean and rust free. Before you touch a new bearing, wash your hands and dip your fingers in clean motor oil to remove acids from your fingers that can corrode the bearing.

Whenever you replace a tapered roller or ball bearing that has a separate outer race, you must replace the race as well. Even though an old outer race may appear to be in perfect condition, it has been subjected to the same stress and metal fatigue as the old bearing. An old race also has a certain wear pattern from running against the old bearing. If you install a new bearing in an old race, it will not fit properly, and premature bearing wear and failure will result.

Whenever you service any type of wheel bearing, install new grease seals. Seals wear with age just like bearings, and are often damaged during bearing removal. Seals keep lubricant in the bearing or axle where it belongs. At the same time, they keep out the dirt, water, and brake dust that can shorten a bearing's or axle's service life.

Cleaning Wheel Bearings

When an unsealed wheel bearing is removed, wipe away as much old grease as possible using dry rags or paper towels. Inspect the grease on the towels for metal chips or other indications of bearing wear or damage. Next,

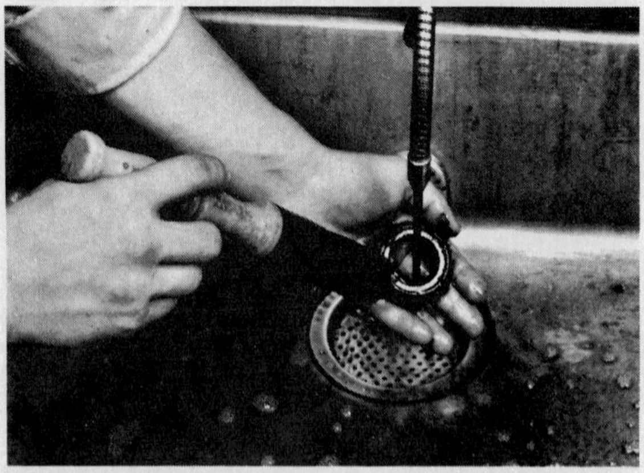

Figure 12-29. Wash unsealed bearings in clean solvent, and use a long-bristled brush to clean them.

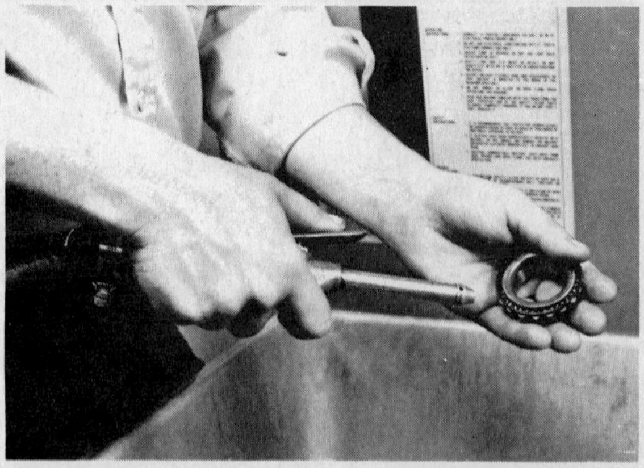

Figure 12-30. You can use compressed air to dry wheel bearings, but do not spin the bearings with air.

clean the bearing in a parts washer using *clean* petroleum-base solvent. A long-bristled brush is helpful to flush out hard to reach grit, figure 12-29. Wash each bearing individually, and keep bearings with detachable outer races separated so you can replace them in the same races from which they were removed. Clean unsealed rear axle ball bearings while they are still on the axle.

Once all of the old grease has been washed out, flush the bearings with a non-petroleum-based brake cleaning fluid. This removes any traces of oil and solvent that can contaminate the new grease and lead to premature bearing failure. Allow the bearings to air dry, or hold the bearings by their cages and dry them with clean compressed air. Direct the air through the bearing so it travels across the balls or rollers from side to side, figure 12-30. *Do not* spin a bearing with air while drying it. The bearing is not properly lubricated at this time, and spinning it at high speed can cause rapid wear and damage. In addition, if the air pressure or bearing speed is great enough, the balls or rollers may fly out of the cage, which can be dangerous.

Inspecting Wheel Bearings

Once the bearings are clean and dry, inspect them for the problems shown in figure 12-31. Although the illustration depicts tapered roller bearings, most of the same problems occur with straight roller and ball bearings as well. To inspect a bearing, rotate it carefully in good light so the complete surface of each ball, roller, and race can be fully checked. If any problems are apparent, or if you have a question about its condition, replace the bearing.

Lubricating Wheel Bearings

Whenever you pack a new or used wheel bearing with grease, it is extremely important that you use the type of grease recommended by the vehicle manufacturer. *Always* clean away every trace of old grease before you repack a wheel bearing, and *never* add new grease to old grease, or bearing damage may result.

When you pack a wheel bearing, work the grease into the cage and races and between the balls or rollers, so that no air spaces remain. The most effective and efficient way to do this is to use a bearing packer, figure 12-32. These devices may be manually operated, or they may use air or hydraulic pressure to force new grease through the entire bearing in one quick operation.

If a bearing packer is unavailable, you can pack wheel bearings by hand, although this method is slower and messier. To hand pack a tapered roller bearing, fill your palm with grease, and place the large end of the bearing down. Draw the bearing across your palm, taking little bits of grease at a time. Force grease into the cage and rollers until it oozes out the opposite side, figure 12-33. Repeat this process all around the bearing until it is completely filled with grease. Finish by spreading a medium coating of grease around the outside of the bearing. When you pack a bearing by hand, work the grease in from both sides until the space between the inner and outer races is completely filled with grease.

ADJUSTABLE DUAL WHEEL BEARING SERVICE

Adjustable dual wheel bearings are used at the front or rear non-driven axle. Almost all late-model cars with adjustable dual wheel bearings use tapered roller bearings. Except for

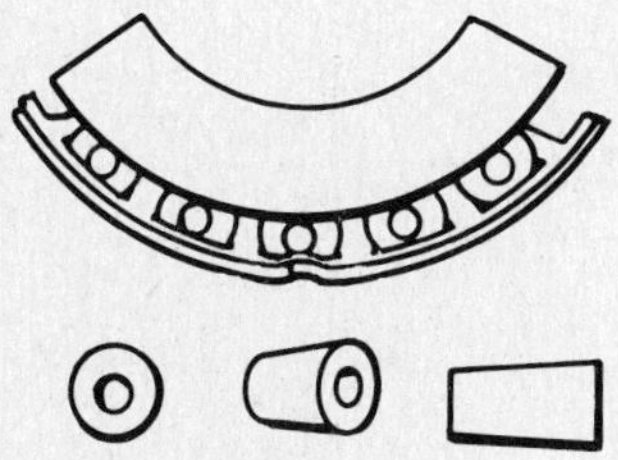

BENT CAGE

CAGE DAMAGE CAUSED BY IMPROPER HANDLING OR TOOL USE

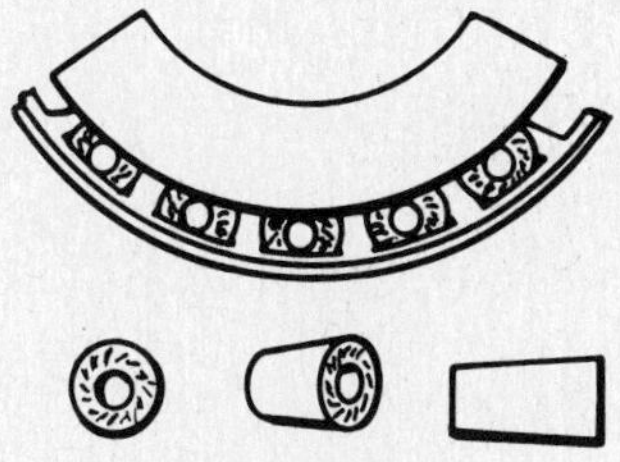

GALLING

METAL SMEARS ON ROLLER ENDS CAUSED BY OVERHEATING, OVERLOADING, OR INADEQUATE LUBRICATION

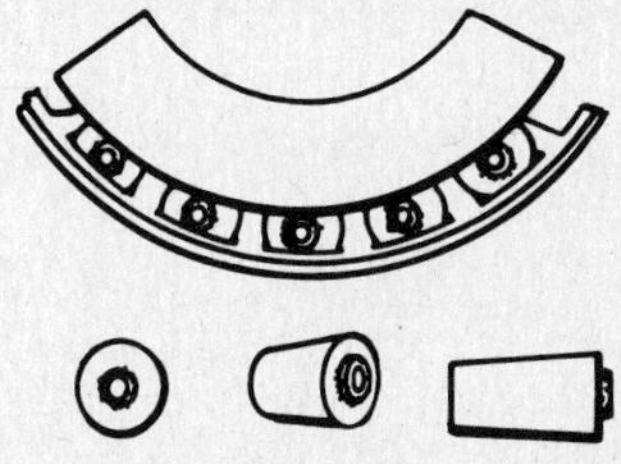

STEP WEAR

NOTCHED WEAR PATTERN ON ROLLER ENDS CAUSED BY ABRASIVES IN THE LUBRICANT

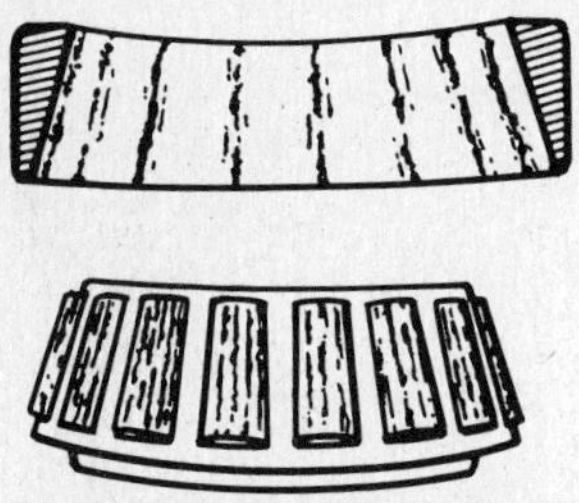

ETCHING AND CORROSION

EATEN AWAY BEARING SURFACE WITH GRAY OR GRAY-BLACK COLOR CAUSED BY MOISTURE CONTAMINATION OF THE LUBRICANT

PITTING AND BRUISING

PITS, DEPRESSIONS, AND GROOVES IN THE BEARING SURFACES CAUSED BY PARTICULATE CONTAMINATION OF THE LUBRICANT

SPALLING

FLAKING AWAY OF THE BEARING SURFACE METAL CAUSED BY FATIGUE

MISALIGNMENT

SKEWED WEAR PATTERN CAUSED BY BENT SPINDLE OR IMPROPER BEARING INSTALLATION

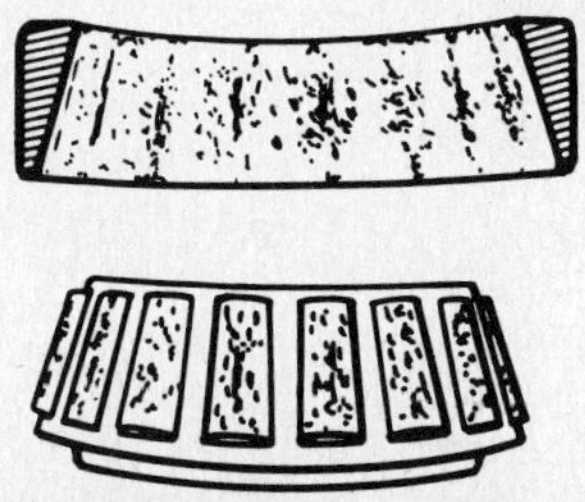

HEAT DISCOLORATION

FAINT YELLOW TO DARK BLUE DISCOLORATION FROM OVERHEATING CAUSED BY OVERLOADING OR INADEQUATE LUBRICATION

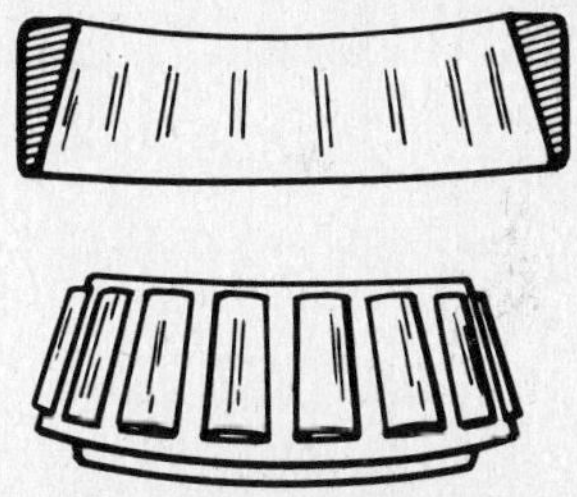

BRINELLING

INDENTATIONS IN THE RACES CAUSED BY IMPACT LOADS OR VIBRATION WHEN THE BEARING IS NOT TURNING

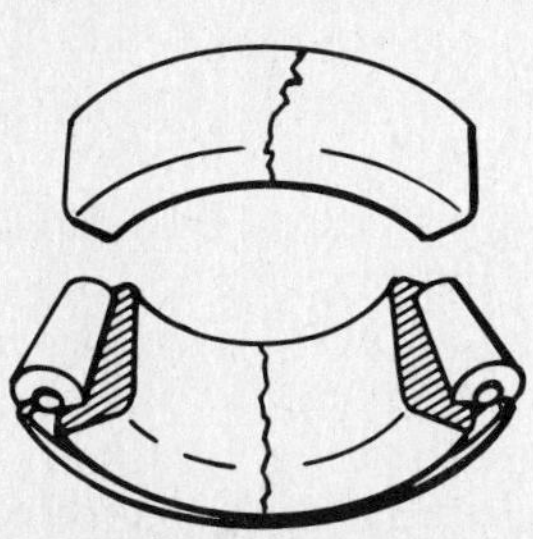

CRACKED RACE

CRACKING OF THE RACE CAUSED BY EXCESSIVE PRESS FIT, IMPROPER INSTALLATION, OR DAMAGED BEARING SEATS

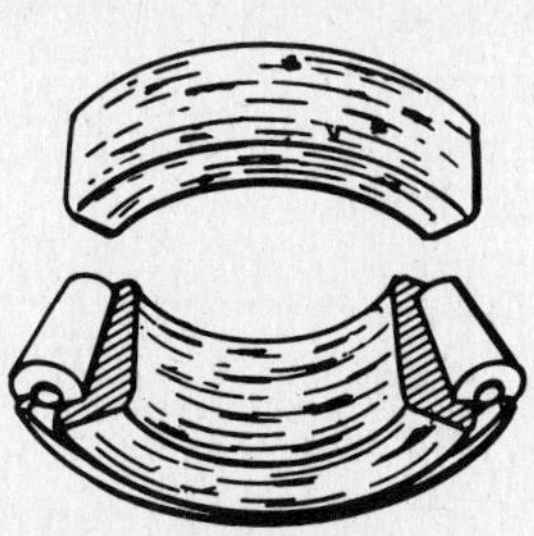

SMEARING

SMEARED METAL FROM SLIPPAGE CAUSED BY POOR FIT, POOR LUBRICATION, OVERLOADING, OVERHEATING, OR HANDLING DAMAGE

FRETTAGE

ETCHING OR CORROSION CAUSED BY SMALL RELATIVE MOVEMENTS BETWEEN PARTS WITH NO LUBRICATION

Figure 12-31. This chart shows common bearing wear patterns and probable causes.

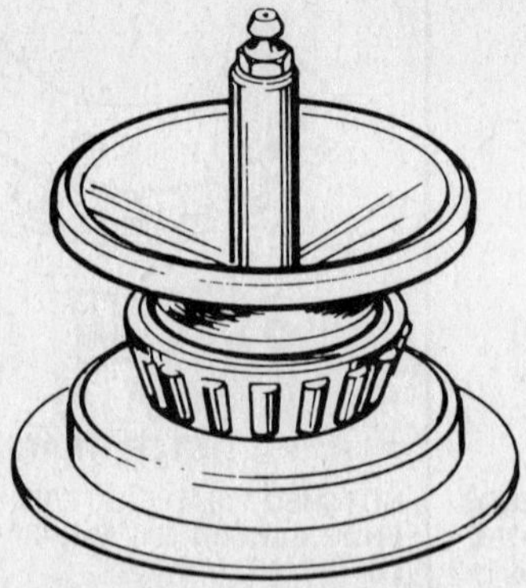

Figure 12-32. Bearing packers do an excellent job of lubricating wheel bearings.

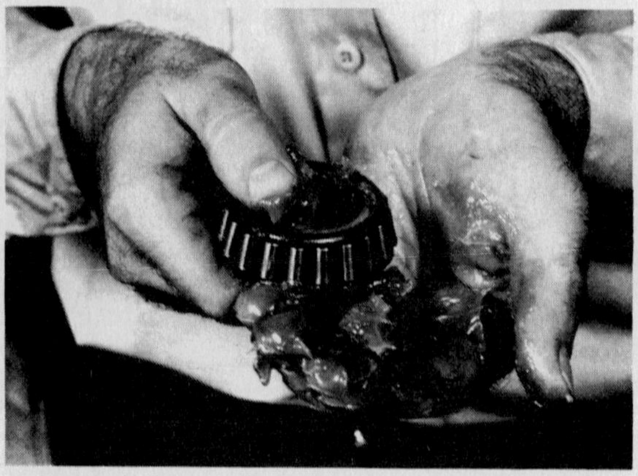

Figure 12-33. You can pack wheel bearings by hand, but it is a rather messy job.

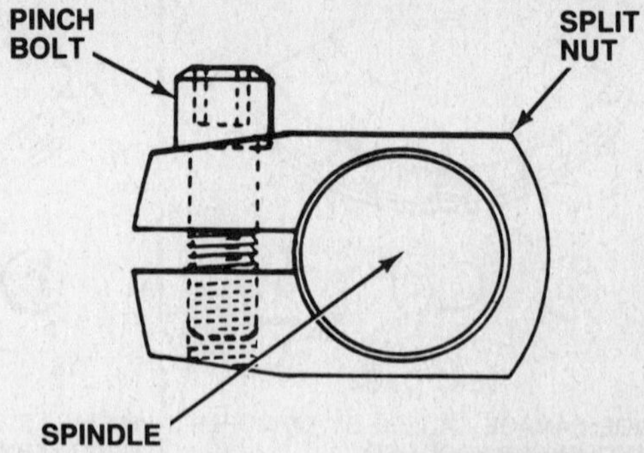

Figure 12-34. A split nut and pinch bolt are used to secure the wheel bearings on some imported vehicles.

Figure 12-35. Remove the outer wheel bearing and related parts.

removing and installing the bearing races, the procedure is basically the same whether tapered roller bearings are being replaced or simply repacked. The procedure is also the same whether the bearings are fitted in a brake drum hub or a brake rotor hub.

To service a set of adjustable dual wheel bearings:

1. Position the vehicle on a lift, and ensure that components will not bend under the weight of the car when it is raised. See Chapter One of this *Shop Manual* for lifting a vehicle. Remove the wheels.
2. If the vehicle has disc brakes, remove the brake caliper, as needed, so that you can remove the rotor. Instructions for removing brake system components are given later in this chapter.
3. Pull the dust cap from the center of the hub to expose the adjusting nut.
4. Loosen the adjusting nut, and back it off about 1/2 inch (13 mm). On most cars, the nut is secured with a castle nut and cotter pin. Remove the pin, and unscrew the nut. Some imported cars have a split nut with a pinch bolt, figure 12-34. Unscrew the pinch bolt, and remove the nut.
5. Pull the drum or rotor outward to free the thrust washer and outer wheel bearing. Then push the drum or rotor inward on the spindle.
6. Remove the adjusting nut, thrust washer, and outer wheel bearing from the hub, figure 12-35, and set them aside.
7. Pull the drum or rotor outward and slide it off the spindle, taking care not to drag the inner wheel bearing across the adjusting nut threads. To remove some brake drums, you may have to loosen the brake adjuster as described later in this chapter.
8. If the inner wheel bearing sticks on the spindle and makes the drum or rotor difficult to remove, use a puller or slide hammer. Once the drum or rotor is off the axle, use a puller or a pair of pry bars to carefully remove the inner bearing and grease seal.

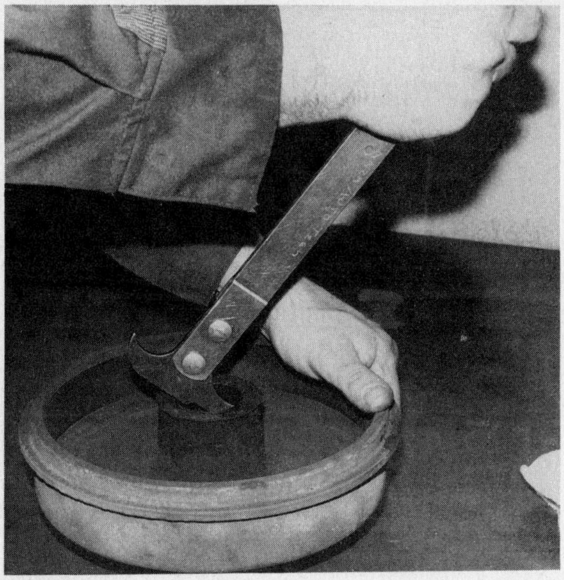

Figure 12-36. Use a seal puller tool to remove the seal from the drum or rotor.

Figure 12-37. You can use a non-metallic drift to remove the inner wheel bearing and grease seal.

9. Support the drum or rotor on the workbench, and remove the grease seal. This can be done in two ways:
   - Hook the claw of a seal puller under the metal retaining ring of the seal, and lever the seal out, figure 12-36. You can use a large screwdriver to pry out the seal.
   - Position a non-metallic drift about three-quarters of an inch (19 mm) in diameter through the outer bearing opening so it contacts the inner race of the inner bearing, figure 12-37. Strike the drift with a hammer until the bearing and seal are driven from the hub.
10. Clean and inspect the bearings and bearing races, as described earlier. Also clean the inside of the drum or rotor hub.

Figure 12-38. Install the grease seal into the hub over the inner bearing.

11. If you must install a new bearing, remove the old race from the drum or rotor hub. This can be done in two ways:
    - Use a special bearing race puller.
    - Insert a metal drift through the hub and position it against the backside of the race. Use a drift that is softer than the hub metal to prevent damage. Strike the drift with a hammer while moving it around the race to drive the race from the hub.
12. Use a bearing race driver or a suitably sized socket to install the new race in the drum or rotor hub. The tool should only contact the outer edge of the race. Drive the race in until it is fully seated. The pitch of the sound made when striking the driver will change when the bearing is seated.
13. Clean and inspect the spindle for rust, nicks, and scratches. Remove any blemishes with fine sandpaper, and make sure the inner races of both bearings slip easily over the spindle. If the spindle is badly scored or cracked in any way, you must replace it.
14. Very lightly coat the spindle with grease. This prevents rust and allows the inner bearing races to creep slightly so wear is distributed evenly around the bearings.
15. Pack the wheel bearings with grease as described earlier.
16. Place the drum or rotor outer side down on the workbench. Coat the inside of the hub with grease to prevent rust.
17. Put a medium coating of grease on the inner bearing race, and place the inner bearing in the race.
18. Install the grease seal with a seal driver, figure 12-38. Put a light coat of grease on the seal lip.

Figure 12-39. A dial indicator allows you to precisely adjust the wheel bearings.

19. Turn the drum or rotor over and put a medium coat of grease on the outer bearing race.
20. Install the drum or rotor onto the spindle, being sure to keep it centered so the inner bearing or outer bearing race does not drag on the adjusting nut threads.
21. Install the outer bearing into the hub, place the thrust washer over it, and install the adjusting nut finger tight.
22. Adjust the wheel bearings as described in the next procedure.
23. If the axle is equipped with disc brakes, install the brake caliper as described later in this chapter. If the axle has drum brakes, and the brake adjustment was loosened earlier, adjust the brakes as described later in this chapter.
24. Install the wheel and tighten the lug nuts to the correct torque in the proper sequence.

Adjusting Tapered Roller Bearings

Tapered roller wheel bearings require adjustment whenever you service them. Before you adjust the bearings, make sure the adjusting nut turns freely on the spindle threads. If the nut does not turn easily by hand, remove any nicks or burrs from the spindle threads with a thread file. If the adjusting nut still binds, run a tap through it or replace the nut. On cars with drum brakes, back off the brake adjustment as needed so the brakes do not drag.

There are three ways to adjust tapered roller bearings:

- By hand
- With a torque wrench
- With a dial indicator.

Always use the adjusting method recommended by the vehicle manufacturer. Once you set the proper axial play, lock the adjusting nut in place, and install the dust cap with a plastic hammer. If the adjusting nut is secured with a cotter pin, place the castle nut over the adjusting nut so the slots in the lock align with the cotter pin hole in the spindle. Insert a new cotter pin through the hole, and wrap the ends around the castle nut. Cut off the extra length. If the car is equipped with a split adjusting nut, hold it in position with a suitably sized wrench, and tighten the pinch bolt to the torque specified by the vehicle manufacturer.

Hand adjustment

To adjust the wheel bearings by hand, rotate the wheel and snug up the adjusting nut with a wrench to seat the bearings. As you continue to rotate the wheel, back off the adjusting nut 1/4 to 1/2 turn, or until it is just loose. Then tighten the nut with your fingers to a snug fit. Lock the adjusting nut in place.

Torque wrench adjustment

To adjust the wheel bearings with a torque wrench, rotate the wheel and torque the adjusting nut to the value specified by the vehicle manufacturer, typically 12 to 25 ft-lb (15 to 35 Nm). Back off the adjusting nut 1/3 turn, then retighten it to the value specified by the vehicle manufacturer, typically 10 to 15 in-lb (1 to 1.5 Nm). Lock the adjusting nut in place.

Dial indicator adjustment

To adjust the wheel bearings with a dial indicator, rotate the wheel and torque the adjusting nut to 12 to 25 ft-lb (15 to 35 Nm). Back off the adjusting nut 1/4 to 1/2 turn, or until it is just loose. Then tighten the nut with your fingers to a snug fit. Mount a dial indicator on the drum or rotor and position the plunger against the end of the spindle, figure 12-39. Move the drum or rotor in and out to measure the axial play, and turn the adjusting nut as needed to obtain the clearance specified by the vehicle manufacturer, typically between 0.001 and 0.005 inch (0.025 to 0.127 mm) end play. Lock the adjusting nut in place.

SEALED WHEEL BEARING SERVICE

As described earlier, the sealed double-row wheel bearing assemblies, used on the front wheels of many FWD cars and the non-driven rear wheels of some cars, are serviced by replacing them when their axial play exceeds the manufacturer's specifications. The replacement procedure varies depending on whether the bearing is on a front or rear axle. Procedures for both follow.

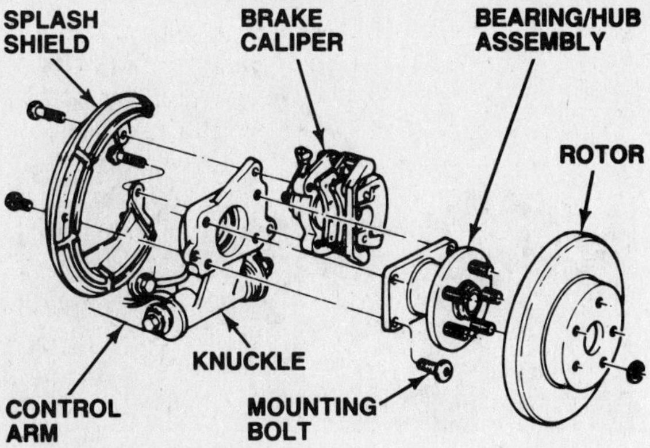

Figure 12-40. In this example, the wheel bearing and hub are an integral assembly.

Replacing Rear Axle Sealed Bearings

General Motors uses sealed rear wheel bearings on some vehicles. Their design combines the bearings with the wheel hub assembly, figure 12-40, which makes replacement a relatively simple job. To replace one of these bearing/hub assemblies:

1. Position the vehicle on a lift, and ensure that components will not bend under the weight of the car when it is raised. See Chapter One of this *Shop Manual* for lifting a vehicle. Remove the wheel at the corner of the car with the defective bearing.
2. If the vehicle has disc brakes, remove the caliper as described later in this chapter.
3. Remove and discard any speed nuts, retaining bolts, or screws. Then remove the brake drum or rotor from the hub.
4. Remove the four bearing/hub retaining bolts, and remove the assembly.
5. Install the new bearing/hub assembly, and tighten the four retaining bolts to the torque specified by the vehicle manufacturer.
6. Position the brake drum or rotor over the wheel studs.
7. If the axle is equipped with disc brakes, install the brake caliper as described later in this chapter.

Replacing Front Axle Sealed Bearings

Replacing a sealed wheel bearing assembly on a driven front axle is much more involved than the same job on a non-driven rear axle. On Chrysler and Ford cars you must remove the steering knuckle from the vehicle before you can replace the wheel bearing. General Motors bearings can be replaced in this manner as well. However, special tools are available that allow the combined bearing/hub assembly to be replaced without removing the steering knuckle. The bearing replacement procedure below applies to domestic FWD cars. The same basic steps apply to most imports as well. To replace a sealed wheel bearing assembly on a driven front axle:

1. On Chrysler cars, remove the cotter pin from the hub retaining nut. The crimped collar on Ford retaining nuts does not have to be straightened prior to nut removal. General Motors cars have self-locking nuts.
2. Have an assistant apply the brakes while you loosen the hub nut with a socket and breaker bar. Do not use an impact wrench for this job because the shock it delivers can damage constant-velocity joints.
3. Position the vehicle on a lift, and ensure that components will not bend under the weight of the car when it is raised. See Chapter One of this *Shop Manual* for lifting a vehicle. Remove the wheel.
4. Remove the brake caliper, as described later in this chapter, so you can remove the rotor.
5. Remove any speed nuts, retaining bolts, or screws. Then remove the brake rotor.
6. Remove the hub nut and washer. On General Motors cars go to step 13.
7. Remove the nut securing the tie rod end. Then use a puller to separate the tie rod end from the steering arm, as described in Chapter Five of this *Shop Manual*.
8. Loosen the clamp bolt that holds the ball joint stud in the steering knuckle. Disconnect the lower control arm from the steering knuckle. Refer to Chapter Nine of this *Shop Manual*, if needed.
9. Loosen the retaining nuts on the bolts that attach the strut to the steering knuckle. If the bolts are used to set the wheel alignment, mark the position of the adjusting cams before you loosen the nuts.
10. Push the axle shaft out of the steering knuckle, using a puller if necessary.
11. Remove the bolts that attach the steering knuckle to the strut, and remove the steering knuckle from the vehicle.
12. Remove the hub from the center of the sealed bearing assembly:
    - On Chrysler cars, use a drift punch to remove the hub, figure 12-41. If an inner bearing race comes out with the hub, use a puller to remove it.

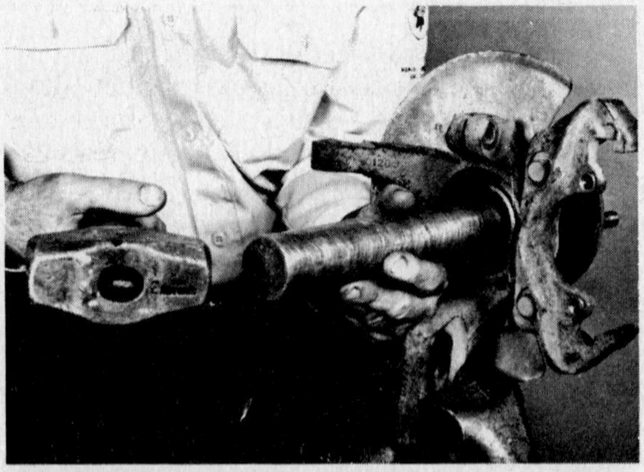

Figure 12-41. You can use a drift to remove the hub.

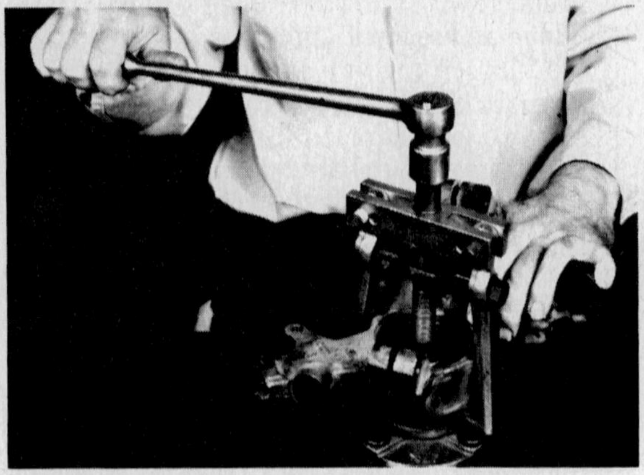

Figure 12-42. On some Fords, remove the hub with a puller.

- On Ford cars, use a puller to remove the hub, figure 12-42.

13. Remove the bearing from the steering knuckle:
    - On Chrysler cars, remove the three bolts securing the bearing retainer. Then press the bearing out of the steering knuckle.
    - On Ford cars, use a screwdriver to remove the snapring that retains the bearing. Then press the bearing out of the steering knuckle.
    - On General Motors cars, remove the bearing/hub retaining bolts and splash shield. Then, use a puller to remove the bearing/hub assembly.
14. Remove the seal:
    - On Chrysler cars, use a screwdriver to pry the seal out of the steering knuckle.
    - On Ford cars, remove the halfshaft and place it in a vise. Use a screwdriver and hammer to remove the seal from the axle.

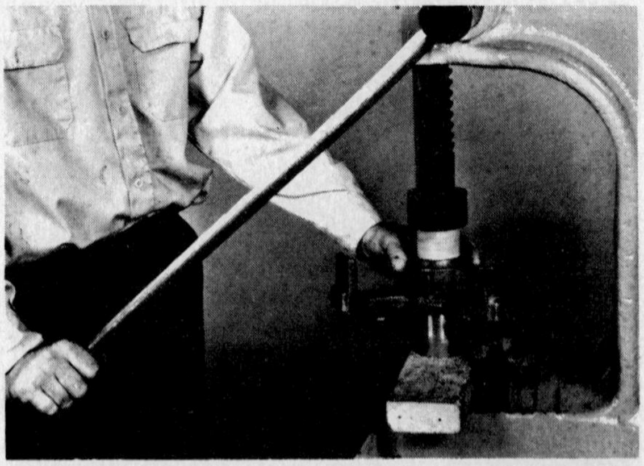

Figure 12-43. Use a press to install the wheel bearing assembly.

    - On General Motors cars, use a punch to drive the seal out of the steering knuckle toward the engine. Then cut the seal off the drive shaft.
15. Clean the bore of the steering knuckle and, on Ford vehicles, the seal mounting surface of the driveshaft. Inspect the bore and mounting surface, and smooth away any nicks or burrs with sandpaper. On General Motors cars go to step 24.
16. Apply a light film of grease to the steering knuckle bore. Then press the new wheel bearing into place, figure 12-43. Apply pressure only on the outer bearing race.
    - On Chrysler cars, install the bearing retainer and tighten the three bolts to the torque specified by the vehicle manufacturer.
    - On Ford cars, install the snapring that retains the bearing into its groove in the steering knuckle.
17. Apply a light film of grease to the inner bearing bore. Then press the hub into the center of the bearing assembly. Support or press *only* on the inner race of the bearing to prevent bearing damage.
18. Install the new seal with a seal driver. Coat the lip of the seal with wheel bearing grease:
    - On Chrysler cars, install the seal in the steering knuckle.
    - On Ford cars, install the seal onto the driveshaft.
19. Align the axle splines with those in the hub. Press the hub and steering knuckle onto the axle. If necessary, use a special tool to fully seat the hub on the driveshaft.
20. Attach the steering knuckle to the strut, but do not tighten the retaining nuts.

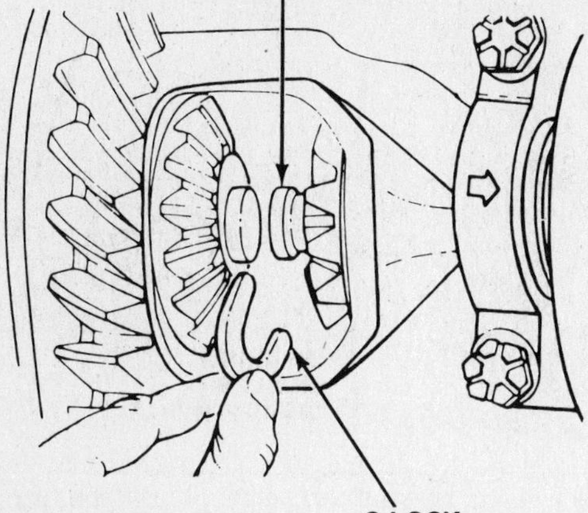

Figure 12-44. Remove the C-lock from its groove in the axle.

21. Position the steering knuckle on the ball joint stud, and tighten the clamp bolt to the torque specified by the vehicle manufacturer.
22. If necessary, align the adjusting cams on the bolts that attach the steering knuckle to the strut with the marks made in step 9. Tighten the retaining nuts to the torque specified by the vehicle manufacturer.
23. Connect the tie rod end to the steering arm. Tighten the retaining nut to the torque specified by the vehicle manufacturer. If you did not remove the bearing/hub assembly, go to step 27.
24. Apply a light film of grease to the steering knuckle bore. Use a seal driver to install the new seal.
25. Install the bearing/hub assembly onto the axle, and install a new washer and hub retaining nut. Tighten the nut until the bearing/hub assembly is fully seated.
26. Install the splash shield and bearing/hub retaining bolts. Tighten the bolts to the torque specified by the vehicle manufacturer.
27. Install the brake rotor over the wheel studs.
28. Install the brake caliper as described later in this chapter.
29. Have an assistant apply the brakes as you tighten the hub retaining nut to the torque specified by the vehicle manufacturer.
    - On Chrysler cars, install the nut lock and a new cotter pin.
    - On Ford cars, use a blunt chisel or similar tool to crimp the retaining nut collar into the groove on the axle.
30. Install the wheel and lower the car.
31. Check and adjust wheel alignment angles.

SOLID AXLE WHEEL BEARING SERVICE

A common reason for removing the wheel bearings from a solid rear axle is to replace leaking grease seals. However, before you undertake this job, first check the axle vent on top of the axle housing. If the vent is plugged, heat can cause a pressure build-up inside the housing that will force lubricant past the seals even if they are in perfect condition. If the vent is not obstructed, replace the leaking seals.

Axle shafts are retained in solid rear axle housings in two ways:

- With C-locks if the axles ride on roller bearings
- With retainer plates if the axles ride on ball or roller bearings.

Because the two types require different service procedures, determine which type of axle you are servicing before you begin work. Begin by positioning the vehicle on a lift. Ensure that components will not bend under the weight of the car when it is raised. See Chapter One of this *Shop Manual* for lifting a vehicle. Remove the wheels. Next, remove the brake drums, or brake calipers and rotors, as described later in this chapter.

Servicing Bearings and Seals — C-Lock Axles

To service the wheel bearing and grease seal on a solid rear axle retained by a C-lock:

1. Clean dirt from around the differential cover. Remove the cover and drain the axle lubricant into a suitable drain pan.
2. Rotate the differential carrier until the pinion shaft lock bolt is accessible. Remove the lock bolt and pinion shaft.
3. Without rotating the axle, push it toward the center of the car until the C-lock clears the counterbore in the differential side gear. Remove the C-lock from its groove, figure 12-44.
4. Return the axle to its original position, and replace the pinion shaft and lock bolt to hold the differential carrier pinion gears in position.

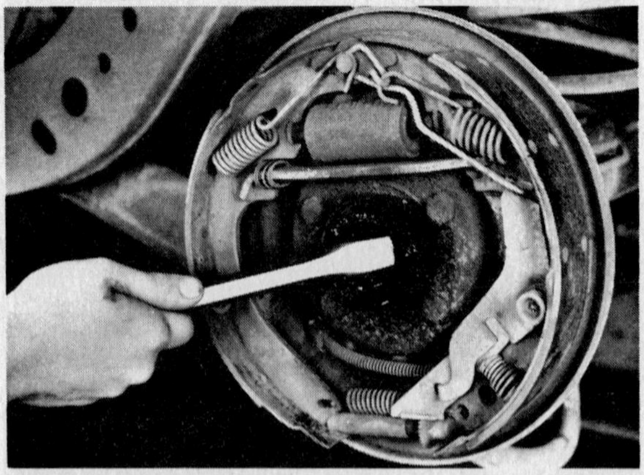

Figure 12-45. Remove the seal from the axle housing with a seal puller or screwdriver.

5. Pull the axle straight out of the housing by hand, making sure it does not drag on the wheel bearing.
6. Pry the seal from the axle housing with a seal puller or large screwdriver, figure 12-45.
7. Use a slide hammer to remove the bearing from the axle housing, figure 12-46.
8. Clean the axle, and inspect the surface where the bearing and seal ride. It should be smooth, dull gray in color, and free of any pitting, indentations, scratches, or flaking metal. Replace the axle and bearing if you find damage.
9. Clean and inspect the bearing as described earlier. Replace the bearing if any problems are present.
10. Clean the axle housing bore where the bearing and seal fit, and check for damage.
11. Lubricate the outside of the bearing with wheel bearing grease. Align the bearing squarely with the opening in the axle housing. Drive the bearing into place with a suitably sized socket or bearing driver that contacts only the outer race of the bearing. The pitch of the sound made when striking the driver will change when the bearing is seated.
12. Install a new grease seal into the axle housing with a seal driver.
13. Lubricate the bearing rollers, seal lip, and axle sealing surface with axle lubricant. Carefully slide the axle into the housing. Do not allow the splines on the axle to damage the seal or bearing during installation. It may be necessary to rotate the axle slightly to align its splines with those in the differential side gear.
14. Remove the lock bolt and pinion shaft from the differential carrier, and push the axle inward until the C-lock groove clears the counterbore in the differential side gear. Do not rotate the axle during this operation or the differential carrier pinion gears may be dislodged, making reassembly more difficult.

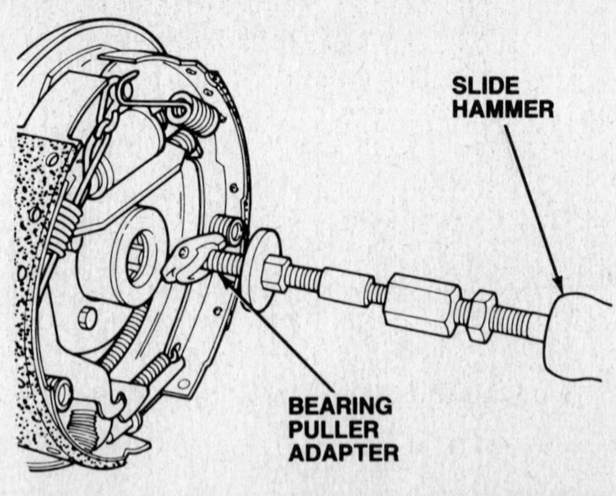

Figure 12-46. Use a slide hammer to pull the bearing from the axle housing.

15. Insert the C-lock into the groove in the inner end of the axle. Pull the axle outward to seat the C-lock in the counterbore of the differential side gear.
16. Install the pinion shaft and lock bolt. Tighten the lock bolt to the torque specified by the manufacturer.
17. Install the differential cover using a new gasket or silicone sealer, as specified by the vehicle manufacturer.
18. With the axle in its normal operating position, fill the differential to the bottom edge of the fill hole with the lubricant specified by the vehicle manufacturer.

Servicing Bearings and Seals — Retainer Plate Axles

On most solid rear axles where the wheel bearing and grease seal are secured by a retainer plate, service the bearing and seal as follows:

1. After removing the necessary brake hardware, as described later in this chapter, remove the nuts that hold the retainer plate to the backing plate, figure 12-47.
2. Slide the retainer plate off the studs. Reinstall two of the nuts finger tight to keep the brake backing plate in place.
3. Use a puller or slide hammer to remove the axle shaft and bearing from the axle housing.
4. Remove the seal from the axle housing.
5. Clean the axle housing bore where the bearing and seal fit, and check for damage.

Figure 12-47. You can loosen the retainer plate nuts through a hole in the axle flange.

6. If the wheel bearing is unsealed, clean and inspect it as described earlier. If the bearing is sealed, turn it by hand to feel for roughness, listen for noise, and make sure the metal side plates are not dented. Replace the bearing if you question its condition.
7. Remove the retaining ring that secures the bearing on the axle using one of these methods, depending on the hardness of the ring and how tightly it fits.
   - Mount the axle in a vise, and use a hammer and chisel to notch the retaining ring in several places around its outer edge. The ring may loosen enough to remove it from the axle.
   - Use a bearing splitter to grip the bearing, and push it off with a hydraulic press or puller. Alternatively, use a grinder to cut away the ring almost to the axle. Use a hammer and chisel to split the ring and remove it from the axle. Take care not to damage the axle in any way.
8. Using a bearing splitter or other adapter that contacts the *inner* race, press the old bearing off the axle. Never attempt to remove a wheel bearing by pressing against the outer race. The bearing may come apart and cause an injury.
9. Remove the retainer plate and inspect the area on the axle where the bearing and seal make contact. It should be smooth and have a dull gray appearance. Replace the axle if it is rough or worn.
10. Clean all traces of the old gasket from the retainer plate and the brake backing plate. If the retainer plate contains an outer seal, install a new seal and lubricate the seal lip with wheel bearing grease. Replace the retainer plate, and install a new gasket on the axle.
11. If the new wheel bearing is unsealed, pack it with grease as described earlier.
12. Press the new wheel bearing onto the axle. Apply pressure only to the inner race of the bearing.
13. Press a new bearing retaining ring onto the axle. Never attempt to press a bearing and retaining ring onto the axle at the same time.
14. Install a new seal in the axle housing with a seal driver, and lightly lubricate the lip of the seal with axle lubricant.
15. Carefully slide the axle into the housing so as not to damage the seal. Rotate the axle as necessary to align its splines with those in the differential side gear.
16. Gently tap the end of the axle with a hammer until the bearing is fully seated in the axle housing.
17. Remove the two nuts holding the backing plate in place. Slide the retainer plate onto the axle housing studs.
18. Install the retainer plate nuts, and tighten them to the torque specified by the vehicle manufacturer.

BASIC DRUM BRAKE SERVICE

Brake shoes may contain asbestos, which is a health hazard for technicians. Chapter One of this *Shop Manual* describes safe methods for servicing brake systems that use asbestos brake shoes. Use the information in this section as reference material when you service the wheel bearings.

Removing Brake Drums

Before you remove a brake drum, you must loosen the parking brake cable and service brake adjustment. You may have to break accumulated rust and corrosion to remove the drum. Whenever you remove a drum, mark it with an L (left) or an R (right) so you can replace it in the same position.

The removal procedures differ for fixed and floating drums, and they are described in the following sections.

Removing fixed drums

Fixed brake drums that are cast as a unit with the hub assembly, are common on the rear axles of FWD cars and the front axles of older RWD cars with front drum brakes. There are two kinds of fixed drums: those whose hubs contain wheel bearings, and those whose hubs

Figure 12-48. You may need a puller to remove a fixed drum if the inner wheel bearing is stuck on the spindle.

have a tapered opening that fits directly onto the axle. Both types are held in place by a single large retaining nut.

In most applications, the nut that retains a fixed brake drum also secures the wheel bearings in the hub. When tapered-roller bearings are fitted, the retaining nut is used to adjust bearing end play and is typically finger tight or torqued to only a few inch-pounds. If ball bearings are used, a spacer and shim pack between the inner and outer bearings sets the bearing clearance, and the retaining nut is tightened to a moderately high torque.

Removing fixed drums with bearings

To remove a fixed drum that contains wheel bearings:

1. Use a pair of dust cap pliers to remove the cap from the center of the hub and expose the retaining nut.
2. Remove the retaining nut. Most cars have a castle nut and cotter pin. Remove the pin along with any other locking devices. Then unscrew the nut. Some imported cars have a split nut with a pinch bolt. Loosen the bolt, and unscrew the nut.
3. Pull outward on the drum and slide it off the spindle. Take care not to let the thrust washer and outer wheel bearing fall on the ground as they clear the end of the spindle. Also, do not drag the inner wheel bearing across the retaining nut threads on the spindle.

Sometimes, the inner race of the inner wheel bearing will stick or have a light press fit on the spindle. In this case, the drum will be difficult to remove unless you use a puller, figure 12-48. When doing this, the inner bearing may remain on the spindle, forcing the grease seal out of the hub. Once the drum is off the car, use a puller or a pair of pry bars to carefully remove the inner bearing and grease seal.

Inspect the grease in the hub and on the wheel bearings. If the grease appears to be in good condition, set the drum on the bench, open side down, and cover the outer bearing and hub opening so the bearings will not become contaminated. If the grease appears old and dirty, repack the wheel bearings before you reinstall the drum.

Removing taper-hub fixed drums

Some cars and light trucks have fixed drums that install directly on the rear driven axles. The hubs of these drums have a tapered opening that fits over a matching taper on the axle. A groove in the hub fits over a locating key in the axle to prevent drum slippage. Once the drum retaining nut is removed, you will need a special puller to remove the drum because it fits tightly on the axle.

The retaining nuts of fixed drums with tapered hubs are tightened to a high torque and require a great deal of force to remove or install. Unless you take certain precautions, this force may damage the axle and drivetrain, or cause personal injury. Always loosen the retaining nuts, or do final tightening, with the car on the ground and the weight of the vehicle on the wheels. Firmly apply the parking brake, or have an assistant apply the service brakes to reduce torque on the axle and drivetrain as you remove or install the retaining nut.

To remove a fixed drum that has a taper fit on the axle:

1. Remove the cotter pin from the drum retaining nut.
2. Use an impact wrench or large breaker bar to loosen the retaining nut.
3. Raise and properly support the vehicle. Remove the wheels from the brake drums.
4. Unscrew the retaining nut from the axle. Loosen the parking and service brake adjustments to ensure that the drum will clear the brake shoes.
5. Attach the drum puller. Use a hammer and the special lever to tighten the shaft against the axle until the puller is under heavy pressure.
6. Strike the end of the shaft two or three times with a three- to five-pound (2-kg) hammer, figure 12-49.
7. Check the pressure on the puller shaft. If it is reduced, the drum is moving. Retighten the shaft and strike it several more times with the hammer. Continue to tighten and strike the shaft until the drum pops free.

Figure 12-49. Strike the puller shaft with a hammer to dislodge the drum from the axle.

If using the puller and hammer does not loosen the drum, and you are trained to use an oxy-acetylene torch, heat the hub where the drum fits on the axle, but do not heat the parts until they are red-hot. Again strike the shaft and check for drum movement. Continue to heat the hub. Tighten and strike the puller shaft until the drum pops free.

Removing floating drums

Floating brake drums are held in place by the wheel and lug nuts during normal operation. However, they may also be retained by speed nuts over the wheel studs, or bolts or screws threaded through the drum into the hub or axle flange. To remove speed nuts, grasp them with a pair of pliers and unscrew them from the studs. On most cars, you can discard the speed nuts because they are unnecessary once the car has left the factory. However, certain Ford cars have high shoulders on the wheel studs that can catch the drum and hold it in a cocked position when the wheel is installed. To prevent the possibility of drum damage in these applications, replace the speed nut after completing brake work. To remove bolts or screws that retain a drum, use the appropriate tool. Set aside bolts or screws so you can reinstall them when the drum is replaced.

After removing retaining devices, the drum should move freely on the hub or axle and slip off over the brake shoes. If necessary, loosen the parking and service brake adjustments until the drum will clear the shoes. If the drum does not move on the hub or axle after you have loosened the brake adjustment, the drum is rusted or corroded in place.

If the drum is frozen to the hub assembly on a non-driven axle, it is often faster and easier to remove the drum and hub as a unit, like a fixed drum, rather than attempt to separate the drum from the hub. However, if the drum is frozen on a driven axle, or if you must remove it from the hub for replacement, use this procedure:

1. Use a pointed probe to scribe the joint where the drum mounts on the hub or axle. This breaks the surface tension, and like a glass cutter, helps crack the joint free.
2. Strike the outer edge of the drum several times with a three- to five-pound (2-kg) hammer at approximately a 45-degree angle. A "dead-blow" urethane hammer loaded with lead shot works very well for this purpose. Striking the drum around the bolt circle area with a hammer may also loosen it.
3. If the drum does not come free, spray penetrating oil around the stud holes and the joint where the drum mounts on the hub or axle. Allow the oil to soak in for a few minutes. Then strike the drum several more times with the hammer.
4. If the drum still does not come free, and you are trained to use an oxy-acetylene torch, heat the drum in the area where it mounts on the hub or axle. Strike the drum again with the hammer until it comes free.
5. Once the drum is off the car, clean all traces of penetrating oil from the drum and hub or axle flange to prevent contamination of the brake linings.

The methods of drum removal described above should be exercised with caution. Excessive hammer blows, puller force, or heat can warp a drum in the process of removing it. After removing a drum, check it for damage and distortion before you reinstall it. In addition, high heat on a hub that contains wheel bearings can melt the grease and lead to seal and bearing failure. Be sure to repack bearings and replace grease seals before installing the drum.

Brake Adjustment

Most drum brakes have automatic adjusters that eliminate the need for periodic manual adjustment. As the linings wear, the adjusters automatically reposition the shoes farther outward on the backing plates to maintain the

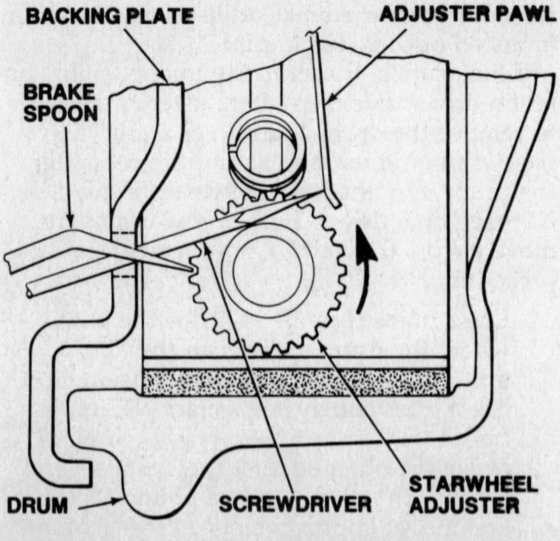

Figure 12-50. Use a screwdriver to push the automatic adjuster pawl out of the way when the access hole is in the backing plate.

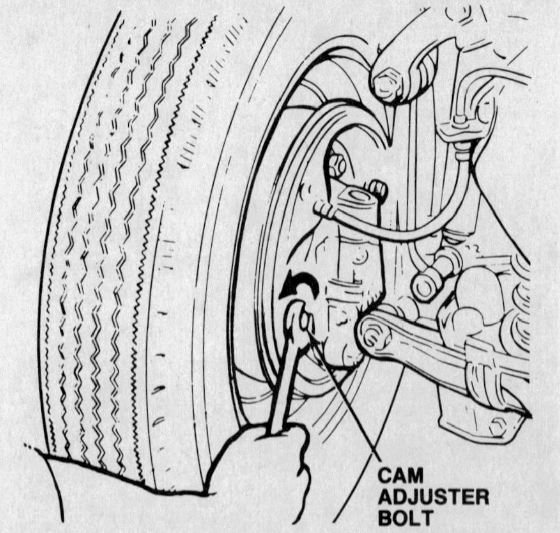

Figure 12-51. Rotate a cam adjuster bolt with a wrench to adjust the brakes on some vehicles.

proper lining-to-drum clearance. Manual adjustment of the brakes on cars with automatic adjusters is unnecessary unless scoring makes the brake drum difficult to remove, the brake shoes are replaced, or the automatic adjusters fail.

Drum brakes without automatic adjusters were once used on all cars, and can still be found on the rear axles of some models. These brakes require periodic manual adjustment of the lining-to-drum clearance. The frequency of these adjustments varies with the type of friction material, the size of the brakes in relation to the weight of the car, the type of roads on which the car is driven, and the habits of the driver.

The exact procedure for adjusting brakes varies depending on whether the vehicle uses starwheel, cam, or wedge adjusters. Refer to the vehicle manufacturer's service manual for specific instructions. However, there are common steps that apply for most brake adjustments:

1. Raise and properly support the vehicle so the wheels hang free.
2. Try to turn each tire and wheel. Unless the brake adjusters or parking brake has been overtightened, the wheels should rotate with no apparent drag. Loosen the adjustment, if needed, and try to rotate the tire again. If loosening the adjustments does not free up the wheels, locate and repair the problem before you adjust the brakes.
3. Reduce the lining-to-drum clearance by adjusting the starwheel, cam, or wedge adjuster.
   a. For starwheel adjusters, remove the plug covering the access hole in the backing plate or brake drum, and insert a screwdriver or other tool through the hole to rotate the starwheel, figure 12-50. When the brake starts to drag, loosen the adjustment until the wheel just turns free, or rotate the starwheel the number of clicks specified by the vehicle manufacturer.
   b. For cam adjusters, rotate the hex head of the adjuster stud that extends through the backing plate, figure 12-51. Some adjuster studs have arrows on them to indicate the tightening direction.
   c. For wedge adjusters, adjust both brake shoes at the same time by rotating a single adjusting bolt that extends through the backing plate, figure 12-52.
4. As you tighten the adjuster, rotate the wheel in the direction that energizes the shoe or shoes being adjusted. This keeps the shoes centered in the drum and properly seated against their anchors. When adjusting a dual-servo or double-leading brake, rotate the wheel in the direction of forward travel. On a leading-trailing brake with a single adjuster, rotate the wheel

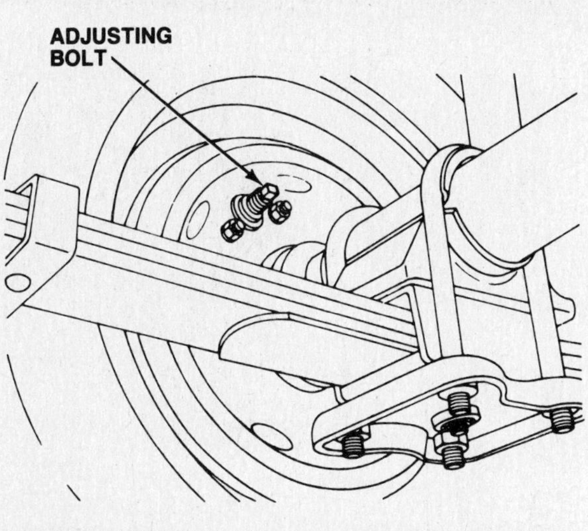

Figure 12-52. To adjust brakes that use a wedge-type adjuster, rotate the adjusting bolt in a clockwise direction to reduce the lining-to-drum clearance.

Figure 12-53. Use a clamp to bottom the piston in its bore.

back and forth as you tighten the adjuster. If a leading-trailing brake has a separate adjuster for each shoe, rotate the wheel in the direction of forward travel to adjust the leading shoe and in the direction of reverse travel to adjust the trailing shoe.

5. Continue to tighten the adjuster until there is a consistent heavy drag on the wheel or until the brake locks. Back off the adjustment the amount specified by the vehicle manufacturer, or until the wheel just turns free. Repeat the adjustment procedure at the other wheels.

Figure 12-54. After removing the protective boots, unscrew and remove the caliper mounting bolts.

BASIC DISC BRAKE SERVICE

Brake pads may contain asbestos, which is a health hazard for technicians. Chapter One of this *Shop Manual* describes safe methods for servicing brake systems that use asbestos brake pads. Reference this section as needed when you are servicing wheel bearings. For in-depth brake service procedures, refer to the vehicle manufacturer's service manual.

Removing and Installing Calipers

Before removing the calipers, remove some of the brake fluid from the master cylinder reservoir with a syringe. When removing brake fluid from the reservoir, wear safety glasses to prevent eye injury from accidental fluid spray. Also, do not get brake fluid on the vehicle finish. To remove a caliper:

1. Raise the car until the wheels hang free. Remove the tire and wheel assemblies to gain access to the calipers.
2. Remove the protective boot from the caliper, if there is one.
3. On some vehicles, you must use a C-clamp to bottom the caliper piston in its bore, figure 12-53, before removing the caliper.
4. Unscrew and remove the caliper mounting bolts, figure 12-54. Remove anti-rattle springs, figure 12-55, and disconnect the parking brake cable from the caliper lever, figure 12-56, if they are used.

Figure 12-55. After unbolting the retaining clips, remove the anti-rattle springs.

Figure 12-56. Remove the parking brake cable clip and pin with pliers.

5. Lift the caliper off the rotor. Use a piece of wire to hang the caliper from a suspension component, figure 12-57, so there is no strain on the brake hose while you are servicing the wheel bearings.
6. After servicing wheel bearings, position the caliper over the rotor.
7. Install the caliper mounting bolts, and torque them according to the manufacturer's specifications. Install the parking brake cable and anti-rattle springs if you removed them earlier.

Figure 12-57. Use wire to hang calipers from the suspension while you are working on the wheel to prevent brake line damage. Never allow the caliper to hang by the brake line.

8. Install the protective rubber boots, if they are used.
9. Adjust the brakes, as described later in this chapter.

Removing and Installing Rotors

After removing the caliper, removing the rotor is not difficult. Remove the rotor as follows:

1. Position the vehicle on a lift, and ensure that components will not bend under the weight of the car when it is raised. See Chapter One of this *Shop Manual* for lifting a vehicle. Remove the tire and wheel.
2. Remove the brake caliper, as described earlier in this chapter, and use a piece of wire to suspend the caliper from a suspension component. Do not allow the caliper to hang by the brake line.
3. If the rotor and hub are an integral component, remove the grease cap from the center of the hub. Pull the cotter pin from the castle nut, and remove the castle nut and thrust washer.
4. If the rotor is separate from the hub, simply unbolt it from the hub.
5. Remove the outer wheel bearing. Pull the rotor off the spindle, being careful to not drag the inner wheel bearing across the spindle.

6. Remove the inner wheel bearing from the hub, using the procedures described earlier in this chapter.

Brake Adjustments

Disc brakes are adjusted by ensuring that fluid is available to the caliper pistons. Fill the master cylinder reservoir, and apply the brake pedal. The pedal will probably go all the way to the floor. Stroke the pedal as many times as necessary to get a high, firm brake pedal. When this happens, the caliper pistons have moved out of their bores far enough to bring the new brake pads into firm contact with the rotor.

Note: Failing to make this initial brake adjustment may cause an accident because the brakes may not work the first time you step on the pedal.

If you cannot get a high, firm pedal, bleed the brakes. If bleeding fails to restore a good brake pedal, and no other cause can be found for the problem, it is possible that rust and corrosion in the caliper bore have disturbed the piston's seal. When this happens, air continually enters the hydraulic system, and the caliper soon starts to leak fluid. The calipers should be rebuilt if this is the case.

Once the pedal feels firm, recheck the fluid level in the master cylinder, and fill it if needed. Turn the ignition ON, release the parking brake, and make sure the brake system warning light does not come on. Install the wheels, lower the vehicle to the ground, and test drive the car to make sure the brakes operate properly.

Chapter

13

Wheel and Tire Runout, Balance, and Rotation

Some vehicle handling problems are caused by excessive wheel and tire runout or imbalance. Excessive wheel and tire runout or imbalance affects ride quality and tire wear, even if wheel alignment angles are within specifications. Check and correct runout and balance problems, as described later in this chapter, before performing a wheel alignment. Recommended tire rotation procedures are also included in this chapter.

Usually, you cannot detect the cause of runout or imbalance with a visual inspection. If a wheel is out-of-round by a fraction of an inch (runout) or if there are small differences in weight distribution around the wheel (imbalance), the tire and wheel will vibrate at moderate highway speeds. When looking for the cause of a vibration problem, first check runout, then balance the wheel and tire assembly. Finally, rotate tires periodically to prolong tire life.

RUNOUT

The procedures in this section describe how to check for radial and lateral wheel and tire runout, figure 13-1, and how you can repair some causes of runout. As discussed in Chapter Fourteen of the *Classroom Manual*, radial runout is the result of an off-center wheel and tire assembly that does not rotate in a true circle. This causes the wheel to hop when rotating at moderate highway speeds. Radial wheel runout is measured at the wheel bead seat, and total radial runout is measured at the tire tread. A tire and wheel assembly has lateral runout when it does not rotate in a single vertical plane, causing the wheel to wobble when rotating at high rpm. Lateral wheel runout is measured at the wheel flange, and total lateral runout is measured at the tire sidewall.

Runout is almost always measured with the wheel and tire mounted on the vehicle, which checks the condition of the entire assembly. You can remove the wheel and tire from the vehicle, mount it on a wheel balancing machine, and take the runout readings there. These readings will tell you if runout problems are due to the tire and wheel or if they are due to other problems on the vehicle, such as misadjusted wheel bearings or a bent hub.

Because runout is measured in fractions of an inch (or millimeters) with a dial indicator, check and correct all other sources of possible error before taking a runout measurement:

- Check tire inflation pressures, and adjust them if needed. Procedures for doing this are in Chapter Twelve of this *Shop Manual*.

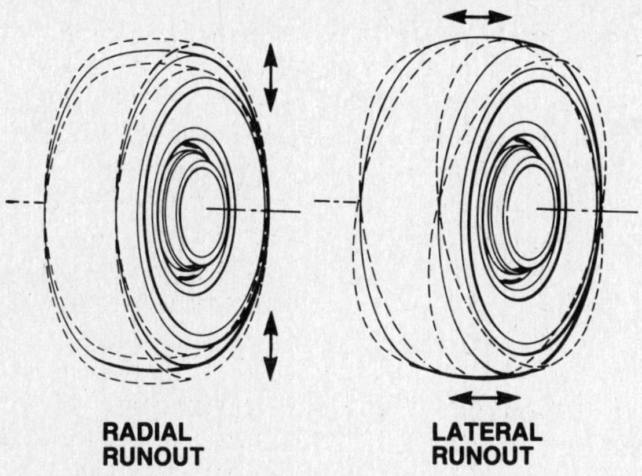

Figure 13-1. A tire and wheel assembly with excessive radial runout will hop. If the assembly has excessive lateral runout, it will wobble.

- Check and adjust wheel bearing play. Chapter Twelve of this *Shop Manual* has procedures for adjusting wheel bearings.
- Check lug nut torque against the vehicle manufacturer's recommendations. Overtightening these nuts may warp the hub or rotor.
- Be sure that the wheel lies flat against the hub. If a technician does not tighten the lug nuts in a star pattern, the wheel may not lie flat and runout problems will result.
- Inspect the tire and wheel for obvious signs of damage that could cause excessive runout, such as bent wheels or abnormal bulges in the tire sidewall.
- Clean away dirt and rocks from the wheel and tire, which may interfere with the dial indicator and affect runout readings.
- Before measuring runout, drive the vehicle for a few minutes to warm the tires. When a vehicle has been parked for a few hours, the tires develop flat spots on the tread. Warming the tires evens out these areas.

After taking runout measurements, as described later, compare the results with the manufacturer's specifications. Ideally, runout should be zero, but this is not likely. For this reason, vehicle manufacturers develop runout tolerances for each model based on vehicle weight, tire size, and tire type. As a rule of thumb, radial or lateral runout of 0.030 inches (0.760 mm) is perceptible to a sensitive driver, 0.060 inches (1.525 mm) to most drivers, and 0.090 inches (2.285 mm) to virtually every driver, as mentioned in Chapter Fourteen of the *Classroom Manual*.

If the vehicle's handling problems are like those described for runout, but runout measurements are within specifications and the tire and wheel are balanced, check the problem further. One possible cause is a tire construction problem. For example, if one part of the tire is stiffer than the rest, the effects of the forces between the tire and the road will not be even as the tire rotates. The result is bouncing, similar to the bounce caused by radial runout. The easiest way to check a possible tire construction problem is to substitute another tire for the one that is possibly damaged, and see if the problem goes away. If the problem persists after switching tires, other parts of the driveline may be faulty, such as a bent axle, brake rotor, or hub.

Measuring Radial Runout

After checking for other possible sources of runout, measure radial wheel runout and total radial runout as follows:

1. Park the vehicle with the brake disengaged and the transmission in neutral.
2. Block the vehicle's rear wheels, and use a floor jack to raise the front wheels off the ground about two inches. You should be able to rotate the front tires with your hands.
3. Support the dial indicator on a stand near the wheel and tire, and place the indicator tip against the bead seat or other horizontal surface on the wheel, figure 13-2. Be sure that the dial indicator and stand do not tip over when you rotate the tire.
4. Zero the dial indicator.
5. Rotate the tire until the lowest reading shows on the indicator. Zero the dial indicator at this low point.
6. Rotate the tire again, and read the dial indicator at the same time. Note if there are areas on the wheel where *radial wheel runout* exceeds the manufacturer's specifications.
7. Reposition the indicator tip on the center of the tire tread, figure 13-3, and zero the indicator.
8. Rotate the tire until the lowest reading shows on the indicator. Zero the indicator at this low point.
9. Rotate the tire again, and read the dial indicator at the same time. Note if there are areas on the tire where *total radial runout* exceeds the manufacturer's specifications.

As a general rule, total radial runout at the front and rear wheels on a vehicle with radial tires should not exceed 0.060 inches (1.5 mm). Wheel runout should not exceed 0.030 inches (0.75 mm) for alloy wheels and 0.040 inches

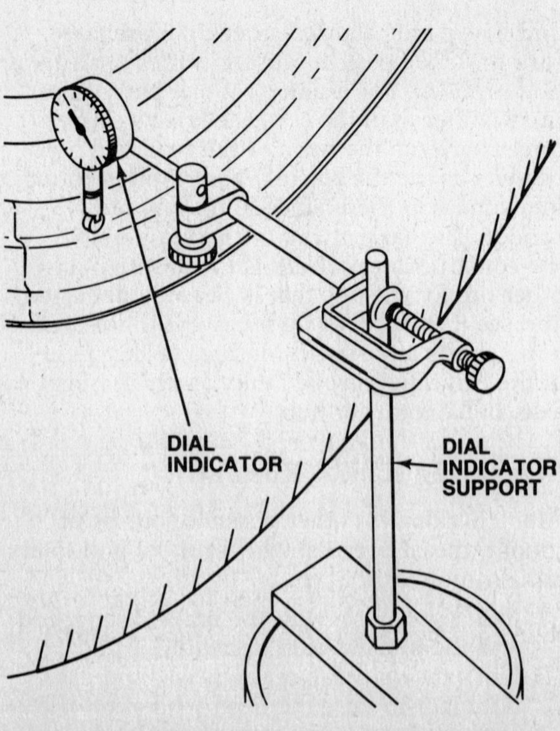

Figure 13-2. Check wheel radial runout with the dial indicator tip against the bead seat or other horizontal surface of the wheel.

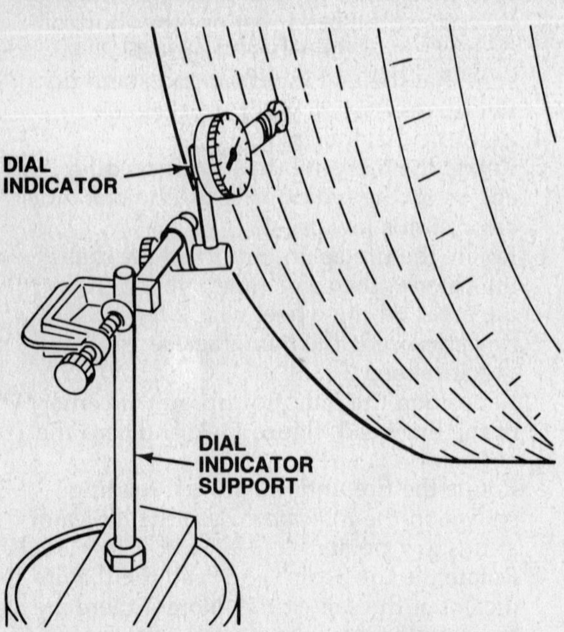

Figure 13-3. Check total radial runout with the dial indicator tip against the center of the tread.

(1.0 mm) for steel wheels, but you should always check readings against the manufacturer's specifications. If the readings are out of

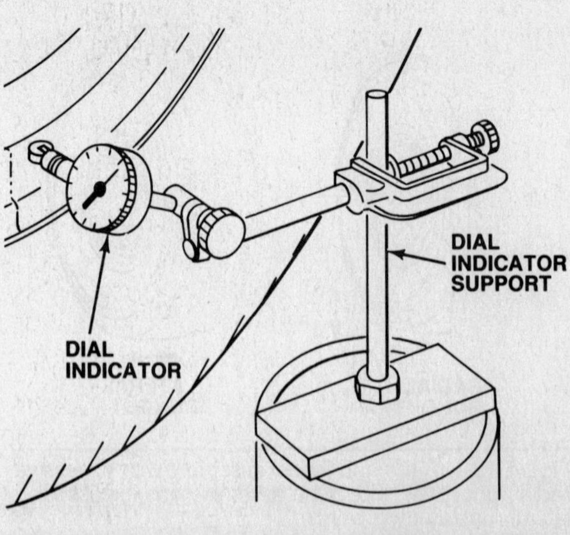

Figure 13-4. Check lateral runout with the dial indicator tip against the wheel flange or other vertical surface on the wheel.

specifications, you have likely found the source of wheel vibration problems. Ways to reduce runout are described later in this chapter.

Measuring Lateral Runout

After checking for other possible sources of runout, as described earlier in this chapter, measure lateral wheel runout and total lateral runout as follows:

1. Park the vehicle with the brake disengaged and the transmission in neutral.
2. Block the vehicle's rear wheels, and use a floor jack to raise the front wheels off the ground about two inches. You should be able to rotate the front tires with your hands.
3. Support the dial indicator on a stand near the wheel and tire, and place the indicator tip against the wheel flange or other vertical surface on the wheel, figure 13-4. Be sure that the dial indicator and stand do not tip over when you rotate the tire.
4. Zero the dial indicator.
5. Rotate the tire until the lowest reading shows on the indicator. Zero the dial indicator at this low point.
6. Rotate the tire again, and read the dial indicator at the same time. Note if there are areas on the wheel where *lateral wheel runout* exceeds the manufacturer's specifications.
7. Reposition the indicator tip on the tire sidewall, figure 13-5, and zero the indicator.

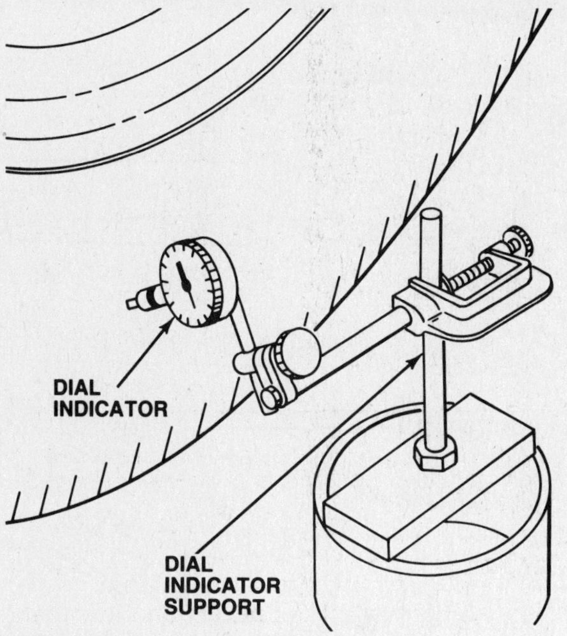

Figure 13-5. Check total lateral runout with the dial indicator tip against the tire sidewall.

8. Rotate the tire until the lowest reading shows on the indicator. Zero the indicator at this low point.
9. Rotate the tire again, and read the dial indicator at the same time. Note if there are areas on the tire where *total lateral runout* exceeds the manufacturer's specifications.

In general for radial tires, total lateral runout should not exceed 0.080 inches (2.0 mm). Wheel lateral runout should not exceed 0.030 inches (0.75 mm) if the vehicle has alloy wheels and 0.045 inches (1.15 mm) if it has steel wheels. Be sure to check the readings against the manufacturer's specifications. If the readings are out of specifications, you have likely found the source of wheel vibration problems. Methods to reduce runout are described in the next section.

Reducing Radial and Lateral Runout

You may be able to reduce radial runout by changing the position of the wheel on the hub. Simply remove the tire and wheel assembly, and move the assembly over two lug holes. Then tighten the assembly using the manufacturer's recommended torque and tightening sequence. Recheck runout and test drive the vehicle, checking for vibration.

To help pinpoint the cause of vibration, you can switch the suspected tire and wheel assembly to another hub. If the vibration remains with the tire and wheel assembly, you have found the defective parts. However, if vibration continues at the original position, even after *balancing* the tire and wheel, a bent hub or axle shaft may be the cause.

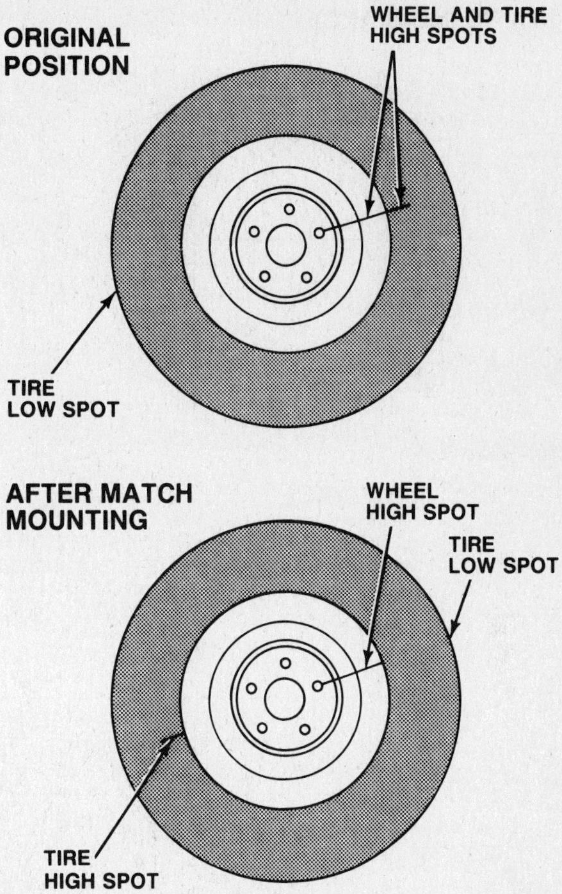

Figure 13-6. Match mounting involves positioning the high spot on the wheel 180 degrees from the high spot on the tire.

If total runout measurements are out of specifications, you may be able to correct vibration problems by changing the position of the tire on the wheel, called "match mounting", figure 13-6. To match mount the tire and wheel, measure runout at the wheel, and mark the area where the runout reading is the highest. Then measure total runout, and mark the area on the tire where runout is the highest. Remove the tire and wheel assembly from the vehicle, and dismount the tire, as described in Chapter Twelve of this *Shop Manual*. Remount the tire, positioning the high spot on the wheel 180 degrees from the high spot on the tire. Measure total runout again, and compare the readings with the manufacturer's specifications.

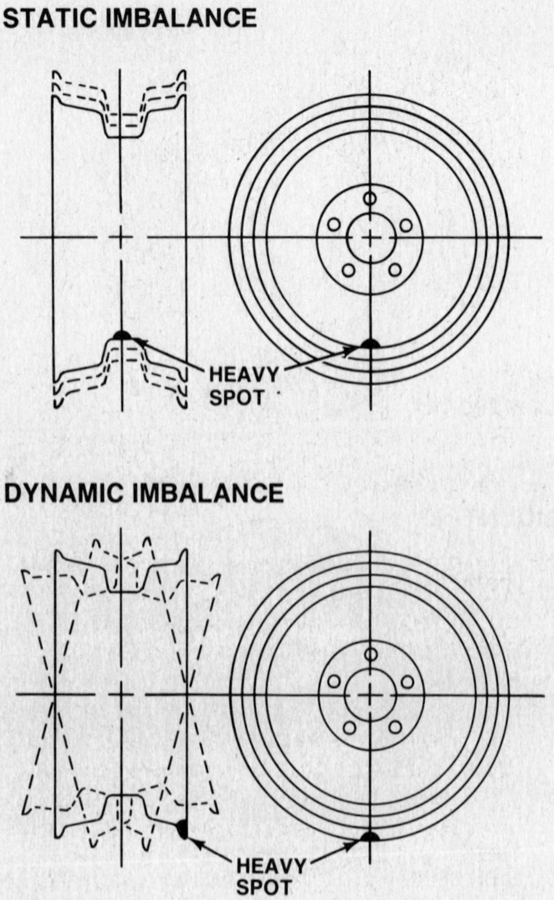

Figure 13-7. If the tire or wheel has a heavy spot, the wheel will not rotate smoothly. Static and dynamic wheel balancers detect heavy spots on the tire and wheel, and most balancers tell the technician where to place weights to offset this imbalance.

If you cannot correct total runout with match mounting, and runout at the wheel is out of specifications, the wheel is bent and should be replaced. If wheel runout is within specifications, but total runout remains out of specifications even after match-mounting, the tire is likely damaged and should be replaced.

BALANCING

If you work at a service station or a wheel-and-tire store, you will likely balance wheels quite a few times. Wheel balancing is a part of every tire replacement procedure, and is recommended as part of a wheel alignment as well.

Balancing ensures that weight is distributed evenly around the tire and wheel so that the wheel rotates smoothly, figure 13-7. Wheel balancers detect heavy spots on the tire and wheel assembly, and most indicate where you should clip or glue weights on the wheel to

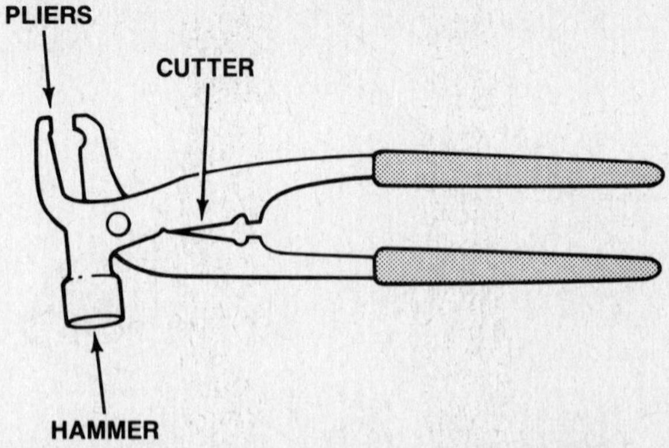

Figure 13-8. This tool is used to remove and install clip-on weights. Use the hook and pliers to remove weights and the small hammer on the other side to install them.

counteract this weight imbalance. When you balance a tire and wheel, never add more than six ounces (170 grams) of weight. If the wheel requires that much weight for balance, it is likely damaged and should be replaced.

There is a special tool that you should use for removing and mounting clip-on weights, figure 13-8. The tool is like a pair of pliers with hooked jaws for grasping weights and prying them off the wheel. The other side of the tool has a small hammer for mounting clip-on weights. Adhesive-backed weights come in a long strip, figure 13-9, and you can cut off as much weight as you need. Simply pull the paper or plastic backing tape off the weights to expose the adhesive, and stick the weights in place. Use coated weights when balancing aluminum wheels. This prevents electrolysis, which is an electrochemical reaction that occurs when you place two different metals together for a long period of time. Electrolysis will ruin the appearance of the wheel.

To prepare a tire and wheel assembly for balancing, you must:

- Clean caked-on mud from the tire and wheel. If you neglect to do this, the wheel will not be balanced when the mud comes off. Remove stones or other objects imbedded in the tread for the same reason.
- Check that the brake shoes or pads are not dragging on the drum or disc. This applies only if you are balancing the wheel on the vehicle. After raising the vehicle, spin the wheel by hand. Listen for a scraping sound, which indicates brake drag.
- Check the wheel for old weights. Remove and discard them. You should not reuse clip-on weights because the spring clips

Figure 13-9. These weights have adhesive on the back. After removing the backing tape from the weights, stick them on the wheel.

are weak after being used once. Do not reuse adhesive-backed weights either. The glue on these weights will not stick a second time.

As discussed in Chapter Fourteen of the *Classroom Manual,* there are two types of wheel balancing methods — static and dynamic, figure 13-7. The object of static balancing is to distribute weight evenly around the wheel hub to prevent wheel hop. The object of dynamic balancing is to distribute weight evenly between the inside and outside of the tire and wheel to prevent wobble.

Most shops use computerized wheel balancers that check static and dynamic balance, and we will cover that procedure in detail. Some shops use balancers that allow you to balance the wheels while they are mounted on the vehicle, which we will also discuss in this chapter. A few shops use the bubble balancer, which is a simple procedure that checks only static balance.

Bubble Balancer

The bubble balancer is for *static* balancing only, and is not a widely used method. Balance a tire and wheel assembly using a bubble balancer, figure 13-10, as follows:

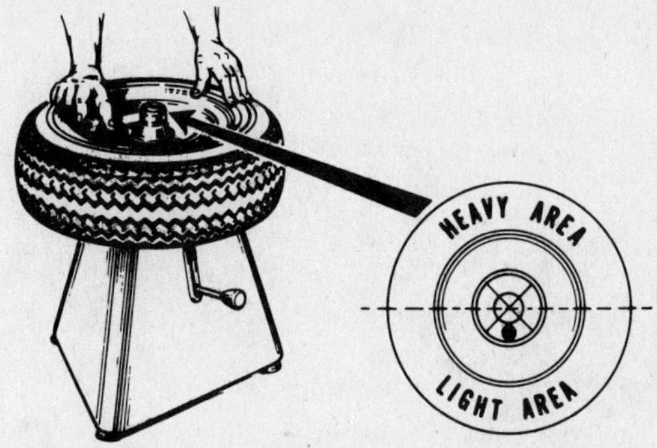

Figure 13-10. The bubble balancer is an "old-fashioned" wheel balancing method that is not used in many shops because it is capable of static balancing only. Spin balancers have almost completely replaced the bubble balancer.

1. Remove the tire and wheel assembly from the vehicle using the procedures in Chapter Twelve of this *Shop Manual,* if needed.
2. Use the guidelines described earlier in this chapter to prepare the tire and wheel for balancing. This includes removing old wheel weights and caked-on mud from the wheel.
3. Before mounting the wheel on the balancer, check the balancer's level indicator to be sure that the bubble is centered in the cross hairs. If it is not, use the adjusters at the base of the balancer to level it and center the bubble.
4. Mount the tire on the balancer. Move the balancer switch to the ON position.
5. Check the bubble. If it moved out of the cross hair region, the wheel is out of balance.
6. Rotate the tire and wheel until the lightest side of the tire is closest to you. (The bubble will also be closest to you.) Then turn the balancer off.
7. Place weights on top of the wheel on the lightest side of the tire. Turn the balancer on to check the position of the bubble. Then turn the balancer off, and add more weight if needed. Continue doing this until the bubble is centered in the cross hairs.
8. Use chalk to mark where the weights should be placed on the wheel, and how much weight should be placed there.
9. Remove the weights, and remove the wheel and tire from the balancer.
10. Mount the weights at the chalk marks. Rather than adding all of the weight to the outside of the wheel, divide the weight in half, and add half of the weight to the outside wheel flange and half of the weight to the inner flange. Doing this distributes the weight evenly so that dynamic balance will

Figure 13-11. This is a typical spin balancer. Spin balancers balance the tire and wheel both statically and dynamically.

not be upset. Use clip-on, adhesive-backed, or coated weights, as described earlier in this chapter.

11. Mount the wheel and tire assembly on the balancer, and check the bubble's position in the cross hairs. Adjust the position of the weights if needed.

Spin Balancers

Most shops use spin balancers because they are more accurate than bubble balancers. Besides checking static balance, spin balancers also check dynamic balance, which is especially important for stability in performance vehicles. There are two types of spin balancers — off-the-vehicle and on-the-vehicle. Procedures for using both types are described in the following sections.

Off-the-vehicle spin balancers

Off-the-vehicle spin balancers are the most commonly used type of balancer, figure 13-11. They are easy to operate, compact, and accurate. Any technician who works in a shop and changes tires will eventually use a spin balancer.

As discussed in Chapter Fourteen of the *Classroom Manual*, spin balancers balance the tire and wheel while they are in motion. A computer detects wheel imbalance and determines where weights should be added to balance the wheel and tire. This information is shown on the balancer's display screen.

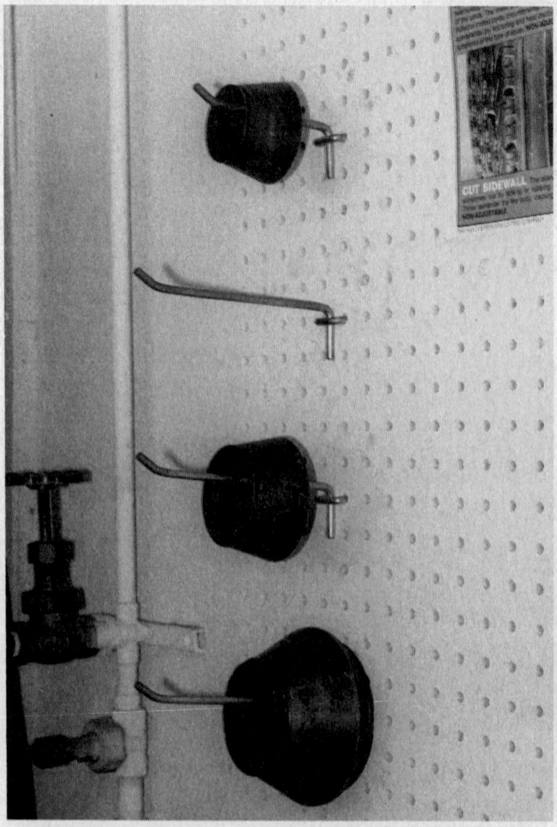

Figure 13-12. These cone-shaped adapters ensure that the wheel fits snugly on the balancer's driveshaft. Wheel balancers come with a wide variety of wheel adapters.

Because the computer can detect both static and dynamic imbalance, the display tells you exactly where to place the weights, including whether you should place them on the inner or the outer wheel flange. You do not have to divide the weight evenly between the top and bottom of the wheel, as you do with a static bubble balancer.

Perform an off-the-vehicle spin balance as follows:

1. Remove the tire and wheel assembly from the vehicle using the procedures in Chapter Twelve of this *Shop Manual*.
2. Use the guidelines described earlier in this chapter to prepare the tire and wheel for balancing. This includes removing old wheel weights and caked-on mud from the wheel.
3. Lift the safety cover on the balancer.
4. Choose an adapter that fits the center of the wheel, figure 13-12, and place the adapter on the balancer's driveshaft.
5. Place the wheel on the driveshaft, and secure it with the threaded clamp, figure 13-13. The clamp may have reverse threads.

Figure 13-13. Place the adapter and wheel on the driveshaft, and secure the wheel with the clamp. Be sure that the tire and wheel assembly is secure before turning on the balancer.

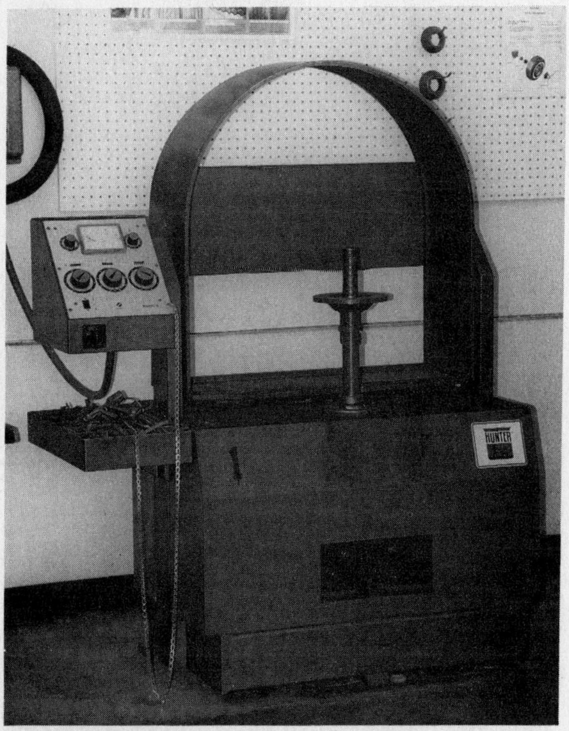

Figure 13-14. On this Hunter wheel balancer, turn the dials on the control panel to set wheel dimensions before starting an alignment.

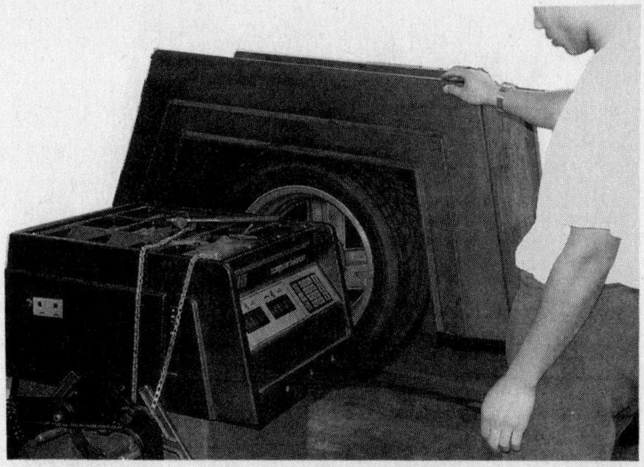

Figure 13-15. This Sun wheel balancer has a numeric keypad that you use to input wheel dimensions.

6. Input wheel dimensions so that the computer can correctly calculate the weight needed to balance the tire and wheel assembly. Some manufacturers have designed the equipment so that you turn dials to set wheel dimensions, figure 13-14. On other balancers, you input the information with a numeric keypad, figure 13-15. Most balancers need the following information to balance the tire and wheel assembly:
   - Wheel diameter
   - Wheel width
   - Distance between the wheel and balancer cover or balancer. This information allows the computer to approximate wheel offset.

   You can determine the wheel diameter by reading the tire sidewall information. Enter the diameter into the computer. For wheel width, use the calipers that come with the balancer, figure 13-16. Place the tips of the caliper on each side of the wheel, and read the wheel width on the caliper scale. Enter the information into the computer. Finally, measure the distance between the wheel and balancer cover or balancer using the gauge that is attached to the balancer, figure 13-17, and input the information in the computer so it can calculate wheel offset.
7. Some balancers allow you to choose between a static or dynamic balance. In most cases, you should use the dynamic setting. However, if the tire and wheel are extremely large or if the tire and wheel are severely out of balance, start with a static balance, and finish with a dynamic balance.

Figure 13-16. Use the calipers that come with the balancer to measure wheel width. The scale at the inner end of the calipers indicates wheel width.

Figure 13-17. Pull out the gauge, and measure the distance from the balancer to the wheel flange. This information helps the computer calculate wheel offset.

8. Some balancers have special settings for alloy wheels and steel wheels. (The alloy wheel setting calculates the weight needed at the inside of the wheels only, so the weights will be hidden.) Choose the appropriate setting.
9. Close the safety cover on the balancer, and push the start button. The balancer will spin the tire, and determine how much weight you should add and where you should add it.

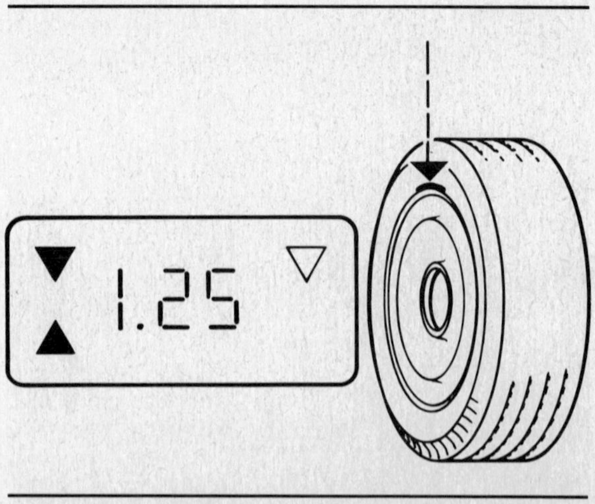

Figure 13-18. On this type of display, rotate the tire by hand until the two arrows on the left light up. Add the weight indicated on the display. The arrow on the right tells you that you must add the weight to the outer wheel flange.

Figure 13-19. Use the small hammer to place the clip-on weights on the wheel flange. Be sure the weights are secure.

10. When the wheel stops spinning, open the safety cover, and read the display to determine where you should add weights, figure 13-18. Remember, unlike the static balancer, you do not have to divide the weight between the inside and outside of the wheel. The computer tells you how much weight to add to the outer and inner wheel flange and where to add it so that the wheel is in static and dynamic balance.
11. If you are using clip-on weights, use the small hammer that comes with the balancer to secure the weights to the wheel, figure 13-19. If you are using adhesive weights, use a cleaning solution, such as window cleaner, to remove dirt and grease

Figure 13-20. When balancing the non-driven wheels with an on-the-vehicle balancer, place the motorized drum under the wheel to turn it.

from the wheel. Dry the wheel before removing the adhesive backing and putting the weights in place.

12. Close the safety cover on the balancer, and recheck wheel balance. If needed, add more weight as indicated by the computer. Do not use more than six ounces (170 grams) of weight, as mentioned earlier. If the wheel requires more than six ounces (170 grams) for balance, it is likely damaged and should be replaced.
13. Lift the safety cover, remove the tire and wheel assembly from the balancer, and install the tire and wheel on the vehicle.

On-the-vehicle spin balancers

Overall, an on-the-vehicle spin balancer can do a more accurate job of balancing the wheels than an off-the-vehicle balancer. Besides simply balancing the tire and wheel assembly, an on-the-vehicle balancer corrects imbalance at the hub or brake components as well. Disadvantages of on-the-vehicle balancers are that it takes longer to balance the wheels, and when you rotate the tires and wheels, you must rebalance them to match the hubs.

A number of manufacturers made these balancers, but they are rarely used now. Most shops use off-the-vehicle spin balancers instead. Use the instructions that come with your balancer because procedures and equipment vary. General instructions for using an electronic on-the-vehicle balancer follow:

1. Use the guidelines described earlier in this chapter to prepare the tire and wheel for balancing. This includes removing old wheel weights and caked-on mud from the wheel.
2. If you are balancing the driven wheels, determine if the vehicle uses a locking or limited-slip differential because there are special service procedures for each. To check the type of differential used, raise the vehicle, and use the following method:
   a. Place an automatic transmission in park, or a manual transmission in gear.
   b. Try to turn one of the wheels by hand. If you cannot turn it, the vehicle uses a locking or limited-slip differential. If you can turn it, the vehicle uses a conventional differential. Note that this method may not always determine the differential type, and you should check with the vehicle manufacturer to be sure.
3. If you are balancing the non-driven wheels, raise one wheel using a floor jack, and place the motorized drum under the wheel, figure 13-20. Lower the wheel until it contacts the drum. Block the other wheels. If you are working on a FWD vehicle, be sure that the suspension is at the proper ride height and the wheels are steered straight ahead. Neglecting to do this will damage the CV joints and transaxle on some vehicles.
4. For driven wheels with a locking or limited-slip differential, raise both wheels, and remove the tire and wheel from the side of the axle that you are not balancing, figure 13-21. For a conventional differential, raise one wheel, but be sure that the wheel on the other end of the axle is firmly on the ground and does not turn. In both cases, block the non-driven wheels.
5. Connect the electronic vibration sensor to the control or trailing arm or to a sturdy brake component on the side of the vehicle that you are balancing. Check the equipment manufacturer's instructions for special procedures.
6. Connect the meter to the electronic vibration sensor, figure 13-22. The meter tells you approximately how much weight to add. Be sure that wires are clear of the wheels and axle.
7. Use white chalk to place a reference mark on the tire, figure 13-23, or use the valve stem as a reference point.
8. Spin the wheel. If you are balancing a non-driven wheel, use the motorized drum to turn it. If you are balancing a driven wheel, use the engine to turn it. The person in the vehicle should not allow the speedometer reading to exceed 25 to 30 mph (48 km/h) for safety purposes. (When

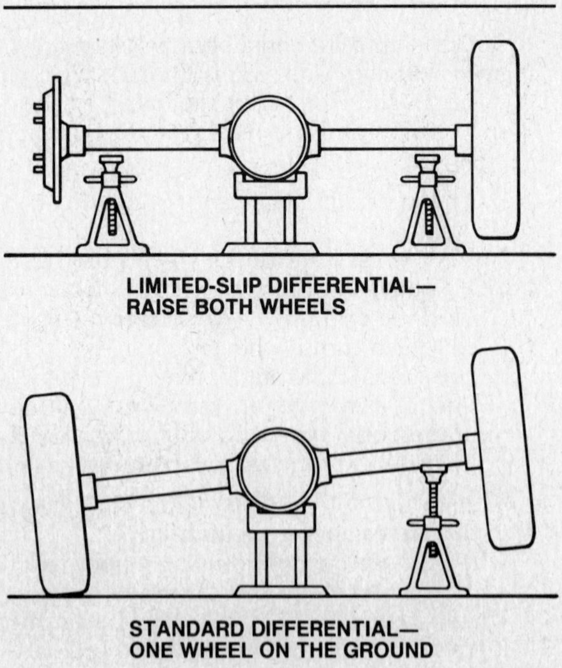

Figure 13-21. When balancing the driven wheels with an on-the-vehicle balancer, use the engine to turn the wheel. For a limited-slip differential, remove the tire and wheel that is not being balanced. For a conventional differential, be sure that the wheel not being balanced is firmly on the ground.

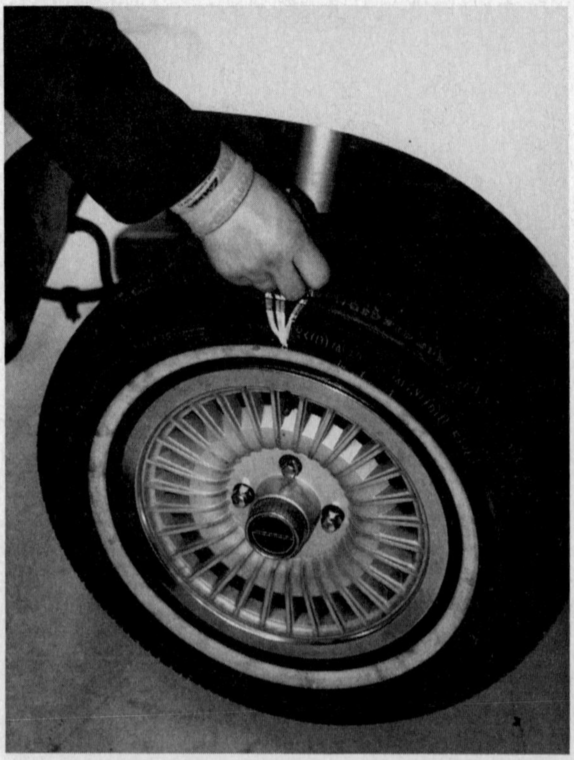

Figure 13-23. Make a reference point on the tire using white chalk. This mark will be noticeable when the strobe light flashes.

Figure 13-22. The electronic sensor that is connected to the control arm detects vibration when the wheel turns, and sends a signal to the meter. This triggers the strobe light, which flashes when vibration is detected.

you use the engine to drive one wheel, the wheel will be spinning twice as fast as the speed indicated on the speedometer.)

9. Point the meter at the tire and electronic sensor that is connected to the control arm. The strobe light flashes when the sensor detects vibration. When the light flashes, the heavy part of the tire is at the bottom and the light part at the top. Note the position of the chalk reference mark when the light flashes. (This works like a timing light.)
10. For driven wheels, stop the tire by turning off the engine and allowing the wheel to coast to a stop. For non-driven wheels, use the brake that is connected to the motorized drum.
11. Turn the tire by hand to position the reference mark as you saw it when the strobe light was flashing. In this position, the heavy part of the tire is at the bottom, and the light part at the top.
12. Check the meter for the approximate weight to be added, and add it to the top of the tire. Divide the weight in half, and add half to the inside and half to the outside of the wheel to keep the assembly in dynamic balance.
13. Check wheel balance again, and add or reposition weights as needed.

TIRE ROTATION

Rotating the tires regularly prolongs tire life because it helps even out wear. Use the schedule and rotation pattern suggested by the vehicle manufacturer. A common rotation interval for radial tires is 7,500 miles (12,100 km) for

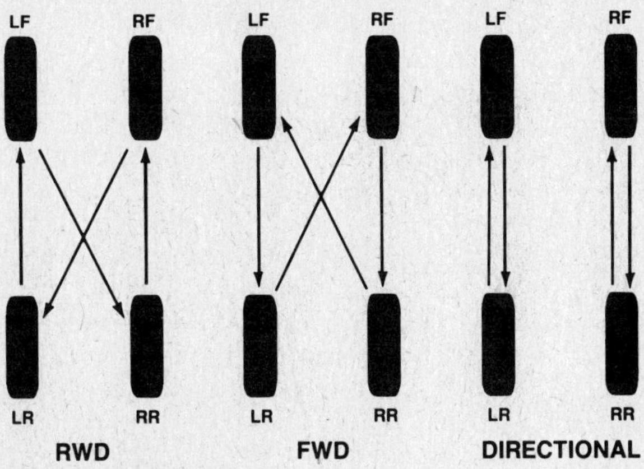

Figure 13-24. Use these standard tire rotation patterns if the vehicle does not have a full-size spare.

the first rotation, and then every 15,000 miles (24,150 km) thereafter. Bias belted tires should be rotated every 6,000 miles (9,660 km).

The standard four-tire rotation pattern for radial and bias belted tires is to cross the non-driven tires and mount them on the other axle. Then move the driven tires to the other axle, but on the same side of the vehicle, figure 13-24. For example, for FWD vehicles, cross the rear tires to the front, and move the front tires straight back. For RWD vehicles, cross the front tires to the rear, and move the rear tires straight forward. In the past, some manufacturers believed that crossing radial tires would cause internal tire damage. Today however, crossing radial tires is an acceptable tire rotation method.

If the vehicle uses directional tires, you cannot cross them because this changes their direction of rotation. As described in Chapter Twelve of this *Shop Manual*, changing the rotation of directional tires can cause hydroplaning in rainy weather. Instead, rotate them on the same side of the vehicle only if the same size tires are used at all four wheels. If the front and rear axles use different size tires, like on

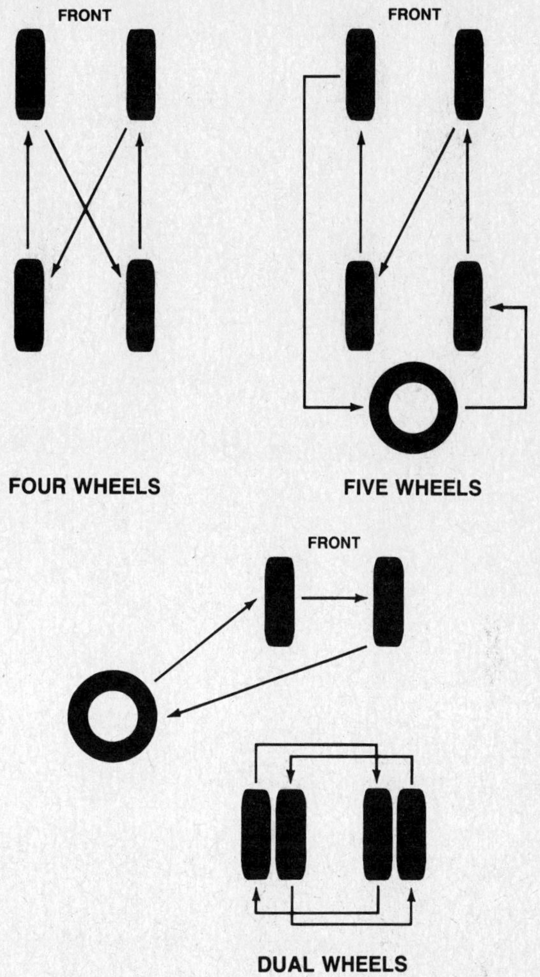

Figure 13-25. Most trucks come with a full-size spare, and the spare is included in the rotation pattern.

the Porsche 911 and the Corvette ZR-1, you cannot rotate them.

Most trucks come with a full-size spare and have special rotation patterns that include the spare, figure 13-25. If the vehicle has a compact spare, use the four-tire rotation pattern.

Chapter

14 Wheel Alignment

A wheel alignment corrects the position of wheel and suspension components, which go out of specifications over time as parts wear. Of course, parts that are extremely worn should be replaced. Many manufacturers recommend that you align the wheels after completing steering and suspension system repairs.

As described in Chapter Fourteen of the *Classroom Manual*, a wheel alignment involves determining and adjusting angles between suspension and steering parts, the wheels, and the road surface. Before measuring and adjusting these angles, check the steering and suspension systems, as well as the tires and wheels. Chapters One through Thirteen of this *Shop Manual* summarize nearly all of the inspection and repair procedures you will need to prepare for a wheel alignment.

This chapter begins with a pre-alignment inspection, which is how every professional wheel alignment starts. If needed, refer to the previous chapters in this *Shop Manual* for detailed inspection procedures. A preliminary inspection includes a visual check, a test drive, and a few simple measurements, which can quickly indicate the source of alignment problems.

After the inspection, measure alignment angles. Procedures for four-wheel and two-wheel alignments using various types of alignment equipment are included in this chapter. Although the equipment in your shop may not be identical to the equipment described in these procedures, the principles and methods are similar.

Methods for *adjusting* alignment angles (camber, caster, and toe) vary from vehicle to vehicle. Refer to the adjustments section of this chapter while you are performing a wheel alignment to set angles to the vehicle manufacturer's specifications.

The last section of this chapter describes special circumstances that you may encounter while performing a wheel alignment. Included are methods for determining if special repairs are needed, as well as tips for custom alignments when steering, suspension, tires, or wheels have been modified.

PRE-ALIGNMENT INSPECTION

The pre-alignment inspection is important because it helps you determine if an alignment will correct a vehicle's handling problems or if the wheels can be aligned at all. Most shops use a pre-alignment checklist, figure 14-1, to help the technician perform a thorough inspection. The technician uses this sheet to note the

PRE-ALIGNMENT CHECKLIST

Steering System (Chapters Four through Seven)

____ Check the steering system and related components for any modifications from the stock design. For example, lowering the vehicle ride height may affect the position of tie rods and other steering components.

____ Check the steering wheel for free-play and centering.

____ Check the inner and outer tie rod ends for play. If there is play, replace the tie rod ends before aligning the wheels.

____ Check the tie rod adjusting sleeves for rust or damage, and replace them if needed. If you cannot turn the sleeves, you will not be able to set toe.

____ Check the center link and other steering linkage parts for play.

____ Check for play between the idler and pitman arms and the center link. The idler arm should mount securely to the vehicle frame, and the pitman arm should be secured to the steering gear output shaft.

____ Check the steering knuckles and arms for bends, which will change the steering geometry.

____ Check the steering gear or rack bushings for wear. The gear or rack should be secured to the vehicle frame or crossmember.

____ Check for steering gear play.

____ Check the steering shaft for bends, and check shaft joints for play.

____ Check the steering damper for leaks, if the vehicle is so equipped.

____ For a power steering system, check for leaks at the fluid lines, steering gear, and pump.

____ For a power steering system, check the pump drive belt for damage and proper tension.

____ For a power steering system, if the wheels turn sharply when you start the engine (while the vehicle is on a frame-contact lift) the steering gear control valve may be damaged. This will likely cause pulling during a test drive.

____ If the vehicle has a four-wheel-steering system, refer to Chapter Seven of this *Shop Manual*. (Service precautions and inspection procedures vary depending on the type of four-wheel-steering system used.)

Suspension System (Chapters Eight through Eleven)

____ Bounce the vehicle at each corner to check the overall condition of the springs and shocks.

____ Check the suspension system for any modifications from the stock design, such as a change in ride height. Look for special height-adjusting shocks and struts, and check for spring inserts that temporarily "fix" sagging springs. Also, check the springs for signs of having been heated with a blow torch.

____ Check shock absorbers and struts for leaks. Check shock absorber mounts and the strut upper bearing for wear.

____ Check coil springs, leaf springs, and torsion bars for sagging. Cracked bump stops are one indication of worn springs.

____ Check load-carrying and non-load-carrying ball joints for wear. If they do not have wear indicators, you must measure wear with a dial indicator, and record the amount of play on the checklist. (This is required by law in some states.)

____ Check kingpins and kingpin bushings for wear, if the vehicle is so equipped.

____ Check bushings at the control arms, trailing arms, strut rods, antiroll bar, stabilizer links, and track bar. Replace cracked, broken, or missing bushings.

____ Check bump stops for cracks or other damage.

____ If the vehicle has an electronic suspension system, refer to Chapter Eleven of this *Shop Manual*. (Service precautions and inspection procedures vary depending on the type of electronic suspension system used.)

Wheels and Tires (Chapters Twelve and Thirteen)

____ Check the tire and wheel size for any modifications from the stock design.

____ Check tire pressure with a gauge, and adjust if needed.

____ Check tire wear patterns. Note possible alignment problems, as indicated by the wear patterns.

____ Check tires for damage. Note if they should be replaced or rotated.

____ Check the sidewall for tire size and construction. Tires on the same axle should be the same size and construction. Tread patterns between tires should match, and directional tires must be installed so that they rotate in the right direction.

____ Compare wheel width and diameter with tire size. The tires and wheels should match closely.

____ Check wheels for dents, bends, cracks, and other damage. If you suspect that the wheel is out-of-round, check tire and wheel runout.

____ If the customer has purchased new tires, balance them before starting the alignment.

Wheel Bearings and Brakes (Chapter Twelve)

____ Rotate the wheel by hand. Listen and feel for roughness, which may indicate worn bearings or dragging brake pads or shoes. To distinguish between the two, test drive the vehicle, and press on the brakes. If the vehicle pulls to the side that had the roughness, the brake pads or shoes are dragging on that side. If the bearings are worn, replace them.

____ Check the bearings for play, and adjust if needed.

Miscellaneous

____ Check CV joint boots for tears or leaks.

____ Check the vehicle body and frame for signs of collision damage. These signs include cracked undercoating, misaligned body panels, fresh rust on undercarriage components, and flaking paint.

____ Be sure that the vehicle is loaded according to the manufacturer's recommendations before starting the alignment.

____ Check for torque steer, and determine the possible cause. If it is due to vehicle design, you cannot repair it. However, torque steer caused by faulty motor mounts, engine cradle design, or CV joints may be repairable.

Figure 14-1. Use this checklist or one provided by the shop to visually inspect a vehicle before the alignment.

condition of steering and suspension parts and to recommend needed repairs. Many computerized wheel aligners have a checklist on the computer screen where the technician can type in information about vehicle condition.

A pre-alignment inspection includes a visual check, a test drive, and a few measurements that will help you determine if the wheels can be aligned or if you must make repairs first.

Visual Inspection

Drive the vehicle onto an alignment rack, and lift it. Use the checklist or one provided by the shop for the visual inspection. Refer to the other chapters in this *Shop Manual* for more information on inspecting the steering system, suspension system, wheels, tires, wheel bearings, brakes, and related components.

Test Drive

After a visual inspection, test drive the vehicle. A test drive gives you the opportunity to check the customer's complaints about handling problems. It also allows you to check other possible problems that you discovered during the visual inspection.

The test-drive course should include a straight-away where you can accelerate and brake a few times. Note pulling, leading, torque steer, or other handling characteristics while accelerating or braking. Also, include 90-degree turns to the left and right. During the turns, check the steering wheel for excessive play or binding. On a power steering system, check for pulling, which may indicate a damaged valve in the steering gear.

See Chapter Two of this *Shop Manual* for instructions on how to test drive, gather information, and diagnose steering and suspension system problems.

Preliminary Measurements

If you determine that the vehicle does not need repairs after the visual inspection and test drive, take the following preliminary measurements. Ride height, engine cradle alignment, toe-out on turns, and setback all should be within specifications before starting the alignment. Use the vehicle manufacturer's recommended procedures, if they are available, when taking these measurements. Many wheel alignment equipment manufacturers include procedures for preliminary measurements in their wheel alignment procedures manual.

Ride height

Ride height or "chassis height" is the distance from a point on the vehicle's suspension to the ground. Proper ride height is determined by the vehicle manufacturer and is usually listed with wheel alignment specifications in the manufacturer's service manual.

Ride height usually lowers with time because the suspension springs wear and begin to sag. When the ride height is lower than specified, wheel alignment angles change. For example, camber decreases when suspension springs sag. Because alignment angles change with ride height, vehicle manufacturers specify the ride height at which you should perform an alignment.

Some points to consider when checking ride height prior to a wheel alignment follow:

- If vehicle ride height has been raised or lowered for handling or other reasons, factory alignment specifications will not apply. See the last section of this chapter, titled "SPECIAL CONSIDERATIONS", for information on how changes to the steering and suspension systems affect alignment angles.
- If vehicle ride height was raised or lowered using equipment or techniques not recommended by the manufacturer, then you should not perform an alignment. Heating the suspension springs or using rubber inserts between the spring and spring seat are examples of "unorthodox" ride height alterations. If the suspension system is impossible to align or if it fails at a later time because of modifications, you may be held responsible in a lawsuit. Some shops require that the customer sign a waiver releasing the shop of liability before starting an alignment on a modified vehicle.
- Most vehicle manufacturers have specific instructions for loading the vehicle before checking ride height. Usually the gas tank, radiator, and crankcase must be full. Also, the spare tire, jack, and other equipment that came standard with the vehicle should be in place. Some vehicle manufacturers call this the "unladen" or "unloaded" condition. However, some alignment equipment manufacturers recommend that you load the vehicle as it is usually driven. This will ensure that the wheels are aligned to the correct position when the vehicle is in use.
- Compare the tires and wheels that are on the vehicle to stock equipment. Many truck owners replace stock tires with higher profile tires, which changes ride height

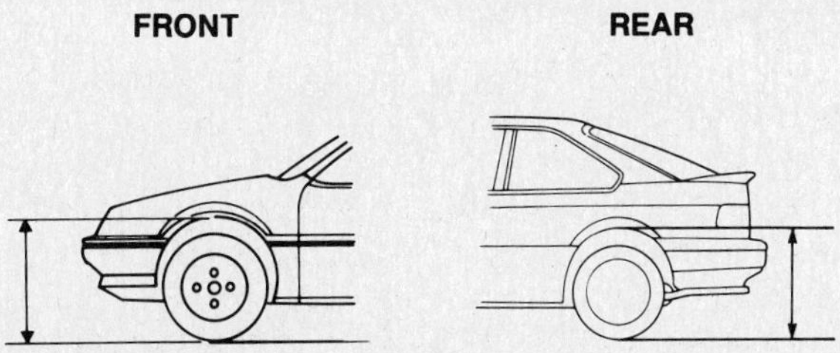

Figure 14-2. To check ride height on Acuras, measure from the center of the fender to the ground at the front and rear wheels.

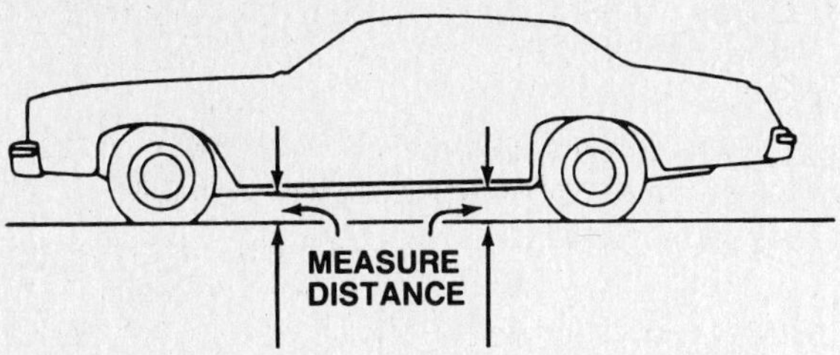

Figure 14-3. To check ride height on some Buicks, measure from the lower edge of the rocker panel to the ground.

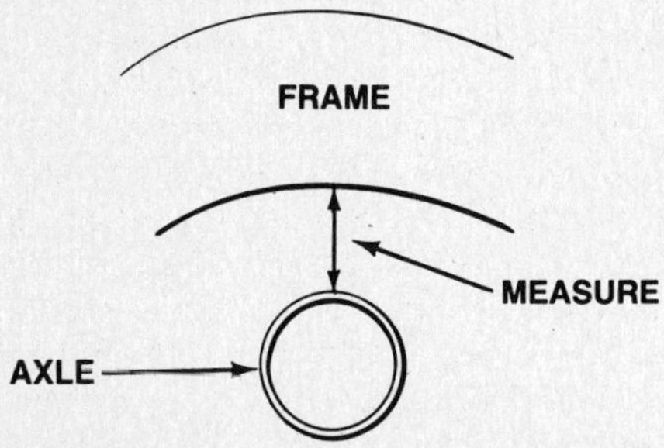

Figure 14-4. To check ride height at the rear of some RWD Cadillacs, measure from the axle to the vehicle frame.

measurements. Some car owners prefer low-profile tires.

- Check and adjust tire pressure. Pressures should be even and within specifications at all four tires.

Measure ride height as follows:

1. Park the vehicle on a level surface. Use a wheel alignment rack if one is available.
2. Bounce the vehicle a couple of times to settle it.
3. Using the vehicle manufacturer's service manual or the instructions that come with the wheel alignment equipment, locate the points on the suspension or body from which you measure ride height.

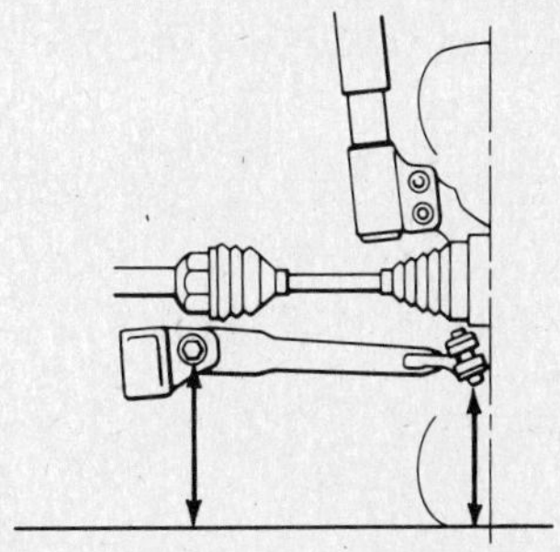

Figure 14-5. To check ride height on some Cadillacs with strut suspensions, measure the distance between the inner control arm pivot and the ground and the outer control arm pivot and the ground. Then subtract the distances for the ride height specification.

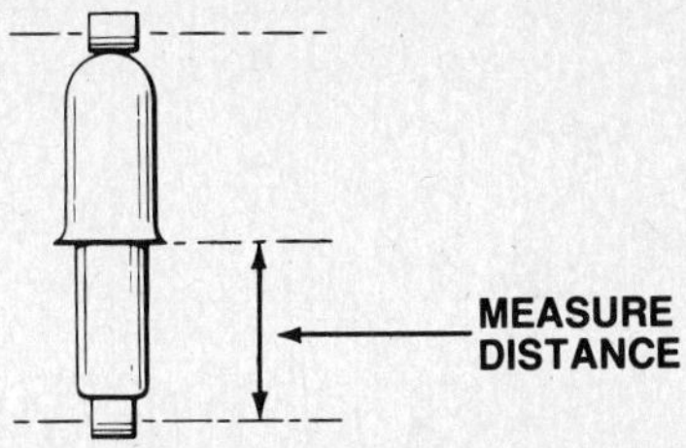

Figure 14-6. To check ride height on some Cadillacs, measure the distance the shock is compressed.

4. Use a tape measure or ruler to measure ride height, and compare the measurements to the manufacturer's specifications.

You will find instructions for measuring ride height in the vehicle manufacturer's service manual or the wheel alignment equipment manufacturer's instruction manual. Some examples of ride height measurements follow:

- Measure from the center of the fender to the ground at the front and rear, figure 14-2.
- Measure from the lower edge of the rocker panel to the ground, figure 14-3.
- Measure from the axle to the vehicle frame, figure 14-4.
- Measure the difference between the inner control arm pivot and the ground, and the outer control arm pivot and the ground, figure 14-5.
- Measure the distance that the shock absorber is compressed, figure 14-6.
- Measure the distance between the inner control arm pivot and the ground, figure 14-7.
- Measure the distance between the center of the spindle and the ground, figure 14-7.

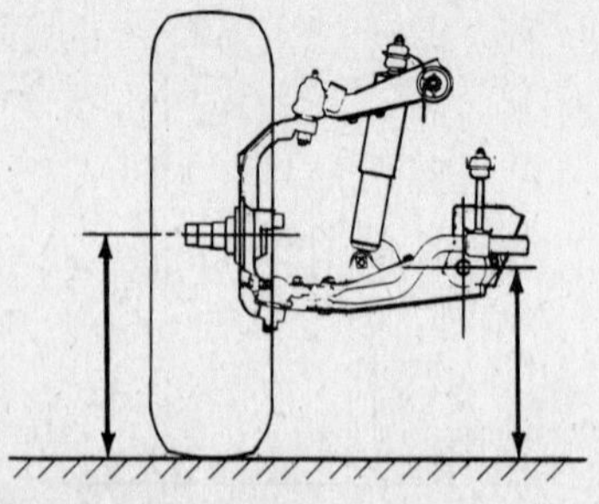

Figure 14-7. To check ride height on some vehicles, measure the distance between the inner control arm pivot and the ground, and between the center of the spindle and the ground.

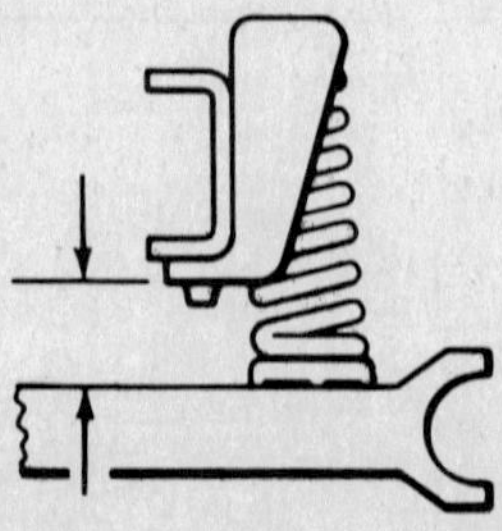

Figure 14-8. To check ride height on some Ford trucks, measure the distance from the lower spring seat to the axle.

- Measure from the lower spring seat to the axle, figure 14-8.
- Measure the distance between the center of the wheel (spindle) and the top of the fender, figure 14-9.
- If the vehicle uses an electronically controlled suspension, use the manufacturer's recommended procedures and cautions when checking ride height. Chapter Eleven of this *Shop Manual* has example procedures for checking ride height.

If ride height measurements vary considerably from the vehicle manufacturer's specifications, make needed repairs before starting a wheel alignment. Replace sagging springs or bent steering and suspension parts using the procedures given in earlier chapters of this *Shop Manual*.

Engine cradle alignment
Many vehicle manufacturers have procedures for checking engine cradle alignment to help determine if the vehicle has been hit hard enough to bend the frame. If the frame is bent, then aligning the wheels may not correct handling problems. In fact, if the frame is bent, you may not be able to align the wheels at all.

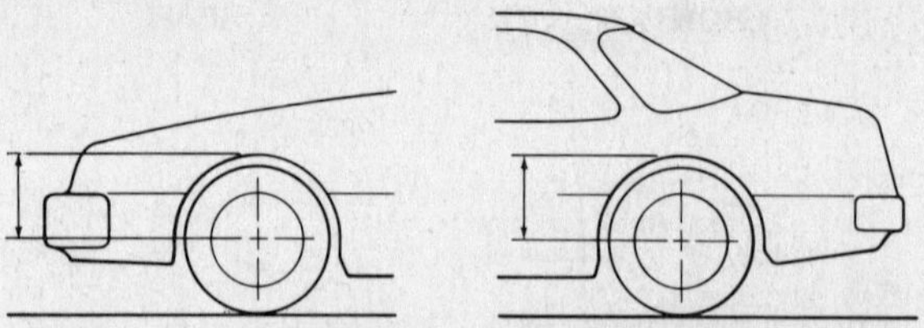

Figure 14-9. To check ride height on most Mazda cars, measure the distance between the center of the wheel (spindle) and the top of the fender.

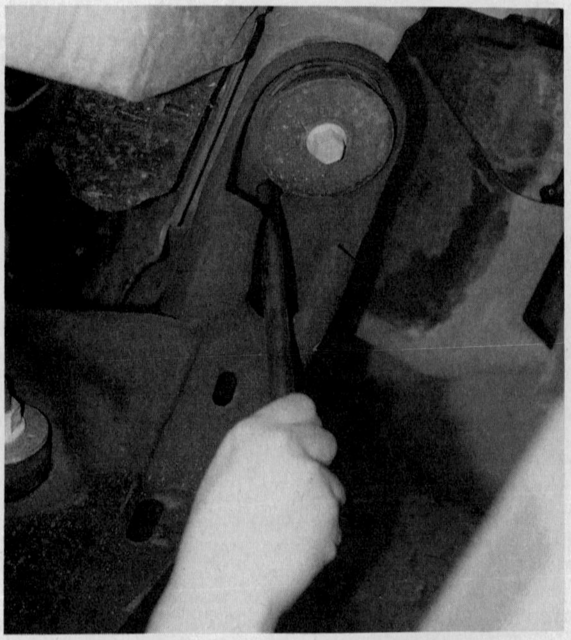

Figure 14-10. An engine cradle alignment check will ensure that the engine is properly positioned. On some vehicles, check cradle alignment by inserting a screwdriver or other tool into the alignment hole.

The vehicle manufacturer's service manual contains the information you need to check engine cradle alignment. Usually it involves simply looking through a hole in the frame that should line up with a mark on the engine or engine cradle. Another method involves sticking a straight tool, such as a "lady finger" or screwdriver, into a hole in the frame, which should line up with another hole in the engine block or cradle, figure 14-10. If the engine cradle is out of alignment, the vehicle needs frame work or the engine must be repositioned before the wheels can be properly aligned.

Toe-out on turns
When a vehicle goes into a turn, the inner and outer wheels turn at different angles, figure 14-11, as described in Chapters Five and Fourteen of the *Classroom Manual*. This angle, when measured for wheel alignment purposes, is

ACKERMANN EFFECT

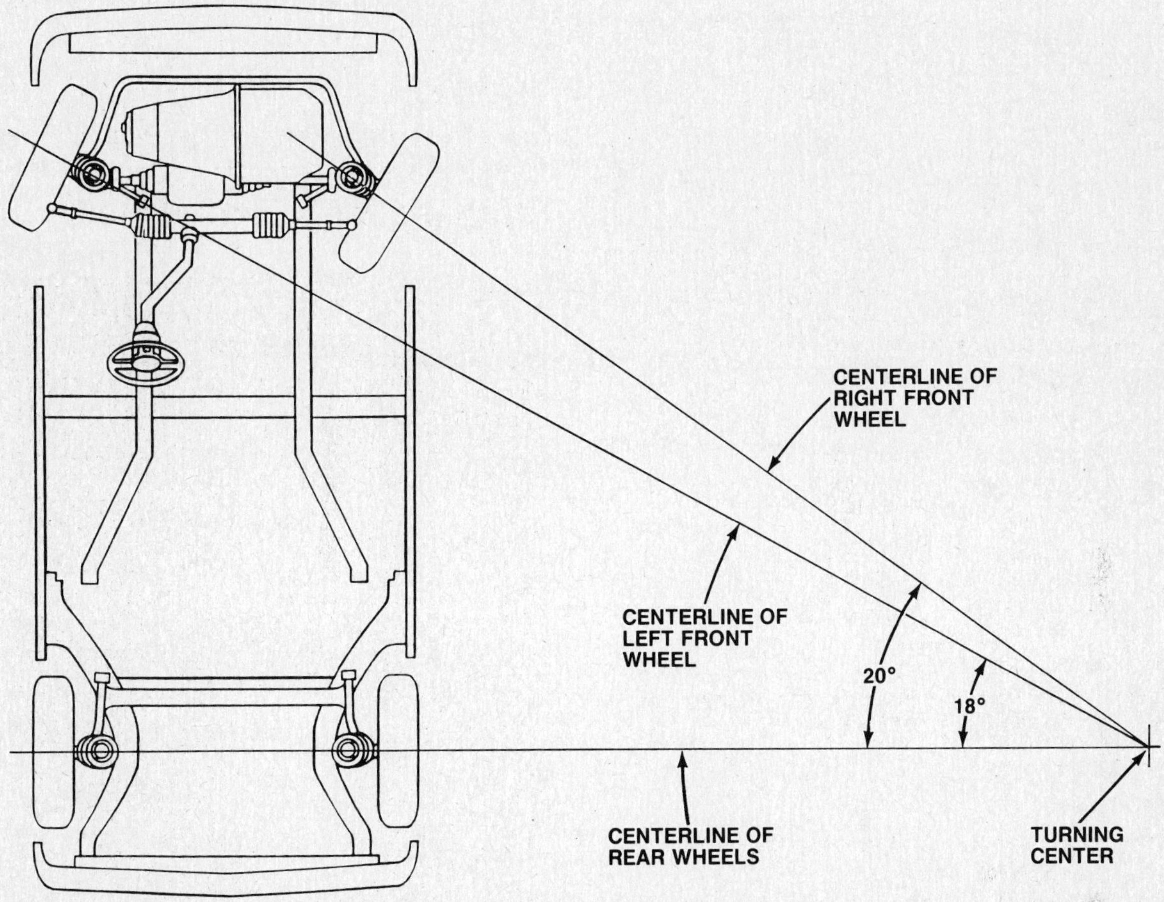

Figure 14-11. The steering and suspension systems are designed so that when the vehicle turns, the outer wheel can travel in a wider arc than the inner wheel. Toe-out on turns is the angle that allows the wheels to turn this way.

called "toe-out on turns", and it determines the vehicle's turning radius.

Toe-out on turns is not an adjustable alignment angle, but it is a good preliminary measurement. If it is out of specifications, steering arms or other steering and suspension components are likely bent. Tire scuffing, squealing, and side slip result when toe-out on turns is out of specifications.

After checking to be sure that toe is within specifications, measure toe-out on turns as follows:

1. Be sure that the turntables on the alignment rack are locked. (The pins should be inserted.) When the turntables are locked, the pointer on the turntable's degree wheel will be at zero.
2. Drive the vehicle onto the alignment rack, and position the front wheels on the centers of the turntables.
3. Set the vehicle's parking brake.
4. Remove the pins from the turntables, figure 14-12.
5. Turn the steering wheel until one wheel is turned the number of degrees specified by the manufacturer. The number of degrees that the wheel is turned is read on the turntable degree wheel. Many manufacturers say to turn the wheel 20 degrees.
6. Read the number of degrees that the other wheel has turned, and compare it to the manufacturer's specifications. A common reading would be 18 degrees, although this number depends on vehicle steering linkage design. The difference in degrees between the inner and outer readings is the actual toe-out on turns measurement. The angle at the inner wheel will be larger than the angle at the outer wheel.

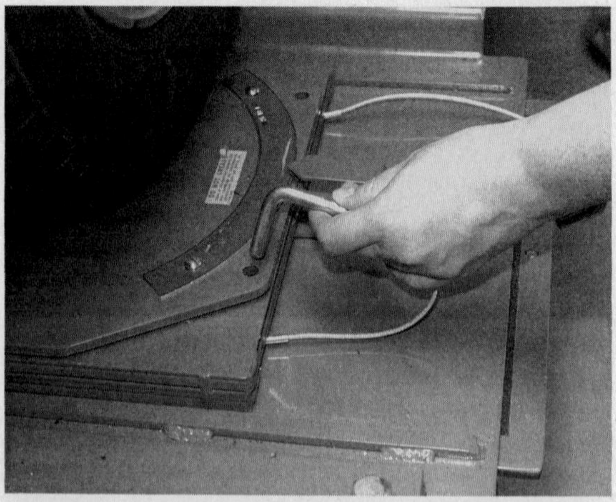

Figure 14-12. The turntables on the alignment rack are marked so that you can read the number of degrees the wheel has turned. After driving the vehicle onto the rack, remove the pins to unlock the turntables.

If toe-out on turns varies from the vehicle manufacturer's specifications, but toe is within specifications, then steering and/or suspension components are damaged and must be repaired before starting the wheel alignment.

Setback

A vehicle has setback if one wheel is farther back than the opposite wheel on the same axle, figure 14-13. Causes of setback vary. Some manufacturers design the vehicle with a slight amount of setback to help alleviate torque steer. However, if the vehicle is not supposed to have setback, torque steer may be increased. Many unit-body vehicles have a small amount of setback because of imperfections in the vehicle manufacturing process. On some vehicles, setback is caused by worn suspension components, such as control arm bushings. A bent frame due to a collision, such as hitting a curb, may also be the cause of setback. If you use a computer wheel aligner and cross caster is out of specifications, the vehicle may have setback, as well.

Setback causes the tie rods to change position, resulting in pulling, bump steer, and unresponsive steering during turns. Check the steering linkage for parallelism, as described in Chapter Five of this *Shop Manual*. During a wheel alignment, if you cannot set front-wheel toe and maintain a centered steering wheel, the vehicle may have setback. This is because of the way a computer aligner measures toe. If the right tire is set back slightly, and the wheel aligner does not compensate for setback, the computer will read right wheel toe-in, even if toe is actually correct. If possible, you should

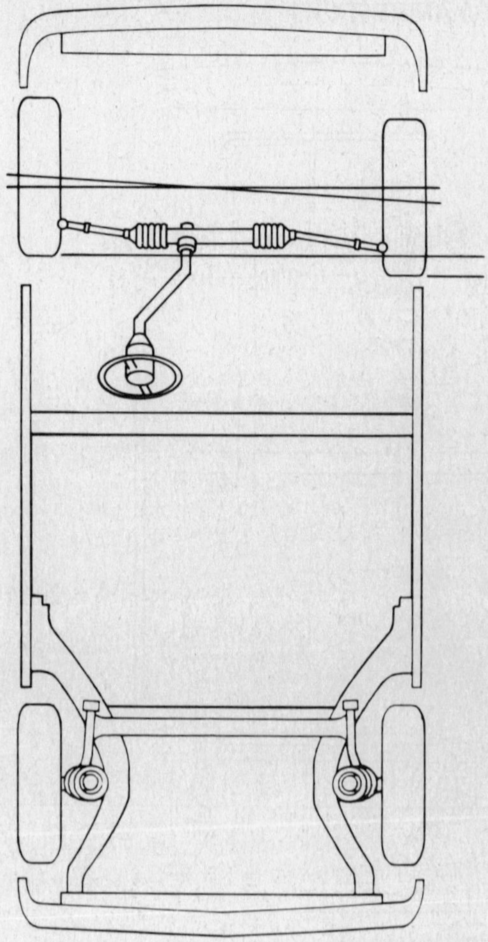

Figure 14-13. If one of the front wheels has been pushed back in a collision so that the front axle is not parallel to the rear axle, the condition is called "setback".

inspect for and repair any damage that causes setback before starting the alignment.

Check for setback at the front axle as follows:

1. Park the vehicle on a level surface. Drive the vehicle onto the alignment rack, block the wheels, and set the parking brake. Be sure to position the front wheels straight ahead. If you do not, the toe-out on turns setting may affect the setback measurement.
2. You can measure setback either manually or with the computer aligner:

   - To measure setback manually, use a tape measure to measure the distance from the front of the rear wheel rim to the front of the front wheel rim. Record the distance. Then take the same

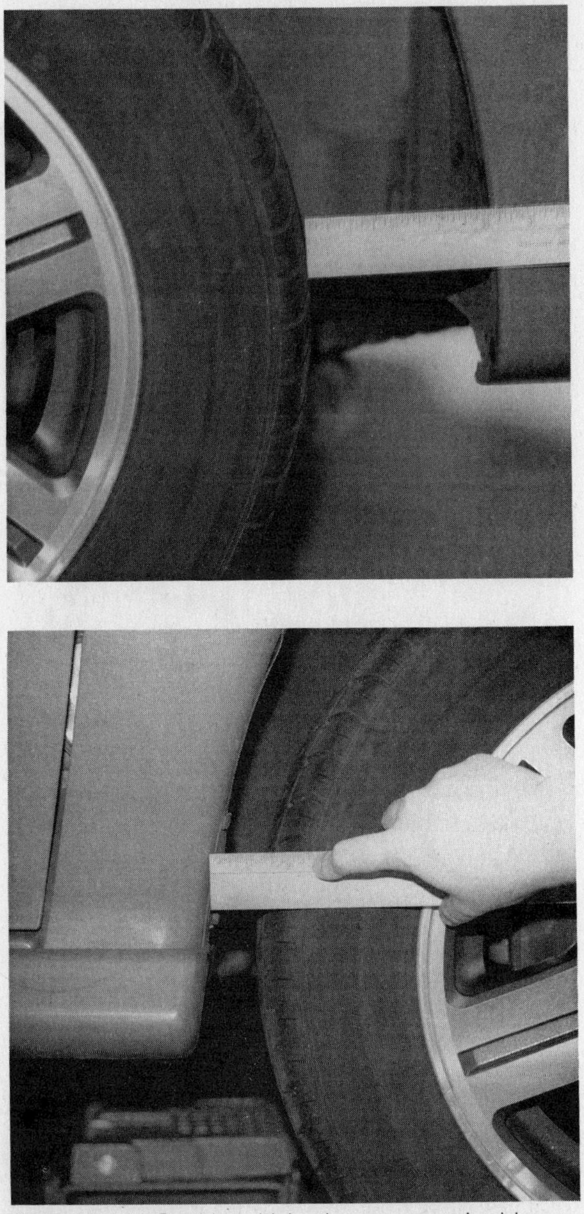

Figure 14-14. On this vehicle, the passenger's side front tire is not centered in the fender well. There is plenty of space at the front of the tire (top) and not much at the rear (bottom). The vehicle probably hit a curb.

measurement on the other side of the vehicle. If the readings vary, find the difference between them and compare the difference with the manufacturer's specifications for setback in inches or millimeters. On some vehicles, setback is easy to detect. Simply check the distance from the front and back of the tire to the fender on both sides of the vehicle, figure 14-14.

- Some computer aligners automatically check setback when you perform an alignment. If the wheel alignment equipment you are using does this, proceed to the section later in this chapter titled "Four-Wheel Alignment". Compare the setback readings with the manufacturer's specifications.

3. If setback measurements do not meet the manufacturer's specifications, repair the vehicle before continuing with the alignment. If the vehicle manufacturer does not provide setback specifications, a general rule of thumb is that more than 1/2 inch of difference from side to side or more than 1/2 degree from centerline is excessive setback.

MEASURING ALIGNMENT ANGLES

A wheel alignment is the final step in many steering and suspension repairs. It checks the position of key steering and suspension components by measuring the angles formed between them. Vehicle manufacturers list specifications for camber, caster, toe, and other angles in their service manuals.

Before an alignment, perform a pre-alignment check, which includes visually inspecting the steering and suspension systems, test driving the vehicle, and taking preliminary measurements, as described earlier in this chapter.

Included in this section are procedures for performing four-wheel and two-wheel alignments. The computerized four-wheel alignment is considered the most accurate method. Chapter Fourteen of the *Classroom Manual* describes the theory of four- and two-wheel alignments in detail. After checking alignment angles, adjust them using the guidelines in the sections titled "ALIGNMENT ANGLE ADJUSTMENTS" and "SPECIAL CONSIDERATIONS" later in this chapter.

Some things to note before starting a professional wheel alignment are:

- Always use the wheel alignment equipment manufacturer's instructions. The procedures in this chapter are to illustrate basic wheel alignment theory and practice, and should only be used as a *guide*. These procedures do not apply to all types of alignment equipment on the market.
- Computerized alignment equipment should be calibrated periodically, as determined by the equipment manufacturer. Also, the alignment rack should be level. Do

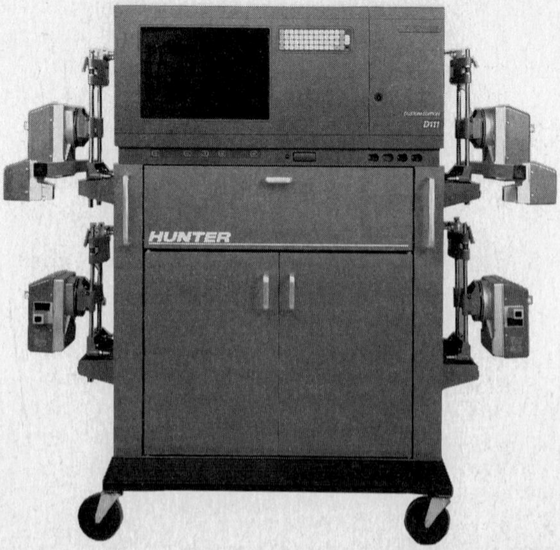

Figure 14-15. The Hunter model D-111 wheel aligner is the equipment we use in our example of how to operate a computerized four-wheel aligner.

Figure 14-16. After driving the vehicle onto the alignment rack, chock the rear wheels.

not attempt to calibrate the equipment or level the rack unless you have been trained to do so.

- Before starting an alignment, *always* check the vehicle for an electronic suspension system. If it has one, check the manufacturer's instructions before lifting the vehicle or aligning the wheels. Chapter Eleven of this *Shop Manual* has general guidelines for identifying electronic suspension systems.
- Familiarize yourself with the steering and suspension systems before you start working on them. For example, identify load-carrying components and the parts that compress and secure suspension springs.
- Always set the parking brake and chock the wheels after driving the vehicle onto the alignment rack.

Computerized Four-Wheel Alignment

As discussed in Chapter Fourteen of the *Classroom Manual*, a four-wheel alignment is considered superior to a two-wheel alignment because it involves measuring and adjusting the angles at the front and rear wheels, rather than measuring and adjusting only the front wheels. If the rear wheels are not adjustable, a computerized four-wheel aligner will calculate the "thrust angle" that is formed by rear wheel toe. Then the front wheels can be aligned to compensate for rear wheel thrust. This is called a "thrust angle alignment". The effects of "dog-tracking", a condition where the vehicle body travels down the road at a slight angle, can be corrected with a four-wheel thrust angle alignment.

Use the procedures in this section to perform a computerized four-wheel alignment using Hunter's D-111 digital wheel aligner, figure 14-15.

Preparing the vehicle for an alignment

Use the following steps to prepare the vehicle for a wheel alignment:

1. Check the turntables at the front of the alignment rack and the slip plates at the rear. They should be locked in place with lock pins.
2. Drive the vehicle onto the alignment rack. Be sure that the front wheels are centered on the turntables. Chock the rear wheels, and set the parking brake, figure 14-16.
3. Check and adjust tire pressures if you did not do so during the pre-alignment check.
4. Be sure that the vehicle is loaded according to the manufacturer's instructions, and that ride height is within specifications, as described in the pre-alignment check procedures.
5. Check for wheel bearing and suspension parts wear. Also check that wheel runout is within specifications, as described in the pre-alignment check procedures.

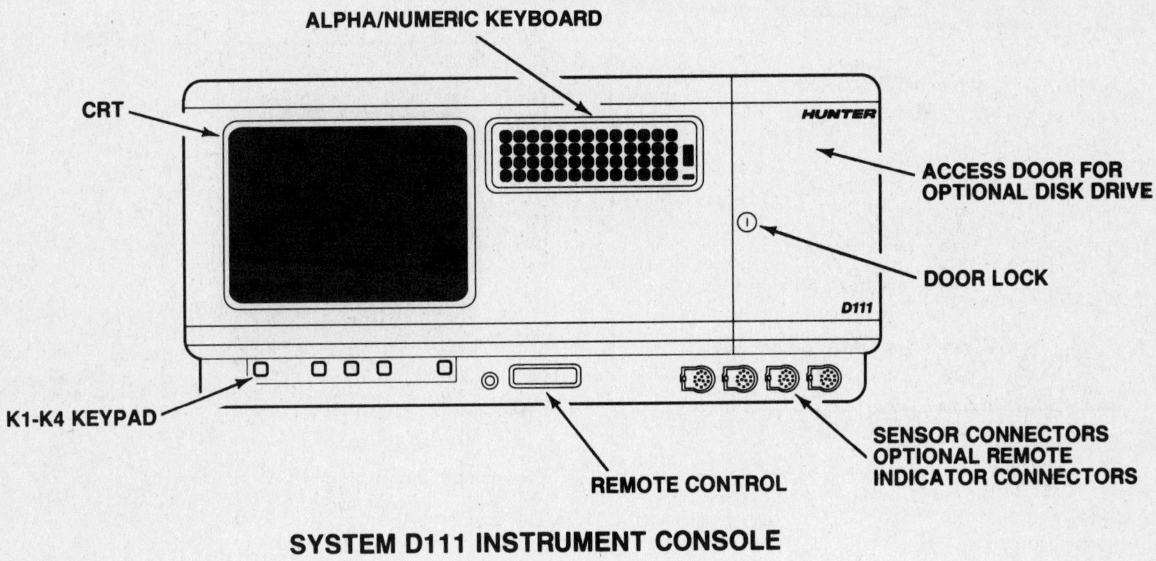

Figure 14-17. This is the layout of the Hunter D-111 instrument panel.

Figure 14-18. The Hunter D-111's multi-function or "K" keys are located below the computer screen.

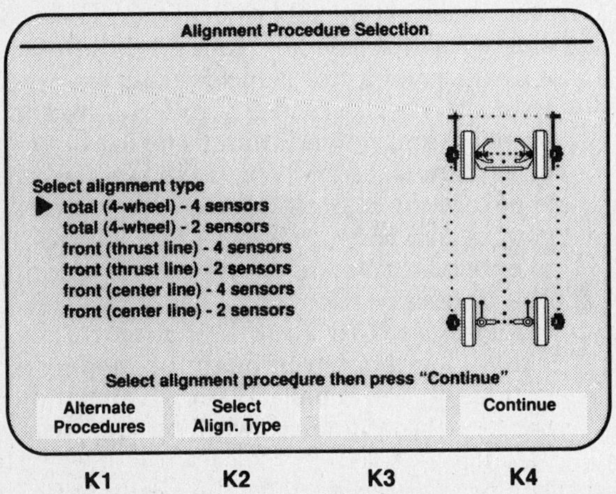

Figure 14-19. Use the "Alignment Procedure Selection" screen to choose a four-wheel, thrustline, or centerline alignment.

Turning on the equipment

Use the following steps to turn on the aligner and set up the computer using the console keypads, figure 14-17:

1. Press the power switch to turn on the computerized aligner, and allow the computer to warm up.
2. Press "Reset" (R) quickly two times, which initializes the wheel alignment program. The message "Ready for Alignment" will appear on the computer screen.
3. Press "Begin Alignment" (K4), figure 14-18.
4. The computer screen will display six alignment types for you to choose from, figure 14-19. The choices are:

   - Total (four-wheel) — 4 sensors
   - Total (four-wheel) — 2 sensors
   - Front (thrustline) — 4 sensors
   - Front (thrustline) — 2 sensors
   - Front (centerline) — 4 sensors
   - Front (centerline) — 2 sensors

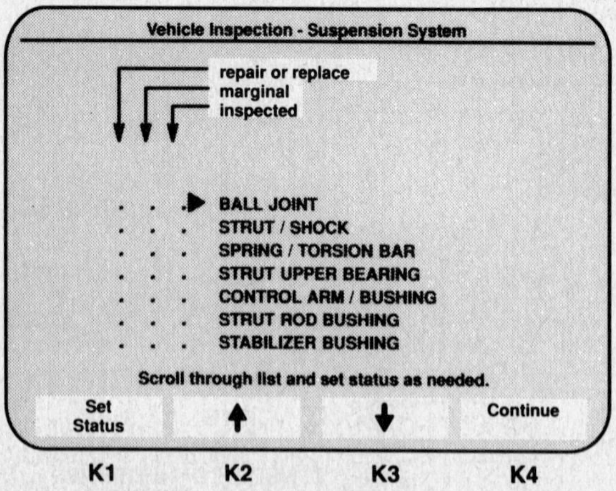

Figure 14-20. This computerized aligner has a checklist on the computer screen that you can use during the pre-alignment inspection.

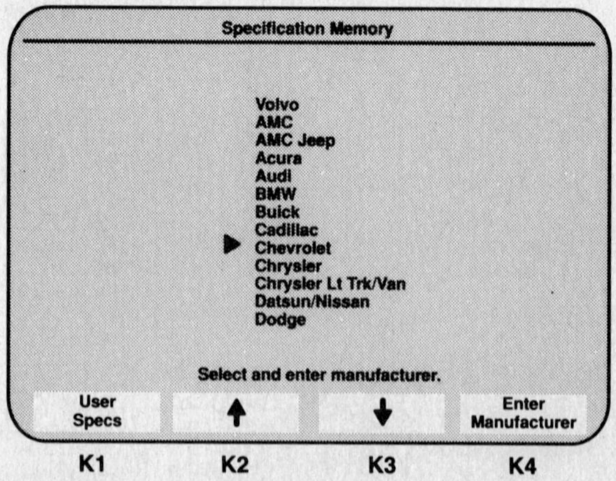

Figure 14-21. The first step in retrieving the wheel alignment specifications is to choose the vehicle manufacturer from the computer screen.

Pressing ''Select Alignment Type'' (K2) toggles a black arrow between the six choices. When the black arrow is next to the alignment type you want, press ''Continue'' (K4), which selects the alignment type and continues to the next screen. For this alignment procedure, select ''Four-wheel alignment''.

Note: A ''four-wheel alignment'' uses four sensors. First the rear wheels are adjusted so that the thrustline is nearly equal to the vehicle's centerline. Then the front wheels are adjusted to the thrustline. A ''thrustline alignment'' uses four sensors. If the rear alignment angles are not adjustable, the thrustline is determined, and the front wheels are aligned to the thrustline rather than the vehicle's geometric centerline. A ''centerline alignment'' simply adjusts the front wheels according to the vehicle's geometric centerline without compensating for rear wheel thrust.

5. Filling out the next screen is optional. It is for recording customer information such as name, phone number, vehicle, license plate number, mileage, and the technician's name. After you finish, press ''Continue'' (K4).
6. The following screen is a vehicle inspection checklist that you can use for the pre-alignment inspection, figure 14-20. After examining each listed item, check ''Inspected'', ''Marginal'', or ''Repair or Replace'', depending on the condition of each part. If you check ''Repair or Replace'' for any parts, do not perform the alignment. Instead discuss repairs with the customer. After completing this screen, press ''Continue'' (K4).

Setting up the specifications screen

The Hunter D-111 computerized aligner has factory specifications in its memory for most domestic and import models. By entering the appropriate information about the vehicle, you can retrieve the manufacturer's wheel alignment specifications for the vehicle you are working on.

1. After you fill out the inspection checklist and press ''Continue'' (K4), the computer will display a list of vehicle manufacturers, figure 14-21. Use the K2 and K3 buttons to scroll up and down through the list of vehicle manufacturers. When the black arrow is next to the manufacturer of the vehicle you are servicing, press ''Continue'' (K4) to enter the manufacturer.
2. The next screen contains a list of vehicle models. Use the K2 and K3 buttons to scroll up and down through the model selections. When the black arrow is next to the model that you are working on, press ''Continue'' (K4).
3. The next screen displays the alignment specifications and tolerances for the vehicle you are servicing, figure 14-22. To tighten the tolerances, press ''Reduce Tolerances'' (K2).

Installing the sensors

After setting up the computer, install the sensors on the wheels. Sensors detect wheel position and feed the information back to the computer. Because the sensors have delicate

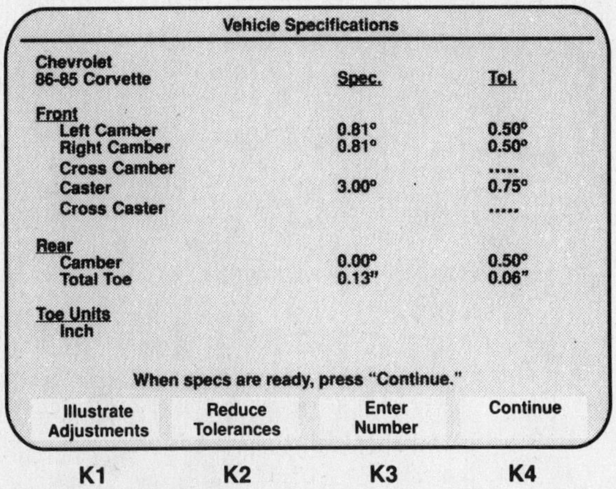

Figure 14-22. The "Vehicle Specifications" screen displays the wheel alignment specifications and tolerances for the vehicle you are servicing.

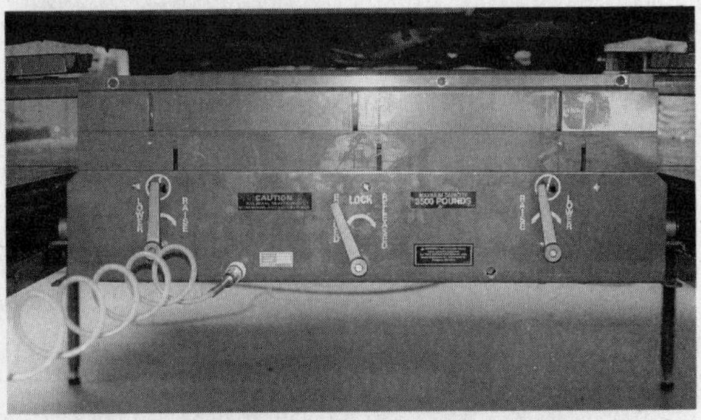

Figure 14-23. Alignment racks have air jacks at the front and rear axles. Use these to raise the vehicle when you install the sensors and when you are making alignment adjustments.

electronic circuitry that measure wheel position, handle them carefully. If you drop one, contact the equipment manufacturer so the sensor can be checked.

There are two types of sensors that can be used with the Hunter D-111 wheel alignment system. The most recent technological trend is to use "optical" sensors that measure wheel position with infrared beams. Other sensors use strings, called "toe lines", to measure the position of the wheels.

There are advantages and disadvantages to both types of sensors. The advantage of the optical sensor is that you do not have to worry about bumping into strings while you are working on the vehicle. The advantage of toe line sensors is that if you step between them and disrupt the readings, the string will remind you that you should allow the computer a few seconds to readjust. You may not realize

Figure 14-24. Mount the sensors on the wheels, fitting them securely into the rim.

that you have disrupted the readings when you step between the optical sensors.

Both types of sensors are installed using the following procedure. Be sure to install and calibrate the sensors properly, or alignment readings will be incorrect.

1. Use the air jacks at the front and rear axles to raise the vehicle, figure 14-23. Be sure to place the lift supports at the vehicle manufacturer's recommended lift points. Raise the vehicle high enough so that all four wheels are off the alignment rack and can turn freely.
2. Turn each wheel so that the tire valve stem is in the one o'clock position. This will allow you to connect the sensors' safety cables to the valve stems.
3. Mount a sensor on the wheel, figure 14-24. If necessary, use adapters on the sensors, figure 14-25, to ensure that the sensor fits snugly. Position the sensor adapters on the lower part of the wheel rim. Then slide the other adapters up until they contact the upper part of the wheel rim.
4. Turn the sensor's T-handle, figure 14-26, which pushes the adapters against the rim, holding the sensor in place.
5. Screw the end of the safety cable onto the valve stem, figure 14-27. Mount the other three sensors in the same manner.
6. Slide the sensor to the center of the wheel.
7. Connect the main data line from the computer to the alignment rack. Connect a

Figure 14-25. Use sensor adapters, if necessary, to ensure that the sensor fits snugly on the wheel.

Figure 14-26. Turning the T-handle on the sensor locks it in place.

data line from each sensor to its plug on the alignment rack, figure 14-28. The data line hookup is much like a telephone jack. Doing this allows each sensor to be connected to the computer through the circuitry in the alignment rack.

8. If you are using sensors with toe lines, connect them now. There are three lines — one toe line runs between the two front sensors, and one toe line runs from each of the front sensors to the rear sensors. There is not a toe line between the rear sensors. If the lines are slack, tighten them, but do not make them taut as this will prevent the sensors from properly measuring wheel position.

Figure 14-27. If the sensor comes loose and falls off the wheel, the safety cable will keep it from falling to the ground.

Figure 14-28. The computer receives information from the sensor about wheel position through this data cable. Be sure the cable is connected.

Compensating the sensors

Before starting the alignment, you must compensate the sensors. "Compensating" tells the

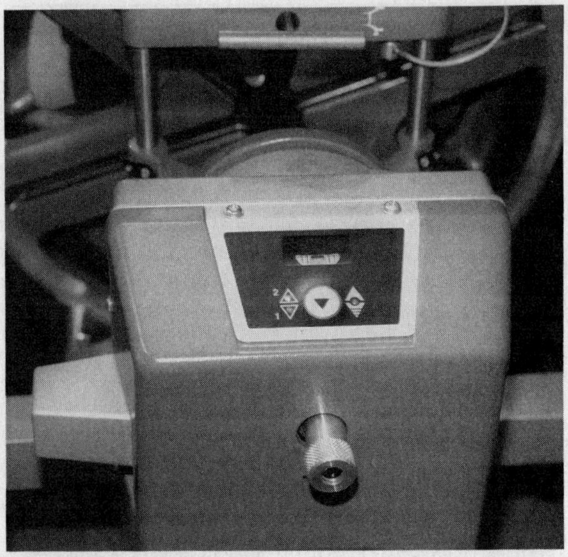

Figure 14-29. A lock knob on the sensor holds it level on the wheel. There is also a bubble gauge, ''compensate'' button, and a light that indicates when the sensor is compensated.

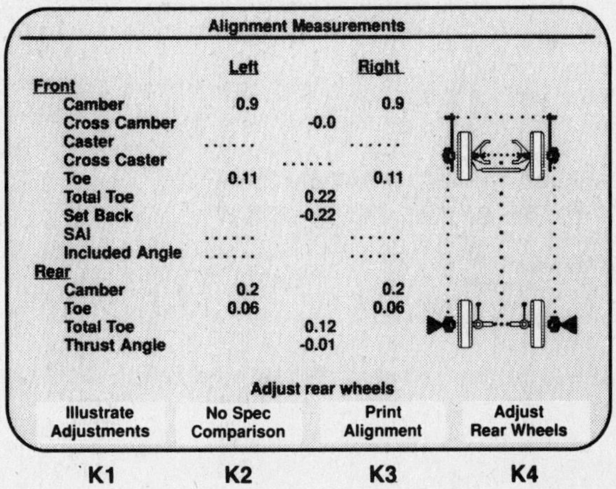

Figure 14-30. This ''Alignment Measurements'' screen displays the initial alignment readings after you have compensated the sensors.

computer how the wheels are positioned, and calculates runout so that the computer can give you accurate alignment readings. If you remove a sensor from a wheel at any time during the alignment, re-compensate it before continuing.

1. Go to any wheel sensor to start the compensation process. Loosen the lock knob at the center of the sensor, figure 14-29.
2. Rotate the wheel 180 degrees. The green VERT or number ''1'' light will illuminate when the wheel is in the right position. Tighten the lock knob.
3. Adjust the wheel, if needed, so that the bubble in the level is at the center of the scale.
4. Press the compensate button, which will start the compensation process. Do not touch the vehicle or sensors until the red light next to the compensate button starts to flash. If you accidentally move the other sensors, repeat the compensation procedure.
5. Loosen the lock knob, and rotate the wheel 180 degrees. The green VERT or number ''2'' light will illuminate when the wheel is in the right position. Tighten the lock knob.
6. Adjust the wheel, if needed, so that the bubble in the level is at the center of the scale.
7. Press the compensate button, which will start the second half of the compensation process. Do not touch the vehicle or sensors until the red light next to the compensate button stops flashing and stays on, which means the sensor is compensated. The computer screen will indicate that the sensor is compensated, as well.
8. Compensate the other three sensors, one at a time, using the same procedure.

Reading the alignment angles

After you have compensated the sensors, the computer will display the alignment angles.

1. Before lowering the vehicle onto the alignment rack:
   - Be sure that the sensors are vertical so that the bubble is centered in the level. Check that the lock knob is tight, and the green VERT lights on each sensor are illuminated.
   - Be sure that the vehicle parking brake is set, and the transmission is in park.
   - Remove the lock pins from the front turntables and rear slip plates.
2. Lower the vehicle onto the alignment rack. Jounce it at the front and rear bumpers to settle the suspension.
3. All alignment angles except for caster, cross caster, SAI, and included angle will be displayed on the computer screen, figure 14-30. To display the other alignment angles, you must perform the procedures in the following sections.

The wheel alignment computer will indicate which readings are out of specifications by displaying them in red. Wheel alignment angles

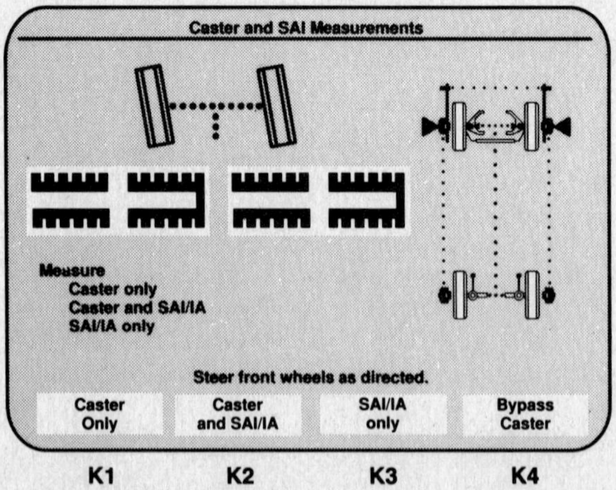

Figure 14-31. Use this screen to help position the wheels when you are measuring caster. The indicator will be at the center of the bar graph when the wheels are properly positioned.

displayed in green are within the manufacturer's recommended tolerances, and do not require adjustment.

Measuring caster and cross caster

The alignment computer cannot measure caster directly. Instead it measures camber roll while you turn the steering wheel, and then calculates caster from the camber roll measurement. The relationship between camber roll and caster is described in Chapter Fourteen of the *Classroom Manual*. Cross caster is simply the difference in caster readings at each front wheel. Measure caster and cross caster as follows:

1. Press the "Measure Caster" button, which brings up the "Caster and SAI Measurements" screen. From this screen, you have the choice of measuring caster, SAI, or both at the same time. Taking both measurements at the same time is not as accurate as taking them separately. This has to do with the way the wheel sensor is designed.
2. For this wheel alignment, measure caster and SAI separately. Press "Caster Only" (K1).
3. The computer will display two bar graphs, one for the right and one for the left wheel. Each bar has an indicator in it that tells you the position of the wheels, figure 14-31.
4. Loosen the lock knobs on the front wheel sensors, and position the front wheels straight ahead. Turn the steering wheel until both indicators are centered in the bar graphs. Hold the steering wheel in position until the computer indicates to continue.
5. The computer screen will instruct you to turn the wheels to the left. Use the steering wheel to turn the wheels until the indicators in the bar graphs are centered again. Hold the steering wheel in position until the computer indicates to continue.
6. Turn the steering wheel to the right, as indicated on the computer screen. Continue turning the steering wheel until the indicators in both bar graphs are centered. Again, hold the steering wheel in place until the computer indicates to continue.
7. Return the wheels to the straight-ahead position and center the indicators in the bar graphs. Hold the steering wheel steady until the computer displays the caster and cross caster readings on the screen.

Measuring SAI and included angle

SAI and included angle are not adjustable in most vehicles, but measuring these angles may give you information about the condition of the suspension. Use the following procedure to check SAI and included angle. The chart in figure 14-32 can help you determine which suspension parts, if any, are damaged based on alignment angle readings.

1. Press the "Measure Caster" button, which brings up the "Caster and SAI Measurements" screen.
2. Press "SAI/IA Only" (K3).
3. Use the air jack at the front of the vehicle to raise the front wheels until they are not touching the turntables. Be sure to place the jack supports at the manufacturer's recommended lift points.
4. Hold down the brake pedal using the brake pedal depressor, figure 14-33. If the vehicle has power brakes, start the engine before pressing the brake pedal. When the brake pedal depressor is in place, turn off the engine. If you wish, remove the brake light fuse so that the battery does not drain, but be sure to install the fuse before returning the vehicle to the customer.
5. Level and lock the sensors at the front wheels.
6. Position the front wheels straight ahead. As in the caster measurements, turn the steering wheel until both indicators are centered in the bar graphs. Hold the steering wheel in position until the computer indicates to continue.
7. The computer screen will instruct you to turn the wheels to the left. Use the steering wheel to turn the wheels until the indicators in the bar graphs are centered again.

DIAGNOSING SAI, CAMBER, AND INCLUDED ANGLE			
SLA AND STRUT/SLA SUSPENSIONS			
SAI	**CAMBER**	**INCLUDED ANGLE**	**DIAGNOSIS**
CORRECT	LESS THAN SPECS	LESS THAN SPECS	BENT STEERING KNUCKLE OR SPINDLE
LESS THAN SPECS	GREATER THAN SPECS	CORRECT	BENT LOWER CONTROL ARM
LESS THAN SPECS	GREATER THAN SPECS	GREATER THAN SPECS	BENT LOWER CONTROL ARM AND STEERING KNUCKLE OR SPINDLE
GREATER THAN SPECS	LESS THAN SPECS	CORRECT	BENT UPPER CONTROL ARM
STRUT SUSPENSIONS			
SAI	**CAMBER**	**INCLUDED ANGLE**	**DIAGNOSIS**
CORRECT	LESS THAN SPECS	LESS THAN SPECS	BENT SPINDLE AND/OR STRUT
CORRECT	GREATER THAN SPECS	GREATER THAN SPECS	BENT SPINDLE AND/OR STRUT
LESS THAN SPECS	GREATER THAN SPECS	CORRECT	BENT CONTROL ARM OR STRUT TOWER OUT AT TOP
LESS THAN SPECS	GREATER THAN SPECS	GREATER THAN SPECS	BENT CONTROL ARM OR STRUT TOWER OUT AT TOP, ALSO BENT SPINDLE AND/OR STRUT
LESS THAN SPECS	LESS THAN SPECS	LESS THAN SPECS	STRUT TOWER OUT AT TOP AND BENT SPINDLE AND/OR BENT STRUT OR CONTROL ARM
GREATER THAN SPECS	LESS THAN SPECS	CORRECT	STRUT TOWER IN AT TOP
GREATER THAN SPECS	GREATER THAN SPECS	GREATER THAN SPECS	STRUT TOWER IN AT TOP AND BENT SPINDLE AND/OR BENT STRUT
KINGPIN TWIN I-BEAM SUSPENSION			
SAI (KPI)	**CAMBER**	**INCLUDED ANGLE**	**DIAGNOSIS**
CORRECT	GREATER THAN SPECS	GREATER THAN SPECS	BENT SPINDLE
LESS THAN SPECS	GREATER THAN SPECS	CORRECT	BENT I-BEAM
LESS THAN SPECS	GREATER THAN SPECS	GREATER THAN SPECS	BENT I-BEAM AND SPINDLE
GREATER THAN SPECS	LESS THAN SPECS	CORRECT	BENT I-BEAM

Figure 14-32. Not all alignment angles are adjustable. SAI is one angle that is not adjustable on most vehicles. By comparing SAI with other alignment angles, you can determine if suspension components are damaged.

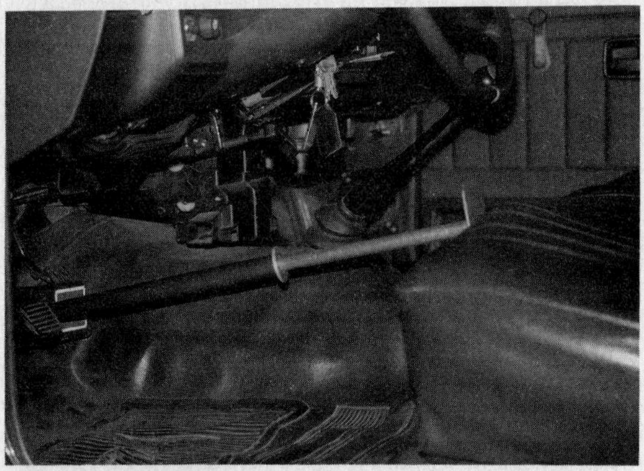

Figure 14-33. The brake depressor is a tool that holds the brake pedal down while you are measuring SAI and included angle.

Hold the steering wheel in position until the computer indicates to continue.

8. Turn the steering wheel to the right, as indicated on the computer screen. Continue turning the steering wheel until the indicators in both bar graphs are centered. Again, hold the steering wheel in place until the computer indicates to continue.
9. Return the wheels to the straight-ahead position and center the indicators in the bar graphs. Hold the steering wheel steady until the computer displays the SAI and included angle readings on the screen.
10. Lower the front wheels onto the alignment rack. The camber and SAI readings will change, but included angle will remain constant.

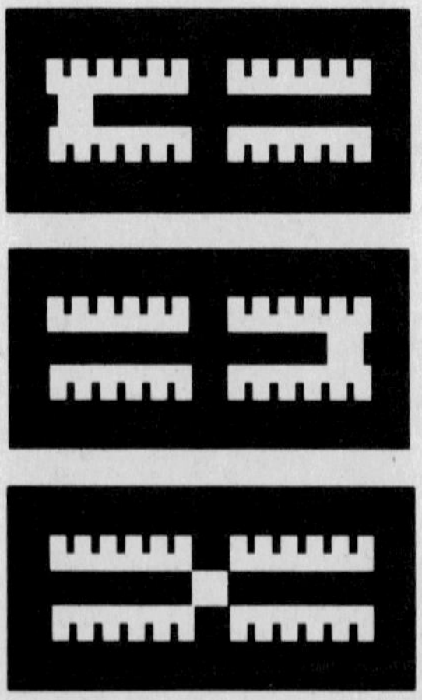

Figure 14-34. When the indicators are centered in the bar graphs, the wheels are properly adjusted.

Adjusting alignment angles
The procedures in this section describe how to use the computer to read alignment angles while you are making adjustments. Adjusting the angles on different types of vehicles is described later in this chapter in the section titled "ALIGNMENT ANGLE ADJUSTMENTS".

The Hunter D-111 and many other aligners have adjustment information for various makes and models stored in the computer's memory. Before starting the adjustments, access adjustment information whenever you need it by pressing the button labeled "Illustrate Adjustments". To return to the adjustment screen, press the K4 button.

The adjustment screens use bar graphs with an indicator that moves along the graph. When the indicator is centered in the bar graph, the angle is within the recommended specifications. As you make adjustments, the indicator will move closer to the center of the graph, figure 14-34.

Angles that are out of specification are flagged on the computer screen. Many aligners will display them in red. As the adjustment moves the angle within the manufacturer's recommended tolerances, the graph will turn green.

Adjust alignment angles in the following order:

- Rear wheel camber (if adjustable)
- Rear wheel toe (if adjustable)
- Front wheel camber
- Front wheel caster (if adjustable)
- Front wheel toe.

An exception to this order of adjustment is when you are working on a vehicle with four-wheel steering, so you should check the vehicle manufacturer's recommendations. For example, when performing a wheel alignment on a Honda Prelude with four-wheel steering, Honda recommends that you first adjust front caster and camber. Then lock the rear steering gear, and adjust rear camber and toe. Check total rear toe. Then adjust front toe, and check total front toe. Finally, check that the steering wheel is centered. If it is off center by more than 0.4 inches (10 mm), remove the steering wheel and reposition it one spline for every 0.4 inches (10 mm) that it is off center.

On some vehicles, you must adjust camber and caster at the same time because a single adjustment affects both angles. Getting both camber and caster within specifications is sometimes a matter of trial-and-error.

Some vehicle manufacturers provide an alternative to the trial-and-error method. A chart in the service manual allows you to determine how far and which way to make adjustments in order to get both camber and caster within specifications, figure 14-35. The chart in this example is for a 1990 Toyota truck, which uses cam bolts to adjust camber and caster. Similar charts exist for other vehicles. You cannot use the same chart for all vehicles. Use this chart when adjusting a 1990 Toyota truck as follows:

1. Use wheel alignment equipment to determine camber and caster angles, as described earlier in this chapter.
2. On the chart, the solid lines represent camber and the dashed lines represent caster. Using the solid lines, locate the camber reading and mark it. Then locate the caster reading on the dashed lines and mark it.
3. Locate the point on the chart where the camber and caster lines meet. Mark this point.
4. To determine the adjustments that you must make, count how far up or down and how far right or left you must go to get to zero. These are your adjustments for the front and rear cam bolts.

For example, if the vehicle that you are working on has a camber reading of 0° and a caster reading of 3°, mark the point where these lines meet on the chart. Then count how far up and how far over you have to go to get to zero. Any movement up or down is an adjustment to the front cam bolt, and any movement right or left is an adjustment to the rear cam bolt. To get to zero, you must move up 3 and to the right 5$^1/_2$. This means that you must increase the front cam bolt by 3 increments and increase the rear cam bolt by 5$^1/_2$

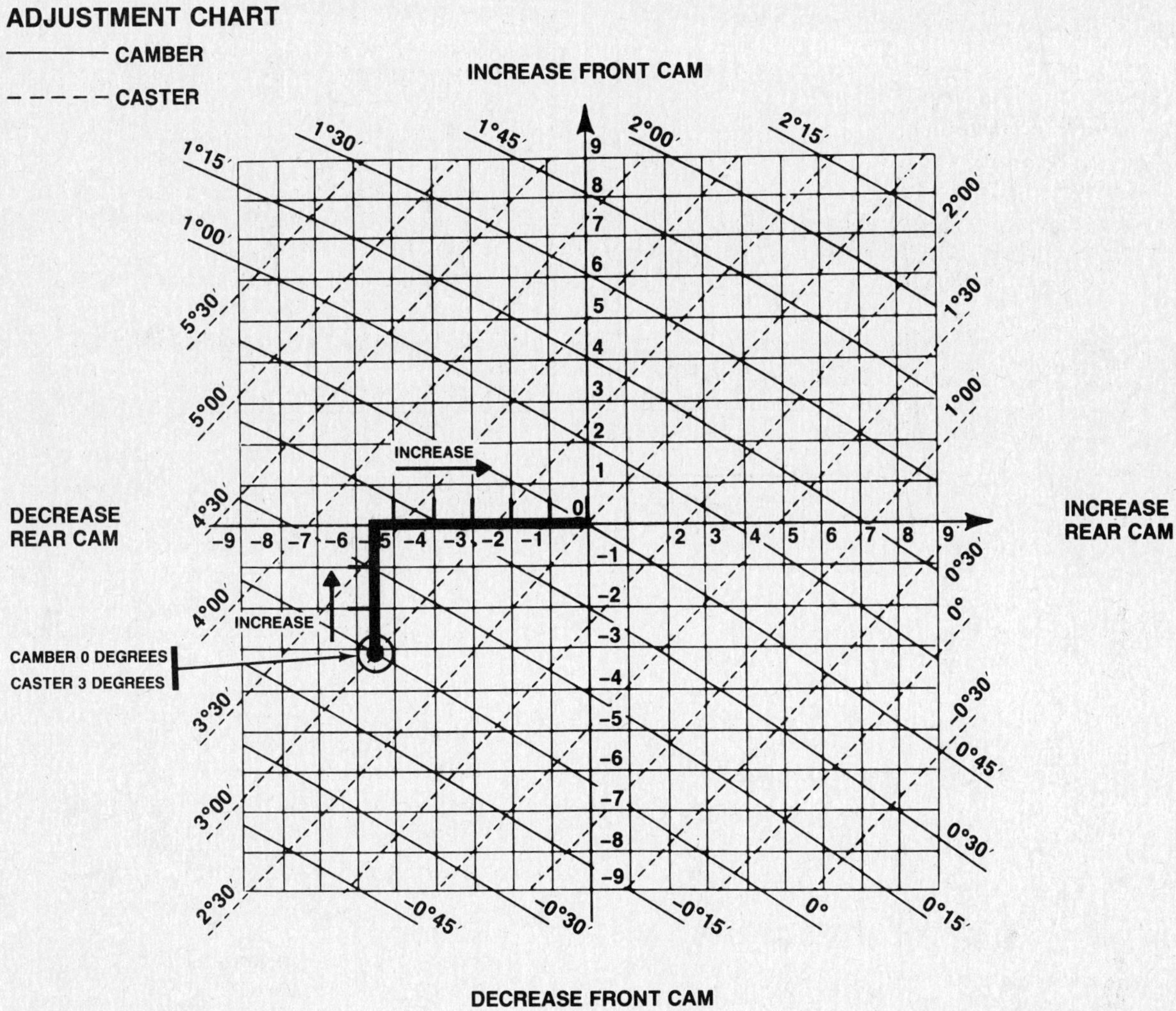

Figure 14-35. This type of chart will help you set camber and caster on a vehicle that uses a single adjustment for both angles. Camber is represented by the solid lines, and caster by the dashed lines.

increments for camber and caster to be within specifications.

Adjusting rear wheel alignment angles
Rear camber and toe are adjustable on some vehicles. If you are not sure whether the rear alignment angles are adjustable on the vehicle you are servicing, use the ''Illustrate Adjustments'' feature mentioned earlier. Illustrations and instructions will be available from the computer's memory if the angle is adjustable. Note that the vehicle manufacturer's service manual may not mention special tools available from the aftermarket for adjustments, but the computer will. Many vehicles that are normally not adjustable from the manufacturer can in fact be adjusted if you have the right aftermarket parts. Wheel alignment aftermarket parts catalogs will also give you access to this information. If you determine that rear alignment angles are adjustable, continue with this procedure. If they are not adjustable, continue with the procedure titled ''Adjusting front wheel alignment angles''.

1. From the ''Vehicle Adjustments'' screen, press ''Select Adjustment'' (K3).
2. Press ''Adjust Rear Wheels'' (K4). This will bring bar graphs on the screen for rear toe and camber adjustments, figure 14-36.
3. Press the shift key twice to start the adjustment procedure.

Note: If you have to raise the wheels to make adjustments, press ''Raise Rear Wheels'' (K1). This lets the computer know that the alignment readings are changing because the rear wheels are being raised.

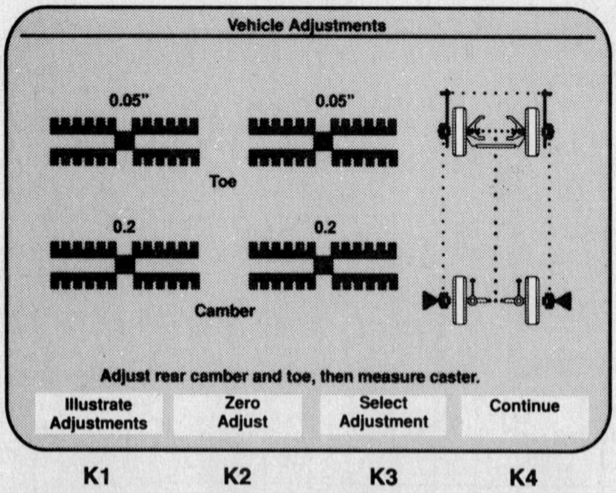

Figure 14-36. The rear wheels are not adjustable on all vehicles. If they are adjustable on the vehicle that you are servicing, use the rear wheel adjustment screen to adjust toe and camber.

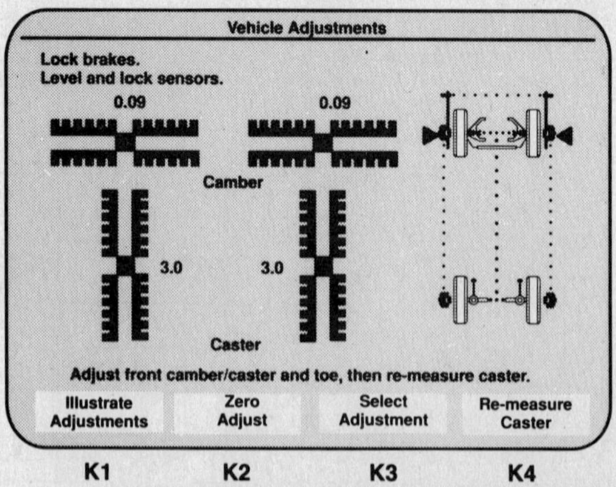

Figure 14-37. Most late-model vehicles have adjustable front camber and caster. Use this screen when making the adjustments. Notice the procedural reminders in the upper left corner of the screen.

4. Use the air jack to raise the rear wheels off the ground. Be sure to place the lift supports at the vehicle manufacturer's recommended lift points.
5. Press "Ready" (K1).
6. Use the instructions from the vehicle manufacturer's service manual or the computer's "Illustrate Adjustments" memory to adjust rear camber. Also, later in this chapter is a section titled "ALIGNMENT ANGLE ADJUSTMENTS" that you can refer to for illustrations and descriptions of rear camber adjustments. Use the bar graphs and indicator as a guide when adjusting camber. When the indicator is in the center of the bar graph, tighten the adjuster so camber will not change.
7. Before lowering the rear axle, press "Lower Rear Wheels" (K1). After lowering the vehicle, bounce it a few times to settle the suspension. Check the rear camber reading to be sure it is within specifications.
8. Press "Adjust Rear Wheels" (K4). Adjust rear toe using the instructions from the vehicle manufacturer's service manual or the computer's "Illustrate Adjustments" memory. Also, you can refer to the section later in this chapter titled "ALIGNMENT ANGLE ADJUSTMENTS" for illustrations and descriptions of various types of rear toe adjustments. Use the bar graphs as a guide when adjusting toe. When the indicator is in the center of the bar graph, tighten the adjuster so the angle will not change.
9. Bounce the vehicle a few times to settle the suspension. Check the rear toe reading to be sure it is within specifications.
10. After making the adjustments, check rear thrust angle, which should be close to zero. As a general rule of thumb, one-tenth of a degree of thrust is considered marginal on a vehicle with adjustable rear alignment angles. If the rear alignment angles are not adjustable, one-fifth of a degree of thrust is considered passable. Align the front wheels if thrust angle is acceptable. If it is not, you should make repairs before continuing with the front-end alignment.

Adjusting front wheel alignment angles

Use this procedure to adjust the front wheels during an alignment. If you are performing a four-wheel alignment and the rear wheels are not adjustable, then you can adjust the front wheels to compensate for rear wheel thrust, as discussed in Chapter Fourteen of the *Classroom Manual*. Use the computer to align the front wheels, as follows:

1. Press the "Adjust Front Wheels" button. The computer will display the bar graphs for front camber and caster, figure 14-37.
2. Set the parking brake. Hold down the brake pedal using the brake pedal depressor. If the vehicle has power brakes, start the engine before pressing the brake pedal. Then turn off the engine. If you wish, remove the brake light fuse so that the battery does not drain, but be sure to install the fuse before returning the vehicle to the customer.

Figure 14-38. Use this tool to lock the steering wheel in place after centering it. If you neglect to center and lock the steering wheel, it may move off-center after you make the toe adjustments.

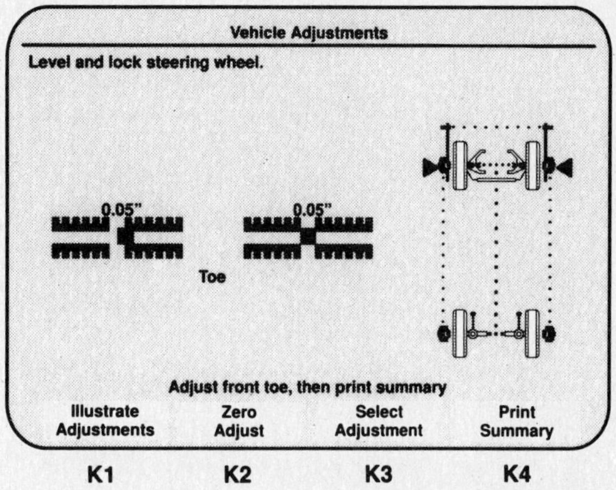

Figure 14-39. This screen is for adjusting toe at the front wheels. Again, notes at the upper left corner of the screen remind you to level and lock the steering wheel before starting the adjustment.

3. Level and lock the sensors.
4. Adjust camber using the instructions from the vehicle manufacturer's service manual or the computer's ''Illustrate Adjustments'' memory. Also, you can refer to the section later in this chapter titled ''ALIGNMENT ANGLE ADJUSTMENTS'' for illustrations and descriptions of various camber adjustments. Use the bar graphs and indicator as a guide when adjusting camber. When the indicator is in the center of the bar graph, tighten the adjuster so the angle will not change.

Note: If you must raise the vehicle to make adjustments, enter this information in the computer. Press the shift key twice. Then press the K1 button (Raise Wheels). Use the air jack to raise the front axle, being sure to put the lift supports at the manufacturer's recommended lift points. Re-level the sensors, if needed, and press ''Ready'' (K1). After making the needed adjustments, press ''Lower Front Wheels'' (K1). Then lower the vehicle.

5. If you had to make a large camber adjustment, re-measure caster. Because caster is determined by measuring camber roll, a large camber adjustment will change the caster reading and may actually put caster within specifications.
6. Press ''Re-Measure Caster'' (K4). Use the procedure given earlier in this chapter titled ''Measuring caster and cross caster''. After completing the caster measurement, new camber and caster readings will be displayed on the screen.
7. If needed, make any final adjustments to caster or camber, and tighten the adjustment. Bounce the front of the vehicle to settle the suspension. Again, re-measure caster by pressing ''Re-Measure Caster'' (K4) to check the final adjustment.
8. The last adjustment is for front toe. First, center the steering wheel, and lock it in position using the special tool, figure 14-38. If the vehicle has power steering, start the engine, center the steering wheel, and place the steering wheel lock tool in position. Then turn off the engine.
9. Press ''Select Adjustment'' (K3), and select ''Toe''. The bar graphs for toe adjustment will be displayed on the computer screen, figure 14-39. Use the instructions from the vehicle manufacturer's service manual or the computer's ''Illustrate Adjustments'' memory to adjust front wheel toe. Also, you can refer to the section later in this chapter titled ''ALIGNMENT ANGLE ADJUSTMENTS'' for illustrations and descriptions of toe adjustments. When the indicator is centered in the bar graph, tighten the adjuster so toe will not change.

Note: If you neglect to center and lock the steering wheel before starting the toe adjustment, the steering wheel may move off-center after adjusting toe.

10. Bounce the front of the vehicle to settle the suspension, and recheck the alignment angle readings. If the readings are within specifications, the alignment is complete. Test drive the vehicle before returning it to the customer.

Printing the alignment report
The Hunter D-111 computer can print the results of the alignment if you need them or if the customer requests them. Print the results as follows:

1. Turn on the printer's power switch.
2. There are four types of print jobs to choose from:

    - Press "Print Inspection" (K4) to print out the pre-alignment inspection.
    - Press "Print Alignment" (K3) to print out alignment measurements. Use this button to print the measurements before making alignment adjustments. Then you will have the "before" and "after" alignment results.
    - Press "Print Specs" (K1) for a printout of the manufacturer's recommended alignment specifications.
    - After making adjustments, press "Print Summary" (K4), which will print the results of the pre-alignment inspection, the manufacturer's recommended specifications, and the actual alignment settings after the adjustment.

Two-Wheel (Front-Wheel) Alignment

You may recall from the procedure given earlier in this chapter that when you use a computerized wheel aligner, you can choose one of three alignment types: a four-wheel alignment, a thrustline alignment, or a centerline alignment. If you were to choose the centerline alignment, you would be performing a two-wheel (front-end) alignment. During a two-wheel alignment, the position of the rear wheels is not taken into consideration when adjusting the front wheels, so there is no compensation for rear wheel thrust, and the front wheels are simply aligned to the vehicle's centerline.

Two-wheel alignments were adequate when nonadjustable, solid-rear axle vehicles dominated the market. However, throughout the 1980s, rear suspension system designs became more sophisticated and sensitive to rear alignment angles, making the four-wheel alignment necessary on many late-model vehicles.

We will not cover two-wheel alignment procedures in great depth because most repair shops use computerized four-wheel aligners. Instead, we will briefly describe the computerized two-wheel alignment and the bubble gauge alignment methods. Before any wheel alignment, perform the pre-alignment inspection, as described earlier in this chapter.

Computerized two-wheel alignment
Perform the computerized two-wheel alignment using the procedures for the four-wheel alignment given earlier in this chapter. However, instead of selecting "four-wheel alignment", select "centerline alignment". Mount and compensate the sensors, take alignment readings, and adjust the alignment angles as described in the four-wheel alignment procedure, except adjust only front-wheel camber, caster, and toe.

Gauge method two-wheel alignment
The gauge alignment method is an old-fashioned way of performing a two-wheel alignment, but it is accurate if performed properly. Camber and caster measurements are made using a bubble gauge that has three levels on it — one for camber, one for caster, and one for positioning the gauge on the wheel, figure 14-40. Toe is measured using a "bar gauge", figure 14-41, which is a long gauge that measures the distance between the fronts and backs of the tires.

To measure camber and caster, drive the vehicle onto an alignment rack or onto two turntables placed on the floor. The turntables should be locked. Be sure that the front tires are centered on the turntables.

Set the parking brake, and hold down the brake pedal using the brake pedal depressor tool. If the vehicle has power brakes, start the engine, put the depressor tool in place, and turn off the engine. You may want to remove the brake light fuse so that you do not drain the battery. Be sure to install the fuse after completing the alignment.

Unlock the turntables. Remove the front wheel hub caps and the dust caps covering the spindles. Wipe off excess grease. Before installing the camber/caster gauge, select the proper template for the vehicle you are servicing, place it over the gauge, and tighten the screws to hold the template in place. These templates have graduated scales that indicate the proper position of the gauge's bubble for the vehicle you are servicing. There are a number of templates available.

Place the camber/caster gauge on the wheel spindle. A strong magnet at the end of the gauge will pull it toward the spindle and hold it in place. Position the gauge horizontally using the small level at the top as a guide.

To read camber, check the camber scale on the template. The bubble in the level should be within the shaded region on the scale. If camber is out of specifications, the bubble will be out of the shaded area, and you must make alignment adjustments.

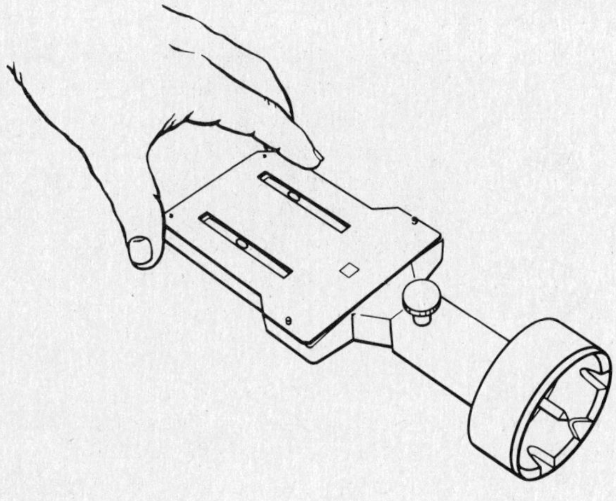

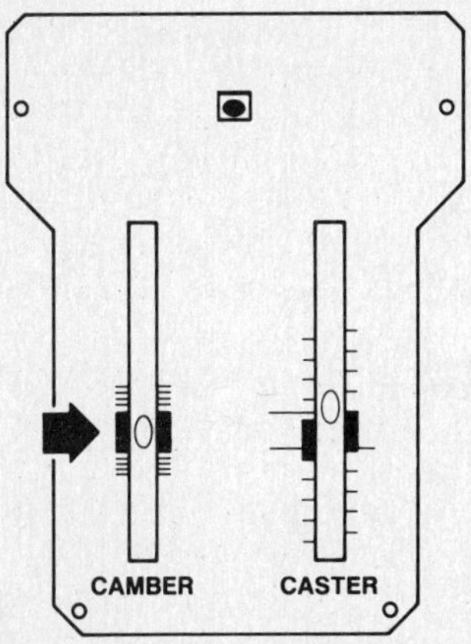

Figure 14-40. On this type of gauge, place the template for the vehicle that you are working on over the levels. The templates mark where the bubbles should be centered when the wheel alignment angles are correct.

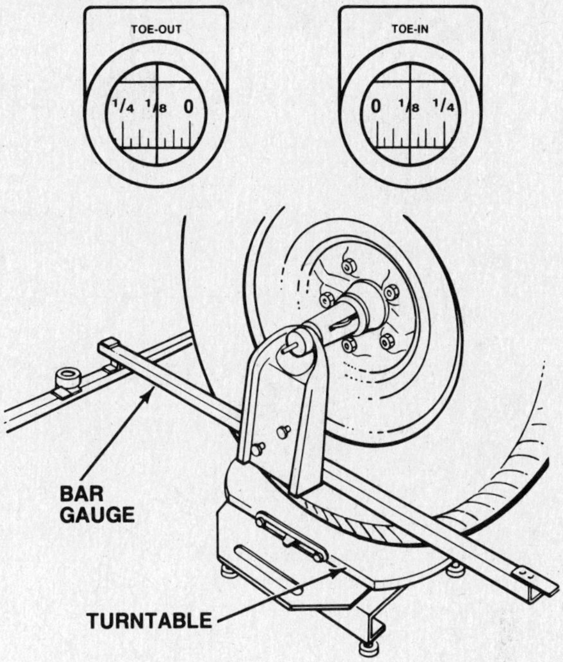

Figure 14-41. This simple tool measures the distances between the fronts of the tires and the backs of the tires. By subtracting one distance from the other, you can determine the vehicle's toe setting.

Then measure caster. Check that the turntable scales are set at zero and the front wheels are positioned straight ahead. Turn the left wheel inward using the steering wheel until the scale on the turntable reads 20 degrees. Adjust the gauge on the left wheel so that the bubble in the caster scale is between the red lines. (You can adjust the position of the gauge by using the screw at the top.)

Turn the steering wheel in the opposite direction until the gauge reads 20 degrees, for a total turn of 40 degrees. The bubble in the gauge should be centered in the shaded area on the caster scale. If it is not, adjust caster. Repeat the measurement to check your adjustment. Check caster on the right wheel using the same method.

Check toe using the "bar gauge" and two parallel bars. Attach one bar to each wheel spindle. Be sure that the bars are parallel to the ground by centering the bubble in the level. The parallel bars will extend past the fronts and backs of the tires.

Attach the bar gauge to the parallel bars to measure the distance between the backs of the tires. Zero the reading on the bar gauge.

Remove the bar gauge, and connect it to the parallel bars at the fronts of the tires. Read the scale on the bar gauge, which will display toe. If needed, adjust front-wheel toe as recommended by the vehicle manufacturer.

ALIGNMENT ANGLE ADJUSTMENTS

A vehicle's suspension system design determines how the alignment angles are adjusted. A good source for wheel alignment information is the vehicle manufacturer's service manual or the alignment computer's memory. Aftermarket parts manufacturers have designed tools and other equipment to make "unadjustable" angles adjustable.

After measuring alignment angles, as described earlier in this chapter, adjust suspension components to bring the alignment angles

Figure 14-42. By placing these shims between suspension components, you can change the alignment angles. Different shim styles are available for the various suspension designs.

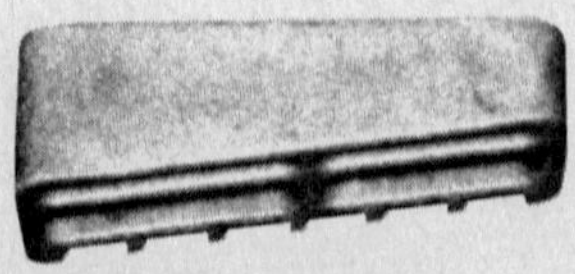

Figure 14-43. Like shims, wedges are made by aftermarket parts manufacturers to fit various suspension types and to change the alignment angles.

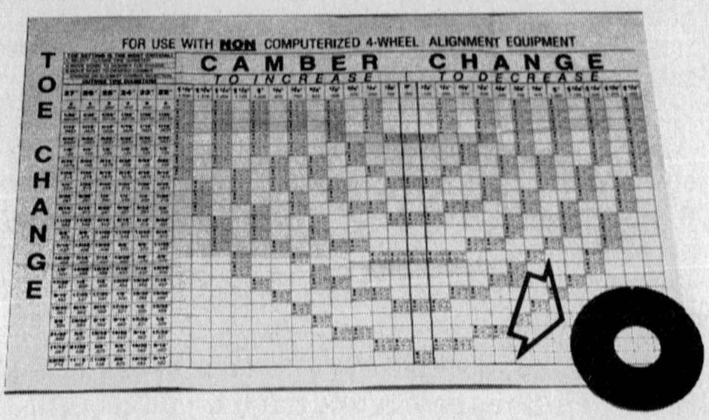

Figure 14-44. This shim is designed for changing toe and camber at the rear wheels. Use the chart that comes with the shim to determine how the shim should be placed on the vehicle to obtain the correct alignment angles.

within specifications. Alignment angles can be adjusted by:

- Placing shims between suspension components and the vehicle frame
- Turning an eccentric cam bolt
- Changing the position of suspension components
- Adjusting the strut rods
- Adjusting ball joint sleeves
- Moving strut mounts
- Bending the axle.

Figure 14-45. This kit contains a variety of shim styles and thicknesses that can be used to adjust alignment angles on a number of vehicles.

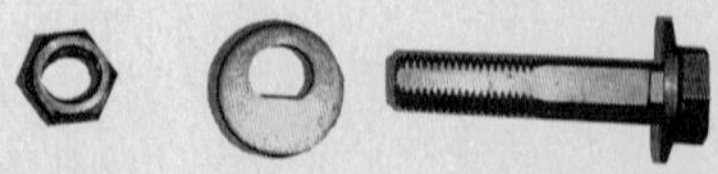

Figure 14-46. The off-center cam on a cam bolt changes alignment angles when you turn the bolt.

Shim Adjustments

Placing shims, figure 14-42, or wedges, figure 14-43, of various thickness between suspension components will change the position of the wheel. This method can be used on front and rear suspensions, figure 14-44. It is best to use one thick shim or wedge rather than many thin ones because thin shims or wedges may slip. Aftermarket manufacturers sell kits that contain shims in a variety of sizes and thicknesses, figure 14-45.

Eccentric Cam Bolt Adjustments

An eccentric cam bolt is a regular bolt with an off-center cam attachment, figure 14-46. When you turn the bolt, the cam turns as well, which moves the suspension component to adjust an alignment angle. For some vehicles, eccentric cam bolts may be purchased from the aftermarket to replace a regular bolt, making an "unadjustable" angle adjustable.

After supporting the rear control arm, remove the bolt that connects the control arm to the frame, figure 14-47. Then use a die grinder to elongate the bolt hole in the frame support, figure 14-48. Insert the replacement eccentric cam bolt in the control arm, and tighten it, figure 14-49. This will allow for rear camber adjustment.

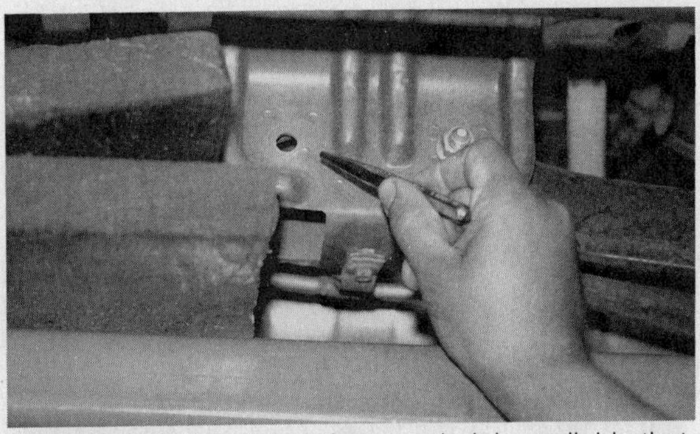

Figure 14-47. An aftermarket cam bolt is available that allows for rear camber adjustment on the Ford Taurus wagon. Start by supporting the rear control arm and removing the bolt that attaches the control arm to the frame.

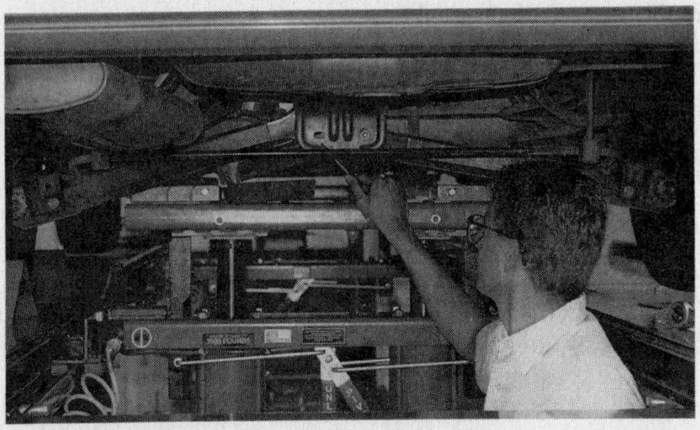

Figure 14-48. Use a die grinder to elongate the bolt hole in the frame support.

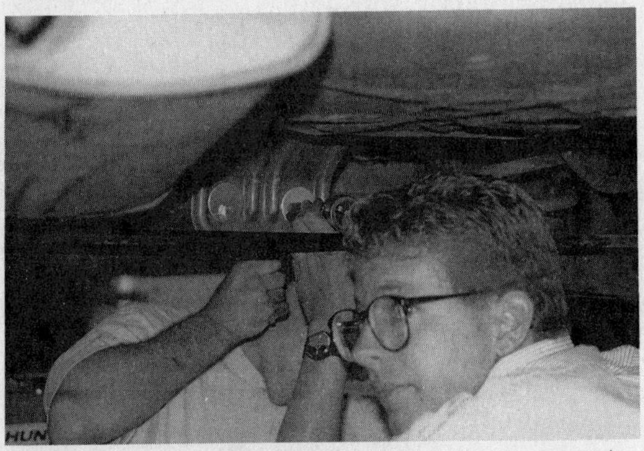

Figure 14-49. Replace the stock bolt with an eccentric cam bolt, which will allow rear camber adjustment.

Suspension Component Adjustment

Changing the position of suspension components will change wheel alignment angles. Some manufacturers use suspension components with slotted bolt holes, which allow you to change the component's position. The vehicle manufacturer may recommend that you

Figure 14-50. These tools are for helping you position the control arms when making alignment adjustments on some Ford vehicles. The small end of this tool fits into a hole in the control arm, and the cross-shaft fits into the tool's grooved fitting. These tools will help you position the control arms before tightening the adjustment.

Figure 14-51. This strut rod bushing allows for adjustment. Turning the bushing changes the position of the strut rod, which changes the alignment angle.

elongate bolt holes yourself with a die grinder to allow for component adjustment. A number of special tools are available from aftermarket parts suppliers to help you position suspension components before tightening the bolts that hold the adjustment, figure 14-50.

Strut Rod Adjustments

Some vehicles have adjustable strut rods. When the strut rod is lengthened or shortened, or if the position of the strut rod is changed by turning a bushing, the control arm moves as well. Aftermarket strut rod bushings with an off-center hole are available for making strut rods adjustable, figure 14-51. By turning the bushing, you can change the position of the strut rod.

Figure 14-52. These sleeves fit into control arms or other suspension components to change the position of the ball joints. This changes the wheel alignment angle.

Ball Joint Sleeve Adjustment

Ball joint sleeves in the control arm allow you to change the position of the ball joint. These sleeves have an off-center hole that the ball joint fits into, figure 14-52. When you turn the sleeve, you change the position of the ball joint and other suspension components. Another type of ball joint sleeve changes the position of the joint, but cannot be rotated for further adjustment.

Strut Mount Adjustment

On some vehicles, the strut mount is movable. Others require modifications in order to change the position of the upper strut mount and adjust camber and caster. Aftermarket manufacturers supply a variety of tools to help you position the strut, such as this one, figure 14-53, that helps you determine where the strut mounting bolts should insert in order to get the correct alignment adjustment. After marking the positions of the mounting bolts, drill the holes, and reposition the strut.

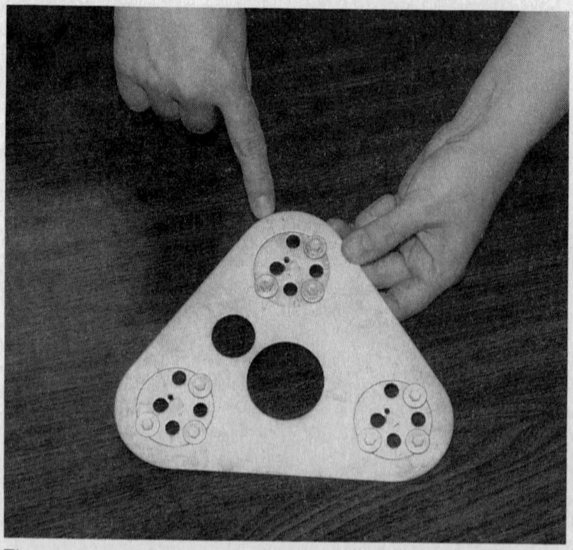

Figure 14-53. This tool is designed to help you position the top of the strut when aligning the front suspension on a Ford Taurus.

Bending the Axle

Although bending the axle seems like an extreme method for adjusting alignment angles, it was used to adjust solid front axle suspensions on some heavy-duty trucks. This adjustment method is not particularly good because bending the axle fatigues the metal. Now many of these vehicles are adjusted with aftermarket parts, such as wedges that fit under the leaf springs.

Camber, Caster, and Toe Adjustments

The following sections include illustrations and brief instructions for adjusting various alignment angles.

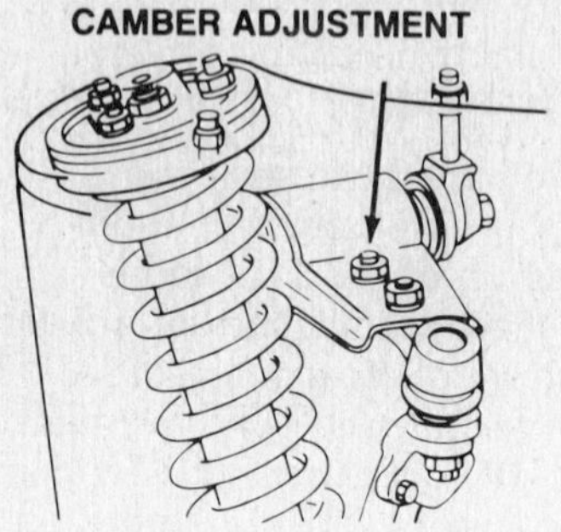

Figure 14-54. Loosen the two nuts on the upper arm, and move the wheel in or out.

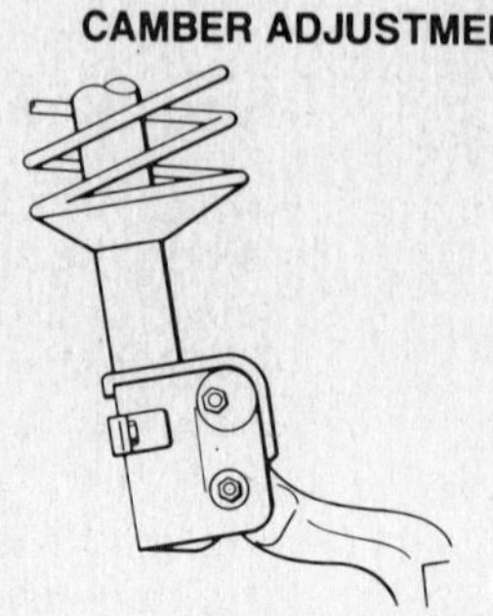

Figure 14-55. Rotate the cam bolt at the strut base.

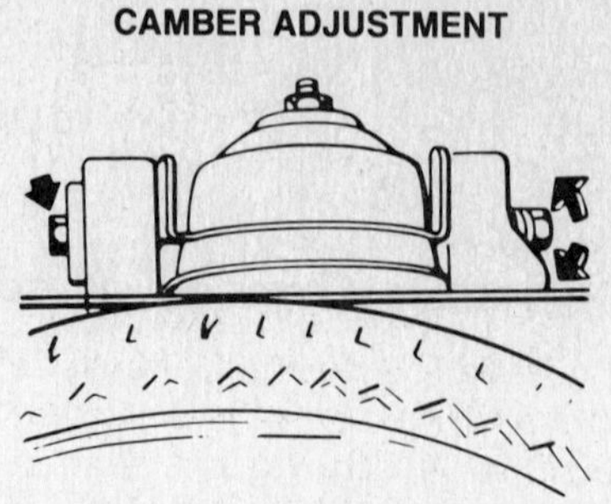

Figure 14-56. Loosen the nut on the upper control arm, and rotate the arm.

CAMBER ADJUSTMENT

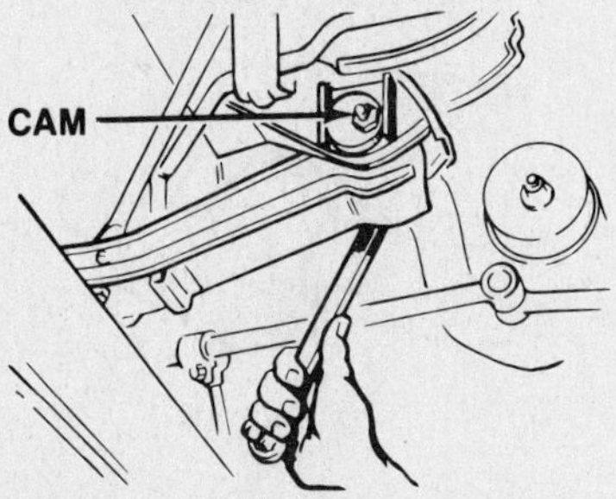

Figure 14-57. Turn the cam at the control arm.

CAMBER ADJUSTMENT

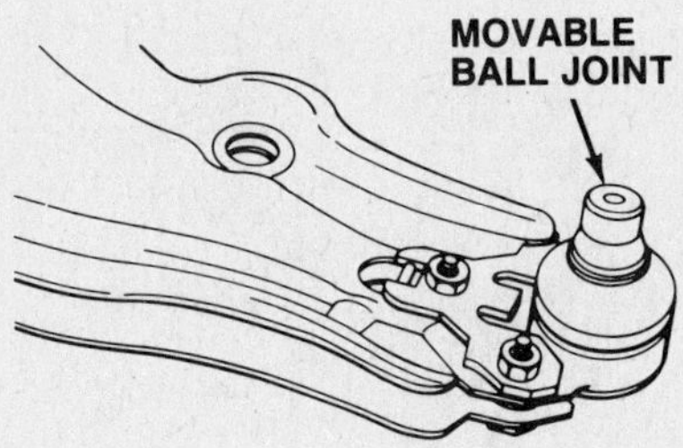

Figure 14-58. Loosen the ball joint mounting bolts, and reposition the joint.

CAMBER ADJUSTMENT

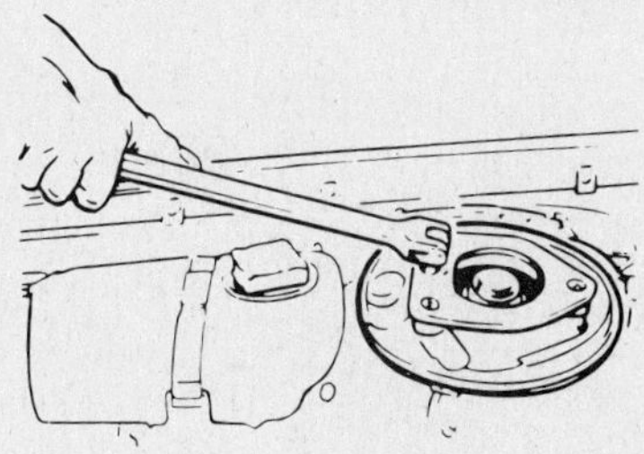

Figure 14-59. Loosen the mounting bolts, and reposition the strut upper mount.

CAMBER ADJUSTMENT

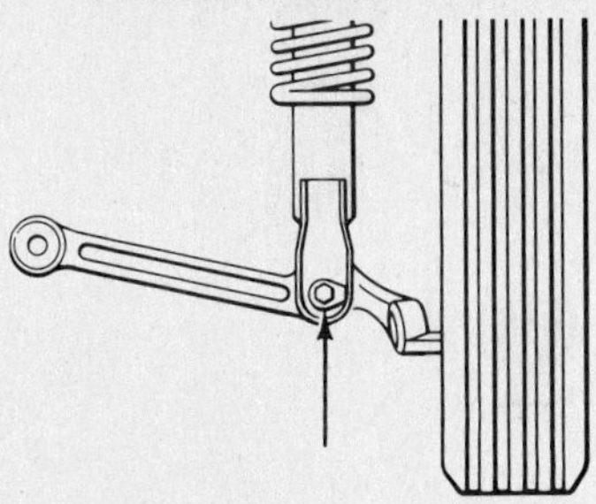

Figure 14-60. Rotate the eccentric at the strut base.

CAMBER ADJUSTMENT

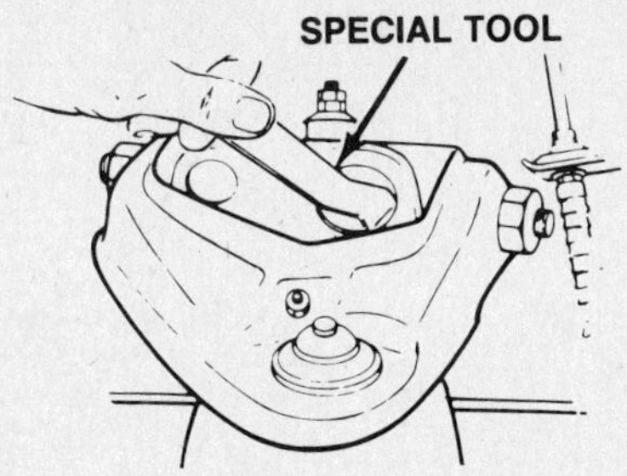

Figure 14-61. At the upper control arm, after adjusting caster, change the shim size to adjust camber.

CAMBER ADJUSTMENT

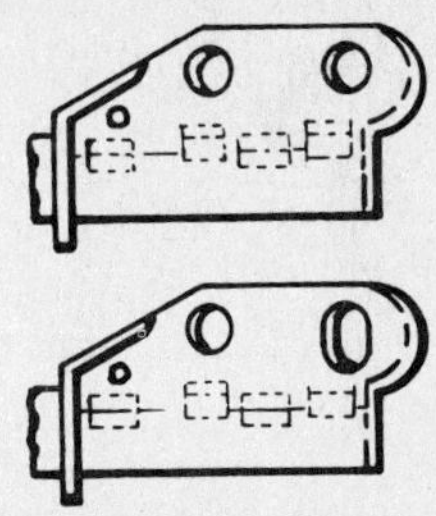

Figure 14-62. Disconnect the strut lower mount, and file the bolt holes so that they become adjustable slots.

CAMBER ADJUSTMENT

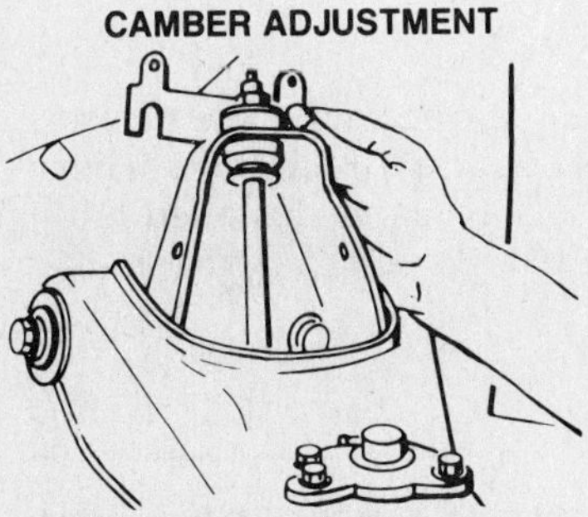

Figure 14-63. Add or remove shims at the control arm.

CAMBER ADJUSTMENT

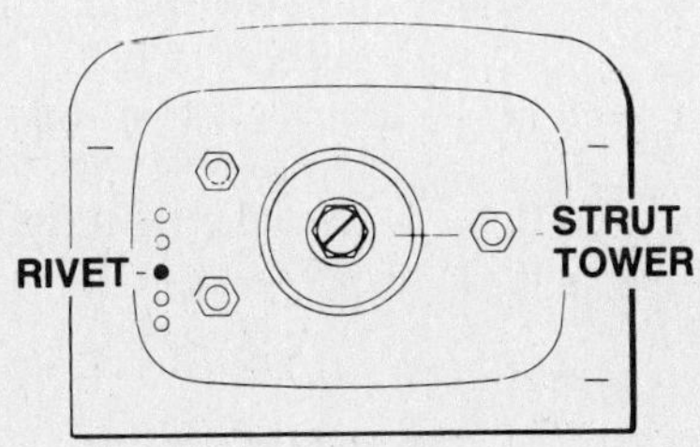

Figure 14-64. Drill out the rivet. Then loosen the strut mounting bolts and reposition the strut.

CAMBER ADJUSTMENT

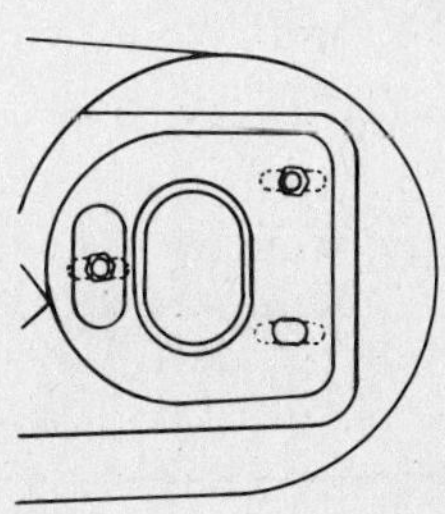

Figure 14-65. File slots in the upper strut mounts and reposition the strut.

Adjusting front wheel camber

Camber is not adjustable on all vehicles. On a strut suspension, camber is often adjusted by repositioning the top of the strut. Other camber adjustments for strut suspensions are made at the base of the strut. On vehicles with control arms, camber may be adjusted in a number of ways. Use figures 14-54 through 14-65 to adjust camber.

CASTER ADJUSTMENT

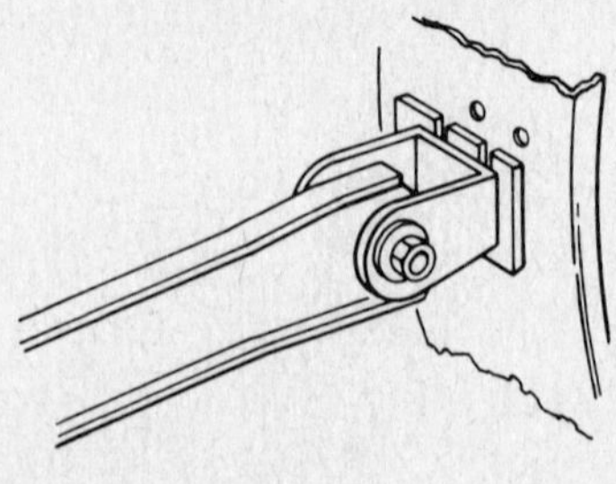

Figure 14-66. Add or subtract shims at the lower control arm.

CASTER ADJUSTMENT

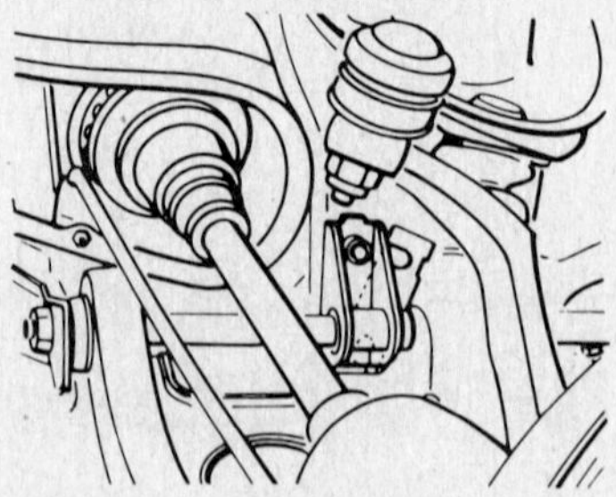

Figure 14-67. Add or remove shims at the torsion bar brace.

CASTER ADJUSTMENT

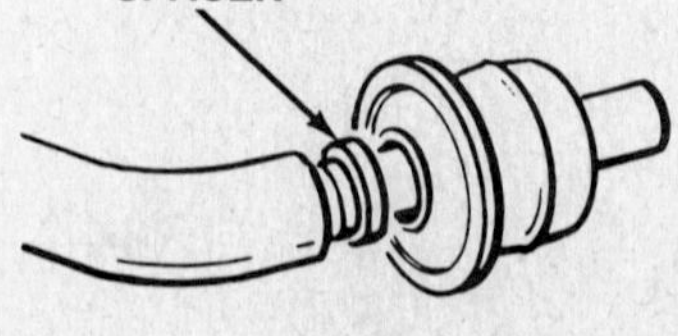

Figure 14-68. Add or remove spacers at the strut rod.

CASTER ADJUSTMENT

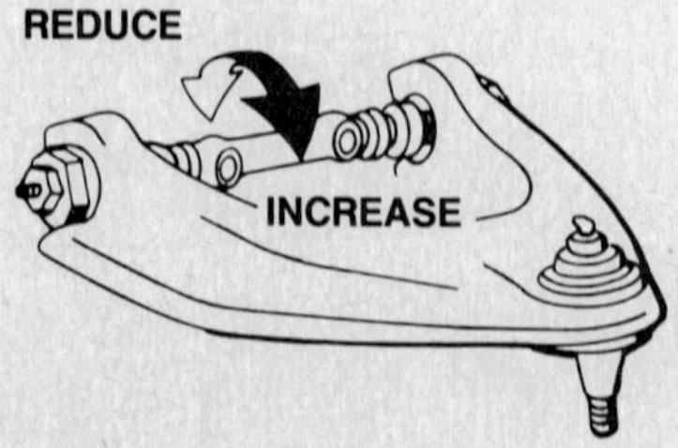

Figure 14-69. Remove the upper control arm from the crossmember, and turn the cross shaft.

CASTER ADJUSTMENT

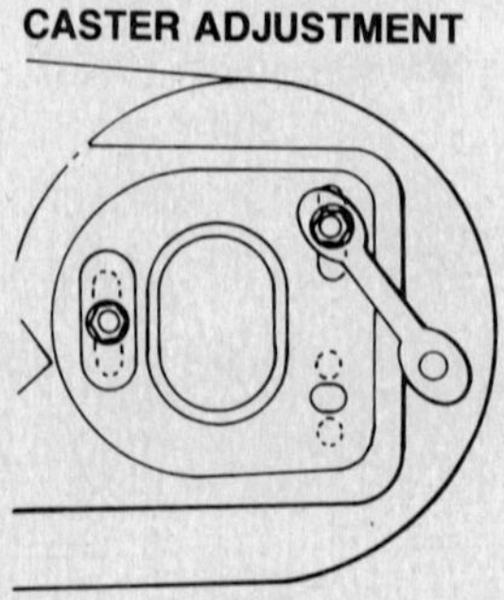

Figure 14-70. File slots in the upper strut mount, and reposition the strut.

CASTER ADJUSTMENT

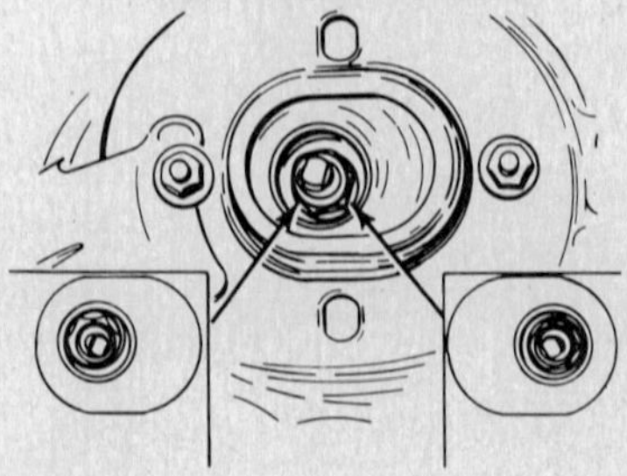

Figure 14-71. Rotate the upper strut mount.

Adjusting caster
Caster is not adjustable on all vehicles, but a few common examples are shown in figures 14-66 through 14-71.

Adjusting front wheel camber and caster simultaneously
On some vehicles, a single adjustment changes both camber and caster. Use figures 14-72 through 14-83 to adjust front wheel camber and caster simultaneously.

Adjusting front wheel toe
Adjust front-wheel toe by turning the tie rod ends to lengthen or shorten the tie rod. There are a number of tools available designed to make turning the tie rods easier, figure 14-84.

CASTER & CAMBER ADJUSTMENT

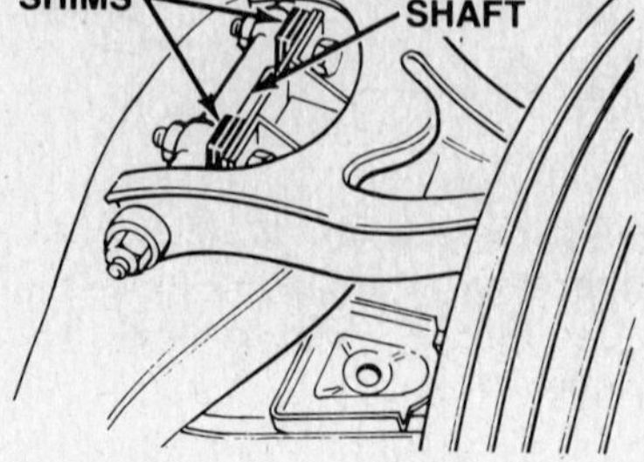

Figure 14-72. Position shims at the upper control arm. To adjust camber only, change shim thicknesses equally.

CASTER & CAMBER ADJUSTMENT

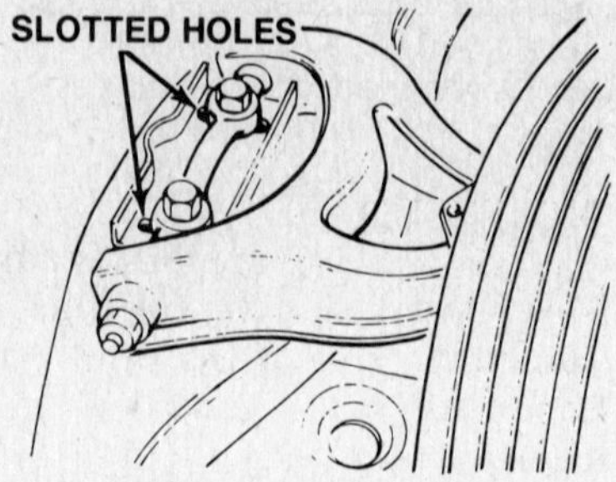

Figure 14-73. Reposition the upper control arm cross shaft in the adjustment slots.

CASTER & CAMBER ADJUSTMENT

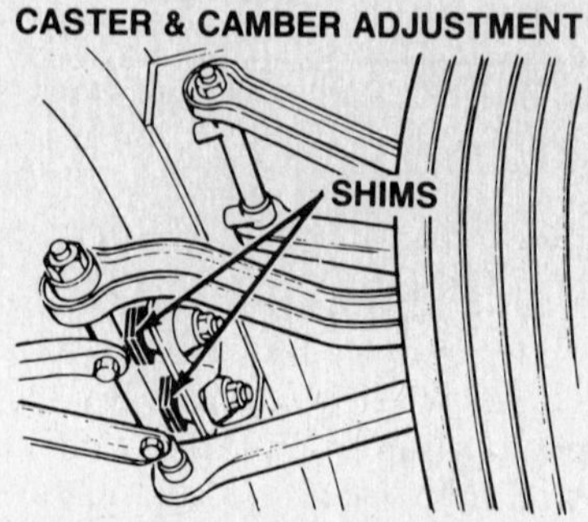

Figure 14-74. Position shims at the upper control arm. To adjust camber only, change shim thicknesses equally.

CASTER & CAMBER ADJUSTMENT

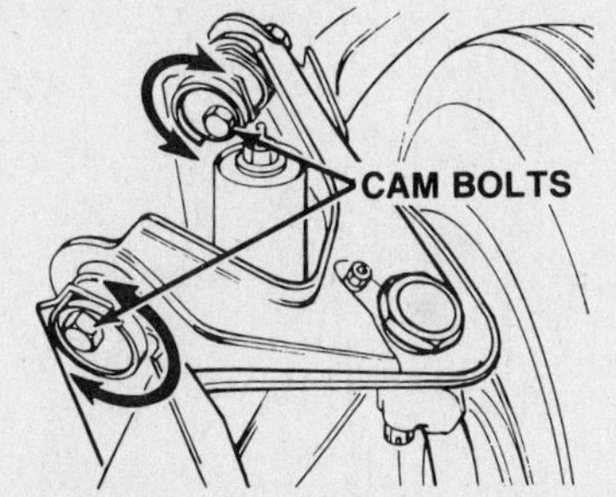

Figure 14-75. Rotate the cam bolts at the upper control arm pivots.

CASTER & CAMBER ADJUSTMENT

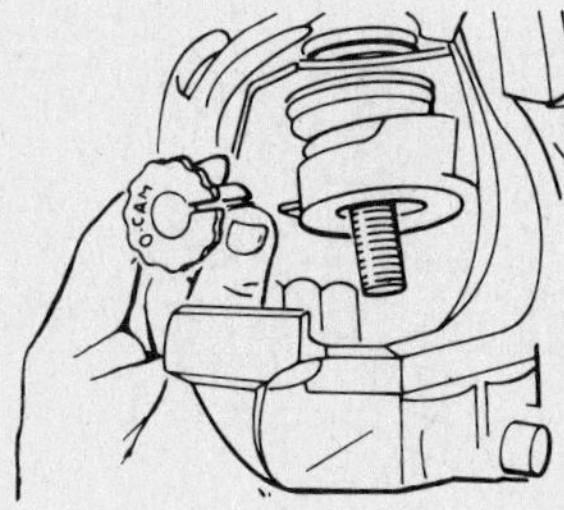

Figure 14-76. Rotate the ball joint adjusting sleeve.

CASTER & CAMBER ADJUSTMENT

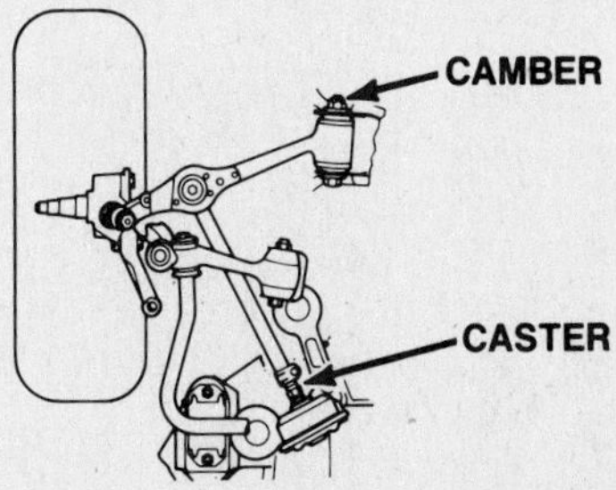

Figure 14-77. For caster, turn the ball pin. For camber, turn the eccentric.

CASTER & CAMBER ADJUSTMENT

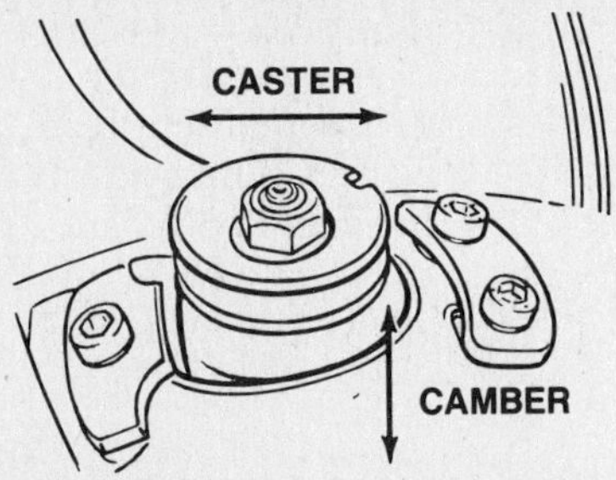

Figure 14-78. Reposition the upper strut mount.

CASTER & CAMBER ADJUSTMENT

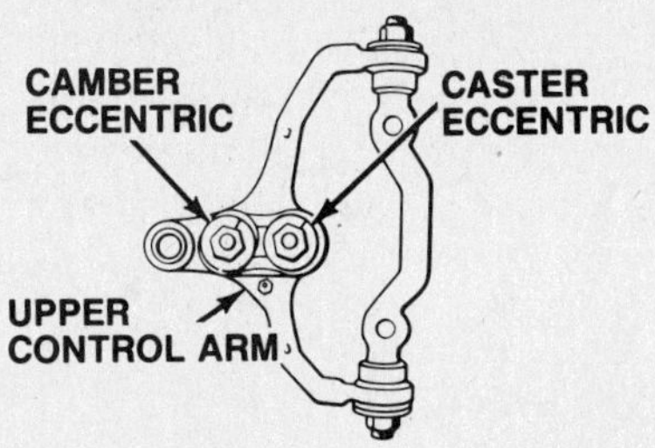

Figure 14-79. Rotate the cams on control arm.

CASTER & CAMBER ADJUSTMENT

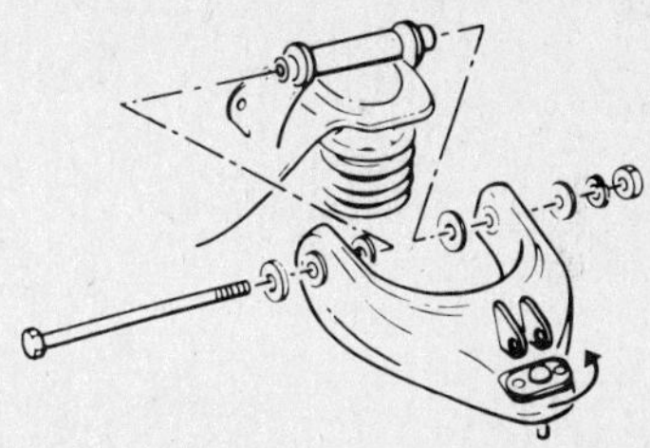

Figure 14-80. For caster, add or remove washers. For camber, rotate the ball joint.

CASTER & CAMBER ADJUSTMENT

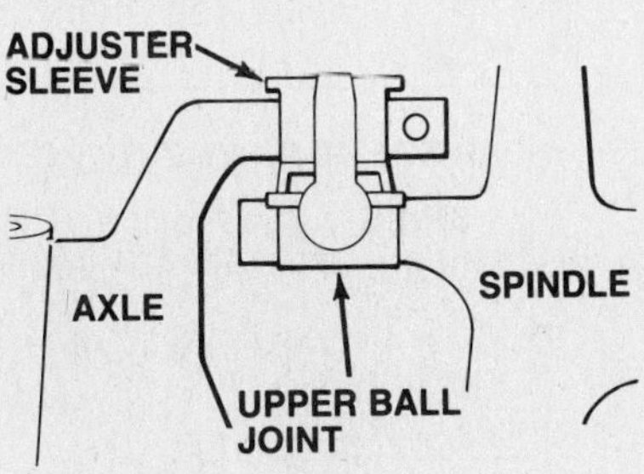

Figure 14-81. Replace the ball joint sleeve with one of the proper size.

CASTER & CAMBER ADJUSTMENT

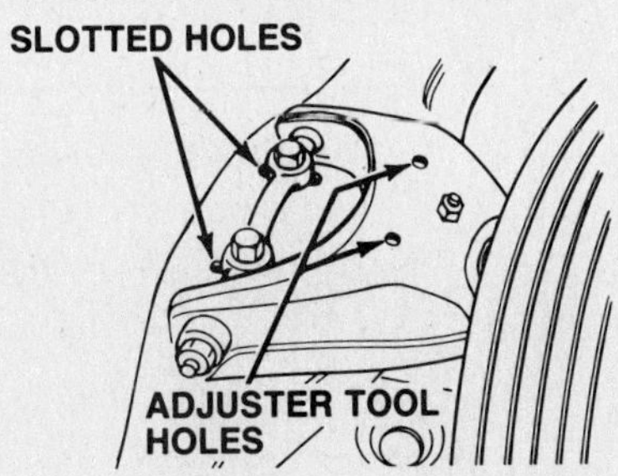

Figure 14-82. Reposition the upper control arm cross shaft on the frame, using the manufacturer's adjuster tool.

CASTER & CAMBER ADJUSTMENT

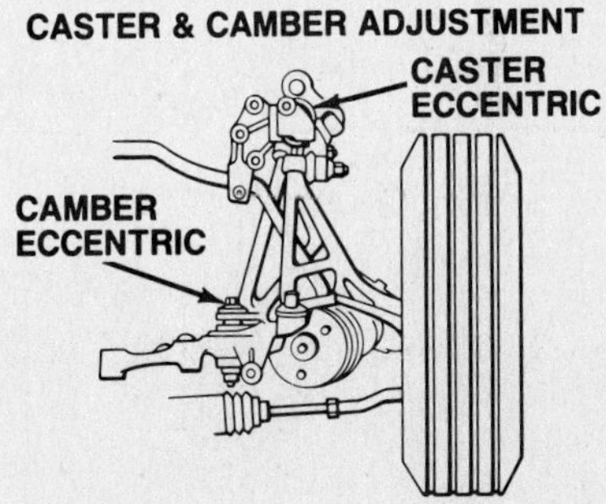

Figure 14-83. Rotate the eccentrics.

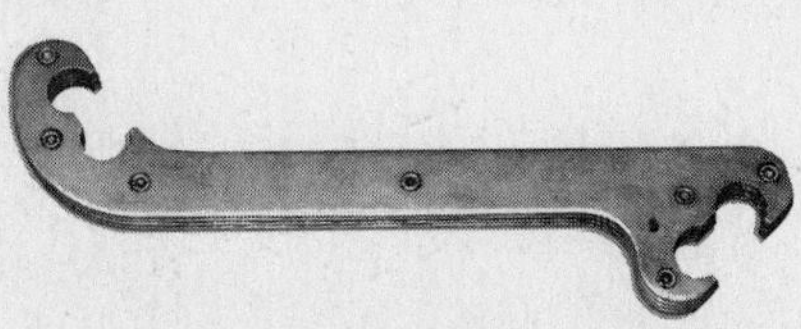

Figure 14-84. This tool is specially designed to grasp tie rods of different diameters so that you can turn them for toe adjustments.

Front toe adjustments are similar from one vehicle to another. As discussed earlier, before adjusting toe you should center and lock the steering wheel. If you forget to do this, the steering wheel will be off-center after you make the adjustments.

REAR CAMBER ADJUSTMENT

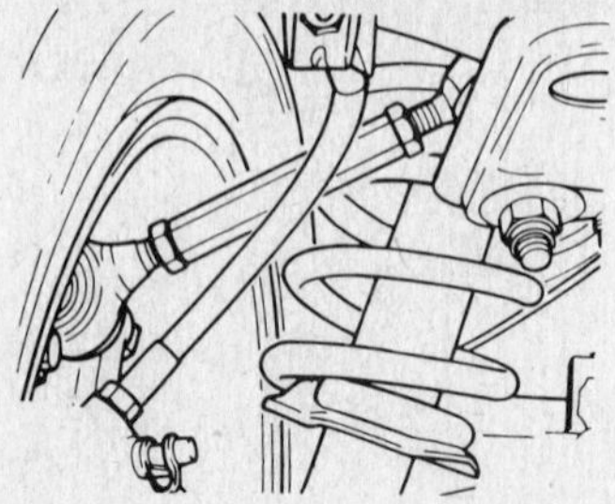

Figure 14-85. Loosen the locknuts, and turn the adjuster.

REAR CAMBER ADJUSTMENT

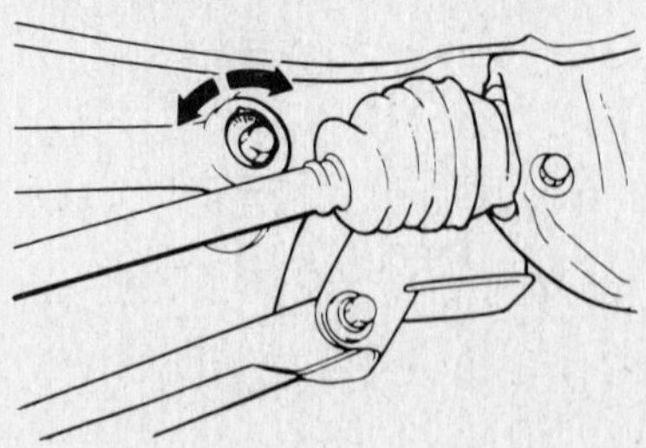

Figure 14-86. Rotate the cam on the inboard side of the upper control arm.

REAR CAMBER ADJUSTMENT

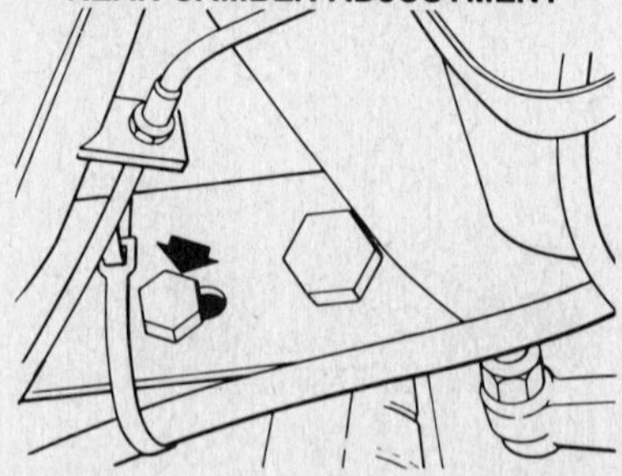

Figure 14-87. Rotate the eccentric bolt at the trailing arm.

REAR CAMBER ADJUSTMENT

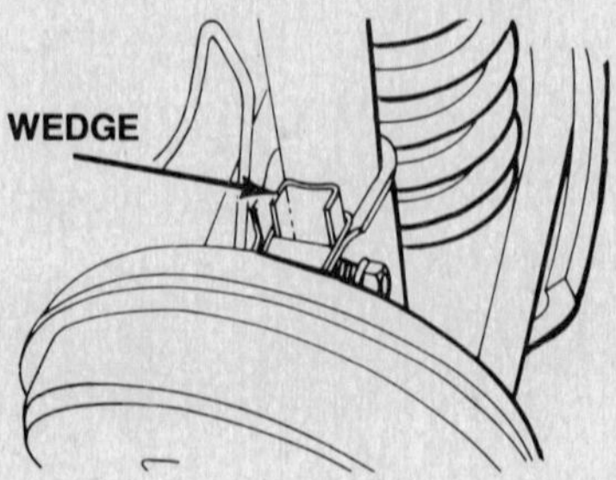

Figure 14-88. Loosen the lower strut mount bolts, and insert a spacer wedge.

Adjusting rear wheel camber

On many vehicles, rear alignment angles are not adjustable or you need aftermarket parts to adjust them. Vehicles with independent rear suspensions normally have rear camber adjustment. Use figures 14-85 through 14-88 to adjust rear wheel camber.

REAR TOE ADJUSTMENT

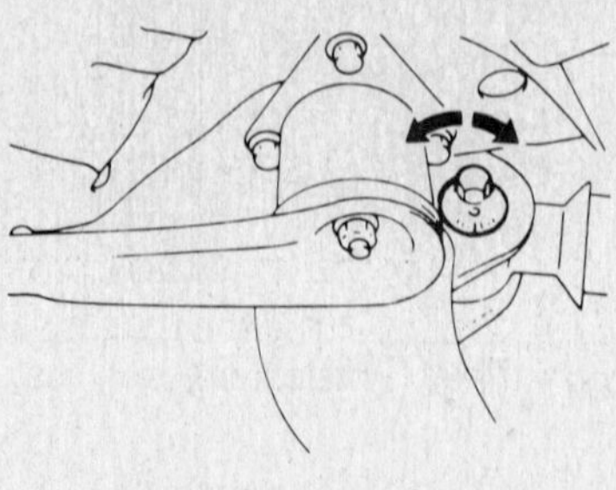

Figure 14-89. Rotate the cam at the control arm.

REAR TOE ADJUSTMENT

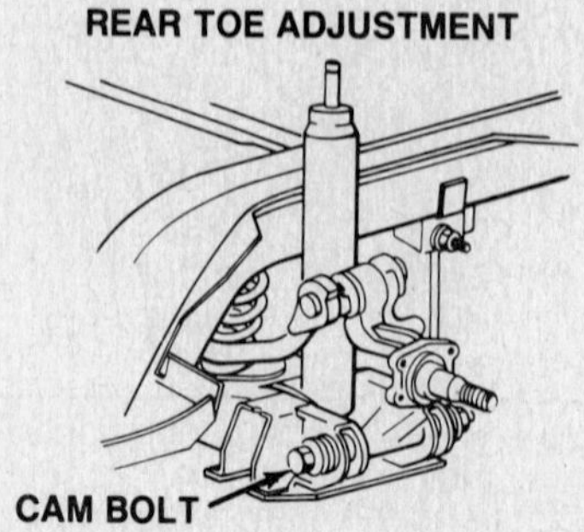

Figure 14-90. Rotate the cam at the control arm.

REAR TOE ADJUSTMENT

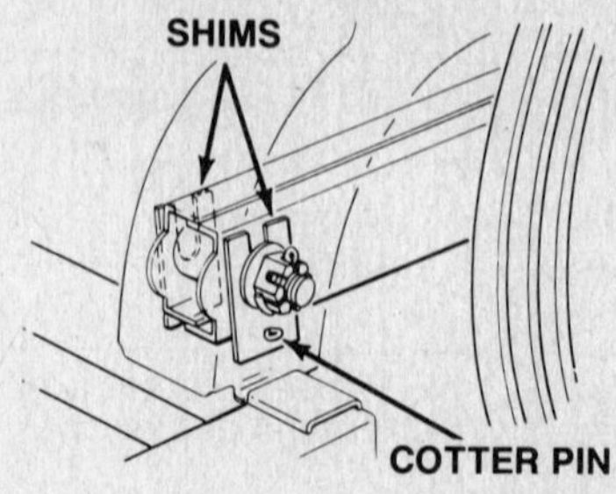

Figure 14-91. Add or remove shims at the control arm.

REAR TOE ADJUSTMENT

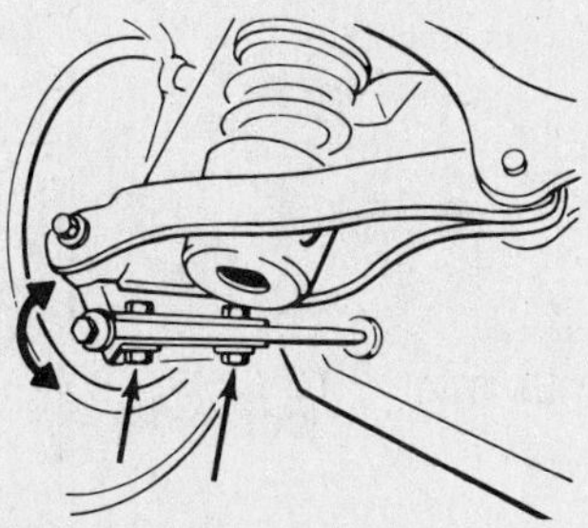

Figure 14-92. Loosen the lock nuts, and turn the eccentrics.

REAR TOE ADJUSTMENT

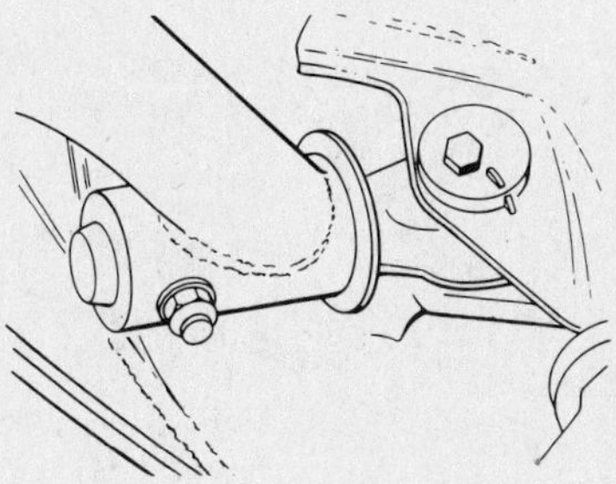

Figure 14-93. Rotate the eccentric at the control arm.

REAR TOE ADJUSTMENT

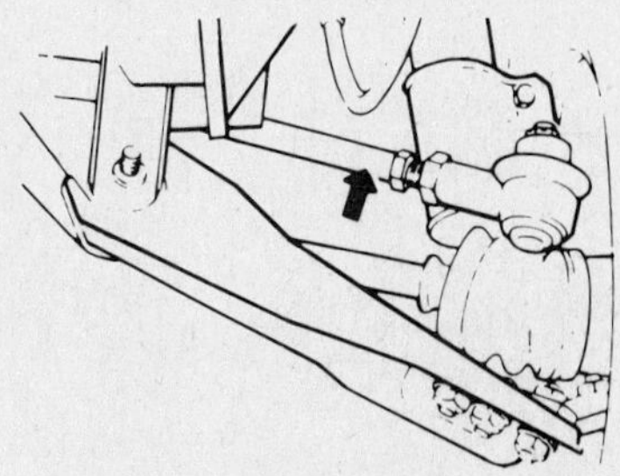

Figure 14-94. Loosen the lock nuts and turn the rear tie rod.

REAR TOE ADJUSTMENT

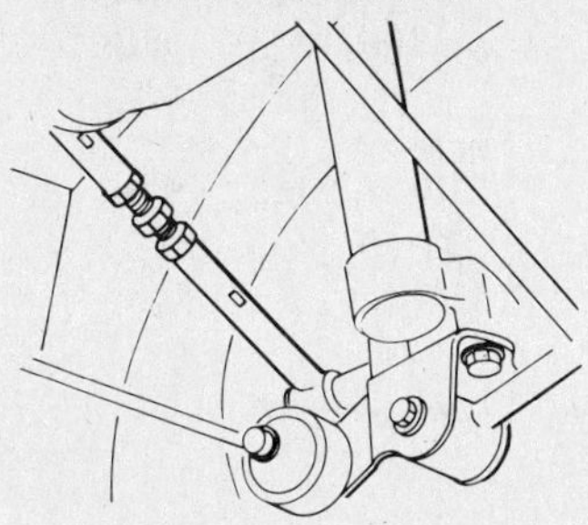

Figure 14-95. Loosen the lock nuts, and turn the center adjuster nut.

REAR TOE ADJUSTMENT

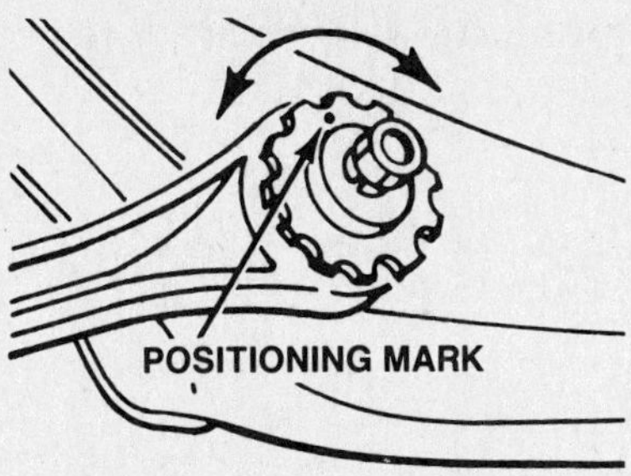

Figure 14-96. Turn the starwheel at the rear control arm.

REAR TOE ADJUSTMENT

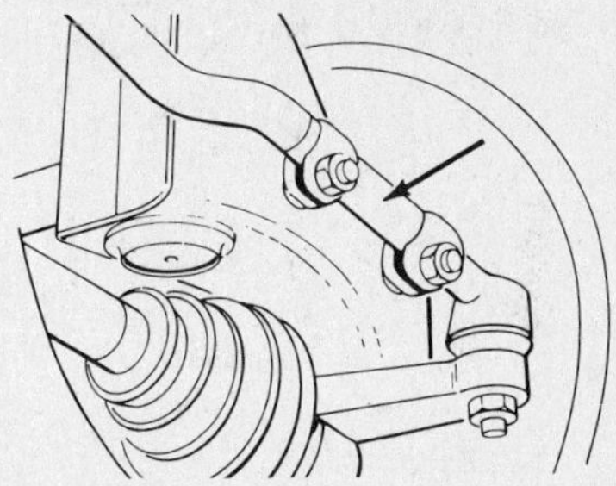

Figure 14-97. Loosen the lock nuts, and rotate the adjuster sleeve.

REAR TOE ADJUSTMENT

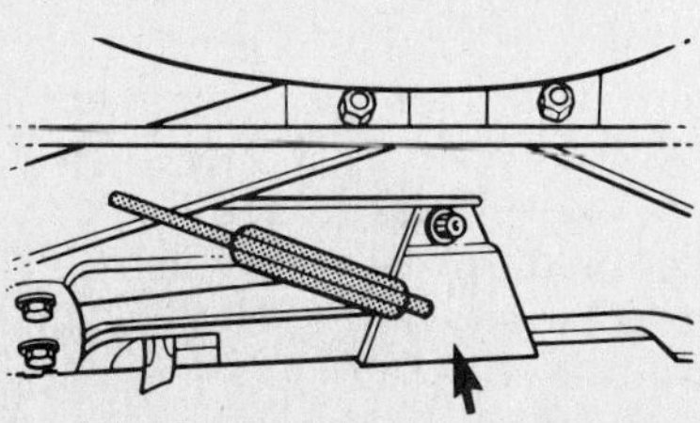

Figure 14-98. Loosen the inboard mounting bolts at the lower control arm, and use a turnbuckle to reposition the control arm before installing the mounting bolts.

REAR TOE ADJUSTMENT

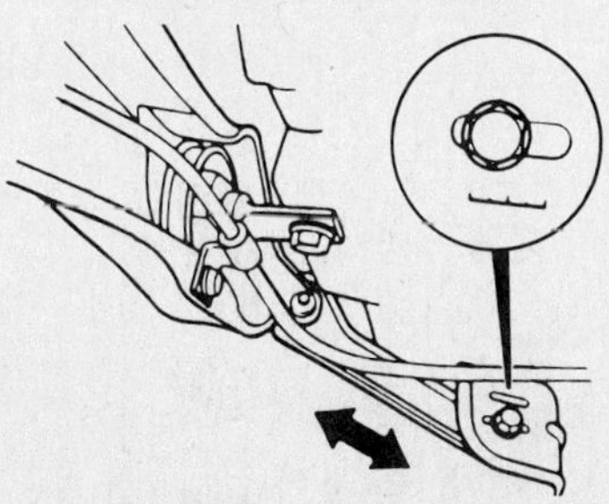

Figure 14-99. Loosen the control arm mounting bolt, and reposition it in the adjustment slot.

REAR TOE ADJUSTMENT

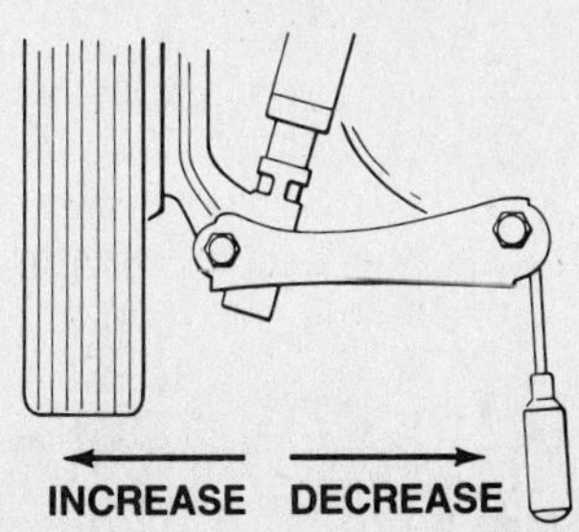

Figure 14-100. Loosen the control arm mounting bolts and pry the control arm to reposition it.

Adjusting rear wheel toe

On some vehicles, you adjust rear toe as you would front toe — by changing the position of tie rod ends. However, if the vehicle has independent rear suspension, you must change the position of the rear control arms to adjust toe. Use figures 14-89 through 14-100 to adjust rear wheel toe.

REAR CAMBER & TOE ADJUSTMENT

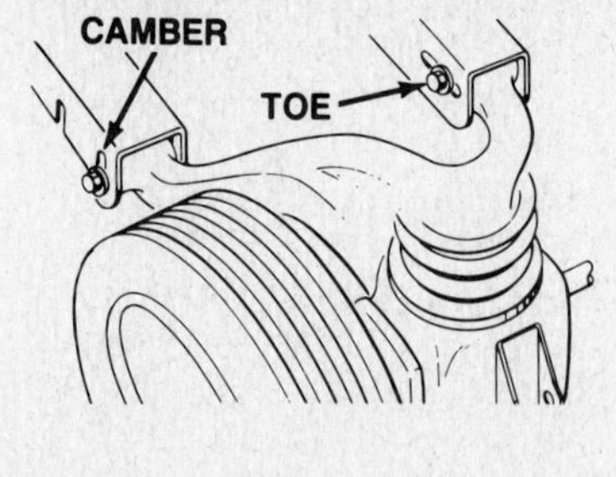

Figure 14-101. Loosen the control arm mounting bolts and pry against the control arm to reposition it.

REAR CAMBER & TOE ADJUSTMENT

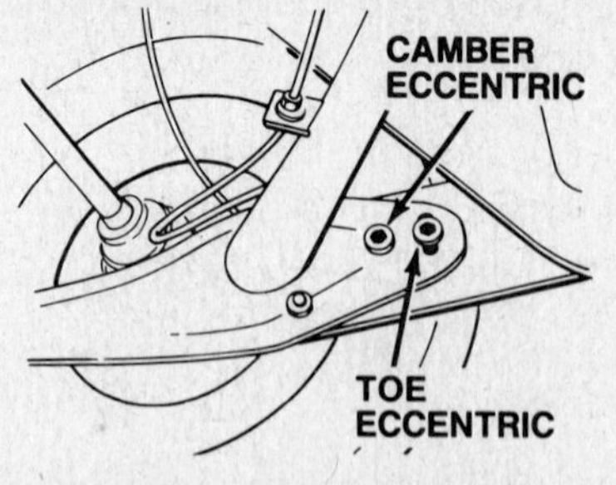

Figure 14-102. Rotate the eccentric bolts at the trailing arm.

REAR CAMBER & TOE ADJUSTMENT

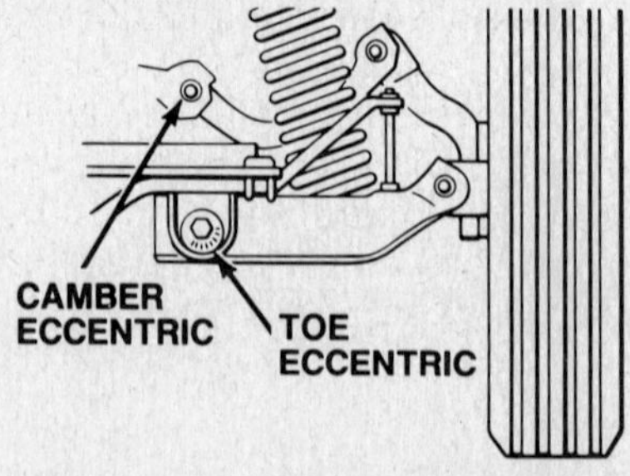

Figure 14-103. Rotate the eccentric bolts at the upper and lower control arms.

REAR CAMBER & TOE ADJUSTMENT

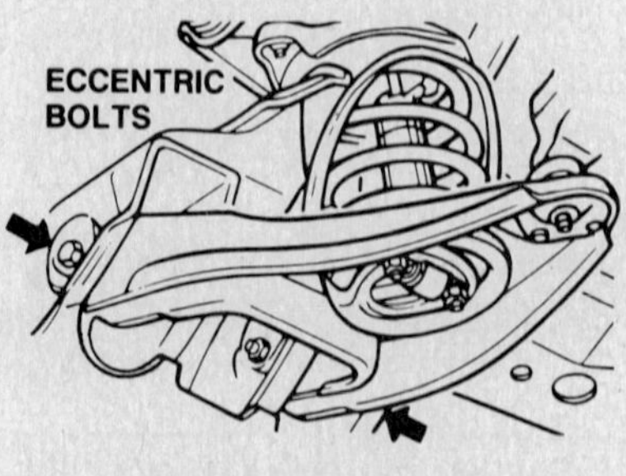

Figure 14-104. Rotate the eccentric bolts at the lower control arm.

REAR CAMBER & TOE ADJUSTMENT

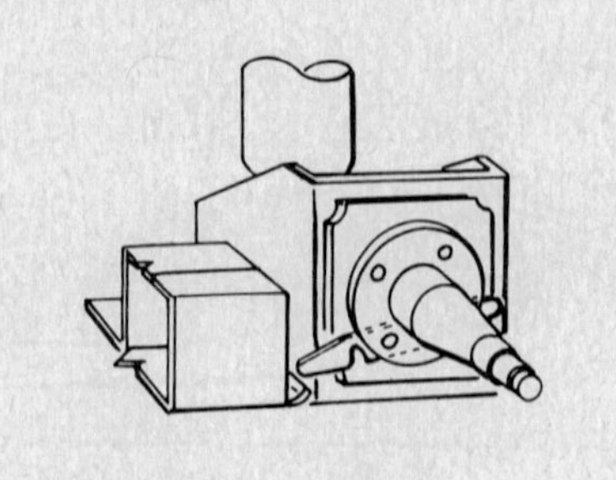

Figure 14-105. Install shims between the rear spindle and hub.

REAR CAMBER & TOE ADJUSTMENT

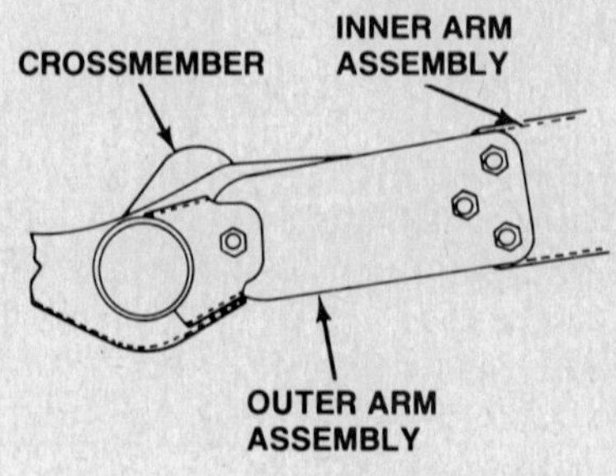

Figure 14-106. Loosen the bolts at the trailing arm, and reposition the inner and outer assemblies.

Adjusting rear wheel camber and toe simultaneously

On some vehicles, rear camber and toe are changed by a single adjustment, figures 14-101 through 14-106.

SPECIAL CONSIDERATIONS

Customers go to an alignment shop for a variety of reasons. Some simply want to fix a vehicle that ''pulls'', while others are looking to improve the handling characteristics of their ''high-performance'' vehicles. Before starting an alignment, talk to the customer and determine his or her needs. If the car pulls or wears tires quickly, setting the vehicle to the manufacturer's recommended alignment specifications should correct the problem.

However, if the vehicle has been modified, setting it to specifications may not correct handling problems. This is because modifications to steering and suspension components change the vehicle's alignment angles and handling characteristics. Many shops refuse to work on modified vehicles because alignment results are difficult to determine. Other shops will work on modified vehicles, but will not guarantee the alignment. In fact, they usually ask the customer to sign a waiver releasing the shop from liability.

When a vehicle comes into the shop for an alignment, ask the customer if it has been modified. Also ask if they normally carry heavy loads in the vehicle. When a vehicle is loaded with extra weight, camber changes in the negative direction. You can slightly modify the alignment settings to compensate for this. Always check for:

- Damage done to the frame, steering, or suspension components due to a collision
- Modifications made for cosmetic purposes; for example, lowering or raising the vehicle, or changing tire and wheel size.
- Modifications made for performance purposes; for example, changing shocks, springs, or tires and wheels.

This section will give you some insight into how steering and suspension system modifications may affect camber, caster, toe, steering-axis inclination (SAI), and included angle (IA). Also included here are hints that you can recommend to customers who are looking for ways to improve vehicle handling by changing alignment angle settings.

Camber

Camber is a tire wear angle. If camber settings are outside of the manufacturer's recommendations, then tire wear will likely result. Excessive positive camber (tire tilts out at the top) will cause wear or scuffing at the tire's outer shoulder. Excessive negative camber (tire tilts in at the top) will cause wear or scuffing at the tire's inner shoulder.

Besides affecting tire wear, camber determines how a vehicle handles because it directly affects how much of the tire's tread contacts the road. A vehicle with zero camber (tire straight up and down) allows for the most tread contact and the best traction. Theoretically, zero camber is what manufacturers try to achieve for the best traction.

However, setting the camber to zero while the vehicle is motionless (static) on the alignment rack will not achieve these results. When the vehicle starts moving, forces due to vehicle movement will change the wheel position and camber angle.

Vehicle manufacturers pick camber settings that will be close to zero when the vehicle is moving. Most production vehicles have a little bit of positive camber at the front wheels and a little bit of negative camber or zero camber at the rear wheels. However, if the steering or suspension systems are modified, the vehicle manufacturer's recommended camber settings may no longer apply.

Camber affects a vehicle's straight-line handling. For example, if camber settings on the front wheels are different, the vehicle will pull to the side with the most positive camber. A vehicle with zero camber will travel straight down a road, provided the road is perfectly flat.

However, most roads are slightly curved or "crowned", causing a vehicle with zero camber at both wheels to drift toward the side of the road. Some vehicle manufacturers make provisions for crowned roads by making camber at the driver-side front wheel more positive than the passenger-side wheel, which pulls the vehicle back toward the center of the road.

If you live in an area that is prone to flooding and the roads are highly crowned, you may suggest a change in camber on the driver's side. If the roads are flat, try to keep camber angles equal.

Manufacturers may use camber angle settings at the front wheels to compensate for torque steer. If the vehicle you are working on has a severe torque steer problem, you may suggest that camber be set so the vehicle will pull in the direction opposite torque steer. For example, if the vehicle torque steers to the right, adjust camber at the driver's side front wheel in the positive direction to steer the vehicle slightly to the left.

You can adjust camber settings based on how the customer loads the vehicle. For example, when a vehicle is loaded with extra weight, camber moves in the negative direction, which reduces traction. If the customer usually uses his or her vehicle to haul heavy loads, a slight positive camber setting may offset the effect of the load and improve handling. You may also recommend installing heavy-duty or helper springs to support the extra weight. Also, when vehicle ride height is lowered due to a modified suspension or spring wear, camber will move in the negative direction. Again, changing camber to a more positive setting may improve handling.

In some cases, camber adjustments may alleviate wheel bearing wear on vehicles that use wide tires and wheels. Wide tires and wheels give the effect of negative camber, changing weight distribution across the wheel's spindle. This can lead to accelerated wheel bearing wear, especially at the outer bearing. By adjusting camber slightly in the positive direction, bearing wear may be reduced. The best way to avoid this problem is to seek professional advice when choosing custom wheels. More information on selecting custom tires and wheels is in Chapter Twelve of this *Shop Manual*.

Some customers may modify their vehicles for racing, so the alignment angle settings depend on the vehicle's suspension and the driver's preference. In general, vehicles used for drag racing will have front camber settings that are zero or very close to zero.

Vehicles used for road racing will emphasize handling during high-speed left and right turns. As discussed in the *Classroom Manual*, during turns, camber at the outside wheel moves in the negative direction and camber at the inside wheel moves in the positive direction. But in high-speed turns, centrifugal force also affects camber. In fact, centrifugal force overcomes the normal turning forces so that camber at the outside wheel moves in the positive direction and camber at the inside wheel moves in the negative direction. To correct for the high positive camber at the outside wheel, many road racers set the front wheels with a little bit of negative camber, depending on vehicle suspension design. Aligning the wheels on this type of vehicle is a balancing act that takes practice.

Vehicles used in circle-track racing always turn to the left. Camber at the outer, passenger's side, wheel moves in the positive direction (again due to centrifugal force) and camber at the inner, driver's side, wheel moves in the negative direction. Giving the outer wheel a slight negative camber setting and the inner wheel a slight positive camber setting will increase traction. Again, aligning the wheels on race cars takes practice.

Because adjusting camber settings changes how much tire tread contacts the road, measuring tire temperature is one way to determine if camber settings are providing optimum traction. This method is used by professional and amateur racing teams alike. They start by driving the vehicle on the track to warm the tires. Then they measure tire temperature at the outer shoulder, center, and inner shoulder. Temperature is measured with a pyrometer, which uses a probe to sense temperature and display it on a meter. If camber settings are good, temperatures will be fairly even across the tread. Large variations in temperature mean that the camber angle should be adjusted so that more of the tread contacts the road. For example, if the tire is much hotter at the outer shoulder compared with the other temperature readings, then camber is too far in the positive direction. Change the wheel to a setting that is slightly more negative, and test the vehicle again.

If you modify camber settings on production vehicles, the front wheels should never vary by more than $1/2°$ camber. At the rear wheels, camber should be as close to equal as adjustments will allow. Also, recheck caster after adjusting camber. Because of camber roll, any large changes to camber will likely affect the caster reading.

Finally, you can use camber readings, along with steering axis inclination (SAI) and included angle (IA) readings, to help determine if suspension parts are bent. Use the chart given earlier in this chapter in the section titled "Measuring SAI and included angle" to determine which suspension parts are bent based on these angle readings.

Caster

Before measuring caster, be sure that camber is within specifications. As described earlier in this chapter, caster is determined by measuring camber roll, so if the camber reading is incorrect, the caster reading may also be off.

When a vehicle has positive caster, the steering axis tilts toward the driver. On a vehicle with negative caster, the steering axis tilts away from the driver. From this information you may realize that caster is not a tire wear angle. It mainly affects straight-line stability and steering feel. If caster is slightly out of specifications, tire wear will not be affected, but vehicle handling may be.

However, there *is* one set of circumstances where caster may affect tire wear. When a vehicle has a high positive caster setting, tire wear at the inner and outer shoulders may result due to camber roll. This wear is even more pronounced if the vehicle uses wide tires and wheels. To decrease tire wear due to camber roll, use narrow tires and wheels, which reduces traction, or decrease caster, which reduces straight-line stability. Finding a compromise is part of the suspension tuning processes.

When a manufacturer determines a vehicle's caster setting, a number of factors are taken into consideration. For example, the effects of increased caster include:

- Hard steering
- Road shock, which is transferred to the passenger compartment
- Straight-line stability
- Good steering wheel returnability.

However, the effects of decreased caster include:

- Reduced straight-line stability
- Reduced steering wheel returnability
- Easy steering.

The manufacturer's recommended caster settings are determined by weighing these handling characteristics and by determining what will work best with the suspension design. For example, high-performance production vehicles that have a high caster setting to increase straight-line stability usually must sacrifice a smooth ride. Vehicles used for drag racing have high positive caster.

Some manufacturers use caster, rather than camber, to compensate for road crown. Vehicles tend to pull to the side with the least amount of caster. A manufacturer may adjust caster on the driver's side to a slightly less positive setting to pull the vehicle away from the side of the road.

If you live in an area that is prone to flooding and the roads are highly crowned, you may suggest a slight decrease in caster on the driver's side. If the roads are flat, try to keep caster angles equal.

If the ride height of the vehicle you are aligning is different from stock due to custom modifications or suspension spring wear, caster will be affected. If you work in a shop that lowers or raises vehicle ride height, be sure to adjust caster to compensate for the change. For example, if the rear suspension is raised or if the front suspension is lowered, caster will move in the negative direction. If the rear suspension is lowered because the springs are weak or because the vehicle is normally overloaded, caster will move in the positive direction. If these changes in ride height noticeably affect vehicle handling, you may suggest to the customer that caster be changed slightly. For example, if the vehicle is lowered in the front, adjust caster in the positive direction to improve ride stability. The following formula allows you to calculate the change in ride height that will change caster by one degree:

$$\text{Wheelbase} \times 2 \times \pi \div 360 = \text{Change in ride height that causes } 1^\circ \text{ change in caster}$$

Whenever possible, keep caster settings even because, as mentioned earlier, the vehicle will tend to steer in the direction with the least amount of caster. If you must have some cross caster, also called "caster spread" (a difference in caster settings) do not exceed $1/2^\circ$. Some vehicles are more sensitive to cross caster than others. In general, if the vehicle has a high caster specification or high SAI, it will tend to be insensitive to cross caster. But if caster and SAI specifications are low, the vehicle will be sensitive to cross caster, resulting in handling problems.

Toe

Manufacturers set toe specifications so that when the vehicle is rolling forward, toe at each wheel is zero and rolling resistance is minimal. Like camber, toe is a tire wear angle. If toe-in settings are higher than specified by the manufacturer, tires will wear faster at the outer shoulder, sometimes creating a feathered edge. If the wheels have excessive toe-out settings, the inner shoulder will wear faster.

In most service manuals, toe specifications are given in terms of total toe. When you set toe at each front wheel, simply use one-half of the total toe specification. For example, if the toe specification calls for $1/2^\circ$ toe-in, then set each wheel to $1/4^\circ$ toe-in. Toe is usually adjusted by lengthening or shortening the tie rods by turning them.

On production vehicles, the rear wheels usually have a slight amount of toe-in or zero toe, if toe is adjustable. Front toe settings vary depending on suspension design, but most vehicles are toed-in at the front wheels as well.

The most common problem resulting from incorrect tie rod position is tire wear, but other problems may result. As discussed in the *Classroom Manual*, bump steer will result if tie rods cannot accommodate suspension travel because they are the wrong length or height. If toe is within specifications and the vehicle bump steers, the problem is likely related to incorrect tie rod height. This problem is usually due to suspension or steering geometry damage, such as front wheel setback or a drastic change in vehicle ride height, as when a vehicle is lowered.

Another toe-related problem is dog-tracking. If rear toe is not set properly, the rear wheels will not follow the front wheels down the road. This is because an incorrect total rear toe setting creates a thrust angle, and the rear wheels will move in the direction of the thrust angle. By setting toe evenly at the rear wheels, dog-tracking is reduced or eliminated.

Finally, if the vehicle is used for drag racing, front toe is usually set to zero for reduced tire drag. For road racing on tight tracks, front toe settings of nearly zero help handling by reducing understeer at the start of a turn. Front toe-out settings usually increase the vehicle's tendency to wander, so on higher speed tracks technicians give the tires a slight toe-in setting.

On FWD race cars, rear toe settings are set to nearly zero to reduce understeer and improve turn-in. On RWD cars, rear toe is usually zero or slightly toed-in, although setting rear toe on these vehicles is more a matter of trial and error.

Unless specified by the vehicle manufacturer, rear toe on production and high-performance street vehicles should always be set at zero or slightly toed-in. Rear toe-out is not recommended.

Steering Axis Inclination (SAI)

Measuring steering axis inclination (SAI), or kingpin inclination (KPI), can tell you a lot about the condition of the suspension system. This angle is determined during the suspension's design, and is normally not adjustable. On some vehicles, SAI is adjustable. When SAI is not adjustable, it can be used in conjunction with camber to help you locate suspension damage.

After a vehicle has had body and frame work, it is normally sent to an alignment shop. Before starting the alignment, check SAI, which will tell you if the frame is straight and if the suspension components are properly positioned. If SAI is out of specifications, the vehicle may need more frame work or may have bent suspension components that need repair before you start the alignment.

Included Angle (IA)

Many specification manuals give the value for included angle rather than the specification for SAI. Included angle and SAI are related in that SAI plus camber equals included angle. Using the included angle measurement as a standard for determining the condition of the suspension system is more precise on vehicles where SAI is an adjustable angle. On some strut suspensions, SAI is adjustable.

ASE Technician Certification Sample Test

This sample test is similar in format to the series of tests given by the National Institute for Automotive Service Excellence (ASE). Each of their exams covers a specific area of automotive repair and service. The tests are given every fall and spring throughout the United States.

For a technician to earn certification in a particular field, he or she must successfully complete one of these tests and have at least two years of "hands on" experience (or a combination of work experience and formal automotive technical training). A person who successfully completes all of the tests is a certified ASE Master Automobile Technician.

The questions in this sample test follow the format of the ASE exams. If you plan to apply for certification later in your career, learning to take this kind of test will help you. The answers for this sample test are on page 332 of this *Shop Manual*.

For test registration forms or additional information on the automobile technician certification program, write to:

National Institute for
AUTOMOTIVE SERVICE EXCELLENCE
13505 Dulles Technology Drive
Herndon, VA 22071-3415

1. Technician A says that on this rear axle setup, the parts labeled X and Y absorb rear axle windup forces.

   Technician B says that part Z is called a Panhard rod.

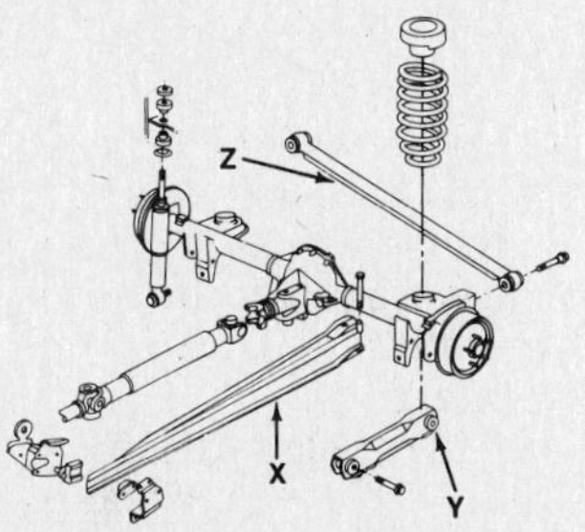

   Who is correct?
   a. A only
   b. B only
   c. Both A and B
   d. Neither A nor B

2. Before leaving for vacation, a driver has the tires checked and the wheels aligned. After driving for a few days, the driver notices abnormal wear on the inner shoulders of the tires. This wear is most likely due to:
   a. Loading the trunk with heavy suitcases
   b. Sustained high-speed driving
   c. Braking during sustained high-speed driving
   d. Driving on highly crowned roads

3. Technician A says that when replacing a rack and pinion steering gear on a vehicle with a driver's side air bag, you must lock the steering wheel before starting the procedure.

   Technician B says that you must center the steering wheel and position the wheels straight ahead before connecting the tie rods.

   Who is correct?
   a. A only
   b. B only
   c. Both A and B
   d. Neither A nor B

4. When adjusting pinion torque on a rack and pinion steering gear, you must:
   a. Tighten the adjuster plug to specifications and back it off a half turn
   b. Tighten the locknut, and then tighten the adjuster plug to specifications
   c. Tighten the adjuster plug, back it off according to the manufacturer's specifications, and tighten the locknut
   d. Check sector shaft endplay first
5. A power steering pump makes a moaning or whining noise when operating. What is the most likely cause?
   a. Low fluid
   b. Air in the system
   c. Either a or b
   d. Neither a nor b
6. If you find power steering fluid in the rack bellows, what is the most likely cause?
   a. Worn pump
   b. Worn bellows
   c. Worn outer tie rod ends
   d. Bad rack seals
7. In the vehicle pictured, if the inside wheel is turned 20 degrees, what is the most likely angle of the outside wheel?

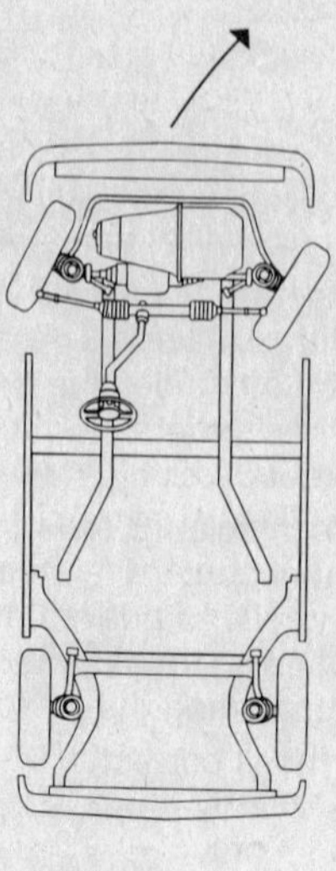

   a. 20 degrees
   b. 25 degrees
   c. 30 degrees
   d. 18 degrees
8. During turns, a vehicle's tires make a squealing noise. What is the most likely cause?
   a. Toe-out on turns out of specification
   b. Toe-in setting out of specification
   c. Tires are too small
   d. Brakes need adjusting
9. When removing a rack and pinion steering gear, you must do all of the following **EXCEPT**:
   a. Disconnect the outer tie rod ends from the knuckles
   b. Disconnect the inner tie rod ends from the rack
   c. Remove the rack mounting bolts from the vehicle frame
   d. Disconnect the steering shaft coupler from the pinion
10. On a vehicle with parallelogram steering, which adjustment may change when you replace the center link?
    a. Toe
    b. Sector shaft endplay
    c. Camber
    d. Caster
11. Technician A says that when you are checking the outer tie rod ends for wear on a rack and pinion steering gear, you must replace any that have torn dust covers.
    Technician B says that worn tie rod ends will allow toe change.
    Who is correct?
    a. A only
    b. B only
    c. Both A and B
    d. Neither A nor B
12. Which illustration shows proper jack placement when checking for ball joint play on these suspensions?

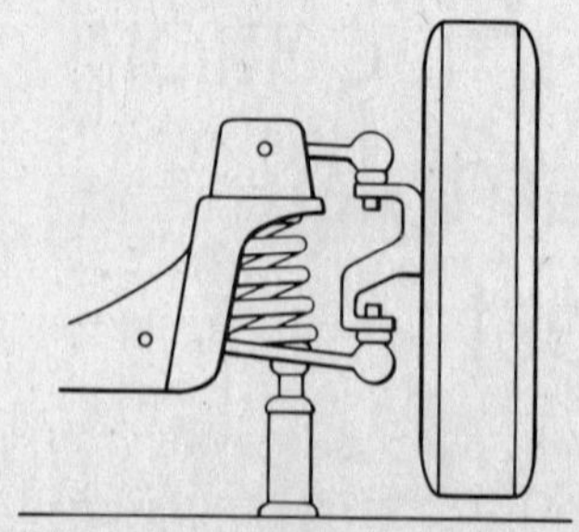

A

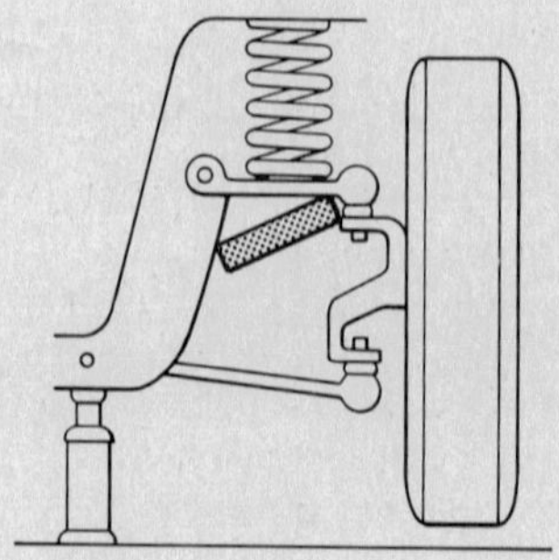

B

    a. A only
    b. B only
    c. Both A and B
    d. Neither A nor B
13. Technician A says that after replacing lower control arm pivot bushings, you must tighten the bolts until the bushings just begin to bulge.
    Technician B says that you must tighten the bushings with the control arm resting against the vehicle frame.
    Who is correct?
    a. A only
    b. B only
    c. Both A and B
    d. Neither A nor B
14. The ride height measurement on a vehicle with MacPherson struts is lower than specified by the manufacturer. What is the most likely cause?
    a. Worn strut springs
    b. Worn strut cartridge
    c. Improperly installed anti-roll bar
    d. Bent strut rod

15. The steering wheel on a vehicle with parallelogram steering is off-center. How do you first attempt to center the steering wheel?
    a. Remove the steering wheel, and reposition it
    b. Lengthen and shorten the left and right tie rods equally
    c. Reposition the idler arm
    d. Replace the steering gear
16. Adjustments made at this component will change:

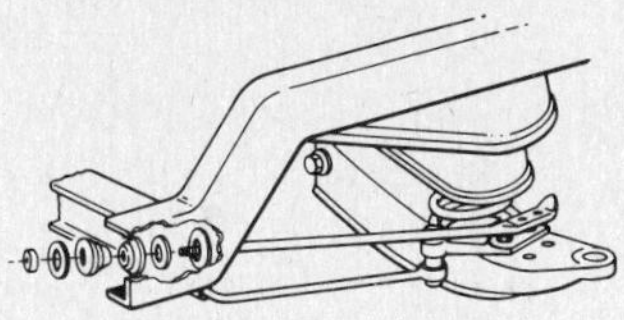

    a. Camber
    b. Caster
    c. Both a and b
    d. Neither a nor b
17. On a FWD vehicle, the spindle nut is used to:
    a. Adjust wheel bearing preload
    b. Seat the CV joint
    c. Adjust axle endplay
    d. None of the above
18. Poor steering wheel return on a rack and pinion steering system is most likely due to:
    a. Low fluid level
    b. Worn pump
    c. Overtightened pinion torque
    d. Worn tie rod ends
19. All of the following indicate that a shock absorber is bad and should be replaced **EXCEPT**:
    a. Drops of oil on shock absorber
    b. Nicks and burrs in shock absorber piston rod
    c. Bent shock absorber housing
    d. Light film of oil on shock absorber
20. A vehicle with new front springs and shocks leans to the left. Technician A says that both rear springs should be checked and replaced.
    Technician B says that the left rear shock is causing the problem, and it should be checked and replaced.
    Who is correct?
    a. A only
    b. B only
    c. Both A and B
    d. Neither A nor B
21. Technician A says that using oversize springs will likely change a vehicle's ride height.
    Technician B says that worn hydraulic shocks will change a vehicle's ride height.
    Who is correct?
    a. A only
    b. B only
    c. Both A and B
    d. Neither A nor B
22. After checking a vehicle's alignment angles on a four-wheel aligner, technicians determine that there is excessive rear axle thrust. What can they do to compensate for it?
    a. Adjust rear wheel caster
    b. Put wider tires and wheels on the rear axle
    c. Install a heavy-duty shock absorber on the side that is pushed back
    d. Adjust front wheel alignment angles
23. The following table shows a vehicle's actual wheel alignment angles and the manufacturer's specifications. How will this vehicle handle?

	Actual L	Actual R	Specifications
Camber	1°	–1/2°	–1/4 ± 3/4°
Caster	1°	11/4°	11/4 ± 3/4°
Toe-in	0°	–1/8°	0 ± 1/4°

    a. Leads to the left
    b. Leads to the right
    c. Wanders at high speeds
    d. Vibrates at speeds exceeding 35 mph
24. If you install the part indicated with an arrow, the vehicle will have:

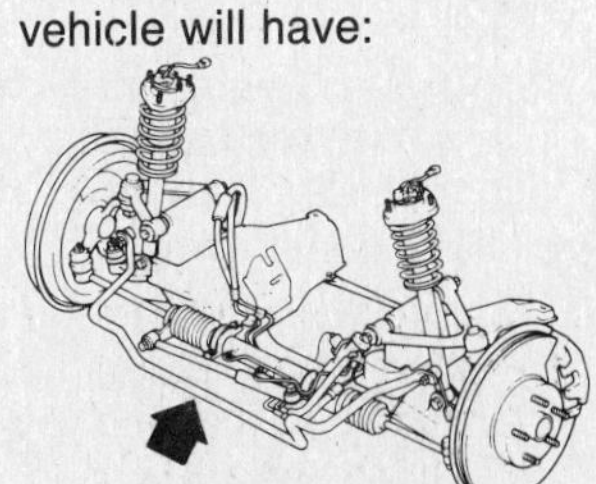

    a. Increased weight-carrying capacity
    b. Decreased steering gear ratio
    c. Decreased body roll
    d. Increased caster
25. Any of the following may be used to connect the steering gear and column **EXCEPT**:
    a. Universal joint
    b. Flexible coupling
    c. Pot joint
    d. Rzeppa joint
26. On a rack and pinion steering system, tightening or loosening the rack support cover adjusts:
    a. Rack torque
    b. Worm bearing preload
    c. Gear mesh lash
    d. Pinion torque
27. Technician A says that Ford's rubber bonded socket (RBS) joints only have to be lubricated every 100,000 miles.
    Technician B says that the RBS joints have memory steer. This means that the steering wheel and linkage must be centered before tightening the joints.
    Who is correct?
    a. A only
    b. B only
    c. Both A and B
    d. Neither A nor B
28. Technician A says that the outer tie rod end must be removed from the steering knuckle and shimmed in order to change toe adjustment.
    Technician B says that you can loosen the lock nut on the tie rod and turn the adjuster sleeve to change toe adjustment.
    Who is correct?
    a. A only
    b. B only
    c. Both A and B
    d. Neither A nor B

29. The function of a steering damper is to:
a. Reduce kickback
b. Reduce shimmy
c. Keep road bumps from affecting steering
d. All of the above

30. If a vehicle's "toe-out on turns" measurement is not within specifications, it is likely due to a:
a. Bent steering arm
b. Bent tie rod
c. Loose center link
d. Loose drag link

31. While inspecting a vehicle's front tires, you notice that when you rub your hand across the tread from the outside to the inside it feels rough, but when you rub your hand from the inside to the outside it feels smooth. What is the most likely cause?
a. Excessive toe-out
b. Excessive toe-in
c. Excessive positive caster
d. Excessive positive camber

32. If the part indicated with an arrow is worn or damaged, which other component should you check for wear?

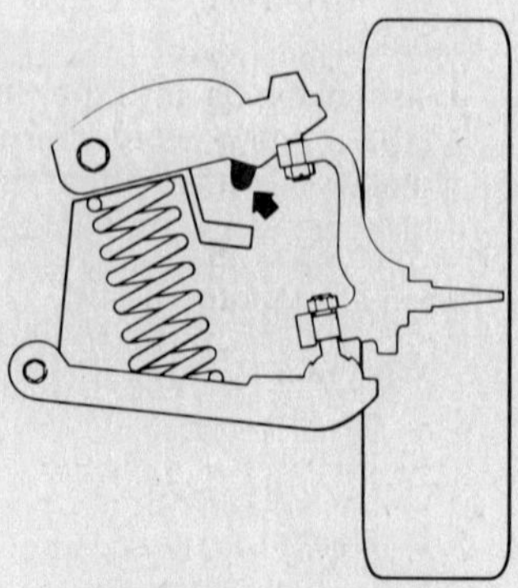

a. Control arm bushings
b. Ball joints
c. Shock absorber
d. Control arm cross shaft

33. A customer brings a vehicle into the shop complaining that it wanders. What is the most likely cause?
a. Excessive positive camber
b. Excessive negative caster
c. Excessive positive toe
d. Excessive negative camber

34. Technician A says that you should clean wheel bearings with soap and warm water to remove grease.
Technician B says that after cleaning the bearings you should spin them with compressed air to dry them thoroughly.
Who is correct?
a. A only
b. B only
c. Both A and B
d. Neither A nor B

35. All of the following handling problems indicate worn shock absorbers **EXCEPT**:
a. Excessive dive during braking
b. Car bounces more than three times during the bounce test
c. Tire cupping
d. Tire wear at the center tread

36. A customer complains that the steering wheel is off-center even when the vehicle is travelling straight down the road. Technician A says that it may be due to excessive rear wheel thrust angle.
Technician B says that adjusting the tie rods at the front wheels while performing a four-wheel alignment will correct the problem.
Who is correct?
a. A only
b. B only
c. Both A and B
d. Neither A nor B

37. If a customer complains of front wheel shimmy, a likely cause is:
a. Static wheel imbalance
b. Dynamic wheel imbalance
c. Both are likely causes
d. Neither are likely causes

38. A tire has the designation "P 275/40 ZR 17". What does the "40" stand for?
a. The aspect ratio
b. The tread width
c. The hardness of the rubber compound
d. The rim diameter

39. All of the following are tire wear angles **EXCEPT**:
a. Camber
b. Toe
c. Caster
d. Toe-out on turns

40. On this type of suspension system, how would you make the camber setting more positive without affecting caster?

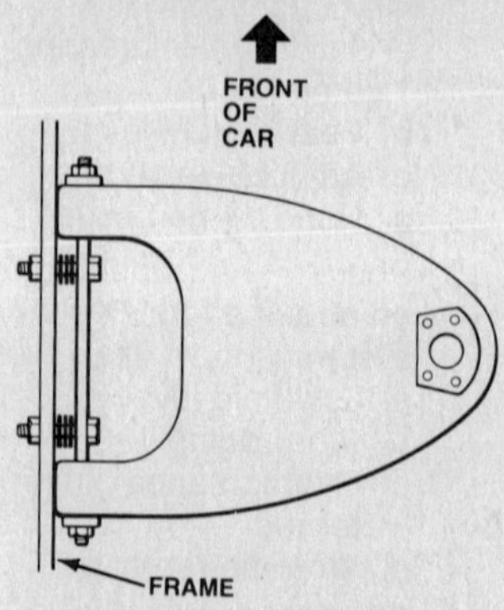

a. Add shims to the front only
b. Add an equal number of shims to the front and rear
c. Remove shims from the front and place them at the rear
d. Remove the same number of shims from the front and rear

Index

Answers to ASE Technician Certification Sample Test

1-C 2-A 3-B 4-C 5-B 6-D 7-D 8-A 9-B 10-A 11-C 12-C 13-D 14-A 15-B 16-B 17-A 18-C 19-D 20-A 21-A 22-D 23-A 24-C 25-D 26-D 27-B 28-B 29-D 30-A 31-A 32-C 33-B 34-D 35-D 36-C 37-B 38-A 39-C 40-B

NOTES

NOTES

NOTES

NOTES

NOTES

NOTES

NOTES

NOTES

NOTES